Topographic Effects in Stra

T0177008

Covering both theory and experiment, this text describes the behaviour of homogeneous and density-stratified fluids over and around topography. Its presentation is suitable for advanced undergraduate and graduate students in fluid mechanics, as well as for practising scientists, engineers and researchers.

Using laboratory experiments and illustrations to further understanding, the author explores topics ranging from the classical hydraulics of single-layer flow to more complex situations involving stratified flows over two- and three-dimensional topography, including complex terrain. A particular focus is placed on applications to the atmosphere and ocean, including discussions of downslope windstorms, and of oceanic flow over continental shelves and slopes.

This new edition has been restructured to make it more digestible, and updated to cover significant developments in areas such as exchange flows, gravity currents, waves in stratified fluids, stability and applications to the atmosphere and ocean.

PETER G. BAINES is an Honorary Senior Fellow in the Department of Infrastructure Engineering at the University of Melbourne. He is a winner of the Priestley Medal and a Fellow and first President of the Australian Meteorological and Oceanographic Society. He is currently Secretary of the Royal Society of Victoria.

Established in 1952, the *Cambridge Monographs on Mechanics* series has maintained a reputation for the publication of outstanding monographs, a number of which have been re-issued in paperback. The series covers such areas as wave propagation, fluid dynamics, theoretical geophysics, combustion and the mechanics of solids. Authors are encouraged to write for a wide audience, and to balance mathematical analysis with physical interpretation and experimental data, where appropriate. Whilst the research literature is expected to be a major source for the content of the book, authors should aim to synthesise new results rather than just survey them.

A complete list of books in the series can be found at www.cambridge.org/mathematics.

RECENT TITLES IN THIS SERIES

Theory and Computation of Hydrodynamic Stability (Second Edition)
W. O. CRIMINALE, T. L. JACKSON & R. D. JOSLIN

Magnetoconvection
N. O. WEISS & M. R. E. PROCTOR

Waves and Mean Flows (Second Edition)
OLIVER BÜHLER

*Turbulence, Coherent Structures, Dynamical Systems
and Symmetry (Second Edition)*
PHILIP HOLMES, JOHN L. LUMLEY,
GAHL BERKOOZ & CLARENCE W. ROWLEY

Elastic Waves at High Frequencies
JOHN G. HARRIS

Gravity-Capillary Free-Surface Flows
JEAN-MARC VANDEN-BROECK

Topographic Effects in Stratified Flows

SECOND EDITION

PETER G. BAINES
University of Melbourne

CAMBRIDGE
UNIVERSITY PRESS

University Printing House, Cambridge CB2 8BS, United Kingdom

One Liberty Plaza, 20th Floor, New York, NY 10006, USA

477 Williamstown Road, Port Melbourne, VIC 3207, Australia

314–321, 3rd Floor, Plot 3, Splendor Forum, Jasola District Centre, New Delhi – 110025, India

103 Penang Road, #05–06/07, Visioncrest Commercial, Singapore 238467

Cambridge University Press is part of the University of Cambridge.

It furthers the University's mission by disseminating knowledge in the pursuit of education, learning, and research at the highest international levels of excellence.

www.cambridge.org
Information on this title: www.cambridge.org/9781108481526
DOI: 10.1017/9781108673983

First edition © Cambridge University Press 1995
Second edition © Peter G. Baines 2022

First published 1995
Second edition 2022

Printed in the United Kingdom by TJ Books Limited, Padstow Cornwall

A catalogue record for this publication is available from the British Library.

Library of Congress Cataloging-in-Publication data
Names: Baines, Peter G., author.
Title: Topographic effects in stratified flows / Peter G. Baines, University of Melbourne.
Description: [Revised edition]. | Cambridge, UK ; New York, NY : Cambridge University Press, 2022. | Series: Cambridge monographs on mechanics | Includes bibliographical references and index.
Identifiers: LCCN 2021029978 | ISBN 9781108481526 (hardback)
Subjects: LCSH: Fluid dynamics. | Stratified flow. | Geophysics. | BISAC: SCIENCE / Mechanics / Fluids
Classification: LCC QC151 .B24 2022 | DDC 532/.053–dc23
LC record available at https://lccn.loc.gov/2021029978

ISBN 978-1-108-48152-6 Hardback
ISBN 978-1-108-72290-2 Paperback

In memory of Adrian Gill and Angus McEwan.

Some books are to be tasted, others to be swallowed,
and some few to be chewed and digested.
FRANCIS BACON, 1561–1626; *Essay 1, Of Studies.*

Contents

Preface to the Second Edition   *page* xi
Preface to the First Edition   xiii

1   Background   1
   1.1   Equations for fluid motion   3
   1.2   Boundary conditions   8
   1.3   Conservation relations   10
   1.4   Terminology   13

2   Non-linear single-layer flow: classical hydraulics   16
   2.1   Basic equations   16
   2.2   Flows with small obstacle height   17
   2.3   One-dimensional non-linear hydrostatic flow   31
   2.4   Downslope flows with frictional drag   48
   2.5   Granular flows   54

**3   Non-linear single-layer flow past obstacles: jumps, bores and wave
dispersion**   56
   3.1   Non-linear waves   56
   3.2   The QRS framework   61
   3.3   Application to hydraulic jumps and undular bores   66
   3.4   Single-layer flow over topography with non-linearity and
       dispersion   71
   3.5   Non-linear flow past three-dimensional obstacles   78

4   Two-layer flow with jumps and topography   95
   4.1   Basic equations   95
   4.2   Linear waves   97
   4.3   Equations for one-dimensional non-linear hydrostatic flow 98
   4.4   Two-layer hydraulic jumps   102
   4.5   Hydrostatic flow over topography   109

Contents

4.6	Non-linear waves and internal bores	120
4.7	Topographic forcing with non-linearity and dispersion	124
4.8	Downstream effects	126

5 **Two-layer and stratified flow through contractions** **129**
5.1	Two-layered flow through contractions with a free upper surface	130
5.2	Two-layered flow through contractions with a rigid upper boundary	132
5.3	Non-linearity with dispersion in contractions	141
5.4	Multi-layered flow through contractions	142
5.5	Continuously stratified flow through contractions	145

6 **Exchange flows** **148**
6.1	Two-layer exchange flow in a uniform channel over topography	149
6.2	Two-layer exchange flow through contractions	149
6.3	Exchange flows through doorways and windows	163
6.4	Multi-layer and continuously stratified exchange flows	164

7 **Gravity currents, downslope and anabatic flows, and stratified hydraulic jumps** **172**
7.1	Gravity currents over horizontal terrain in uniform environments	172
7.2	Gravity currents in density-stratified environments	176
7.3	Gravity currents down slopes	179
7.4	Hydraulic jumps in stratified flow	189
7.5	Anabatic flows	192

8 **Waves in stratified fluids** **196**
8.1	Waves in multi-layered models	196
8.2	Continuously stratified fluids: equations	202
8.3	Waves in finite-depth systems	205
8.4	Waves in infinitely deep stratified fluids	211
8.5	Trapped and leaky modes	217
8.6	The effects of molecular viscosity and diffusion on internal waves	220
8.7	Energy and momentum transport in a non-uniformly moving fluid	221
8.8	The "slowly varying" or WKB approximation	226
8.9	Critical layers	229
8.10	Wave-overturning and saturation	238
8.11	Wave propagation in three dimensions	239

9 **The stability of stratified flows** **243**
| 9.1 | Stability of stratified shear flow: a general criterion | 244 |

9.2 The process and products of the instability of shear flows 245

9.3 Instability in laminar boundary-layers: Tollmien–Schlichting waves 254

9.4 The stability of internal waves 259

10 Stratified flow over two-dimensional obstacles: linear and near-linear theory 261

10.1 Observations of flows of infinite depth 263

10.2 Infinite-depth flows: theory for small Nh/U 275

10.3 Infinite-depth flows: finite-amplitude topography and "Long's model" 286

10.4 Infinite-depth flows with $Nh/U > (Nh/U)_c$: numerical studies 294

10.5 Linear theory for small Nh_{m}/U: finite depth 297

10.6 Comparison between linear theory, and observations and numerical results for finite depth and small Nh/U 304

10.7 Long-model solutions for finite depth 313

11 Stratified flow over two-dimensional obstacles: non-linear hydraulic models with applications 321

11.1 Models with non-linearity and dispersion 321

11.2 Non-linear hydraulic flow theory for finite depth 325

11.3 Applications of the hydraulic theory 336

11.4 The approach to continuous stratification 344

11.5 Observations and numerical results for finite Nh/U in finite depth: short obstacles 358

11.6 Application of the hydraulic model to infinite-depth flows 365

11.7 Flows with large Nh/U: deep blocked flow, topographic drag and clear-air turbulence 368

11.8 Details of the dynamics of downslope windstorms 372

11.9 Flow across valleys 378

12 Stratified flow over three-dimensional topography: linear theory 390

12.1 Linear theory for small-amplitude topography, with the lower boundary as a stream surface 391

12.2 Linear theory for trapped lee waves 414

12.3 Atmospheric lee waves 419

12.4 Limitations and extensions of linear theory 424

13 Three-dimensional stratified flow over finite obstacles 433

13.1 The topology of the flow field on the surface of an obstacle 433

13.2 Observations of the flow past three-dimensional obstacles 439

13.3 Flow properties for finite Nh/U: theoretical aspects 473

13.4 The drag force on isolated obstacles: $Nh/U > 1$ 484

14　Flow over complex and realistic terrain in the atmosphere and ocean　　486

14.1　Flow over complex terrain　　486

14.2　Some atmospheric examples in the troposphere　　489

14.3　Internal waves in the upper atmosphere　　499

14.4　Internal waves in the deep ocean　　502

14.5　Topographic effects in coastal oceanography　　502

14.6　Oscillating ocean flows and tides　　504

15　Applications to practical modelling of flow over complex terrain　　505

15.1　Laboratory modelling　　505

15.2　The natural ventilation of buildings　　510

15.3　Parametrisation of the effects of sub-grid-scale orography in large-scale numerical models　　510

References　　520

Index　　542

Preface to the Second Edition

The main objective of this second edition is to bring the topic up to date at time of writing (~2019 to early 2020). The material covered is somewhat broader: every topic described in the first edition is included here, and a number of new topics such as downslope flows and waves in the upper atmosphere have also been added. In addition, some corrections have been made, and more emphasis has been given to applications. These are becoming more apparent as observations in the ocean and atmosphere improve. The effects of rotation have still largely been omitted, though the Coriolis force/frequency does get an occasional mention. There is a finite limit to everything.

To some extent, the subject is a closed book (or at least, more so than for the first edition), but the type of analysis described here may (or does) have application to other fields. Two under-developed topics that have been proposed are non-linear optics and Bose–Einstein condensates. The latter is probably relevant to the missing mass of the universe.

I would also like to express my appreciation of the work of the staff of Cambridge University Press, particularly David Tranah, for his support and professionalism, and to Leon Chan and Jimmy Philip for assistance with some simulations and figures.

Preface to the First Edition

This project was conceived about 10 years ago, but the incentive to pursue and complete it was hampered until recently because of several fundamental unresolved questions about the nature of stratified flow around topography. Within the last few years it has become possible to answer these questions, as a result of the efforts of several people, and the answers are embodied in the synthesis presented here.

Who will benefit from purchasing, reading or thumbing through this book? It is primarily addressed to fluid dynamicists, meteorologists, oceanographers, engineers, physicists and mathematicians who wish to learn more about the dynamics of stratified fluids. Some background in fluid dynamics is probably necessary, but the subject is treated from first principles and is developed from simple situations toward more complex ones. Overall, the order of presentation is based on logic rather than the historical development. There is balance between theory and experiment, where the comparison is made whenever possible, and a consistent attempt has been made to provide a physical understanding of the phenomena involved.

I have gone to some length to make the material easily assimilable, as the number of figures testifies. As I see it, a book such as this is the next step in the scientific process of the documentation of a subject, following the initial "source" material in journals. It is a documented attempt to digest such material, and should therefore be easier to read. However, much of the material presented here is new, as part of the process of filling gaps and providing a (more) complete picture. A number of new experiments have been carried out at Aspendale specifically for this volume.

I have attempted to give an adequate list of references so that readers can delve deeper into the subject, but it is not exhaustive, and some relevant work may have been omitted. I apologise in advance to any colleagues to whom due reference has not been given.

In its present form, this book has been made possible by the dedicated and professional efforts of several people at Aspendale: most notably David Murray, who has played a major part in most of my experimental studies over the past 10 years; David Whillas, whose talents are evident in several photographs; and Sean Higgins, who has skillfully adapted and created many of the line drawings. Thanks are also due to others who have provided continual background support.

I am grateful to several colleagues who have contributed photographs or figures, and to various copyright holders for permission to use some figures, and these are acknowledged in the captions. I am also grateful to others who have taken the time to read and comment on drafts of chapters in varying degrees of imperfection, and specifically these include Jim Rottman in particular, and Ian Castro, Terry Clark, Jack Katzfey, Peter Killworth, Greg Lawrence, Mike Sewell, Bill Snyder, Larry Armi and Sharan Majumdar. Thanks are also due to numerous colleagues for informative discussion on the material of this book over many years, to George Batchelor for his advice and support, and to Alan Harvey, Brian Watts and the staff of Cambridge University Press for their cooperation and attention to detail.

1

Background

I shall not be satisfied unless I produce something
that shall for a few days supersede the last fashionable
novel on the tables of young ladies.
LORD THOMAS B. MACAULAY, on his *History of England.*

The term "stratified flow" is commonly used to denote the flow of "stratified fluid", or more correctly "density-stratified fluid", and it is so used here. In such fluids the density (mass per unit volume) varies with position in the fluid, and this variation is dynamically important. Normally this density variation is stable with lines of constant density oriented nearly horizontally, with lighter fluid above and heavier fluid below. The density variation may be continuous, as occurs in most of the atmosphere and ocean, or be concentrated in discontinuous interfaces, such as at the surface of the ocean. In many situations the variation in density is very small, but such variations may have a dominant effect on the flow if the small buoyancy forces are given sufficient time to act. This book is about the motion of such fluids caused by their flow over topographic features. There has been substantial progress in recent years in this area. It is now possible to view the subject as a whole, and to understand the relation between, for example, the flow of a river over a ridge or weir, and the flow of the atmosphere over a mountain range. Whilst some details remain to be resolved, a corner has been turned, and the subject may now be viewed from a new and broader perspective.

The terms "topography", used in the title, and the equally common "orography" are not equivalent. The first may be taken to mean any departure from a level surface, such as lumps and bumps in a laboratory experiment for example. "Orography", on the other hand, implies obstacles on the large scale such as mountain ranges, which affect stratified processes in the atmosphere, and excludes rocks, buildings and the like, which don't. Topography therefore includes orography, and we will use the first term here for generality and consistency, although the second may be more appropriate when relating to atmospheric flows.

The subject of stratified flows is relevant to meteorology, oceanography and environmental engineering in the broad sense, with specific applications ranging from flow around hills to flow under ice keels. However, the impetus for the recent progress has been driven by atmospheric considerations more than others. This is for two reasons. First, it has become clear that a lack of understanding of the effects of mountain ranges and smaller topographic features on the atmosphere has been a significant impediment to the improvement in weather forecasting. The form drag (as distinct from surface frictional drag) of the topography is known to constitute about 50% of the total drag on the atmosphere (e.g. Palmer et al., 1986), and this drag is manifested in stratified effects such as internal gravity waves. The correction of this deficiency requires adequate parametrisation of sub-grid-scale topographic effects in large-scale models for numerical weather prediction. The conspicuous deficiencies have stimulated atmospheric field research programmes such as ALPEX, PYREX, MAP and T-REX, as well as many smaller studies. Second, and on a smaller scale, increasing concerns about environmental issues such as atmospheric pollution and air quality have instigated a number of mesoscale field and other studies which aim to describe the motion of air in or around mountains, hills and valleys, collectively known as "complex terrain", in considerable detail (see Blumen, 1990). However, field programmes do not usually provide mechanistic answers by themselves, because of the general sparseness, paucity and ambiguity of most field data sets. Consequently, most of the improved understanding in dynamics has instead come from analytical, numerical and laboratory studies, with the field data supplying confirmation of applicability to the atmosphere.

The effects of the Earth's rotation (i.e. the Coriolis force) have been excluded from the whole of this monograph, and this places some restrictions on the applicability of the material described to atmospheric flows. Specifically, it strictly applies to atmospheric flows that last for a few hours or less (significantly less than a pendulum day), and have length-scales of a few tens of kilometres or less. In the ocean, where fluid velocities are typically smaller by a factor of 100, the maximum relevant length-scale is smaller by the same factor and is typically several hundred metres. However, when the flow is constrained to a channel, as in narrow straits, estuaries and rivers, the effect of rotation on flow in the downstream direction can be small, even over quite large distances.

"Text books" that aim to cover particular areas of current research are not that common, but I would like to mention two that concentrate on areas that are outside the main theme of this book, but overlap with it to some extent. The first is *Rotating Hydraulics: Nonlinear Topographic Effects in the Ocean and Atmosphere*, by Pratt & Whitehead, which includes rotation and the Coriolis force in almost all sections.

The second is *Buoyancy-Driven Flows*, edited by Chassignet et al., which contains 10 chapters by separate distinguished authors, where the effects of buoyancy are paramount, and any effects of rotation or topography are incidental. There is a complementary relationship between all three.

The remainder of this chapter is concerned with the basic equations and their applicability to the atmosphere, their relevant boundary conditions, some conservation equations that will be needed later, and some comments on terminology. Chapters 2 and 3 are concerned with the properties of topographically forced flows of a homogeneous fluid layer with a free surface, which is applicable to flow in a river or channel. This system constitutes, in a sense, the simplest example of a stratified fluid, but there is still plenty of scope for non-linear complexity here, and it provides a useful prototype for the more general stratification considered in subsequent chapters. In Chapters 4, 5 and 6 the additional effects present when the layer is surmounted by a second layer are discussed. These effects on the lower layer may only amount to "reduced gravity" due to the density of the overlying layer if the latter is deep, the motion is inviscid, and it has long horizontal length-scales. On the other hand, if the upper layer is shallow and in motion, the character of the flow may be quite different from that of a single layer. An example of this is the important case of "exchange flows", where the two layers are flowing in opposite directions. Chapter 7 describes gravity currents and flows down slopes, notably into stratified environments, and in Chapters 8 and 9 we proceed from two layers to "many", and discuss the behaviour of disturbances in continuously stratified fluids *per se*, without considering the effect of topography specifically. For the most part, this concerns the general properties of small-amplitude internal gravity waves in stratified flows with shear. Chapters 10–14 are devoted to the effects of continuously stratified flow over two- and three-dimensional topography, respectively, and constitute the heart of the book, whereas Chapter 15 is concerned with the application of the material of the preceding chapters to laboratory modelling of flow over complex terrain, and the parametrisation of sub-grid-scale orographic effects in numerical models of the atmosphere.

1.1 Equations for fluid motion

The equations governing the motion of a stratified fluid are (see, for example, Batchelor, 1967)

$$\frac{D\mathbf{u}}{Dt} = -g\hat{\mathbf{z}} - \frac{1}{\rho}\nabla p + \nu\nabla^2\mathbf{u}, \tag{1.1}$$

$$\frac{1}{\rho}\frac{D\rho}{Dt} + \nabla \cdot \mathbf{u} = 0. \tag{1.2}$$

where $\mathbf{u} = (u, v, w)$ is the fluid velocity, ρ the fluid density, p the pressure, g the acceleration due to gravity, and the kinematic viscosity $\nu = \mu/\rho$ (assumed spatially uniform), where μ is the viscosity. The term D/Dt denotes the Lagrangian derivative with respect to time t, and $\hat{\mathbf{z}}$ is the unit vector vertically upwards. In most situations the viscous term is small, and may be neglected for present purposes. If the density of each very small particle of fluid always remains constant as it moves, in spite of variations in pressure, the fluid is *incompressible* and we have

$$\frac{D\rho}{Dt} = 0, \tag{1.3}$$

which in conjunction with (1.2) implies

$$\nabla \cdot \mathbf{u} = 0. \tag{1.4}$$

An important quantity that characterises continuously stratified fluids is the *buoyancy frequency* N (formerly known as the Brunt–Väisälä frequency), which for incompressible fluids at rest is defined by

$$N^2 = -\frac{g}{\rho}\frac{d\rho}{dz}, \tag{1.5}$$

where N is the frequency of local unforced vertical oscillations of small amplitude, and is the highest frequency that local buoyancy-driven fluctuations may have. It therefore gives a characteristic time-scale for these motions (see, for example, Gill, 1982, for more details). This frequency N has maximum values of about 10^{-2} rad/s in both the atmosphere and ocean, and this sets the minimum time-scale for the motions there that we will be considering. Under these circumstances, it is often appropriate to consider these motions as variations about a basic state, which is in hydrostatic equilibrium. The pressure and density fields may be expressed as

$$p = p_0(z) + p'(x, y, z, t), \qquad \rho = \rho_0 + \rho'(x, y, z, t), \tag{1.6}$$

where p_0 and ρ_0 represent the values in hydrostatic equilibrium, and are related by

$$\frac{dp_0}{dz} = -\rho_0 g. \tag{1.7}$$

The inviscid equation of motion for the perturbations may then be written

$$(\rho_0 + \rho')\frac{D\mathbf{u}}{Dt} = -g\rho'\hat{\mathbf{z}} - \nabla p'. \tag{1.8}$$

For most purposes water is effectively incompressible, but in general air is not. The equilibrium state of the atmosphere, specified by $p_0(z), \rho_0(z)$, is obtained from (1.7) and the equation of state for a uniform ideal gas, $p = \rho\mathcal{R}T$, for a specific temperature profile $T(z)$. Here $\mathcal{R} = \mathcal{R}^*/M$, where $\mathcal{R}^* = 8314.3$ joule/(Kelvin kilomole) is the *Universal Gas Constant*, and M is the molecular weight of the gas in question. Since air is a mixture of gases, we require its mean or effective molecular weight, which for dry air is 28.97; and this gives $\mathcal{R} = 287$ joule/(Kelvin kilogram). (For a more detailed discussion of the thermodynamics, including the effects of water vapour, the reader is referred to Wallace & Hobbs, 1977.) If $T(z)$ (in Kelvin) is taken to be uniform with height, as a simple approximation to the mean atmospheric profile, then $p_0(z) = \rho_0(z)\mathcal{R}T$, and (1.7) gives

$$p_0(z) = p_0(0)e^{-z/H_s}, \qquad \rho_0(z) = \rho_0(0)e^{-z/H_s}, \qquad (1.9)$$

where H_s is the *scale height*, defined by $H_s = \mathcal{R}T/g$. If T is 280 K, then $H_s = 8.2$ km. The equations (inviscid for simplicity) for the adiabatic motion of an ideal gas are (1.1) omitting viscous terms, (1.2) and

$$\frac{D}{Dt}(p/\rho^{\gamma}) = 0, \quad \text{or} \quad \frac{D\theta}{Dt} = 0, \qquad (1.10)$$

where γ is the ratio of specific heats ($\gamma = 1.4$ for air) and θ is the *potential temperature* defined by $\theta = T(p_r/p)^{1-1/\gamma}$, where T is the actual temperature at pressure p, p_r is the pressure at the reference level, usually 1000mb, and θ is the temperature an air parcel would have if transported adiabatically to the level where $p = p_r$. For a compressible ideal gas the buoyancy frequency N is given by

$$N^2 = \frac{g}{\theta}\frac{d\theta}{dz}. \qquad (1.11)$$

There are circumstances under which air in motion satisfying these equations may be regarded as incompressible. These have been discussed in general terms by Batchelor (1967), and we consider them here in the special context of flow forced by topography. From (1.2) and (1.10) we may obtain

$$\frac{1}{\gamma p}\frac{Dp}{Dt} = -\nabla \cdot \mathbf{u}, \qquad (1.12)$$

and for the fluid to be effectively incompressible we require

$$\left|\frac{1}{\gamma p}\frac{Dp}{Dt}\right| \ll \left|\frac{\partial \mathbf{u}}{\partial x}\right|, \qquad (1.13)$$

where the x-direction is chosen to be the principal direction of fluid flow (if there is one), and $\partial \mathbf{u}/\partial x$ is taken to be representative of the constituent terms in $\nabla \cdot \mathbf{u}$. If the variations in $\mathbf{u}$ due to the topography have magnitude U and horizontal length-scale L, and the variables are expressed relative to a hydrostatic basic state as in (1.8), then (1.3) becomes

$$\left|\frac{1}{\rho c_s^2}\frac{\partial p}{\partial t} - \frac{1}{2c_s^2}\frac{D\mathbf{u}^2}{Dt} - \frac{g'w}{c_s^2}\right| \ll \frac{U}{L}, \qquad (1.14)$$

where $c_s = (\gamma p/\rho)^{1/2}$ is the speed of sound in an ideal gas, and $g' = \rho'g/\rho$. Now if the pressure varies with a frequency ω, the magnitude of these pressure variations will be of order $\rho L U \omega$, so that the first term on the left-hand side will be of order $L U \omega^2 c_s^2$, and the second of order $\omega U^2 / c_s^2$. If the maximum local frequency of oscillations forced by flow over topography is the buoyancy frequency N, there are two possible choices for ω, namely N and the advective frequency, U_T / L, where U_T is the total fluid speed relative to the ground (and by implication, $U < U_T$). Admitting both of these magnitudes, it follows that the three terms on the left-hand side of (1.14) will all be individually smaller than the right-hand side if the following (not respective) conditions are met:

$$\frac{U_T^2}{c_s^2} \ll 1, \qquad \frac{N^2 L^2}{c_s^2} \ll 1, \qquad \frac{g'w}{c_s^2} \ll \frac{U}{L}. \qquad (1.15)$$

For internal gravity waves forced by topography with frequency comparable with N, the appropriate length-scale is $L \sim U_T / N$; on the other hand, if L is a topographic length-scale that is much longer than U_T / N, the relevant frequency is not N but U_T / L. Either way, the second criterion reduces to the first, and since $c_s \sim 330$ m/s, these two conditions will be met provided $U_T < 100$ m/s. This is normally the case in the atmosphere, with the exception of jet streams in the upper troposphere.

For the last of (1.15), we may assume that $U/L \sim W/H$, where W is a typical magnitude for the vertical velocity and H is a typical vertical length-scale for its variation. We also have $g' \sim N^2 H$, so that the third requirement becomes

$$\frac{N^2 H^2}{c_s^2} \ll 1. \qquad (1.16)$$

This criterion is satisfied in the atmosphere if $H < 10$ km, which is normally the case.

Approximation (1.16) may be avoided if $\partial \rho / \partial t$ is small, so that (1.2) may be approximated by

$$\nabla \cdot (\rho \mathbf{u}) = 0. \qquad (1.17)$$

This equation then takes the place of (1.4), and this approximation is termed the *anelastic approximation* (Ogura & Phillips, 1962), because the system no longer supports sound waves. For its validity we require the first two of (1.15), namely $U_T^2 / c_s^2 \ll 1$, as discussed above, and the assumption that the motion is buoyancy driven so that no frequency is greater than N. This approximation has been used in a number of numerical studies that will be described later. The solutions of this system are very similar in character to those of the incompressible system.

In addition to anelasticity and incompressibility, two further approximations will often be made, namely the Boussinesq approximation and the hydrostatic approximation. Each of these approximations provides a useful simplification of the

dynamical equations in appropriate circumstances, without significantly affecting
the character of the motion being studied. The conditions for their validity are as
follows. We consider motion in a layer of fluid of total depth D and express ρ_0 as
$\rho_0 = \overline{\rho} + \triangle\rho_0(z)$, where $\overline{\rho}$ is a mean density in the fluid layer. For incompressible
and anelastic motions, ρ' will be of the same order of magnitude as $\triangle\rho_0(z)$. If
we have $\triangle\rho_0/\overline{\rho} \ll 1$ in this layer, we may replace $\rho_0 + \rho'$ in (1.8) by its mean
value $\overline{\rho}$, and incur an error of relative magnitude $\triangle\rho_0(z)/\overline{\rho}$. The density variations
are thereby neglected in the inertia term, but retained in the buoyancy-force term
where they are multiplied by g. This constitutes the *Boussinesq approximation*. It
is normally a good approximation for all watery fluids in geophysical situations,
and the analytic simplification that it provides is valuable. For this approximation
to be valid in the atmosphere we also require $D \ll H_s$, the scale height, which
restricts us to the lowest 1 or 2 km of the atmosphere. This is not as restrictive as
it sounds, however. Bretherton (1966) has shown that it is possible to transform
the anelastic equations to a form in which the Boussinesq approximation remains
valid for *linear* motions over a much deeper layer, provided that the vertical scale
H of such motions is much less than H_s. Throughout this book, we will mostly be
concerned with the Boussinesq form of the equations, since these apply to the lower
part of the atmosphere and contain the essence of the topics being covered, without
surplus complexity. For readers who are interested in upper atmosphere phenomena,
most of these results are at least qualitatively valid in the non-Boussinesq system.

The *hydrostatic approximation* is valid when

$$\frac{\partial p'}{\partial z} \cong -\rho' g, \tag{1.18}$$

implying that the dynamical variations about the mean state are in hydrostatic
balance, in addition to the mean flow state itself (from (1.7)). For this to hold, we
require

$$\left|\frac{\partial w}{\partial t} + \mathbf{u} \cdot \nabla w\right| \ll \left|\frac{\partial p'}{\partial z}\right|. \tag{1.19}$$

If U and W are taken to be the horizontal and vertical velocity variations on the
length-scales L and H respectively, and with $W \sim UH/L$ and $H \sim U_T/N$, it may
be readily shown that (1.19) is satisfied if

$$\frac{U_T}{NL} \ll 1. \tag{1.20}$$

For small-amplitude motions L is the horizontal scale of the forcing topography,
but for large-amplitude motions we may have $L \sim U_T/N$ also, so that hydrostaticity
depends on amplitude, or steepness of the streamlines.

Leaving aside the effect of pressure, the principal factor causing the variation in the density of air is temperature, and the density of fluid particles may alter because of the molecular diffusion of heat from one to another. The density of sea water is affected by both temperature and salinity, which diffuse at different rates. For present purposes we will assume that the density of the fluid concerned is affected by the diffusion of a single component, notionally the temperature T, which satisfies

$$\frac{DT}{Dt} = \kappa_T \nabla^2 T, \tag{1.21}$$

where κ_T is the thermal diffusivity. For an incompressible fluid with an equation of state of the form $\rho = \rho(T)$, we may write

$$\frac{D\rho}{Dt} = \frac{d\rho}{dT}\frac{DT}{Dt} = \kappa \nabla^2 \rho, \tag{1.22}$$

where κ is the diffusivity of density. If $\rho(T)$ is effectively linear over the range of interest with the form

$$\rho(T) = \rho_0[1 - \alpha(T - T_0)], \tag{1.23}$$

where ρ_0, α and T_0 are constant, then $\kappa = \kappa_T$. If molecular diffusion is important, therefore, (1.22) replaces (1.3) in the equations for the motion of an incompressible fluid.

In many cases it is convenient to approximate a continuously stratified fluid with a fluid that is made up of a superposition of distinct layers, each of uniform density. Normally, each of these homogeneous layers is thin relative to the length-scale of the motion, and the horizontal velocity within the layer may be supposed to be uniform through the local depth of the layer. Layered models are often used in conjunction with the hydrostatic approximation. Many phenomena of general interest and applicability may be described with a system consisting of only one or two layers, and Chapters 2 to 5 are concerned with such systems.

1.2 Boundary conditions

Most flows considered will be established by time-dependent development from some known initial state, so that one can see how a given state may be established. In many situations with stratified fluids, it is possible to obtain steady-state flow solutions by assuming known flow conditions upstream (or downstream). However, in two-dimensional (or nearly two-dimensional) situations, there are restrictions on the properties of the steady upstream velocity and density profiles that may exist for an obstacle of given height (see Chapter 5), and for those profiles that are permissible, it may not be obvious how such a flow could be established. It is not difficult to obtain steady solutions that appear to be unrealistic or unphysical. For this reason, we concentrate on initial-value problems where the motion is commenced from simple initial states.

We will mostly be considering isolated topography; that is, topography where the lower surface becomes horizontal at large distances from the origin (though not necessarily at the same level). Boundary conditions on the flow at large distances from the topography will embody the assumption that there is no inward-propagating energy from infinity, other than that specified (which in the cases considered here is nil).

Since we are mainly concerned with inviscid equations, we will mostly omit viscous stresses and diffusive effects here in the boundary conditions. In general, the lower boundary will be a rigid surface specified by $z = h(x, y)$, where $h \to 0$ as $x^2 + y^2 \to \infty$. For inviscid flow, the boundary condition is that the velocity component normal to this surface must vanish. For viscous flow, the tangential component must also vanish. For an isolated obstacle, the maximum value of $h(x, y)$ is denoted by h_m, but this will be abbreviated to h in dimensionless ratios such as Nh/U. The upper boundary condition may take one of two main forms. The first is the "finite-depth" form, where the stratified fluid is bounded above by either a rigid horizontal surface – a "rigid lid", or alternatively by an infinitely deep homogeneous layer. For the rigid lid, the boundary condition is

$$w = 0, \qquad \text{at } z = D. \tag{1.24}$$

There is no existing term for the upper boundary of a fluid surmounted by a deep homogeneous layer, and it is defined here to be a *pliant surface* or *pliant boundary*. The appropriate boundary conditions for the stratified fluid with a pliant surface (a material surface of fluid particles) at $z = d = D + \eta(x, y, t)$, with homogeneous fluid above it, are

$$\left. \begin{matrix} w = D\eta/Dt, \\ \\ p\text{-continuous} \end{matrix} \right\} \qquad \text{at } z = D + \eta, \tag{1.25}$$

where here "p-continuous" implies that the pressure at the boundary depends on its elevation in the static upper layer. If the fluid above is immiscible with the fluid below, surface tension forces are present, but these are ignored in this work. If the density of the upper homogeneous fluid is (effectively) zero the pliant surface is termed a *free surface*, and (1.25) becomes $p = 0$.

The second form of upper boundary condition, the "infinite depth" form, applies to an infinitely deep stratified fluid where internal wave energy may propagate to great heights. This condition is a "radiation condition", which specifies that the waves radiate "out the top" without reflection. This means that there is no downward propagation of wave energy above a certain level ($z = D$, say).

1.3 Conservation relations

For the purposes of this section we take more general forms of the equations for incompressible flow by adding external forcing and heat gain (or loss) terms, so that (1.1) and (1.3) become

$$\frac{D\mathbf{u}}{Dt} = -g\hat{\mathbf{z}} - \frac{1}{\rho}\nabla p + \mathbf{F}, \qquad (1.26)$$

$$\frac{D\rho}{Dt} = -\dot{H}, \qquad (1.27)$$

and (1.2) still applies. The term $\mathbf{F}$ represents some general and unspecified forcing on the otherwise inviscid flow, and $\dot{H}$ (in suitable units) represents some heating process that affects the density without changing the total mass. Both of these may represent internal processes, such as turbulence or viscous effects, or external ones such as additional body forces or radiative heating. In the applications to be considered in this book, $\mathbf{F}$ is only non-zero in certain regions, and $\dot{H}$ is always zero and is only included in this section for generality.

1.3.1 Total head and energy density conservation

The *total head* or *Bernoulli* function R is defined by

$$R(x, y, z, t) = \frac{1}{2}\mathbf{u}^2 + gz + \frac{p}{\rho}, \qquad (1.28)$$

and from (1.26), (1.27) and (1.2) we obtain

$$\frac{DR}{Dt} = \frac{1}{\rho}\frac{\partial p}{\partial t} + \mathbf{u}\cdot\mathbf{F} + \frac{p}{\rho^2}\dot{H}. \qquad (1.29)$$

In steady flow without forcing or heating, therefore, R is constant along a streamline (the Bernoulli integral). On the other hand, if the fluid passes through a region where there are significant frictional or turbulent stresses (represented by $\mathbf{F}$), such as may occur in a hydraulic jump (see §2.3.1), the flow on each streamline will be affected by these stresses, which will reduce R. The same equations may also be used to derive the equation for the energy density per unit volume, e, where

$$e = \frac{1}{2}\rho\mathbf{u}^2 + \rho gz, \qquad (1.30)$$

in the form

$$\frac{\partial e}{\partial t} = -\nabla\cdot(\rho\mathbf{u}R) + \rho\mathbf{u}\cdot\mathbf{F} + \frac{p\dot{H}}{\rho}. \qquad (1.31)$$

Therefore $\rho\mathbf{u}R$ constitutes the total energy flux, with the additional terms providing energy production or dissipation.

1.3.2 Potential vorticity conservation and flux

We define the vorticity by $\boldsymbol{\Omega} = \nabla \times \mathbf{u}$, and taking the curl of (1.26) gives the vorticity equation

$$\frac{D\boldsymbol{\Omega}}{Dt} = \boldsymbol{\Omega} \cdot \nabla \mathbf{u} + \frac{1}{\rho^2} \nabla \rho \times \nabla p + \nabla \times \mathbf{F} - \frac{1}{\rho} \dot{H} \boldsymbol{\Omega}. \tag{1.32}$$

On the right-hand side, the first term denotes vortex stretching, the second is the baroclinic generation term, the third denotes vorticity generation by direct forcing, and the fourth denotes vorticity generation by divergence due to heating. The baroclinic generation term only affects components of vorticity that lie in the surfaces of constant density. This suggests that a simpler equation may be found for the component of vorticity that is perpendicular to the density surfaces, and this is embodied in the Ertel *potential vorticity*, $\zeta = \nabla \rho \cdot \boldsymbol{\Omega}/\rho$. From (1.2), (1.27) and (1.32) we then obtain the equation for conservation of potential vorticity

$$\frac{D\zeta}{Dt} = \frac{1}{\rho} \nabla \rho \cdot \nabla \times \mathbf{F} - \frac{1}{\rho} \boldsymbol{\Omega} \cdot \nabla \dot{H}. \tag{1.33}$$

In the absence of the effects of $\mathbf{F}$ and $\dot{H}$, therefore, ζ is conserved for each fluid particle even for unsteady motions; the vortex stretching term has been incorporated through the density gradient in ζ.

Equation (1.33) may be expressed in a flux form for the density-weighted potential vorticity, namely

$$\frac{\partial}{\partial t}(\rho \zeta) = -\nabla \cdot \mathbf{J}, \quad \text{where } \mathbf{J} = \rho \zeta \mathbf{u} + \nabla \rho \times \mathbf{F} + \dot{H} \boldsymbol{\Omega}, \tag{1.34}$$

so that $\mathbf{J}$ is the total flux of $\rho \zeta = \nabla \rho \cdot \boldsymbol{\Omega}$, relative to fixed axes, incorporating both $\mathbf{F}$ and $\dot{H}$ (Haynes & McIntyre, 1987). In fact, by suitable manipulation of the above equations $\mathbf{J}$ may be expressed entirely in terms of flow variables, in the form (Schär, 1993)

$$\mathbf{J} = \nabla \rho \times \nabla R + \nabla \rho \times \frac{\partial \mathbf{u}}{\partial t} - \boldsymbol{\Omega} \frac{\partial \rho}{\partial t}, \tag{1.35}$$

even though $\mathbf{F}$ and $\dot{H}$ are non-zero and the direction of $\mathbf{J}$ no longer coincides with $\mathbf{u}$. For steady flow, therefore, $\mathbf{J} = \nabla \rho \times \nabla R$, so that the flux vectors of $\rho \zeta$ lie along the intersections of the surfaces of constant ρ and R. Only if $\mathbf{F}$ and $\dot{H}$ are zero (or their effects cancel) do these intersections coincide with streamlines.

The addition of processes represented by $\mathbf{F}(t)$, such as wave breaking, dissipation or turbulence, occurring in a localised region, can therefore affect the local distribution of potential vorticity, whose flux may be carried by either the advective or productive terms, or a combination of both. However, from (1.34) the total amount of $\rho \zeta$ within a region is unchanged if $\mathbf{J}$ on the boundary is not altered.

1.3.3 Integral relations

We assume that the lower boundary is at $z = h(x, y)$ and the upper boundary is at $z = d$, which may be rigid or pliant. If the equation of conservation of mass is expressed in the form

$$\frac{\partial \rho}{\partial t} + \nabla \cdot (\rho \mathbf{u}) = 0, \tag{1.36}$$

then integrating vertically gives (see, e.g., Phillips, 1977)

$$\frac{\partial}{\partial t} \int_h^d \rho \, dz = -\frac{\partial Q_\alpha}{\partial x_\alpha}, \quad \text{where } Q_\alpha = \int_h^d \rho u_\alpha \, dz, \quad \alpha = 1, 2, \tag{1.37}$$

where the use of the same index twice denotes summation. The term Q_α is the *mass flux* in the direction of x_α, where $x_1 = x$, $x_2 = y$. Similarly, integrating (1.26) vertically gives

$$\frac{\partial}{\partial t} Q_\alpha = -\frac{\partial}{\partial x_\beta} S_{\alpha\beta} - p(x, h, t)\frac{\partial h}{\partial x_\alpha} + \int_h^d \rho F_\alpha \, dz, \quad \alpha, \beta = 1, 2, \tag{1.38}$$

where $S_{\alpha\beta}$ is the *momentum flux tensor*, given by

$$S_{\alpha\beta} = \int_h^d \rho u_\alpha u_\beta + p \delta_{\alpha\beta} \, dz, \tag{1.39}$$

and where $\delta_{\alpha\beta}$ is the Kronecker delta function. The corresponding equation for the conservation of energy may be obtained by the vertical integral of (1.31)

$$\frac{\partial}{\partial t} E = -\frac{\partial}{\partial x_\alpha} \int_h^d \rho u_\alpha R \, dz + \int \left(\rho \mathbf{u} \cdot \mathbf{F} + \frac{p\dot{H}}{\rho} \right) dz, \tag{1.40}$$

where E is the vertically integrated energy density, given by

$$E = \int_h^d \rho(\mathbf{u}^2/2 + gz) \, dz = \int_h^d e \, dz. \tag{1.41}$$

These equations may be used to derive useful properties of flows when the latter are known to have a particular vertical structure.

When the flow is steady and independent of the y-coordinate, the drag force F_D on a localised obstacle is given by

$$F_D = \int p(x, h, t)\frac{dh}{dx} \, dx = -\Delta S_{11}, \tag{1.42}$$

where ΔS_{11} denotes the difference in S_{11} across the obstacle. Also, for two-dimensional (x, z) flows we may define a stream function Ψ by

$$\rho u = -\frac{\partial \Psi}{\partial z}, \quad \rho w = \frac{\partial \Psi}{\partial x}, \tag{1.43}$$

and the horizontal energy flux $\mathcal{F}(x)$ may be expressed as

$$\mathcal{F}(x) = \int_h^d \rho u R \, dz = -\int_h^d \frac{\partial \Psi}{\partial z} R(\Psi) \, dz = \int_0^Q R(\Psi) \, d\Psi = Q\overline{R(\Psi)}, \quad (1.44)$$

where $Q = Q_1$ and $\overline{R(\Psi)}$ is the average of R over Ψ. If the fluid density is constant, we may write $Q = \rho \mathcal{Q}$, where $\mathcal{Q}$ is the associated *volume* flux. The flux $\mathcal{F}$ is independent of horizontal position in the absence of external forcing, dissipation or heating (i.e. $\mathbf{F} = \dot{H} = 0$). These results have been derived for fluid between upper and lower boundaries at $z = d, h$, but they also hold for fluid between two material lines or surfaces.

1.4 Terminology

The present general state of the terminology in this subject is unsatisfactory. The most important dimensionless parameter in any study in the field is normally called the Froude (pronounced "Frood") number, F (though often with varying subscripts). This typically has the form $F = U/\sqrt{gl}$, where U is the fluid speed and l is a length. For flow of a stream past an obstacle there are several possible choices for l: it may be the horizontal length L or the vertical height h of the obstacle, or it may be the depth of the stream, d. These different choices have different dynamical significance, as is shown in later chapters. To say that "the Froude number" is "large", "small", or "equal to one" conveys no information unless one specifies how the Froude number is defined. This contrasts with the Reynolds number $R_e = Ul/\nu$, which may be defined with different lengths l, but the dynamical significance of the parameter in each case is really the same: viscously-dominated flow if R_e is small, turbulent if it is large, etc. The quantities for continuously stratified fluids that correspond to the above single layer parameters are U/NL, U/Nh and U/ND, where the depth of the fluid is now D, and these, and sometimes their reciprocals and squares, have all been termed "the Froude number" by someone at sometime. In general, all of these various quantities are important, and in order to choose a sensible and appropriate terminology for them, and in particular to escape from the "Froude number for everything" syndrome, it is helpful to review the historical background.

The Froude number is named after William Froude (1810–1879), who developed towing tank techniques and made numerous experiments on modelling the wave and frictional drag of ships. His work was continued by his son, Robert Edmund Froude (1846–1924). William Froude used dynamical similarity based on the quantity $U/\sqrt{gL}$, where L is the horizontal length of, say, a ship, and he is justifiably known as the father of laboratory modelling. However, contrary to popular belief, Froude did not discover this dynamical principle, as it was enunciated somewhat earlier by a Frenchman, Ferdinand Reech (Rouse & Ince, 1957). The quantity $U/\sqrt{gL}$ relates to the wave drag produced by a ship or other obstacle. But in the hydraulics of streams the term denoted "Froude number" is usually $F = U/\sqrt{gd}$, although this appears never to have been used by William Froude. When compared with unity, this parameter $U/\sqrt{gd}$ specifies to what extent long linear gravity waves may propagate against the stream, and the result was first derived by Lagrange. In spite of this, the use of the term "Froude number" for both parameters in their separate contexts has been common since the work of Moritz Weber (1919). In single-layer flows here, however, we will only use the term "Froude number" for the quantity $U/\sqrt{gd}$, which represents a fluid speed divided by the significant wave speed. This conforms with the most common general acceptance, and gives significance to the numerical value. We will also take this latter property to be the basis for generalising the definition of "Froude number" to other more complex systems.

For a continuously stratified fluid, the parameters U/NL, U/Nh and U/ND are all important for stratified flow over topography, and for different reasons. The parameters U/NL and U/ND relate to wave drag and wave propagation, respectively, in the same manner as their single-layer counterparts, and (multiplied by appropriate constants) they may both, with some justification, also be termed Froude numbers. To extend this terminology further to include U/Nh, however, is inappropriate for two reasons. Firstly, U/Nh has different dynamical significance from U/NL and U/ND; it is not a fluid speed divided by a wave speed, nor is it simply related to the wave drag. As will be discussed later, it relates to non-linear factors, such as the steepness of the wave field and blocking of low-level fluid. Secondly, it extends the recognition of one man's contribution too far beyond his sphere of activity, especially when measured against that of many other eminent contributors. Since it does not help the comprehension of the subject to have several different parameters with different dynamical significance having the same name, and in order to distinguish this last parameter from the others, we will write it as Nh/U. A suitable name might be "Nhu" or, following Flanders and Swann, "Gnhu", with the symbol G. The important parameters would then be F and G.

A new approach by Mayer & Fringer (2017) with a rigorous scale analysis has shown that Nh/U is equivalent to the square of a Froude number for stratified flow over small obstacle heights, based on the internal variables (see Chapter 10). Here the length-scale in the Froude number is the vertical scale of the lee waves, and there is no connection with critical flow.

To be consistent with single-layer flows, the term "Froude number" for general stratified horizontal wave guides will here be used to denote the mean fluid speed divided by the relevant internal wave speed (usually the fastest). For unsheared uniformly stratified flow, this would be $U/(ND/n\pi)$, corresponding to the third of the above four possibilities. The flow may then be said to be subcritical if $F < 1$, critical if $F = 1$, and supercritical if $F > 1$, (with implications as described in succeeding chapters), in quite general circumstances.

Miles (1969) has suggested calling Nh/U the Russell number R_u after Scott Russell, who discovered solitary waves in a canal in Scotland, and made numerous other contributions to nautical engineering. The chief virtue in this would be that the five fundamental parameters in geophysical fluid dynamics would form a vowel-ordered set: R_a, R_e, R_i, R_o, (after Rayleigh, Reynolds, Richardson and Rossby) and R_u. However, Russell did not work on stratified fluids, and furthermore he opposed the funding of the construction of Froude's large tank, with which Froude did much of his most significant work, so that this name seems inappropriate also.

Underlying the above remarks is the strong feeling that the value of a named dimensionless parameter should mean something, so that the parameter is not just a vague assemblage of terms. The definition of a Reynolds number, for example, is standardised to the extent that if $R_e = 1, 100$ or $10\,000$, then the character and properties of the flow are specified. The value of the Froude number, as defined and used in this work, has the same sort of meaning.

There are other items for which the current terminology seems inappropriate or is non-existent, and new terminology is proposed here. These include the *pliant boundary* of §1.2 above and the *supercritical leap* of Chapter 4. A term that has sometimes been used for the latter and related phenomena in the past is *hydraulic drop*. Dr G.A. Lawrence and I have suggested that the term *hydraulic drop* be used to represent a hydraulic jump that is inverted (as on a thin upper layer in two-layer flow, for example), to avoid confusion, so that the interface (or streamlines) are depressed on passing through the drop, rather than rising as in a jump. The *supercritical leap* is a very different phenomenon, as described in §4.8.

2

Non-linear single-layer flow: classical hydraulics

> A classic is something that everybody wants to have read,
> and nobody wants to read.
> MARK TWAIN, *The Disappearance of Literature.*

In this chapter and the next we discuss the flow of an incompressible homogeneous fluid with a free surface past isolated topography. This system is a (relatively) simple surrogate for stratified flows in general. All of the phenomena described here have their counterparts in stratified flows, and a familiarity with the behaviour of the single layer is very helpful in understanding more complex flows. The converse does not apply: stratified flows contain many phenomena that are not represented in the flow of a single layer.

2.1 Basic equations

The equations for the motion of an incompressible, homogeneous inviscid fluid with density ρ are

$$\frac{D\mathbf{u}}{Dt} = -g\hat{\mathbf{z}} - \frac{1}{\rho}\nabla p, \tag{2.1}$$

$$\nabla \cdot \mathbf{u} = 0, \tag{2.2}$$

where $\mathbf{u} = (u, v, w)$ with components in the cartesian directions (x, y, z) respectively, with $\hat{\mathbf{z}}$ directed vertically upwards. In general we will assume that some bottom topography is present, which has the form $z = h(x, y)$, with a base level at $z = 0$. The boundary condition of zero velocity normal to this surface may then be expressed as

$$w = \mathbf{u} \cdot \nabla h, \quad \text{on} \quad z = h(x, y). \tag{2.3}$$

The fluid has an upper free surface at the mean level $z = d_0$, with a displacement η. (This notation is illustrated in Figure 2.2, page 22.) The mass of any fluid situated above this surface is negligible, so that the pressure at the surface is constant, and may be taken to be zero for convenience. Consequently we have

$$p = p_0 = 0, \quad \text{on} \quad z = d_0 + \eta.$$
$$w = D\eta/Dt.$$
$$(2.4)$$

The curl of (2.1) gives the vorticity equation $D\omega/Dt = \omega \cdot \nabla \mathbf{u}$, and if $\omega = 0$ initially it remains so, and the motion is irrotational throughout. If the vertical accelerations Dw/Dt are everywhere much less than gravity, the motion is hydrostatic, so that

$$-\frac{1}{\rho}\frac{\partial p}{\partial z} - g = 0. \qquad (2.5)$$

If the horizontal scale of the fluid motions is L, simple scale analysis shows that this approximation holds if $(d_0/L)^2 \ll 1$. We then have $p = p_0 + \rho g(d_0 + \eta - z)$ within the fluid, and the horizontal pressure gradient is independent of z. Hence, if u and v are initially independent of z, they will remain so. The equations of motion may then be expressed as (e.g. Whitham, 1974, p. 454)

$$u_t + uu_x + vu_y = -g\eta_x, \qquad (2.6)$$
$$v_t + uv_x + vv_y = -g\eta_y, \qquad (2.7)$$
$$\eta_t + (du)_x + (dv)_y = 0, \qquad (2.8)$$

where the suffices denote derivatives, and $d = d_0 + \eta - h$ is the thickness of the fluid layer.

In order to ensure that we are considering physically realisable flows, and that we know how to realise them, we discuss time-dependent problems with given initial conditions. These may take several forms, such as the sudden onset of a uniform flow past a stationary obstacle, or the sudden insertion of an obstacle in a moving stream (by upward movement of the bottom surface), or the sudden commencement of the motion of a towed obstacle through stationary fluid. These all imply the sudden onset of equivalent steady forcing, but there are differences in the mathematical form of the initial conditions for different cases. In most situations the final solution obtained in the large-time limit is not sensitive to the details of the initial conditions, but there are some significant exceptions, as discussed below.

2.2 Flows with small obstacle height

We may formally write down the solution of the above non-linear initial-value problems as a power series expansion for each of the dependent variables about the undisturbed state, where the expansion parameter ϵ is proportional to the maximum value of h, h_m say. The first term of this series is the "linear solution", and this usually dominates the total solution if h_m (and hence ϵ) is sufficiently small. Equations governing the linear solution may be obtained by substituting the power series for the dependent variables into the governing equations above, and omitting all terms which are $O(\epsilon^2)$. As we proceed through the chapter we consider the solutions to initial-value problems for flow over topography where the governing equations have increasing complexity.

2.2.1 Linear hydrostatic flow

We consider a uniform stream of velocity U in the x-direction, and insert a long obstacle with small height of the form $z = h(x, y)$ into this stream, by uplift from below. Linearising (2.6) to (2.8) about this initial state gives $u = U + u'$, and

$$u'_t + U u'_x = -g\eta_x, \quad v_t + U v_x = -g\eta_y, \quad \eta_t + U\eta_x + d_0(u'_x + v_y) = U h_x. \quad (2.9)$$

Eliminating u' and v then gives the equation for η:

$$\left(\frac{\partial}{\partial t} + U\frac{\partial}{\partial x}\right)^2 \eta - g d_0 \frac{\partial^2\eta}{\partial x^2} - g d_0 \frac{\partial^2\eta}{\partial y^2} = U^2 \frac{\partial^2 h}{\partial x^2}, \quad (2.10)$$

applicable for $t > 0$, with the initial conditions

$$\eta = h(x, y), \quad \frac{\partial\eta}{\partial t} = 0, \text{ at } t = 0. \quad (2.11)$$

The sudden introduction of the obstacle results in an instantaneous "potential flow", which causes a deformation of the interface but does not immediately alter the fluid velocity. This system has one dimensionless parameter, the Froude number F_0 for the undisturbed flow, defined by

$$F_0 = U/\sqrt{g d_0}, \quad (2.12)$$

which is the ratio of the flow speed to the speed of long gravity waves on stationary fluid of depth d_0. Flows with $F_0 = 1$ are termed "critical", those for $F_0 < 1$ "subcritical", and for $F_0 > 1$ "supercritical". Clearly, in subcritical flows linear waves (or "information") may propagate both upstream and downstream, whereas in supercritical flow upstream propagation is not possible.

In Chapters 2 and 3, the term "one-dimensional flow" means flow in the x-direction over a two-dimensional obstacle; "two-dimensional flow" means flow in the x–y plane past a three-dimensional obstacle.

One-dimensional flow If h is independent of y, the resulting flow will have $v = 0$ everywhere, and the solution to (2.10), (2.11) may be shown by standard methods to be

$$\eta = \frac{F_0^2}{F_0^2 - 1} h(x) + \frac{1}{2}\left(\frac{h[x - (U + \sqrt{g d_0})t]}{F_0 + 1} - \frac{h[x - (U - \sqrt{g d_0})t]}{F_0 - 1}\right), \quad (2.13)$$

provided that $F_0 \neq 1$. This solution is illustrated in Figure 2.1. It consists of a steady component over the obstacle and two waves, the one with larger amplitude propagating against the stream and the other with smaller amplitude with it. These two wave terms are functions of the characteristic variables of the hyperbolic equation (2.10), and will be different if the initial conditions are changed. The larger wave appears on the upstream side if $F_0 < 1$, and on the downstream side if $F_0 > 1$. All three terms have the form of the obstacle itself, and become long if the obstacle does. In particular, for an obstacle that is semi-infinite on the side $x > 0$, (that is, $h \to h_0 \neq 0$ as $x \to \infty$), the upstream propagating "transient" does not detach from the obstacle, and becomes a permanent part of the flow.

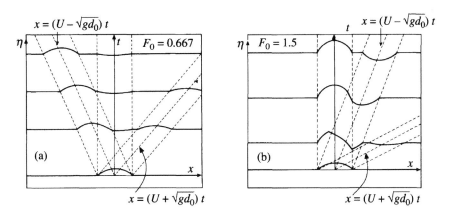

Figure 2.1 (x, t) diagrams showing time-dependent structure of linear hydrostatic flow solutions (2.13) of impulsively started flow over a two-dimensional (x, z) obstacle. The solution consists of a steady part plus two propagating transients. (a) $F_0 = 0.667$, (b) $F_0 = 1.5$.

The solution (2.13) becomes singular as $F_0 \to 1$, and cannot be valid near this point because the linearisation assumption of small amplitude is violated. The solution to (2.10), (2.11) for $F_0 = 1$ is

$$\eta = \frac{Ut}{2} \frac{\partial h}{\partial x} + \frac{1}{4}h(x - 2Ut) + \frac{3}{4}h(x). \tag{2.14}$$

Here the flow over the obstacle grows linearly with time, but the downstream-propagating wave is unaffected. This singular behaviour arises because the forcing by the obstacle resonates with the upstream-propagating wave. When $F_0 = 1$, the latter is stationary relative to the obstacle, and the equation for this mode has the form $\partial \eta / \partial t$ = (topographic forcing). For this resonant case there is an associated drag force on the obstacle, whereas there is no drag in the steady state for $F_0 \neq 1$.

One may ask whether this singular behaviour near $F_0 = 1$ is important in the overall description of one-dimensional flow over a ridge, given that only a small part of the total range of F_0 is involved. The answer is an emphatic yes, as will be seen below. If the initial flow (of speed U) is not directed normally to the ridge $h(x)$ but is instead inclined at an angle θ to this normal (the x-axis), the above analysis still applies but with U replaced by the normal component, $U \cos \theta$. There is also a uniform along-ridge velocity, $v = U \sin \theta$.

Two-dimensional flow We next consider (2.10), (2.11) with $h(x, y)$ a function of both x and y, representing an isolated piece of topography on the infinite (x, y)-plane. For $F_0 \neq 1$, the time-dependent solution again consists of initial transient waves radiating outward, leaving a steady solution. The details of the transients are mathematically complex and are of no particular interest here. When $F_0 < 1$, the steady-state form of (2.10) is elliptic, with the form of a Poisson equation. The steady-state solution for $F_0 < 1$ may be obtained by Green's function methods (see Morse & Feshbach (1953), for example), and is

$$\eta = \frac{F_0^2}{2\pi(1 - F_0^2)^{1/2}} \iint h(x', y') \frac{[(x - x')^2 - (1 - F_0^2)(y - y')^2]}{[(x - x')^2 + (1 - F_0^2)(y - y')^2]^2} dx' dy', \quad (2.15)$$

where the integral is taken over the region of the obstacle. This solution is again localised in the region of the topography, but extends beyond it. If the topography is symmetric about each of the x and y axes, so is the solution. At large distances from the topography where $x^2 + (1 - F_0^2)y^2 \gg (A^2 + B^2)$, and where A denotes the half-width of the obstacle in the x-direction and B the half-width in the y-direction, we may deduce from (2.15) that

$$\eta \approx \frac{F_0^2}{2\pi(1 - F_0^2)^{1/2}} \frac{x^2 - (1 - F_0^2)y^2}{(x^2 + (1 - F_0^2)y^2)^2} \iint h(x', y') dx' dy'. \quad (2.16)$$

The displacement η in the far-field is therefore proportional to the volume of the obstacle, and the surface is raised in the upstream and downstream directions and depressed in the transverse direction for positive topography ($h \geq 0$). Correspondingly, the fluid velocity is decreased upstream and downstream, and increased at the sides. The displacement over the obstacle, on the other hand, depends on its shape. In general terms, the surface is depressed over the topography if $(1 - F_0^2)^{1/2}B > A$, (cf. the one-dimensional solution above), and elevated if the reverse holds. Schematic diagrams of the surface for obstacles which are elongated in the across-stream and downstream directions are shown in Figures 2.2a and 2.2b, respectively. The solution becomes singular as $F_0 \to 1$, but the singularity is weaker than in the one-dimensional case $[(1 - F_0^2)^{-1/2}$ versus $(1 - F_0^2)^{-1}]$, because of lateral spreading of wave energy. The total drag force (that is, wave drag or form drag) on the obstacle is zero.

When $F_0 > 1$, the steady-state form of (2.10) is hyperbolic in the spatial variables.

Waves generated by the topography are all advected downstream. Again, we omit the details of the time-dependent solution. The steady-state solution may be obtained by Green's function methods and may be expressed as (Jiang & Smith, 2000)

$$\eta = \frac{F_0^2}{2(F_0^2 - 1)^{1/2}} \frac{\partial}{\partial x} \{f_-(x, y) + f_+(x, y)\}, \qquad (2.17)$$

where

$$f_-(x, y) = \int_{-\infty}^{y} h[x - (F_0^2 - 1)^{1/2}(y - y'), y'] dy',$$

$$f_+(x, y) = \int_{y}^{\infty} h[x + (F_0^2 - 1)^{1/2}(y - y'), y'] dy'. \qquad (2.18)$$

The integrals in (2.18) are of course confined to the region where h is non-zero. The form of this solution for a single-humped obstacle is shown in Figure 2.3. Away from the obstacle, the solution is confined to two beams extending indefinitely in the directions of the characteristics of the steady-state form of (2.10). These are oriented at the "Mach angle" $\alpha = \arcsin(1/F_0)$ to the flow direction, with η positive in the leading part of the beam and negative in the trailing part.

Figure 2.4 shows plan views of flow of a shallow layer past an obstacle towed at supercritical speed. In frames (a) and (b) the obstacle is dome-shaped, with $h_m/d_0 = 0.58$. The wave beams are quite evident, but are distorted by non-linear effects, with the leading and trailing parts of the beams steepening to form nearly discontinuous hydraulic jumps (see below), corresponding to shock waves in compressible flow. In (a), $F_0 = 1.69$, and the upstream jump is detached ahead of the obstacle, but in faster flow, (b), $F_0 = 3.59$) it is attached. The pattern of disturbance caused by a vertical circular cylinder (c) is similar but more turbulent, because all the fluid must pass around the obstacle.

In these supercritical flows there is a wave drag force on the obstacle, which contrasts with the previous (subcritical) flow situations where there was no drag. This drag may be calculated as follows. From integrating the energy equation derived from (2.9), one sees that the rate of working of the obstacle (via the drag force) is equal to the flux of wave energy into the beams. Further, the total energy flux in the y-direction for each beam is $\rho g d_0 \int \eta v dx$. Since η and v are only functions of ζ in the beams, (2.9) gives $v = -g(F_0^2 - 1)^{1/2} \eta / U$. The drag force F_D is therefore given by

$$U F_D = 2\rho \frac{g^2 d_0}{U} (F_0^2 - 1)^{1/2} \int \eta^2 dx, \qquad (2.19)$$

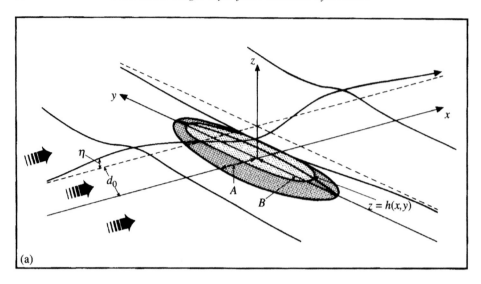

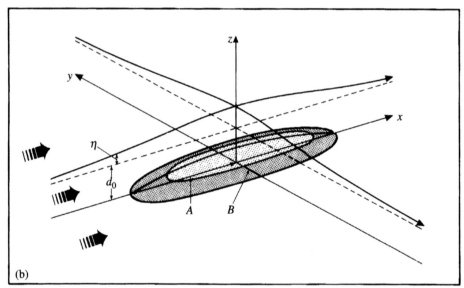

Figure 2.2 The steady-state linear hydrostatic solution (2.15), (2.16) over a three-dimensional obstacle (shaded) for $F_0 < 1$. (a) Across-stream; (b) downstream. The free surface over the obstacle is depressed if $B > A$, and elevated if $B < A$.

which is the total flux of energy away from the obstacle in both beams. This gives

$$F_{\mathrm{D}} = \frac{\rho g F_0^2}{2(F_0^2 - 1)^{1/2}} \int_{-\infty}^{\infty} f_{\pm}(x, y)^2 \mathrm{d}x, \qquad (2.20)$$

where $f_{\pm}(x, y)$ is given by (2.18) and represents either beam. For a single-humped

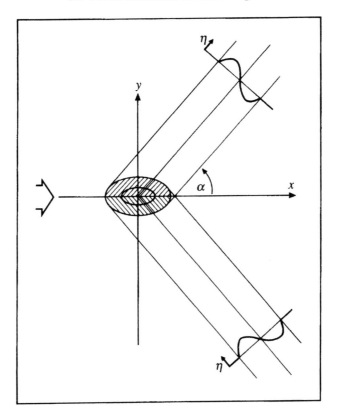

Figure 2.3 Plan view of the steady-state linear hydrostatic solution (2.17), (2.18) over an obstacle (shaded) for $F_0 > 1$. Away from the obstacle the flow disturbance is confined to two trailing beams inclined at the "Mach angle" α to the flow direction, where $\sin\alpha = 1/F_0$, and extending unchanged (in principle) to infinity. The structure of the interface across these beams depends on the obstacle shape, but for a single-humped obstacle it is elevated in the leading part of the beam and depressed in the trailing part.

axisymmetric obstacle of radius A, from (2.18) we may estimate $f_{\pm}(x, y)^2 \propto h_{\mathrm{m}}^2/F_0^4$ within each beam, so that (Jiang & Smith, 2000)

$$F_{\mathrm{D}} \sim \frac{\rho g h_{\mathrm{m}}^2 A}{F_0(F_0^2 - 1)^{1/2}}. \tag{2.21}$$

The supercritical solution (2.17) also becomes singular as $F_0 \to 1$, as $(F_0^2 - 1)^{-1/2}$. An examination of the time-dependent solution for $F_0 = 1$ shows that it grows as $t^{1/2}$ in the vicinity of the obstacle. In two space dimensions, therefore, linear theory also fails for F_0 near unity in the same manner as in the one-dimensional case, although here the singularity is not as strong because the generated wave energy

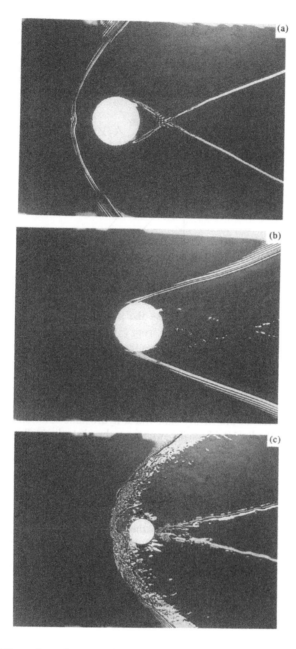

Figure 2.4 Plan view showing the surface features of the flow past obstacles towed through a shallow layer of water of depth 6 mm. (a) Dome-shaped obstacle of radius 5 cm, with $h_m=3.5$ mm, $d_0=6$ mm, $F_0=1.69$; (b) as for (a) but with $F_0=3.59$; (c) circular cylinder of radius 2.5 cm, with $d_0=6$ mm, $F_0=2.27$. Note the leading and trailing hydraulic jumps, or shock waves.

may disperse laterally. As may be seen from (2.20), (2.21), this singularity is also manifested in the drag force on the obstacle. In steady state, this drag is zero for $F_0 < 1$, but increases discontinuously to a theoretically infinite value as F_0 increases above unity.

There is a general analogy here between hydrostatic flow past obstacles moving at subcritical and supercritical speeds, and flow past thin two-dimensional bodies (thin to make the disturbances small, and hence linear) moving at subsonic and supersonic speeds in a compressible fluid. The Mach number, $M = U/c_s$, where c_s is the speed of sound, plays the same role as the Froude number F_0. As $M \to 1-$ the drag (on an aircraft, for example) is observed to increase substantially due to incipient shock wave formation and wave drag, constituting the "sound barrier". Equation (2.21) suggests corresponding behaviour here, and the analogy is discussed further in §3.5.

2.2.2 *Linear non-hydrostatic flow*

We return to the more general equations (2.1) to (2.4). Since the motion is irrotational, the fluid velocity may be written in the form of a velocity potential ϕ which from (2.2) satisfies Laplace's equation, so that

$$\mathbf{u} = \nabla\phi, \quad \text{where } \nabla^2\phi = 0. \tag{2.22}$$

The boundary conditions (2.4) at the free surface are

$$w = \frac{D\eta}{Dt}, \quad \frac{\partial\phi}{\partial t} + \frac{1}{2}\mathbf{u}^2 + g\eta = \text{constant}, \tag{2.23}$$

on $z = d_0 + \eta$, which together with the bottom boundary condition (2.3) and the initial conditions completely specifies the initial-value problem. If we again consider perturbations about an undisturbed state of velocity U in the x-direction, we write $\phi = Ux + \phi'$, and the linearised boundary conditions are

$$\frac{\partial\phi'}{\partial z} = \left(\frac{\partial}{\partial t} + U\frac{\partial}{\partial x}\right)\eta, \quad \frac{\partial\phi'}{\partial t} + U\frac{\partial\phi'}{\partial x} + g\eta = 0, \quad z = d_0, \tag{2.24}$$

$$\frac{\partial\phi'}{\partial z} = \left(\frac{\partial}{\partial t} + U\frac{\partial}{\partial x}\right)h, \quad z = 0. \tag{2.25}$$

One-dimensional flow In stationary fluid ($U = 0$) of uniform depth ($h = 0$), from (2.22)–(2.25) linear waves of the form $\exp i(kx - \omega t)$ satisfy the dispersion relation

$$\omega^2 = gk \tanh kd_0, \quad \text{with} \quad c^2 = \left(\frac{\omega}{k}\right)^2 = \frac{g}{k} \tanh kd_0, \tag{2.26}$$

where c is the phase speed of the waves. The *group velocity* c_g is given by $c_g = d\omega/dk$, and this represents the overall propagation velocity of the energy, or of a *group* of waves with wavenumbers approximately equal to k. In the limits when the depth d_0 is very large or very small, we have

$$kd_0 \gg 1, \quad c_g \approx \frac{1}{2}c \approx \left(\frac{g}{k}\right)^{1/2}, \tag{2.27}$$

$$kd_0 \ll 1, \quad c_g \approx c \approx (gd_0)^{1/2}, \tag{2.28}$$

the latter being the hydrostatic limit. For a general discussion of group velocity in the present context, the reader is referred to Lighthill (1978).

The above time-dependent problem of perturbed flow over a small bump may be solved by Fourier transforms. For simplicity we again confine our attention to the solution for large times, and write the topography in the form

$$h(x) = \frac{1}{2\pi} \int_{-\infty}^{\infty} \hat{h}(k)e^{ikx} dk, \tag{2.29}$$

where $\hat{h}(k)$ is the Fourier transform of $h(x)$. The general solution for ϕ' for large times when $F_0 \neq 1$ is given by

$$\phi'(x, z) = \frac{iU}{2\pi} \int_{-\infty}^{\infty} \hat{h}(k)e^{ikx}\left(\sinh kz + \cosh kz \frac{F_0^2 kd_0 \tanh kd_0 - 1}{\tanh kd_0 - F_0^2 kd_0}\right) dk. \tag{2.30}$$

This integral for the steady-state or large-time solution contains poles on the real k-axis, at points where

$$\tanh kd_0 = F_0^2 kd_0. \tag{2.31}$$

Consideration of how this steady solution is realised as time becomes large shows that the contour of integration should be taken below the poles (Stoker, 1953; Lighthill, 1978). For $F_0 > 1$, there are no poles (excluding the one at $k = 0$, which gives a constant contribution to ϕ' and does not contribute to the flow field), and the solution for η obtained from (2.24), (2.30) is evanescent away from the topography, and is symmetric about $x = 0$ if $h(x)$ is. Consequently there is no wave drag on the obstacle when $F_0 > 1$. For $F_0 < 1$, on the other hand, there is a contribution from the two poles of (2.31), and for $x > d_0$ and downstream of the obstacle this gives

$$\eta \simeq \frac{F_0^2 k_0}{(F_0^2 - 1 + F_0^4 k_0^2 d_0^2)\cosh k_0 d_0} \left[i\hat{h}(k_0)e^{ik_0 x} + \text{complex conj.}\right], \tag{2.32}$$

where k_0 is the positive solution of (2.31). This consists of a sinusoidal standing wave of wavenumber k_0 on the lee side of the obstacle, forced by the Fourier component of $h(x)$ with this wavenumber. Here k_0 is the wavenumber of waves that have the phase velocity U, as may be seen from (2.26), (2.31). On the upstream side the solution is again evanescent. The hydrostatic limit of (2.32) may be realised by taking the limit $d_0 \to 0$ with F_0 held fixed; this implies that $k_0 d_0$ is fixed so that k_0

becomes large; η then vanishes in this limit away from the obstacle because $\hat{h}(k_0)$ vanishes.

The wave drag force F_D on the obstacle may be calculated from energy considerations by equating the work done per unit time by the obstacle, UF_D, with the flux of energy away from it, as in §2.2.1. For such a plane wave with ϕ' given by (2.36), the energy density is $\rho g a^2/2$ (e.g., Phillips, 1977) and the energy flux is $(U - c_g)\rho g a^2/2$, where a is the wave amplitude given by (2.32) and c_g is the group velocity obtained from (2.26). This gives

$$F_D = \frac{\rho g |\hat{h}(k_0)|^2}{4\cosh^2 k_0 d_0} \frac{F_0^2 k_0^2}{(F_0^2 - 1 + F_0^4 k_0^2 d_0^2)}. \tag{2.33}$$

It is somewhat paradoxical that this one-dimensional dispersive system has wave drag for $F_0 < 1$ but not for $F_0 > 1$, and the two-dimensional non-dispersive system described above has wave drag for $F_0 > 1$ but not for $F_0 < 1$. However, this is quite natural given the steady-state waves possible in the two systems.

When F_0 is close to unity, (2.32) has the form

$$\eta \simeq \frac{\sqrt{3}}{2d_0(1 - F_0^2)^{1/2}} \left[i\hat{h}(k_0)e^{ik_0 x} + \text{complex conj.} \right] [1 + O(1 - F_0^2)], \tag{2.34}$$

which is singular at $F_0 = 1$. Further, (2.30) is not applicable at $F_0 = 1$ and gives unbounded values for η at large x. An examination of the full time-dependent solution of (2.22), (2.24), (2.25) (with suitable initial conditions, and the choice makes no difference) shows that $\partial\phi/\partial x$ (i.e. the velocity field) in the vicinity of the obstacle grows as $t^{1/3}$ for large times. Hence the linear solution is again singular at $F_0 = 1$ because the transients cannot escape from the generation region, but the singularity is not as severe as in the hydrostatic case $[(1 - F_0^2)^{-1/2}$ versus $(1 - F_0^2)^{-1}]$ because of the partial dispersion of wave energy. The wave drag F_D remains finite as $F_0 \to 1-$, with

$$F_D \to \frac{3\rho g |h(0)|^2}{8d_0^2}, \quad \text{as } F_0 \to 1-, \tag{2.35}$$

so that the drag is discontinuous at $F_0 = 1$, but is not singular. This is because the nett group velocity relative to the obstacle $(U - c_g)$ decreases as $F_0 \to 1-$, in an inverse manner to compensate for the increase in wave energy density.

Two-dimensional flow In the infinite (x, y)-plane, from (2.22), (2.24), (2.25) with $h = 0$, plane waves in fluid of depth d_0 and mean velocity U have the perturbation velocity potential

$$\phi' = \cosh(k^2 + m^2)^{1/2} z \cdot e^{i(kx+my-\omega t)}, \tag{2.36}$$

and satisfy the dispersion relation

$$(\omega + Uk)^2 = g(k^2 + m^2)^{1/2} \tanh (k^2 + m^2)^{1/2} d_0. \tag{2.37}$$

For flow of mean velocity U past an isolated obstacle, the total flow (excluding the vicinity of the obstacle) may be regarded as an integral of these Fourier components, whose amplitudes are determined by the forcing at the obstacle. An examination of the full initial-value problem (2.22), (2.24), (2.25) shows that there is no singularity at $F_0 = 1$: the combined processes of lateral dispersion and wavenumber dispersion are sufficient to remove the singularity that is present if either process is absent. Consequently, the linear theory is applicable over the whole range of Froude number values. The steady-state solution is comprised of those Fourier components for which $\omega = 0$, so that from (2.37) they must satisfy

$$K^2 = (K^2 + M^2)^{1/2} \tanh[(K^2 + M^2)^{1/2}/F_0^2], \tag{2.38}$$

where

$$K = kU^2/g, \qquad M = mU^2/g. \tag{2.39}$$

With this scaling, the limit of infinite depth $d_0 \to \infty$ is obtained by taking $F_0 \to 0$, so that (2.38) becomes

$$K^2 = (K^2 + M^2)^{1/2}. \tag{2.40}$$

The shapes of the wavenumber surfaces (2.38) for several representative values of F_0 are shown in Figure 2.5. The group velocity of the waves is defined by $c_g = (\partial\omega/\partial k, \partial\omega/\partial m)$, which in this case is obtained by differentiating (2.37). This implies that the directions of c_g in physical space are normal to the surfaces in the wavenumber space of Figure 2.5, if the x-direction is identified with the k-direction and y with m. These directions are indicated by the arrows normal to the surfaces in Figure 2.5. Such figures are known as "Lighthill diagrams", and have the physical interpretation that if one looks in the direction of the c_g arrows in physical space, one sees waves in the far-field that have the wavenumbers k, m corresponding to the location of the base of the arrow in wavenumber space. They are thus very useful for depicting the properties of this and other anisotropic dispersive wave systems encountered in subsequent chapters. The full theory is described in Lighthill (1978), to which the reader is referred for further details.

All the curves of Figure 2.5 are asymptotic to the same curve for large $K^2 + M^2$. There is also little variation with F_0 for $F_0 < 0.8$, where the curves are all very similar to the deep-water curve, $F_0 = 0$. For $F_0 < 1$ the curves cross the K-axis perpendicularly (although this is not apparent for F_0 near unity), but they have a cusp there for $F_0 = 1$. For $F_0 > 1$, for small K, the group velocity vectors (perpendicular to the curves) are aligned at the "Mach angle" of hydrostatic shallow-water waves.

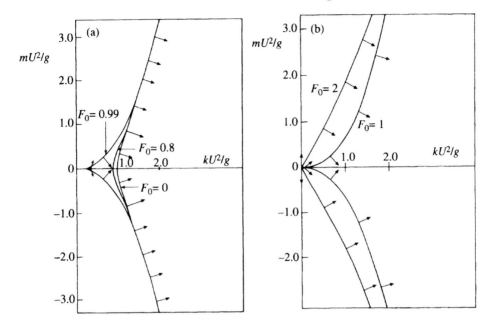

Figure 2.5 Lighthill diagrams for wavenumber surfaces at zero frequency for surface gravity waves (2.38), for a steady moving wave source with Froude number F_0. Arrows normal to the curves denote the directions of group velocity in physical space (with x and z corresponding to k and m) for that wavenumber vector. The case $F_0 = 0$ corresponds to the Kelvin ship-wake pattern, and the $F_0 \neq 0$ curves show the changes resulting from various finite depths. (a) $F_0 < 1$; (b) $F_0 > 1$.

The diagrams of Figure 2.5 show the far-field locations and types of the waves that may be generated, but their amplitudes depend on the properties of the source of the waves. In the case of forcing by topography, the wavenumber (k, m) will be forced if the topography contains Fourier components with this wavenumber. Long obstacles tend to force small wavenumbers and short obstacles large ones.

Such wave fields have been studied at length for the case of generation of waves by moving sources such as ships (Lighthill, 1978). In linearised theory, the forcing of waves by a ship is represented by a moving pressure pattern on the surface. This forcing pressure pattern is that which would be produced by the ship if the surface were constrained to be horizontal, rather than free, and it extends beyond the ship itself. The results are similar to, but different from, the effects produced by forcing due to flow past bottom topography. A given topographic shape on the bottom may be related to a corresponding surface pressure forcing field that will produce the same wake, or wave pattern, although the flow fields close to the forcing will of course be different. Ship wave theory is therefore directly relevant to flow past topography, although most of these studies have concentrated on the deep-water case where F_0 is small or zero. The curve $F_0 = 0$ of Figure 2.5 shows that the waves produced by the ship are confined within a wedge of half-angle $\alpha_K = \arcsin(1/3) = 19.5°$, spanning the ship's track. The term α_K is known as the *Kelvin ship-wake wedge angle*, and is also the angle made with the k-axis by the group velocity vector at the inflexion point of the wavenumber curve; all other group velocity vectors make a shallower angle, so that the corresponding waves lie within the wedge. Waves close to the boundary of the wedge have a wide range of wavenumbers, and the region is an example of a *caustic* – a boundary between a region with no waves and a region where two or more wavetrains are present. The wave field is evanescent outside the wedge with an Airy function structure at the caustic, which implies that the waves extend slightly beyond the Kelvin wedge angle (Lighthill, 1978).

The wave patterns actually produced by boats and ships show considerable variety. Such craft generate waves with a spectrum of different wavelengths, with a dominant wavenumber $k \sim 2\pi/L$, where L is the length of the vessel. For these waves, (2.40) gives

$$m \sim \frac{2\pi}{L} \left[\left(\frac{2\pi U^2}{gL} \right)^2 - 1 \right]^{1/2}. \tag{2.41}$$

If $2\pi U^2/gL \leq 1$, as for an ocean liner or large container vessel, the dominant waves produced have small m and are transverse, behind the ship. On the other hand if $2\pi U^2/gL \ll 1$, as for a speedboat, (2.41) gives $m \ll k$. The waves are then concentrated in a wedge of angular half-width $\sim gL/2\pi U^2$ which is much less than the Kelvin angle α_K, with the phase lines aligned at a similarly small angle to the boat's track.

As F_0 increases, the wedge angle increases and the (theoretically possible) wake broadens, so that wakes noticeably wider than the Kelvin angle of $19.5°$ are obtained for $F_0 \geq 0.7$. For $F_0 \geq 1$, the caustic disappears and the wake resembles the hydrostatic wake described above. The wedge angle is now close to $90°$, but it decreases as F_0 increases further and narrows to the Kelvin wake angle as $F_0 \to 3$. Narrower wakes are found when $F_0 > 3$, where the wake is confined within the Mach angle, which narrows to zero as F_0 increases to infinity. The observed wakes of small craft at high speed with $F_0 \ll 1$, described above, have sometimes been incorrectly attributed to shallow water effects (with $F_0 \gg 1$) in the past. Ship-wake situations where $F_0 > 0.7$ are the exception rather than the rule, but are possible in shallow water (for example, $F_0 = 1$ for ship (or boat) speed $U = 28$ knots in water of 20 m depth). For the case of forcing by topography, where U is the current speed, Froude numbers of order unity are common in shallow, fast-moving streams, and the resulting non-linear effects are discussed in the following sections. But for most of the time, for most rivers, the Froude numbers are small. In the stratified analogues to be discussed in subsequent chapters, and which are our primary concern, Froude numbers of all values are possible. I am not aware of any detailed observational tests of the linear theory for two-dimensional ship wakes, but the latter gives a good qualitative description of the disturbances produced by ducks, boats etc. (Examples are shown in Stoker, 1957, and Lighthill, 1978.) For flow over topography, there is no reason to question the validity of linear theory for sufficiently small h, but this range of validity is dependent on F_0.

2.3 One-dimensional non-linear hydrostatic flow

We now return to one-dimensional flow and investigate non-linear effects. These phenomena are most readily appreciated in the hydrostatic system with its long horizontal length-scales, where dispersive effects are not present. From (2.6), (2.8) we obtain for this system, with $d = d_0 + \eta - h$,

$$u_t + uu_x = -g\eta_x, \tag{2.42}$$

$$d_t + (du)_x = 0, \tag{2.43}$$

or alternatively

$$\eta_t + [(d_0 + \eta)u]_x = (uh)_x, \tag{2.44}$$

and we again take the initial conditions (2.11), namely $u = U$, $\eta = h$, $\eta_t = 0$, at $t = 0$. This is a classic system of hyperbolic diffential equations with a forcing term, which may be expressed in the characteristic form

$$\frac{d}{dt}(u \pm 2\sqrt{gd}) = -g\frac{dh}{dx}, \tag{2.45}$$

on the respective characteristic curves, which are given by

$$\frac{dx}{dt} = u \pm \sqrt{gd}. \tag{2.46}$$

For a discussion of this type of mathematical system, see Whitham (1974); an interesting geometrical approach has been described by Broad et al. (1993). For small times, the solution of this initial-value problem is essentially the same as for the linear system, and it is given by (2.13) and displayed in Figure 2.1. The flow is still governed by the value of the Froude number for the undisturbed flow $F_0 = U/\sqrt{gd_0}$, but as time increases the characteristics over the obstacle do not remain straight but become curved, depending on the values of the flow variables u and d. Representative examples for $F_0 < 1$ and $F_0 > 1$, showing the characteristics of the upstream-propagating waves only, are presented in Figure 2.6. In most situations of interest, the downstream-propagating wave (on the characteristics $dx/dt = u + \sqrt{gd}$)

is little affected by non-linearities, and travels quickly downstream away from the vicinity of the obstacle. Whilst necessary to satisfy the initial conditions, these waves are unimportant otherwise and are not relevant to the following discussion. However, on the upstream side of the obstacle, the equations for the variables on this same family of characteristics (namely (2.45), (2.46) with the plus sign) may be integrated to yield

$$u + 2\sqrt{gd} = U + 2\sqrt{gd_0}, \tag{2.47}$$

since the initial conditions are the same for each member of this family of characteristics on the upstream side. Substituting (2.47) into (2.45), (2.46) then yields, for the other set of characteristics (for waves propagating against the flow),

$$\frac{du}{dt} = \frac{dd}{dt} = 0, \text{ on } \frac{dx}{dt} = U + 2\sqrt{gd_0} - 3\sqrt{gd}, \tag{2.48}$$

on the upstream side of the obstacle. In this region, d is constant on each characteristic which is therefore straight, but the slopes are different for different characteristics. The result is that, for $F_0 > 1$, the characteristics converge and intersect, as shown in Figure 2.6a, because a disturbance of larger elevation travels faster. For $F_0 > 1$ the effect may be large enough to cause the upstream waves over the obstacle to change direction, as shown in Figure 2.6b. The same processes occur, *mutatis mutandis*, on the downstream side.

This phenomenon of wave speed being dependent on wave amplitude is termed *amplitude dispersion*, and is illustrated in a simpler form in Figure 2.7. A wave of elevation, here simplified to a monotonic increase in surface level moving to the left into undisturbed fluid, is also governed by (2.47) and (2.48), which show that the interface steepens with time as shown in Figure 2.7a. On the other hand, a decrease in surface level moving in the same direction and leaving shallower fluid at rest on the right becomes more spread out, or "rarefied", with time, as shown in Figure 2.7b. The upstream wave of elevation produced by the starting motion of the obstacle therefore steepens, to the point where the interface becomes vertical, and, in principle, overturns. Within the framework of the present model, this results in a "hydraulic jump" which may be modelled as a discontinuity, and it is not governed by the above equations *per se*. Hence we must consider its behaviour as an independent entity.

2.3.1 Hydraulic jumps

We consider a simple model of hydraulic jumps based on the principles of mass and momentum conservation, which will be adequate for our present purposes. More detailed properties of jumps will be discussed later in this chapter. We assume that a hydraulic jump may be regarded as a locally steady and compact region with its own

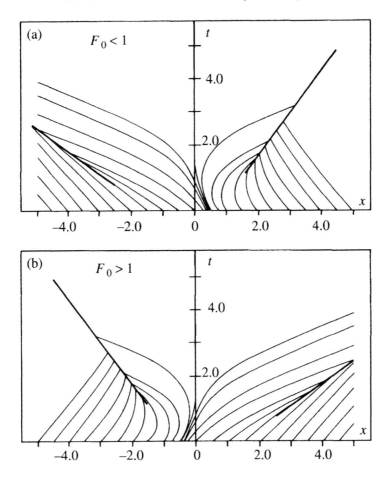

Figure 2.6 Representative characteristics for the upstream-propagating disturbances for non-linear hydrostatic flow forced by an obstacle impulsively set into
motion, when upstream jumps form (cf. Figure 2.1). The obstacle is centred at
$x = 0$, and decreases in height to near zero at $x = \pm 3$. The jumps are denoted by
heavy lines. (a) $F_0 < 1$, (b) $F_0 > 1$. (Adapted from Grimshaw & Smyth, 1986.)

internal dynamics, which may be modelled as a discontinuity between two uniform
streams, and that it is produced as a result of wave steepening as just described. We
then require equations relating conditions on the upstream and downstream sides of
the jump, which must be considered from first principles. At the most fundamental
level, we must have conservation of mass and momentum in the region of fluid
containing the jump. We take coordinates fixed relative to the jump, and in this
frame of reference denote the fluid velocity and depth on the upstream side by
u_u and d_u, and on the downstream side by u_d and d_d, as shown in Figure 2.8.

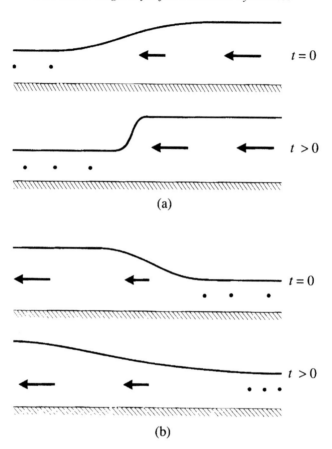

Figure 2.7 The development of a moving surge according to the hydraulic model. (a) A monotonic increase in depth, where the deeper fluid is moving to the left at $t = 0$ into fluid at rest, steepens at subsequent times to form a discontinuity or bore. Arrow length denotes speed of fluid, and dots denote fluid at rest. (b) The reverse situation where the deeper fluid moves away from the depth change, leaving fluid at rest on the right. At later times this disturbance is more spread out, or "rarefied".

Conservation of mass into and out of the jump then gives

$$u_u d_u = u_d d_d = Q, \tag{2.49}$$

where Q is the volume flux relative to the jump. Conservation of momentum applied to a vertical column of fluid implies that the vertically integrated form of (2.42) applies across the jump, and integrating (2.42) vertically and then horizontally gives

$$d_u u_u^2 + \frac{1}{2}g d_u^2 = d_d u_d^2 + \frac{1}{2}g d_d^2. \tag{2.50}$$

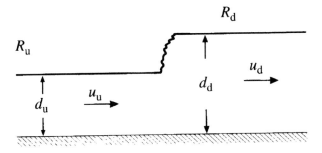

Figure 2.8 Notation in the frame of a steady hydraulic jump (see text).

These two equations completely specify the properties of the jump at this level of approximation, and enable relevant properties to be calculated. One of these is the rate of energy dissipation within the jump. The energy density of a column of fluid is given by

$$E = \frac{1}{2}\rho d u^2 + \frac{1}{2}\rho g d^2, \tag{2.51}$$

and from (2.42), (2.43), over level ground the rate of change of energy is given by

$$\frac{\partial E}{\partial t} = -\frac{\partial}{\partial x}\rho\left(\frac{1}{2}du^3 + gud^2\right) = -\frac{\partial}{\partial x}\rho Q R, \tag{2.52}$$

where $\rho Q R = \rho du(u^2/2 + gd)$ may be identified as the flux of energy, where d is the local depth. Note that the flux of energy is not simply given by uE, because each fluid column is not independent of the others, and account must be taken of work done by the pressure force that acts between them (see (1.28)–(1.31)). The rate of energy dissipation within the jump is then given by the differences between the energy fluxes into and out of the jump, namely

$$\frac{dE_J}{dt} = \rho Q(R_u - R_d) = \frac{\rho g Q}{4}\frac{(d_d - d_u)^3}{d_d d_u}, \tag{2.53}$$

where E_J denotes the energy of the fluid in a region containing the jump and moving with it. The mechanisms for this energy dissipation depend on the detailed internal dynamics of the jump, and (2.53) is effectively a requirement imposed by the external conditions. Note that for the energy dissipation to be positive we must have $d_d > d_u$.

For a jump advancing into water that is at rest, the speed of the jump in this frame, c_J, is given by $u_u = c_J$. Eliminating u_d from (2.49), (2.50) yields Rayleigh's hydraulic equation

$$c_J^2 = \frac{g d_d}{2}\left(1 + \frac{d_d}{d_u}\right), \tag{2.54}$$

a speed which is faster than that of linear waves on fluid of depth d_d (where wave speed $c = \sqrt{gd_d}$), because the fluid following the jump is moving in the same direction in this reference frame.

2.3.2 Flow solutions with topography

The steady-state flow solutions may be specified by the two dimensionless parameters F_0 and $H_m = h_m/d_0$, where h_m is the maximum height of the topography. If h_m is sufficiently small the solution is approximately linear, as described by (2.13). Here, when $F_0 < 1$ the upstream disturbance escapes from the region of the obstacle, and for $F_0 > 1$, the corresponding wave is advected away downstream. In both cases, we are left with a steady disturbance over the obstacle, where the upstream conditions are unchanged from the initial ones of $u = U, d = d_0$. The steady-state forms of (2.42), (2.43) may be expressed as

$$\frac{d}{dx}\left(\frac{1}{2}u^2 + gd + gh\right) = 0, \qquad \frac{d}{dx}(ud) = 0, \tag{2.55}$$

which give

$$ud = Q = Ud_0, \tag{2.56}$$

and

$$\frac{1}{2}u^2 + gd + gh = \frac{Q^2}{2d^2} + gd + gh = \frac{1}{2}U^2 + gd_0, \tag{2.57}$$

which gives $d(x)$ as a function of $h(x)$. This equation may be expressed as

$$\frac{1}{2}\left(\frac{F_0d_0}{d}\right)^2 + \frac{d}{d_0} + \frac{h}{d_0} = \frac{1}{2}F_0^2 + 1, \tag{2.58}$$

and the solutions are qualitatively similar to those obtained from linear theory (2.13) and shown in Figure 2.1. The range of applicability of (2.57), (2.58) is limited, though, and the limit is seen from (2.55) which gives

$$\left(\frac{u^2}{gd} - 1\right)\frac{dd}{dx} = \frac{dh}{dx}. \tag{2.59}$$

This implies that at the crest of a single-humped obstacle where dh/dx vanishes, either dd/dx also vanishes or the local Froude number F, defined by $F^2 = u^2/gd$, is unity. Also, when $F = 1$ we must have $dh/dx = 0$. Now in these steady solutions we have $F = F_0$ upstream, and for $F_0 < 1$, we see that F increases over the obstacle as d decreases and u increases with increasing h. But from (2.59), F can only equal unity at the crest of the obstacle where h has its maximum value, h_m. Hence F cannot exceed unity and h therefore has a maximum value. In these solutions the flow is subcritical everywhere. Similarly for $F_0 > 1$, we see that F decreases over the obstacle and the solutions are again limited by $F = 1$ at the crest, so that here the flow is supercritical everywhere. This maximum obstacle height specified by $F = 1$ is a function of the initial Froude number F_0 and is shown in Figures 2.10–2.12 (see pages 40–43) as the curve BAE. From the above equations this curve is readily shown to be given by

$$H_m = 1 - \frac{3}{2}F_0^{2/3} + \frac{1}{2}F_0^2. \qquad (2.60)$$

This is the boundary of the regions of (F_0, H_m)-space containing solutions of wholly sub- or supercritical flows, with steady upstream flow states unaffected by the obstacle. When H_m exceeds these values the flow is more complicated and, as the above theoretical discussion suggests and laboratory experiments have shown, an upstream hydraulic jump forms on the upstream side, and must be incorporated into the flow solution.

For flow over depressions, where $h < 0$, if $F_0 < 1$ the steady solutions with subcritical flow everywhere apply regardless of the depth of the hole, and the same applies to the solutions with supercritical flow if $F_0 > 1$. These flow states stem from two different solutions to the cubic equation (2.58) for d/d_0, and if F_0 is varied through unity for fixed $H_m(< 0)$, the flow state will jump from one to the other. If $F_0 = 1$, the steady solution is indeterminate, and may take either form. This suggests that this flow is unstable and unsteady, at least for small obstacles, and this property is discussed further in §3.4. Note that flow over depressions of any size may be either sub- or supercritical. Waterfalls are a limiting case. Flow in a waterfall must be supercritical, and hence the flow approaching it must also be supercritical if the depth decreases monotonically in the direction of flow. Subcritical flow may exist upstream if it is separated from this supercritical downstream state by a sill or obstacle, where a transition from sub- to supercritical flow occurs.

The preceding equations may be used to incorporate an upstream jump into the characteristic solutions specified by (2.48). However, a much simpler approach to locally steady flows over topography is to look for solutions that are steady in the vicinity of the obstacle, and assume that an upstream hydraulic jump (moving away from the region) exists *ab initio*. In fact, the region downstream of such a jump must be subcritical in the frame of the obstacle, so that waves may propagate from the obstacle to the jump and vice versa. As discussed above, a jump imposes its own conditions on the flow. In practice, therefore, when an upstream jump forms, the information is transmitted back to the obstacle and affects the flow there; the consequent change is then transmitted back to the jump, and so on, with decreasing magnitude of effects. An equilibration between the two is therefore set up, and this

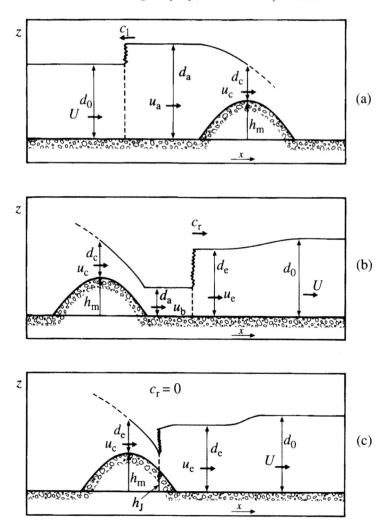

Figure 2.9 Definition sketches for notation for flow over a long obstacle. (a) Upstream flow with a hydraulic jump; (b) downstream flow with a moving hydraulic jump and rarefaction wave; (c) downstream flow with a stationary hydraulic jump and rarefaction wave.

is manifested in the steady-state solutions. In practice, the jump usually forms over, or very close to, the obstacle anyway.

From (2.59), where h_m is sufficiently large for (2.58) not to be applicable, we must have $F = 1$ at $h = h_m$. We therefore consider a model of the flow as shown in Figure 2.9a, where axes are taken in the frame of the topography, and for the present we consider only the upstream side. The upstream jump is moving to the left at a speed c_1 whereas the other parts of the flow are steady. We therefore have five unknown variables (c_1, u_a, d_a, u_c and d_c), to be determined by the five equations

$$(U + c_1)d_0 = (u_a + c_1)d_a, \tag{2.61}$$

$$u_a d_a = u_c d_c, \tag{2.62}$$

$$(U + c_1)^2 = \frac{g d_a}{2}\left(1 + \frac{d_a}{d_0}\right), \tag{2.63}$$

$$\frac{1}{2}u_a^2 + g d_a = \frac{1}{2}u_c^2 + g(d_c + h_m), \tag{2.64}$$

$$u_c^2 = g d_c. \tag{2.65}$$

Here the first two equations come from mass conservation, the third from the jump speed expression (2.54), the fourth from the Bernoulli equation (2.55), (2.56), and the fifth from the critical condition at $h = h_m$. The solution of these equations yields results as shown in Figure 2.10 for the upstream jump speed and its elevation, rendered dimensionless by

$$C_1 = \frac{c_1}{\sqrt{g d_0}}, \quad D_a = \frac{d_a}{d_0}. \tag{2.66}$$

Here, curves of constant values of C_1, D_a are shown in terms of the two fundamental dimensionless parameters of the system, F_0 and H_m. Long (1970, 1972) has made comparisons with laboratory experiments in the regions shown shaded in Figure 2.10, with satisfactory agreement. We note from Figure 2.10 the curious result that solutions to (2.61)–(2.65) are not confined to the right of the curve BAE, but extend to the curve AG when $F_0 > 1$. We determine AG by the criterion $c_1 = 0$, and may be shown to have the equation

$$H_m = \frac{(8F_0^2 + 1)^{3/2} + 1}{16 F_0^2} - \frac{1}{4} - \frac{3}{2}F_0^{2/3}. \tag{2.67}$$

Near $F_0 = 1$, $H_m = 0$, the perturbation to the undisturbed flow in the solutions to (2.61)–(2.65) grows as $H_m^{1/2}$, rather than H_m. As H_m increases, the value of $u_a d_a$ in these solutions progressively decreases, and reaches zero at the curve BC, which is given by

$$F_0 = (H_m - 1)\left(\frac{1 + H_m}{2 H_m}\right)^{1/2}. \tag{2.68}$$

To the right of this curve, there is no flow over the obstacle, as the latter is high enough to block the flow completely. On this boundary $D_a = H_m$, and to the right of it $D_a < H_m$; further increases in H_m with F_0 constant cannot change the upstream flow properties. The flow states pertaining to these various regions of the $F_0 - H_m$ diagram are shown in Figure 2.11. When $F_0 = 4.47$, curve BC intersects AE at a point (not shown) denoted by K, and BC continues and asymptotes to a straight

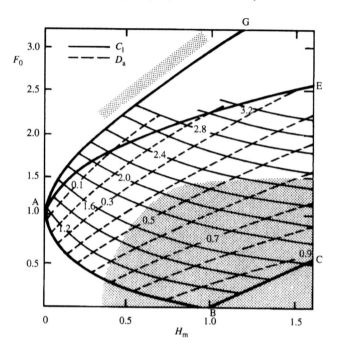

Figure 2.10 Values for upstream jump speed $C_1 = c_1/\sqrt{gd_0}$ and jump amplitude $(D_a = d_a/d_0)$ as functions of F_0 and H_m, from the hydraulic model. The shaded regions denote the parameter ranges covered in Long's (1970) experiments.

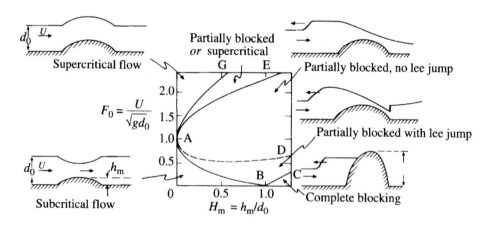

Figure 2.11 Flow regimes on the (F_0, H_m)-diagram for hydrostatic single-layer flow over an obstacle.

line that is parallel to AG. In the region CKE, in which $F_0 > 4.47$, $H_m > 6.90$, the steady flow is either supercritical everywhere, or totally blocked.

Within region EAG, the flow may be either supercritical, as in the region to the left of AG, or controlled by a critical condition at the crest with an upstream jump, as in the region to the right of AE. Although the governing equations have been known for a hundred years or more, the presence of these two different flow states in region EAG is a relatively recent discovery. [Curve BAE was described by Long (1954, 1970), curve BAG by Houghton & Kasahara (1968), and region EAG by Baines & Davies (1980).] The presence of these two states implies that the state obtained in practice depends on the initial conditions. It also implies that there is hysteresis in the system.

If, for example, one starts with a steady supercritical state in EAG and then slowly decreases F_0 so that the flow evolves quasi-statically through a succession of steady states, on reaching AE the flow suddenly changes from supercritical flow to the critically controlled state with the upstream jump. If one then reverses the process and increases F_0, this new state persists until curve AG is reached, where the flow makes a sudden transition back to the supercritical state. Similar transitions occur if F_0 is held constant and H_m is increased and decreased. This hysteresis has been verified numerically by Pratt (1983), and it may be readily demonstrated in a hydraulic laboratory.

In region GAE of Figure 2.11, there is also a third solution of the above equations that consists of a stationary hydraulic jump situated over the forward face of the obstacle (Baines & Whitehead, 2003). This solution is contiguous with the other two, but is unstable, in the sense that a small displacement causes the jump to move away from the location of this solution. Attempts to realise it in the laboratory have confirmed its existence and its instability.

When the flow is critically controlled at the obstacle crest, the downstream flow may have one of the two forms shown in Figure 2.9b,c (Houghton & Kasahara, 1968). Flow on the lee side is supercritical (i.e. $F > 1$) and is followed by a hydraulic jump that may be swept downstream as in Figure 2.9b, or situated over the topography as in Figure 2.9c. However, these structures alone do not permit the downstream flow to be equated to the initial flow state. This connection may be made by adding a time-dependent downstream propagating wave. This is a rarefaction wave propagating into fluid at rest on the one family of characteristics only, and as for (2.47), (2.45) and (2.46) give

$$u_e - 2\sqrt{g d_e} = U - 2\sqrt{g d_0}. \tag{2.69}$$

For the configuration and notation of Figure 2.9b, we again have five undetermined variables, and the five equations

$$u_c d_c = u_b d_b, \tag{2.70}$$

$$\frac{1}{2}u_c^2 + g(d_c + h_m) = \frac{1}{2}u_b^2 + g d_b, \tag{2.71}$$

$$(u_b + c_r)^2 = \frac{g d_e}{2}\left(1 + \frac{d_e}{d_b}\right), \tag{2.72}$$

$$(u_b + c_r)d_b = (u_e + c_r)d_e, \tag{2.73}$$

and (2.69), which enable u_b, d_b, u_e, d_e and c_r to be determined. Here (2.70), (2.73) come from mass conservation, (2.71) from the Bernoulli equation over the obstacle, and (2.72) from the jump speed relation (2.54). These equations apply above the dashed line AD in Figure 2.11, where $c_r = 0$ on this line. Below AD the configuration of Figure 2.9c is applicable, with a stationary jump over the lee side of the obstacle. Here we have seven unknowns: u_-, d_-, and u_+, d_+, which are the flow variables immediately upstream and downstream of the jump respectively, u_e, d_e, and h_J, the height of the topography at the location of the jump. The equations are

$$u_c d_c = u_- d_- = u_+ d_+ = u_e d_e, \tag{2.74}$$

$$u_-^2 = \frac{g d_+}{2}\left(1 + \frac{d_+}{d_-}\right), \tag{2.75}$$

$$\frac{1}{2}u_c^2 + g(d_c + h_m) = \frac{1}{2}u_-^2 + g(d_- + h_J), \tag{2.76}$$

$$\frac{1}{2}u_+^2 + g(d_+ + h_J) = \frac{1}{2}u_e^2 + g d_e, \tag{2.77}$$

and (2.69), from which the details of the flow may be found. Respectively these equations come from conservation of mass, the condition for a stationary jump ($c_r = 0$), and the Bernoulli equations over the obstacle. Curves showing the values of c_r (above AD) and h_J (below AD) are shown in Figure 2.12.

In the blocked flow region to the right of curve BC in Figure 2.11, the downstream jump disappears and the downstream flow is completely governed by the condition (2.69) with $u_e = 0$. This gives

$$\frac{d_e}{d_0} = \left(1 - \frac{F_0}{2}\right)^2, \tag{2.78}$$

so that d_e decreases from $d_0/4$ to zero as F_0 increases from 1 to 2, and must vanish for $F_0 \geq 2$ where the fluid "dries out" on the lee side of the obstacle.

In the region enclosed by EADC (or GADC) of Figure 2.11, the drag force on the obstacle is

$$F_D = \int p\frac{dh}{dx}dx = \frac{\rho g}{2}\frac{(d_a - d_b)^3}{d_a + d_b}, \tag{2.79}$$

where $p = \rho g d$ is the pressure on the topographic surface.

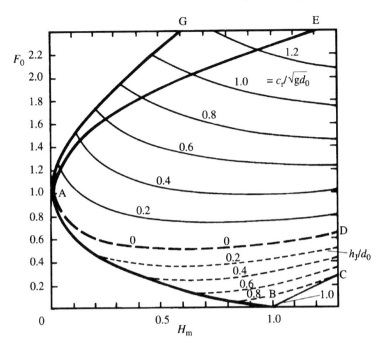

Figure 2.12 Above AD: speed $C_r = c_r/\sqrt{gd_0}$ of the downstream jump (relative to the obstacle), and below AD: the location of the stationary jump as measured by the local topographic height h_J.

2.3.3 Flow through variable cross-sections: contractions and expansions

Single-layer flow through a channel whose cross-section varies on length-scales that are long compared with the channel width and depth, and where η and u across each cross-section are effectively uniform, is still governed by (2.42). (These assumptions of uniformity of the flow across the channel are often not valid for fully supercritical flows: see §3.5.) However, (2.43) may be generalised to

$$\frac{\partial A}{\partial t} + \frac{\partial}{\partial x}(Au) = 0, \qquad (2.80)$$

where $A(x,\eta)$ is the cross-sectional area of the channel occupied by fluid at the point x, and $\eta(x,t)$ is the displacement of the free surface relative to some upstream reference level, as shown in Figure 2.13. For steady flows we then have from (2.80)

$$Au = Q, \qquad (2.81)$$

where Q is the volume flux of fluid in the channel, and from (2.42)

$$\frac{Q^2}{2A^2} + g\eta = \text{constant.} \qquad (2.82)$$

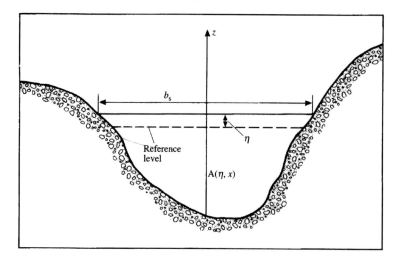

Figure 2.13 Notation for hydrostatic flow through a channel of variable cross-section.

Differentiating with respect to x then gives

$$\left[g - \frac{Q^2}{A^3} \left(\frac{\partial A}{\partial \eta} \right)_x \right] \frac{\partial \eta}{\partial x} = \frac{Q^2}{A^3} \left(\frac{\partial A}{\partial x} \right)_\eta, \tag{2.83}$$

where the subscripts here denote the variables held constant in the partial derivatives. Hence, when $(\partial A/\partial x)_\eta = 0$, we must have

$$\frac{\partial \eta}{\partial x} = 0, \quad \text{or} \quad \left(\frac{\partial A}{\partial \eta} \right)_x = \frac{g A^3}{Q^2} \equiv \frac{g A}{u^2}. \tag{2.84}$$

This is a generalisation of the hydraulic alternative described by (2.59). For given x, $\partial A/\partial \eta$ is equal to the channel width b_s at the surface, and A may be written $A = b_s \overline{d}$, where $\overline{d}$ is the mean depth of the channel at this x value. Hence the second alternative of (2.84) may be written

$$F^2 = u^2 / g\overline{d} = 1, \tag{2.85}$$

defining a generalised Froude number for a channel of non-uniform depth.

The special case where the channel has vertical sidewalls is relevant to later studies. Here, A may be written $A = (d_0 + \eta - h)b$, where $b(x)$ is the channel breadth. If there is no depth variation we have $h = 0$, and at a minimum contraction point $\partial b/\partial x = 0$. At such a point (2.84) implies that

$$\frac{\partial \eta}{\partial x} = 0, \quad \text{or} \quad F^2 \equiv \frac{u^2}{g(d_0 + \eta)} = 1. \tag{2.86}$$

The effects of pure sidewall contractions in channels are similar to those produced by flow over obstacles, and they may be described in the same terms as in §2.3.2. Here we only discuss a special case that does not have a counterpart in the flow over an obstacle. This is the flow with initially specified flux Q from one very (infinitely) wide reservoir of depth d_0 to another, through a region of width $b(x)$ (as shown in Figure 5.2a), and we examine how the flow changes as the minimum gap width b_m is reduced. If the minimum value of b is sufficiently large, (2.81), (2.82) give

$$Q = b(d_0 + \eta)u, \quad \frac{1}{2}u^2 + g\eta = 0, \tag{2.87}$$

and the flow is everywhere subcritical, with the interface depressed in the contraction. If the contraction is then made progressively narrower, the point is reached where $F = 1$ at the narrowest point, and there, equations (2.86), (2.87) give

$$\eta = -d_0/3, \quad u^2 = \frac{2}{3}gd_0, \quad Q = b_m \left(\frac{8}{27}gd_0^3\right)^{1/2}, \tag{2.88}$$

so that the depth of water in the contraction is 2/3rd the depth in the upstream reservoir. The flux Q is determined solely by the width and depth over the weir, which can be easily measured. This is a main reason for having weirs in rivers.

If the gap width is decreased further, (2.86), (2.87) show that the flow alters to reduce the flux Q, but conditions (2.88) at the minimum gap remain unchanged. The change in Q is effected by a hydraulic jump that is sent upstream (in the manner of §2.3.2). This jump weakens as it moves into a wider channel and its amplitude decreases to zero, so that it cannot change conditions in the reservoir. Downstream of the contraction, the supercritical flow adjusts to the subcritical flow in the downstream reservoir via a stationary hydraulic jump. If the flow is driven by a difference in levels between the two reservoirs, Q is controlled by the critical conditions (2.88) in the contraction, where η is now the displacement of the surface from the upstream reservoir level, d_0.

This behaviour may be compared with the case of flow in a channel (with vertical sidewalls) where the *base level* of the depth $d(x)$ is uniform along and across the channel at all locations x, but the channel width $b(x)$ varies. The steady-state flow is still governed by equation (2.57) in the form

$$u(x)^2 + 2gd(x) = u(x_0)^2 + 2gd(x_0), \tag{2.89}$$

where x_0 is a reference location, and the total flux Q is given by

$$Q = u(x)d(x)b(x) = Q_0 = u(x_0)d(x_0)b(x_0). \tag{2.90}$$

If the width of the channel increases downstream of section x_0 at a uniform rate, so that width $b(x) = b(x_0)x/x_0$, from equations (2.89, 2.90) we obtain

$$D^3 - (1 + \frac{F_0^2}{2})D^2 + \frac{F_0^2}{2X^2} = 0, \qquad (2.91)$$

where $D = d/d_0$, and $X = x/x_0$.

One may solve this cubic equation for D as a function of F_0 and X, to obtain the flow properties downstream of location x_0, and their dependence on the Froude number F_0 at that location, as the downstream flow becomes wider or narrower. The results show that, for a widening channel downstream, if $F_0 > 1$, the Froude number increases with downstream distance, and the fluid depth decreases (see also §3.5.4). If $F_0 < 1$, the reverse happens: the Froude number decreases downstream and the fluid depth increases. Correspondingly similar behaviour is found on the upstream side if the channel width varies there.

If the breadth $b(x)$ and depth $d(x)$ vary independently with x but are both uniform across the channel (with vertical sidewalls), the equations governing u and d are (Armi, 1986)

$$\frac{1}{u}\frac{\partial u}{\partial x} = -\frac{1}{1 - F^2}\frac{1}{b}\frac{db}{dx} + \frac{1}{1 - F^2}\frac{1}{y}\frac{dh}{dx}, \qquad (2.92a)$$

$$\frac{1}{d}\frac{\partial d}{\partial x} = \frac{F^2}{1 - F^2}\frac{1}{b}\frac{db}{dx} - \frac{1}{1 - F^2}\frac{1}{b}\frac{dh}{dx}. \qquad (2.92b)$$

An inspection of these equations shows that where $F = 1$, to avoid singular behaviour we must have

$$-\frac{1}{b}\frac{db}{dx} + \frac{1}{d}\frac{dh}{dx} = 0. \qquad (2.93)$$

This implies that if $\frac{db}{dx} = 0$ and $\frac{dh}{dx} = 0$ coincide, critical flow will be found at this location. But if the highest and narrowest sections do not coincide, critical flow must occur somewhere in between these positions. This means that the location of critical flow will depend on the flow rate, as well as the geometry.

However, weirs in rivers or canals are often constructed to have a minimum width or depth, or both, so that critical flow is attained at the appropriate location (Vanden-Broeck & Keller, 1987). The total flux Q can then be readily measured or monitored because it is determined solely by the local width b_c and depth d_c over the weir, as given by $Q = b_c(g d_c^3)^{1/2}$.

2.3.4 Practical limitations of the unidirectional single-layer hydraulic model

The above theoretical treatment assumes that the flow is effectively inviscid except within the hydraulic jumps. In real world situations, bottom friction is present everywhere. Since we are primarily concerned with flow at large Reynolds numbers where $U d_0/\nu \gg 1$, the flow will be turbulent in realistic situations, and the bottom friction is associated with a turbulent boundary layer. In these cases, frictional drag is then most realistically represented by a quadratic drag law, which adds the term $-C_D u^2/d$ to the right-hand side of (2.42). The significance of this term depends on the magnitude of the drag coefficient C_D, which is a function of the roughness of the lower boundary. The term C_D has values that are typically of order 10^{-3}, so that it is small and may be neglected for many purposes. It would, for example, have a small effect on the behaviour of flows for a limited period of time after onset.

The unidirectional single-layer hydraulic model is widely used for practical applications to unidirectional horizontal flows, and it is important to be aware of its limitations. Application to the flow of rivers, or flow through straits with minimal bottom slopes (in the mean), are all subject to the effects of bottom and internal friction. These effects generally have a small effect locally, but they are cumulative and can affect the overall structure of the flow. Pratt (1986) has examined the effect of bottom friction on unforced flow over a level surface, where the friction is restricted to the surface of the topography only. This is an approximate model for the laboratory situation where an obstacle is towed through stationary fluid. The significance of bottom friction in this system is measured by the parameter $\gamma = C_D L/h_m$, where L is the half-width of the obstacle. If γ is small the overall pattern of flow is similar to that of the inviscid case, but as γ increases the hysteresis effect (of §2.3.2) disappears, and stationary upstream jumps over the obstacle become possible.

Garrett & Gerdes (2003) and Garrett (2004) have applied the model to flow through straits, and some numerical results have been computed by Hogg & Hughes (2006), who compare the effects of bottom and internal friction. In general, bottom friction is the more significant, and it causes the location of the critical-flow control point to move downstream of its position in inviscid flow. Paradoxically, internal friction causes it to move in the opposite direction, but bottom drag is the stronger component.

The usefulness of the single-layer hydrostatic model becomes less relevant as the flow becomes more density-stratified, the horizontal length-scales decrease, and the possibility of entrainment of overlying fluid is included. The limiting effects of these factors have been explored by Nielsen et al. (2004). But if the flow is still dominated by long waves, the concept of critical control points (where the wave propagation speed against the mean flow is zero) is still valid and is useful in interpreting flow properties. In most practical applications the mean velocity profile is approximately uniform, under the influence of a bottom boundary layer and associated turbulence throughout the water column, but the effects of departures from uniformity deserve consideration. Analysis of the associated eigenvalue problem for linear waves on non-uniform velocity profiles $U(z)$, with mean $\overline{U}$ in fluid of depth d, shows that long waves travel at the speed

$$c \simeq \overline{U} \pm \sqrt{gd} \left[1 + \frac{\overline{U}^2}{gd} \frac{3}{2d} \int_0^d \left(\frac{U}{\overline{U}} - 1 \right)^2 dz \right], \tag{2.94}$$

taking only the first term in a series expansion. The presence of mean shear increases the long-wave propagation speed. For many practical purposes this extra term is small, but it may not always be negligible when compared with other small effects such as non-linear steepening and dispersion, as discussed in the next chapter. The single-layer model is sometimes applied where there is an inactive overlying fluid, so that the process of entrainment of this fluid may affect the flow. This process has been explored by Gerdes et al. (2002) and Nielsen et al. (2004). Entrainment tends to act in a similar manner to bottom friction. A summary of some of these effects is given in Chapter 1 of Pratt & Whitehead (2008).

Another fluid property affected by frictional stresses is flow separation. Under what conditions does the flow of a homogeneous layer separate from the surface of an obstacle? Flow separation does not normally make the hydrostatic approximation invalid for the layer as a whole, but it usually has local practical significance. In general, boundary layers tend to separate when the fluid moves against an adverse pressure gradient. In the present system, this occurs on the lee side of an obstacle when the flow is everywhere subcritical, and also on the downstream side of a stationary hydraulic jump. Huppert & Britter (1982) have verified experimentally that separation occurs under both of these circumstances if the pressure reduction is sufficiently rapid. Hence, we may expect some degree of separation from the lower boundary in subcritical single layer flows when the fluid depth increases in the downstream direction.

Akers & Bokhove (2008) have made an experimental and theoretical study of horizontal homogeneous flow through a contraction, with no bottom topography, in a laboratory tank of dimensions 1.1 m length by 0.2 m width, with water depths of order 1–2 mm. A variety of oblique hydraulic jumps and multiple steady states were realised, and could be modelled by appropriate analytical and numerical analysis. Despite displaying a variety of novel non-linear features, these results showed good agreement with the inviscid analysis except in the neighbourhood of hydraulic jumps.

2.4 Downslope flows with frictional drag

If steady-state flows with friction are to be realised, there must be some external forcing that balances the bottom drag. In stratified analogues to be discussed in following chapters, this forcing may be due to an externally imposed pressure gradient. In the single-layer case, external forcing may be imposed by wind, but a more common situation is flow down a slope, where the flow is forced by gravity. We will examine this case in more detail, regarding it as a prototype for flows where the mean state is due to an explicit balance between forcing and frictional drag.

If we consider a plane slope at angle θ to the horizontal, with x-axis down the slope and z-axis perpendicular to it, the equations corresponding to (2.42), (2.43) are

$$u_t + uu_x + g\cos\theta\,\eta_x = g\sin\theta - \frac{C_{\mathrm{D}}u^2}{d}, \tag{2.95}$$

$$d_t + (du)_x = 0, \tag{2.96}$$

where u is directed down the slope and d is the depth measured normal to it. For the case of external forcing f_e per unit mass over level ground we would instead have $\theta = 0$, and $g\sin\theta$ would be replaced by f_e.

Equation (2.95) has the steady equilibrium solution

$$u^2 = \frac{gd\sin\theta}{C_{\mathrm{D}}}, \tag{2.97}$$

where u and d are constant. If the flow is not steady but the horizontal gradients are small, so that the terms on the left-hand side of (2.95) are collectively much smaller than each of those on the right, the flow is governed by the simpler, reduced system (2.43), (2.93). This is the *kinematic wave approximation* of Lighthill & Whitham (1955). The resulting equations are applicable to flood waves in rivers, and reduce to

$$d_t + \frac{3}{2}\left(\frac{gd\sin\theta}{C_{\mathrm{D}}}\right)^{1/2} d_x = 0, \tag{2.98}$$

or in characteristic form

$$\frac{\mathrm{d}d}{\mathrm{d}t} = 0 \quad \text{on} \quad \frac{\mathrm{d}x}{\mathrm{d}t} = \frac{3}{2}\left(\frac{gd\sin\theta}{C_{\mathrm{D}}}\right)^{1/2} = c_k = \frac{3u}{2}. \tag{2.99}$$

They describe waves propagating at speed c_k in the direction of the flow velocity u but at a rate 50% faster, and which increases with $\sqrt{d}$. Consequently a wave of elevation will steepen to form a bore or "discontinuity", in the same manner as the long gravity waves described in §2.3. From the more complete equations (2.95), (2.96) it may be shown that a monotonic increase in surface elevation (a flood wave) steepens to a steady profile with width of order $d_0\cot\theta$, provided its amplitude is not too large; for larger amplitudes, a hydraulic jump forms at the leading part of the profile. The mathematical details of the relationships between kinematic waves and gravity waves are lucidly described by Whitham (1974), to which the reader is referred for further details.

A linear stability analysis of a uniform steady flow $u = U$, $d = d_0$, where

$C_D U^2 = g d_0 \sin \theta$, with disturbances governed by the system (2.95), (2.96) shows that the flow is stable if (e.g. Whitham, 1974)

$$\tan \theta < 4C_D, \qquad (2.100)$$

or equivalently

$$F_s = \frac{U}{\sqrt{g d_0 \cos \theta}} < 2, \qquad (2.101)$$

where F_s is the initial slope Froude number (which corresponds to F_0 when $\theta = 0$, but the distinction is made here to avoid confusion), or again equivalently

$$U - \sqrt{g d_0 \cos \theta} < c_k < U + \sqrt{g d_0 \cos \theta}, \qquad (2.102)$$

and unstable otherwise. Relation (2.102) shows that for stability the kinematic wave speed c_k must lie within the range of linear gravity wave speeds, which are the fastest speeds for the propagation of information in this system. If these conditions are not satisfied, the flow is unstable with disturbances of long wavelength growing most rapidly at marginal stability. This instability seems to be due to the onset of disorder in the wave system, rather than to the release of potential energy. It may be attributed physically to the fact that water that is made deeper by a perturbation has lower velocity and hence less frictional resistance; consequently it accelerates under gravity, and conversely for water that is made shallower. Advection then causes steepening and increases the wave amplitude.

For $F_s > 2$, in addition to the unstable uniform flow solution, there are periodic flow solutions that are probably stable. These spatially periodic steady solutions (of (2.92a), (2.92b), (2.43)) are termed *roll waves*, which are essentially a sequence of periodic hydraulic jumps propagating down the slope, as shown in Figure 2.14. Observations of these waves have been described by Cornish (1934), who provided ample empirical evidence for their stability. In each of his examples the roll waves were observed in shallow (several centimetres or less in depth) conduits of several hundred metres in length. Some simple attempts by the author to produce them in the laboratory were not very successful, which suggests that they take time to develop fully and are not easily produced in short channels. Roll-wave solutions of (2.95), (2.96) employing discontinuous jumps were obtained by Dressler (1949), and Figure 2.14 shows an example. Needham & Merkin (1984) have included a frictional term νu_{xx} in (2.95), where ν is kinematic viscosity, and studied roll-wave solutions for this system. Here there is an additional parameter, a Reynolds number, and the roll-wave solutions are continuous without requiring discontinuities, and arise as a Hopf bifurcation. (See Drazin & Reid, 1981, for a description of Hopf bifurcations.)

Otherwise the overall properties of the solutions of the two systems are very

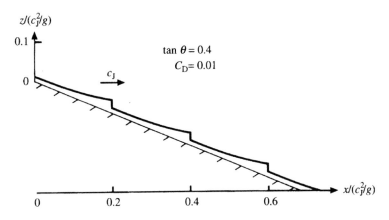

Figure 2.14 A typical roll-wave solution, constructed from periodic hydraulic jumps; c_J denotes the horizontal speed of each jump. (From Dressler, 1949.)

similar. For both systems, roll waves are only possible in the unstable range $F_s > 2$, with amplitude tending to zero as $F_s \to 2+$. For given θ and F_s (and Reynolds number for the extended system), there is a one-parameter family of periodic solutions, where the parameter is the speed, the amplitude or the wavelength; once one of these is fixed, the other two follow and the solution is unique. All roll waves propagate downward at speeds c_r in the range $U < c_r < 3U/2$. Roll-wave solutions do not exist in the limit $F_s \to \infty$, which suggests that they become unstable for sufficiently large F_s, leading to chaos and turbulence. Flows down steep spillways are normally turbulent. Figure 2.15 shows flow down a steep spillway that appears to consist primarily of unstable roll waves.

The effect of obstacles placed in a frictionally balanced uniform downslope flow ((2.97) with $u = U$, $d = d_0$) may be described using (2.102), (2.96), with $d = d_0 + \eta - h(x)$. As for the case of flow over level ground, described above, we look for steady solutions. This requires stable flow, so that we are limited to $0 < F_s < 2$. An examination of the equations shows that solutions that are steady near the obstacle but unsteady away from it (as in Figure 2.9) are not possible here, because the depth cannot be uniform unless $d = d_0$. Spatially periodic solutions can also be ruled out. Hence the disturbance forced by the topography must die out as $|x| \to \infty$, and the horizontal fluid flux is the same everywhere, with the value $Q = U d_0$. Writing

$$\mathcal{D} = \frac{d}{d_0}, \quad X = \frac{x}{d_0}, \quad H = \frac{h}{d_0}, \tag{2.103}$$

Figure 2.15 Unstable roll waves on a steep spillway at the Wellington reservoir in Western Australia, observed looking upwards. (Photo courtesy of J. Imberger.)

and eliminating u from (2.95), (2.96) gives

$$(\mathcal{D}^3 - F_{\rm s}^2)\frac{{\rm d}\mathcal{D}}{{\rm d}X} + \mathcal{D}^3\frac{{\rm d}H}{{\rm d}X} = C_{\rm D}F_{\rm s}^2(\mathcal{D}^3 - 1), \qquad (2.104)$$

which may be integrated to give

$$\frac{F_{\rm s}^2}{2\mathcal{D}^2} + \mathcal{D} + H(X) = \frac{F_{\rm s}^2}{2\mathcal{D}_1^2} + \mathcal{D}_1 + H(X_1) + C_{\rm D}F_{\rm s}^2\int_{X_1}^{X}\left(1 - \frac{1}{\mathcal{D}^2}\right){\rm d}X, \quad (2.105)$$

where $\mathcal{D} = \mathcal{D}_1$ at $X = X_1$, a location that may be chosen arbitrarily (for example, $X_1 = \pm\infty$, or the point given by (2.107)). We also have the boundary conditions

$$\mathcal{D} \to 1 \quad \text{as } |X| \to \infty. \qquad (2.106)$$

Equation (2.104) shows that at the point where

$$\frac{{\rm d}H}{{\rm d}X} = C_{\rm D}F_{\rm s}^2\left(1 - \frac{1}{\mathcal{D}^3}\right), \qquad (2.107)$$

which is "somewhere near" the crest of the obstacle, we must have

$$\frac{{\rm d}\mathcal{D}}{{\rm d}X} = 0, \quad \text{or} \quad \mathcal{D} = F_{\rm s}^{2/3}. \qquad (2.108)$$

The latter is equivalent to $F = 1$, where the local Froude number F is defined by $F^2 = u^2/gd\cos\theta = F_{\rm s}^2/\mathcal{D}^3$. We therefore have the same hydraulic alternative in this frictional case as that for (2.59), at a point that is well-defined by (2.107). Also, examination of (2.104) with $H = 0$ shows that flows that are subcritical everywhere

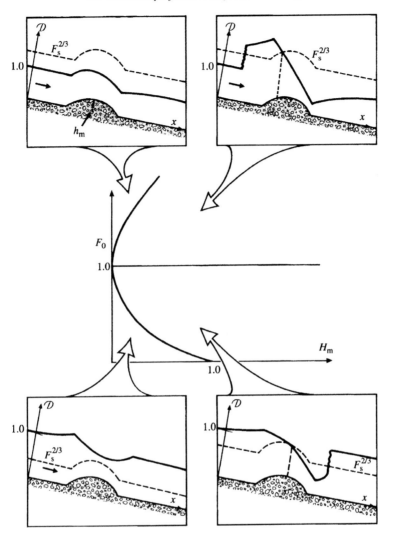

Figure 2.16 Flow regimes in the (F_0, H_m)-diagram for frictionally controlled downslope flow over an obstacle, where $\mathcal{D} = d/d_0$. The dashed line denotes the value of $F_s^{2/3}$ on the same scale as $\mathcal{D}$ (cf. Figure 2.11).

may have $\mathcal{D} \to 1$ continuously as $X \to -\infty$ but not for $X \to \infty$, and the reverse applies for flows that are everywhere supercritical. From these conditions we may infer that there are four types of solution, corresponding to sub- and supercritical flows far upstream with each of the two conditions given by (2.107), (2.108). These are shown schematically in Figure 2.16.

The details may be calculated by integrating (2.104), and incorporating stationary hydraulic jumps where appropriate to achieve $\mathcal{D} = 1$ upstream of the obstacle for supercritical flows and downstream of it for subcritical flows. We assume that these jumps are effectively discontinuous and thereby unaffected by the bottom friction, and that they satisfy (2.54). When the upstream jump occurs it may be located anywhere upstream of the "control point" (2.107), including over the obstacle, and similarly the downstream jump may be anywhere downstream of it. Here the details of the flow are dependent on the shape of the obstacle and on C_D, as well as F_s and H_m, in contrast to the "zero slope" solutions of Figure 2.11, which are dependent only on F_0 and H_m. However, there is no hysteresis region here (since Q and the upstream and downstream conditions cannot be altered), and for a given C_D and obstacle shape the flow is completely specified by F_s and H_m. Further, there is no blocking region. If a tall obstacle is introduced into the flow, the downslope force causes fluid to accumulate on the upstream side of the obstacle until it spills over, and the flow then settles down to the appropriate solution shown in Figure 2.16.

These downslope flows with stationary upstream hydraulic jumps whose location depends on bottom friction have some similarities with those of a single layer flowing over a ridge in an inviscid rotating system (Baines & Leonard, 1989). The presence of rotation causes the amplitude and speed of an upstream-propagating jump to decrease so that the jump becomes stationary, but the details are outside the scope of this work.

2.5 Granular flows

Granular flow is the common term for the flow of material consisting of (effectively) solid particles, possibly down sloping terrain, in a manner similar to the flow of a fluid. As one might expect, there are similarities with fluid flow and also systemic differences, which mostly depend on the size of the grains constituting the flow. In a sense, each granular fluid is different, and the subject is not that well developed, but a representative description of some effects is given here. In particular, granules could be regarded as the obstacles, or part of the flow.

The "father" of this topic is Ralph Bagnold (1896–1990), who combined a theoretical and experimental scientific career on the motion of granular material, particularly in sandy deserts, with a military role in both world wars to the rank of brigadier. Perhaps his best known work is the volume *The Physics of Blown Sands and Desert Dunes*, first published in 1941. But the subject is many faceted, given the variety of sizes and shapes of particles of various materials in (notional) fluids of various types, that may be present in nature, or may have some commercial application.

As an example of the phenomena involved, one can cite experiments by Vreman et al. (2007) in a shallow tank (length 2 m, width $<$ 13 cm, depth $\sim$ 1 cm) of downslope flow of "fluid" consisting of uniform granular (but not necessarily spherical) material particles through a linear lateral contraction. This produced a variety of different flows, depending on the slope and the contraction. These were (i) a smooth supercritical flow throughout, (ii) a steady reservoir with an upstream stationary jump in the contraction, and (iii) non-steady flow with a time-varying upstream jump in the contraction. Friction is obviously important here, but the Froude number F_0 is still the dominant parameter, with non-steady flows occurring for $F_0 > 4$.

3

Non-linear single-layer flow past obstacles: jumps, bores and wave dispersion

It is a truth universally acknowledged,
that a single man in possession of a good fortune,
must be in want of a wife.

JANE AUSTEN, *Pride and Prejudice*

The hydrostatic approximation normally provides a good description of flow over fairly long obstacles, but there are special situations where it breaks down and shorter non-hydrostatic non-linear waves become important, regardless of how long the obstacles are. For shorter obstacles we expect wave motions with smaller wavelengths to be generated, as described above, and both dispersive and non-linear advective effects to be present. We now describe these more complex situations, and begin by discussing equations that incorporate both linear dispersion and non-linear advection as small variations to uniform propagation at constant speed. This leads to a more general description of steady flows in terms of the mass, energy and momentum fluxes of the fluid, termed the "*QRS* framework". The effects of viscosity and entrainment are not described here, but the discussion in Chapter 2 is equally applicable if frictional effects are significant and/or there is overlying fluid that can be entrained.

3.1 Non-linear waves

The dispersion relation for one-dimensional linear gravity waves of the form $e^{i(kx-\omega t)}$ in fluid at rest of depth d_0 is

$$\omega^2 = gk \tanh k\, d_0, \tag{3.1}$$

shown in Figure 3.1. Expanding this in powers of kd_0 gives

$$\omega^2 \simeq c_0^2 k^2 \left(1 - \frac{1}{3}k^2 d_0^2\right), \tag{3.2}$$

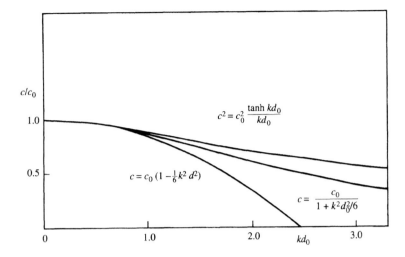

Figure 3.1 The dispersion relation for surface gravity waves (3.1) with $c = \omega/k$, and long-wavelength approximations (3.8) pertaining to (3.9), and to the linear form of (3.13).

where $c_0^2 = g d_0$, as the first approximation incorporating dispersion for long waves. Waves satisfying this dispersion relation also satisfy the equation

$$\eta_{tt} - c_0^2 \eta_{xx} - \frac{1}{3} c_0^2 d_0^2 \eta_{xxxx} = 0, \tag{3.3}$$

since, arguing backwards, this equation gives (3.2). Equivalently, to the same degree of approximation with $k d_0 \ll 1$, we have

$$\eta_{tt} - c_0^2 \eta_{xx} - \frac{1}{3} d_0^2 \eta_{xxtt} = 0, \tag{3.4}$$

which is known as the *linear Boussinesq equation*. In both (3.3) and (3.4), the third term incorporates weak dispersion into the nondispersive wave equation. Solutions to (3.4) satisfy the dispersion relation

$$\omega^2 = \frac{c_0^2 k^2}{1 + k^2 d_0^2/3}. \tag{3.5}$$

This gives real frequencies over the whole range of k, and is a significantly better approximation to (3.1) than (3.2); the latter is seen to be a good approximation to (3.1) for $k d_0 < 1$, and (3.5) similarly for $k d_0 < 2$. These considerations imply that (3.4) gives a better description of dispersive wave motion than (3.3).

An extension of the non-linear hydrostatic equations (2.42), (2.43) in uniform depth which incorporates this dispersion is (Whitham, 1974)

$$u_t + uu_x + gd_x + \frac{1}{3}d_0 d_{xtt} = 0,$$ (3.6)

$$d_t + (du)_x = 0,$$ (3.7)

which are known as the *Boussinesq equations*. This system contains two small parameters, a/d_0 and $(d_0/L)^2$, where a is the amplitude of the surface displacement and L is a characteristic horizontal length-scale. In the limit $a/d_0 \to 0$ the equations reduce to (3.4), and in the limit $(d_0/L)^2 \to 0$, (3.6) reduces to (2.42).

The Boussinesq equations describe waves moving in both directions. When considering wave motions forced by flow over topography with the hydrostatic equations, it is clear that only waves moving in the direction against the stream are important, as those propagating with the stream are advected away relatively quickly after the motion has commenced. Consequently, rather than deal with the analytically intractable Boussinesq equations, and without losing the essence of the phenomena involved, we may simplify them to describe waves moving in one direction only. However, as shown in Figure 2.1 and §2.3.2, the omitted downstream waves do have an effect on the mass of fluid in the region of the obstacle, and this effect is omitted in the one-directional equations. The one-directional form of (3.2) is

$$\omega = c_0 k \left(1 - \frac{1}{6}k^2 d_0^2\right)$$ (3.8)

where c_0 may be taken to be positive or negative. This is compared with (3.1) in Figure 3.1. The equation corresponding to (3.3) is then

$$\eta_t + c_0 \eta_x + \frac{1}{6}c_0 d_0^2 \eta_{xxx} = 0.$$ (3.9)

On the other hand, from the hydrostatic equations [(2.45), (2.46), giving (2.48) or (2.69) with $U = 0$] for waves moving into fluid at rest (with direction depending on the sign of c_0) and of depth d_0, we have

$$u = 2c_0[(1 + \eta/d_0)^{1/2} - 1],$$ (3.10)

and using this to eliminate u from either of (2.42), (2.43) we have

$$\eta_t + c_0 \left(1 + \frac{3}{2}\frac{\eta}{d_0}\right)\eta_x = 0.$$ (3.11)

Both (3.9) and (3.11) describe waves moving with velocity c_0 with a small corrective term, (3.9) depending on a/d_0 being small, and (3.11) on $(d_0/L)^2$ small. It seems plausible that we can incorporate both of these small terms by simply adding them, and this gives rise to the *Korteweg–de Vries equation* (KdV)

$$\eta_t + c_0\eta_x + \frac{3}{2}\frac{c_0}{d_0}\eta\eta_x + \frac{1}{6}c_0 d_0^2\eta_{xxx} = 0. \tag{3.12}$$

Here c_0 may be positive or negative, depending on the direction of propagation of the waves. Since $\eta_t \approx -c_0\eta_x$, to the same degree of accuracy we may express (3.12) in the form

$$\eta_t + c_0\eta_x + \frac{3}{2}\frac{c_0}{d_0}\eta\eta_x - \frac{1}{6}d_0^2\eta_{xxt} = 0, \tag{3.13}$$

for which the linear form has the dispersion relation $\omega = c_0 k/(1 + k^2 d_0^2/6)$, also shown in Figure 3.1. Equation (3.13) bears the same relationship to (3.12) as (3.4) does to (3.3), and has the same relative advantages. Of course, these equations may also be derived on a more formal basis, on the assumption that $a/d_0 = O(d_0/L)^2$ (e.g. Whitham, 1974). Equation (3.12) has been studied at length because it may be solved exactly and has a number of nice analytical properties, whereas (3.13) is sometimes preferred for numerical studies. A generalisation of these equations to channels of arbitrary cross-section has been made by Peregrine (1968).

Detailed comparisons between numerical solutions of (3.13) and waves in a channel in water of depth $d_0 = 3$ cm, generated by a paddle at one end, have been made by Bona et al. (1981), who also summarise the results of earlier qualitative studies. In experiments of this type, viscous effects are quite significant, and Bona et al. included a frictional term $\mu\eta_{xx}$ on the right-hand side of (3.13) for their theoretical comparisons, where the viscosity μ is estimated from the observed damping rate for small-amplitude waves. With this addition, the observations were in good agreement with the computations for values of $\mathfrak{A} = aL^2/d_0^3$ ranging up to about 12. As $\mathfrak{A}$ increased further the agreement progressively deteriorated, apparently because higher harmonics at shorter wavelengths were generated which were not well represented by the model. The conclusions were that (3.13) [and by implication, (3.12)] should give good descriptions of the motion when $aL^2/d_0^3 = O(1)$ and dissipation is small.

Korteweg and de Vries found analytic solutions to (3.12) of the form

$$\eta = a\,\mathrm{sech}^2\left[\left(\frac{3a}{4d_0}\right)^{1/2}\frac{(x-ct)}{d_0}\right], \tag{3.14}$$

where the speed c is given by

$$c = c_0\left(1 + \frac{a}{2d_0}\right) \approx [g(d_0 + a)]^{1/2}. \tag{3.15}$$

This is the well-known *solitary wave* solution, which is a two-parameter family of solutions (two dimensional parameters, one dimensionless parameter) with a/d_0

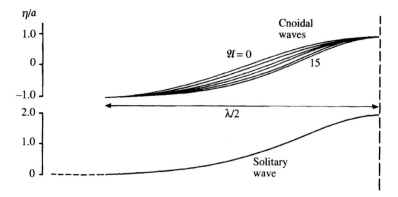

Figure 3.2 Cnoidal wave profiles from (3.16) for fixed wavelength λ and amplitude
a, with $a\lambda^2/d_0^3 = 0$ (sinusoidal profile), 3, 6, 9, 12 and 15. Note that the wave
peak becomes progressively steeper and more isolated. The lower curve shows a
solitary wave profile plotted using (3.14), with the horizontal scale λ chosen to be
the distance within which the surface elevation exceeds 3% of its maximum value.
(From Lighthill, 1978, reproduced with permission.)

as the dimensionless parameter. A profile is given in Figure 3.2. Note that as the
amplitude increases, the horizontal scale of the wave decreases. There is now a
considerable body of literature devoted to solitary wave interactions, including:
overtaking or colliding pairs (which pass through each other intact with only a
phase change); the development of an initial arbitrary disturbance into a number
of solitary waves; and the oblique reflection of solitary waves from barriers. These
details are peripheral for our present purposes, and a succinct summary with many
references has been given by (Miles, 1980).

Korteweg and de Vries also found a three-parameter family of periodic analytic
solutions to (3.12), where the parameters may be taken to be the amplitude a, the
wavelength λ and the minimum depth, d_0. These solutions have the form

$$\eta = a \operatorname{cn}^2 \left[2K(m) \frac{(x - ct)}{\lambda} \right], \tag{3.16}$$

where cn is a Jacobian elliptic function,[1] $K(m)$ is the complete elliptic integral of
the first kind and m is the modulus of the elliptic function, given by

$$mK(m) = \frac{\sqrt{3}}{4} \left(\frac{a\lambda^2}{d_0^3} \right)^{1/2}. \tag{3.17}$$

The two dimensionless parameters may be taken to be any two of a/d_0, d_0/λ and

[1] cn X is defined by cn $X = \cos \phi$, where $X = \int_0^\phi \frac{d\theta}{(1-m\sin^2\theta)^{1/2}}$, and $K(m)$ by $K(m) = \int_0^{\pi/2} \frac{d\theta}{(1-m\sin^2\theta)^{1/2}}$.

m, or $\hat{R}$ and $\hat{S}$ (see below). The wave speed *c* is given by

$$c = c_0 \left[1 + \frac{a}{d_0} \left(1 - \frac{1}{2m^2} \right) \right].$$

(3.18)

The modulus *m* is restricted to the range $0 \le m \le 1$, and its magnitude measures the relative importance of non-linearity to dispersion. There are restrictions on the possible amplitudes and wavelengths which may occur in these solutions. Possible values are shown in Figure 3.5 (see p. 67). As $m \to 0$ the waves become linear and cn $X \to \cos X$, and as $m \to 1$ the wavelength increases and the waves become solitary, with cn $X \to \operatorname{sech} X$. Examples of these cnoidal wave forms are shown in Figure 3.2.

3.2 The QRS framework

The above solitary and cnoidal wave solutions constitute a class of steady flows in a uniform stream with speed *U* and depth d_0 if $U = -c$. In order to discuss more general steady flows that are not restricted to a/d_0 and $(d_0/\lambda)^2$ being small, we consider the full equations of conservation of mass, momentum and energy in a channel where the bottom surface is $z = h(x)$ and the free surface is at $z = d$. From Chapter 1, these may be expressed as

$$\frac{\partial}{\partial t} \rho d = -\rho \frac{\partial Q}{\partial x},$$

(3.19)

$$\frac{\partial}{\partial t} \int_h^d \rho u \, dz = -\rho \frac{\partial S}{\partial x} - p(x, h, t) \frac{dh}{dx},$$

(3.20)

$$\frac{\partial}{\partial t} \int_h^d \rho [\frac{1}{2}(u^2 + w^2) + gz] dz = -\rho \frac{\partial}{\partial x} QR,$$

(3.21)

where *Q* is the *volume flux*, given by

$$Q = \int_h^d u \, dz,$$

(3.22)

ρS is the *flow force* or *momentum flux*, given by

$$S = \int_h^d (u^2 + p/\rho) dz,$$

(3.23)

and ρQR is the *energy flux*, given by

$$QR = \int_h^d u \left(\frac{1}{2}(u^2 + w^2) + gz + p/\rho \right) dz.$$

(3.24)

In steady flow, the equations of motion may be expressed as

$$\nabla \left[\frac{1}{2}(u^2 + w^2) + gz + p/\rho \right] = \nabla R(x, z) = 0, \tag{3.25}$$

so that

$$R = \frac{1}{2}(u^2 + w^2) + gz + p/\rho = \frac{1}{2}(u^2 + w^2)|_{z=d} + gd \tag{3.26}$$

is independent of position in the fluid.

We consider steady motion over level ground ($h = 0$) and define a stream function ψ by

$$u = -\psi_z, \quad w = \psi_x, \tag{3.27}$$

where ψ is a harmonic function (i.e., $\nabla \cdot \nabla \psi = 0$) expressed as

$$\psi = zf(x) - \frac{z^3}{3!}f''(x) + \frac{z^5}{5!}f''''(x) - \cdots, \tag{3.28}$$

where $f(x)$ is an arbitrary function. Successive terms in this series are smaller by a factor of order $(d_0/L)^2$ Substituting (3.28) in the above expressions for Q and S and retaining only the first two terms in the series gives

$$Q = df - \frac{d^3}{3!}f''(x), \tag{3.29}$$

$$S = \int_h^d \left(R - gd - \frac{1}{2}\psi_x^2 + \frac{1}{2}\psi_z^2 \right) dz = Rd - \frac{1}{2}gd^2 + \frac{1}{2}f^2d - d^3(f^2 + ff''), \tag{3.30}$$

and eliminating f to the same order of approximation, we obtain after some rearrangement (Benjamin & Lighthill, 1954)

$$\frac{1}{3}Q^2 \left(\frac{dd}{dx} \right)^2 = \mathcal{B}(d) \equiv Q^2 - 2Sd + 2Rd^2 - gd^3. \tag{3.31}$$

This equation is equivalent to an integral of the steady-state form of (3.12), and it has the same cnoidal wave solutions. The difference is that the coefficients of (3.31), and hence its solutions, are expressed in terms of the fundamental physical quantities Q, R and S. The right-hand side of (3.31) is shown in Figure 3.3, and the curve must be non-negative for solutions to exist. The three curves shown correspond to different values of Q with the same values of R and S. An analysis of the behaviour of solutions of (3.31) near the points where $\mathcal{B}(d) = 0$ shows that the solutions "reflect" from a simple zero such as $d = d_5$, but do not reflect from a tangent point such as $d = d_1$. Hence it is clear that the curve marked "sub-" corresponds to a constant solution $d = d_3$, the curve "cnoidal" to a cnoidal wave solution oscillating

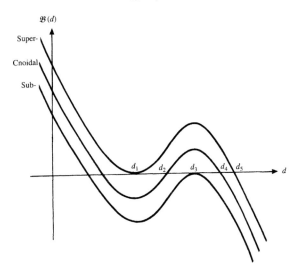

Figure 3.3 The cubic expression for the right-hand side of (3.31), shown for three different values of Q.

between d_2 and d_4, and the curve "super-" to a solitary wave solution with the depth taking the maximum value d_5 and then tending to the constant value d_1.

For a *uniform stream* of velocity u and depth d, the values of Q, R and S are given by

$$Q = ud, \quad R = \frac{1}{2}u^2 + gd, \quad S = u^2 d + \frac{1}{2}gd^2, \tag{3.32}$$

and the Froude number F is of course given by $F^2 = u^2/gd$. For steady flows in a channel, Q is independent of horizontal position, so that it is natural to consider the possible flows in terms of R and S with Q constant. For this purpose we may scale R and S by their values for critical flow ($F = 1$) for given Q, which from (3.32) are given by

$$R_c = \frac{3}{2}(gQ)^{2/3}, \quad S_c = \frac{3}{2}(gQ^4)^{1/3}. \tag{3.33}$$

Fixing Q also determines length and velocity scales d_c and u_c, where $u_c^2 = gd_c$, and

$$d_c = (Q^2/g)^{1/3}, \qquad u_c = \sqrt{gd_c} = (gQ)^{1/3}. \tag{3.34}$$

The possible values for R and S for uniform stream flows are then given parametrically in terms of F by

$$\hat{R} = R/R_c = \frac{1}{3}F^{4/3} + \frac{2}{3}F^{-2/3}, \quad \hat{S} = S/S_c = \frac{2}{3}F^{2/3} + \frac{1}{3}F^{-4/3}. \tag{3.35}$$

These equations specify two curves in the $(\hat{R}, \hat{S})$-plane with F as a parameter, a subcritical one with $F < 1$, and a supercritical one with $F > 1$, and they are shown plotted in Figure 3.4, where they form a cusp at the origin of the diagram. These curves are quite general, in that they have been derived completely independently of (3.31) and the assumptions behind it. Equation (3.31) has solutions only within the cusp, and these assumptions are only formally valid close to the vertex where R/R_c, $S/S_c < 1.02$. Uniform flows on the subcritical branch therefore correspond to the curve "sub-" in Figure 3.3, and those on the supercritical branch to the curve "super-". Benjamin & Lighthill (1954) hypothesised that there would be unique steady solutions of the full equations (2.1)–(2.4) in the cusp region of Figure 3.4 for $\hat{R}, \hat{S} > 1.02$, which would be qualitatively similar to the solutions of (3.31). This has been vindicated by the computations of Cokelet (1977), with the exceptions given below. Between the two curves periodic wave solutions may be found, and these tend to sinusoidal form and zero amplitude as the subcritical branch is approached, and to the "long-wave" solitary wave limit as the supercritical branch is approached. No steady solutions to the full equations with uniform Q are possible outside the cusped region.

Figure 3.4 also contains another boundary denoted by the heavy dashed line, which marks the limit of steady non-linear wave solutions for fixed Q. This boundary has been interpolated from the results of Cokelet, who has computed the properties of these periodic waves beyond the cnoidal wave region ($\hat{S} < 1.02$), and it is an envelope of his computed solutions. It marks the onset of a zone where the waves are unsteady. There is a very thin region to the left of this curve, where there is more than one solution for given $R/R_c, S/S_c$, with the additional solution(s) having greater amplitude. These additional solutions and their properties are not known precisely due to the inaccuracies in the numerical methods used to compute them. On the supercritical side of the cusp, however, the highest wave is known to occur at smaller F (=1.288) than the boundary at F=1.296 (i.e. $\hat{S}$=1.028), which shows the narrowness of this non-unique region. The additional higher-amplitude steady solutions are apparently unstable (see Schwartz & Fenton (1982) for details and references). Apart from these solutions, the flows to the left of the dashed curve are believed to be uniquely specified by Q, R and S.

This QRS framework is very useful in interpreting steady flows where there are localised regions of energy dissipation (such as hydraulic jumps), or localised regions of momentum loss such as drag due to flow over obstacles. From (3.20), (3.21) the energy flux associated with a non-linear wavetrain is given by $\rho Q \Delta R$, where ΔR is the difference in R between the point denoting the wavetrain and the subcritical uniform stream with the same value of S, and the drag force on an obstacle is equal to $\rho \Delta S$, where ΔS is the difference in S across the obstacle. In these cases the flow changes from one steady state to another as it passes over the

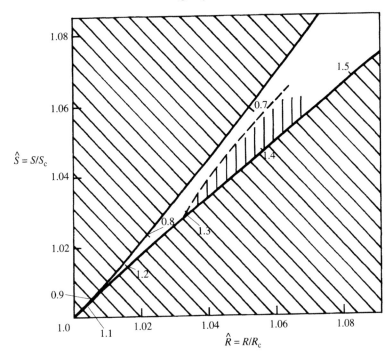

Figure 3.4 The energy flux-momentum flux diagram (in terms of the variables $\hat{R}$ and $\hat{S}$) for steady flow of a single layer with given volume flux Q. Possible flow states lie within the cusp, with subcritical uniform streams on the upper boundary, supercritical uniform streams on the lower boundary, and steady non-linear "cnoidal type" wavetrains in the interior. The numbers on the cusp give the value of the Froude number, F, of the stream. The dashed line marks the boundary of steady wave solutions; in the shaded region to the right of this boundary, steady periodic wave solutions are not possible and the motion must be unsteady (or uniform, on the supercritical curve).

obstacle or through the jump. The QRS framework may also be applied to situations where the flow state is continuously altered, as by bottom friction for example. If S (or R) is varying continuously with x but the flow is steady, we may interpret the flow as passing through a succession of steady states on the (R, S) diagram in a quasi-static fashion as the fluid flows along (i.e. as x increases). Both of these types of interpretation of the diagram are made below. If one attempts to establish steady flows in the region to the right of the dashed curve of Figure 3.4, the flow is unsteady or unstable so that the waves presumably break. This dissipates energy, reducing R to (at least) the corresponding value on the dashed curve with the same S value.

It is difficult to see the structure inside Figure 3.4 because of its shape, and the diagram has been replotted in Figure 3.5 in terms of the variables $\hat{T}$ and $\hat{S}$, where

$\hat{R} = R/R_{\rm c}$, $\hat{S} = S/S_{\rm c}$, and where $\hat{T}$ is defined by

$$\hat{T} = \frac{1}{2} + \frac{\hat{R}^3 - \frac{3}{2}\hat{R}\hat{S} + \frac{1}{2}}{2(\hat{R}^2 - \hat{S})^{3/2}}. \tag{3.36}$$

Note $\hat{T} = 0$ on the subcritical (left-hand) side of the cusp curve, and $\hat{T} = 1$ on the supercritical (right-hand) side, as does m of (3.17). Some curves showing amplitude and wavelength of the steady periodic waves are plotted in Figure 3.5 with the scaling of (3.34).

3.3 Application to hydraulic jumps and undular bores

In §2.3 on hydrostatic flows, a hydraulic jump over a level surface with negligible bottom friction was modelled as a discontinuity between supercritical and subcritical uniform streams, with a concomitant energy dissipation rate given by (2.53). In the *QRS* framework, this hydraulic jump "solution" may be seen as a transition across the cusp of Figure 3.4 from the supercritical to the subcritical side, along a line of constant S. This is because R has been decreased by energy dissipation (by turbulence or viscous effects) and the momentum flux is unchanged. The loss of energy flux is given by $\rho Q \Delta R$, where ΔR is the change in R, which is consistent with (2.53). We can also see that if R were decreased by an amount less than this, the result would be a cnoidal-type wavetrain, with properties depending on position within the cusp. If $F < 1.295$, a very tiny amount of dissipation would convert the supercritical stream into a train of large-amplitude cnoidal waves, which would constitute an undular bore. This might occur, for example, if the initial wave of the bore "spilled over" slightly on a continuous basis but the other waves did not, giving a semi-infinite wavetrain rather than a subcritical stream. If instead $F > 1.295$, there must be a more substantial dissipation of energy to cross the "unsteady zone", which would correspond to a more visibly turbulent bore structure. This sensitivity of the structure of undular bores to the dissipation partly explains the considerable scatter in the observed properties of wavetrains in undular bores.

It is obvious from the foregoing that steady inviscid bores are not possible, and that steady flows which change from one uniform stream to another, or to a periodic wavetrain (which is steady in some frame of reference) are not possible unless R or S is changed in some way. An exception to this, of course, is the solitary wave on the supercritical stream, on the supercritical side of the cusp of Figure 3.4, where the stream returns to its initial state after the passage of the wave. The time-dependent behaviour of an *inviscid* system where the initial state consists of a monotonic change in depth from one uniform stream to another has been studied numerically (with the KdV or Boussinesq equation) by Peregrine (1966) for a gradual initial change, and by Fornberg & Whitham (1978) for a discontinuous one. An example

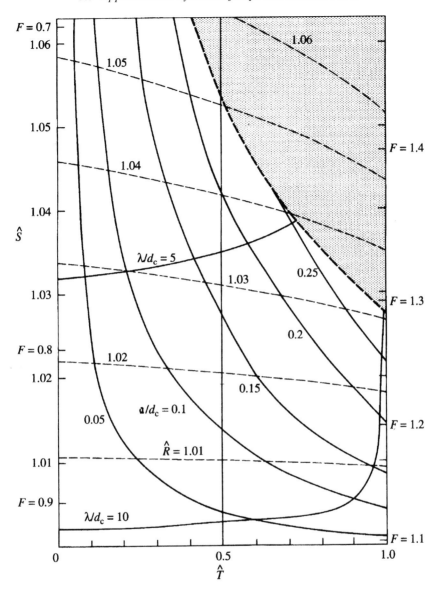

Figure 3.5 The diagram of Figure 3.4 replotted in terms of the variables $\hat{S}$ and $\hat{T}$, which shows the non-linear wave properties more clearly. The two sides of the cusp are now the lines $\hat{T} = 0$ and $\hat{T} = 1$ respectively, and the values of F on these boundaries are shown with labelled marks. The region of unsteady motions is shown shaded. Contours of $\hat{R}, a/d_c$ and λ/d_c are shown. These curves for λ and a were interpolated from the values computed by Cokelet (1977). Linear sinusoidal waves are found at $\hat{T} = 0$, and solitary waves at $\hat{T} = 1$. The waves have cnoidal form for $\hat{S} < 1.02$. Wavelength increases as one moves downward and to the right in this diagram, whereas wave amplitude a/d_c increases as one moves upward and to the right.

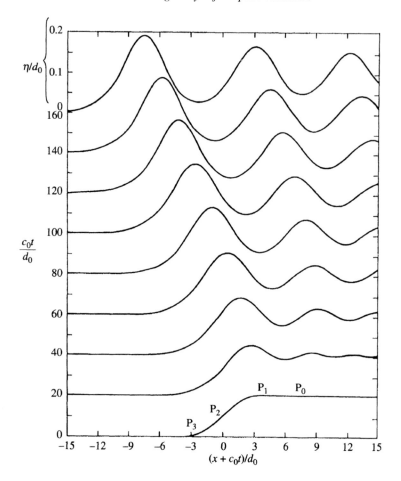

Figure 3.6 The growth of an inviscid undular bore with time from an initial monotonic form, via the Boussinesq equations. (Modified from Peregrine, 1966.) Time t is given in units of d_0/c_0, and the initial fluid velocity $u(x,0)$ is negative and given in terms of η by $(u/2c_0)^2 + u/c_0 = \eta/d_0$.

of the results using the Boussinesq equations is shown in Figure 3.6. Here the horizontal length-scale is small enough for vertical accelerations to be significant in the vertical equation of motion, and the initially smooth profile at $t = 0$ is seen to develop an undular structure.

The reason for this may be seen by examining the deviation of this flow from hydrostatic balance. In the initial profile, the fluid between points P_2 and P_3 has an upward acceleration, implying that the pressure there is greater than the hydrostatic value. Similarly, the fluid between P_1 and P_2 has upward *deceleration*, implying that the pressure is reduced below hydrostatic. Hence the pressure field at $t = 0$ oscillates about the hydrostatic profile, and these pressure gradients cause convergence toward the region between P_1 and P_2. There is also an associated divergence of fluid away from the region between P_0 and P_1. These perturbations and others that develop subsequently result in the wavy structure shown. However, the flow does not reach a steady state. The leading wave, having the largest amplitude, eventually detaches, moves ahead and forms a solitary wave, and this subsequently happens to the second and following waves in turn. The "inviscid bore" structure therefore takes the form of a continual source of solitary waves, spreading over a progressively longer region of the fluid. The parameter m, (3.17), has the value of unity at the leading part of this unsteady bore, but if one moves through the bore (at a given time), m progressively decreases to zero, where the wave amplitude tends to zero and the waves are sinusoidal.

This behaviour of inviscid bores implies that two inviscid streams, separated by a considerable distance, may be connected by a *time-dependent*, spreading, inviscid bore with accumulating energy which is equivalent to energy dissipation in a steady bore. This undular bore will eventually extend to reach any given upstream or downstream location. By a suitable choice of frame of reference, the two uniform streams (with different depths) may have the same Q value. The energy accumulating in the growing undular bore is then given by $\rho Q \Delta R$. The same time-dependent bore may also connect two other steady flows, such as a uniform stream and a fully periodic uniform cnoidal wavetrain.

The *observed* structure of bores in nature or in the laboratory takes three possible forms, depending on the amplitude:

(a) a smooth steady undular bore, with wave amplitudes decreasing with distance to give a uniform stream on the downstream side (if the bore is long enough);

(b) a "broken" undular bore, with turbulence on the crest of the first wave, and possibly also on subsequent waves; and

(c) a fully turbulent bore, with an initial "roller" followed by a uniform stream with some irregularities.

Figure 3.7a shows observed bore speeds and amplitudes compared with Rayleigh's hydraulic equation, (2.54); agreement is generally good, except that the observed bore amplitudes are slightly less than predicted at larger amplitude. Type (a) undular bores are found for $F_u \approx 1.3$ (upstream Froude number) (i.e. $d_d/d_u = 1.4$), which agrees with QRS theory and Figure 3.6. Type (b) bores are found for $1.3 < F_u < 1.55$ (i.e. $d_d/d_u = 1.75$), and the fully turbulent type (c) for $F_u > 1.55$. The corresponding observed maximum wave amplitudes in bores are shown in Figure 3.7b; the wave amplitude progressively decreases with the increasing degree of turbulence as F_u increases from 1.3 to 1.55. However, these boundaries between the bore types are only approximate, because bottom (and sidewall) friction may have a significant

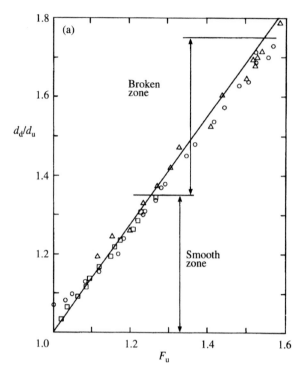

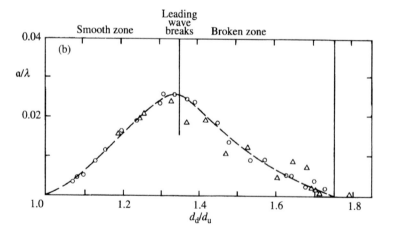

Figure 3.7 (a) The speed–amplitude relationship for undular bores observed in a laboratory channel, compared with Rayleigh's equation (2.54). (b) Observed amplitude a of the leading wave, scaled with wavelength λ, as a function of jump amplitude. In the figure, $\triangle$: $d_u = 6.35$ cm; $\circ$: $d_u = 11.5$ cm; $\square$: $d_u = 20.3$ cm. Note a/λ increases with d_d/d_u up to the point where the first wave breaks, and then decreases as the bore becomes more turbulent. (From Binnie & Orkney, 1955.)

effect on their structure in a practical situation (as demonstrated in the book by Henderson, 1966).

Why do the observed waves in an undular bore decrease in amplitude with distance from the leading edge? One possibility, discussed above, is that the bore is time-dependent and is behaving in an essentially inviscid manner, with the character of a modulated cnoidal wavetrain; here the modulus m of the cnoidal waves (3.16), (3.17) varies from 1 (solitary waves) at the leading edge, to a value m_0 (with $0 \le m_0 < 1$) at the trailing edge, where the waves have smaller amplitude. However, if dissipation is present (as it always is in practice) the bore will eventually reach a steady state. We may identify two main types of steady bores:

(i) bores moving into fluid that is effectively at rest (as would apply to many tidal bores in estuaries where the river flow is weak); and

(ii) stationary bores, such as those on the downstream side of topographic features.

The situation with each of these types is shown schematically in Figure 3.8. For a bore moving into fluid at rest, R/R_c and S/S_c both *increase* as the fluid progresses through the bore (Sturtevant, 1965), due to the effect of bottom stress which, in *the frame of the bore*, acts in the direction of the flow. The flow therefore progresses along the curve in the $(\hat{R}, \hat{S})$ diagram shown in the inset in Figure 3.8a, causing the wave amplitudes to decrease to zero. Steady solutions of this type, obtained by incorporating a viscous boundary layer into the inviscid model, have been obtained by Byatt-Smith (1971). For a stationary bore, on the other hand, the bottom stress under the bore is opposed to the fluid motion, and both R and S progressively *decrease* as the fluid passes through the bore. The flow then follows the route shown in the inset in Figure 3.8b. The reason for the change in bore type at $F_u = 1.3$ is therefore attributable to the absence of stable, steady cnoidal wave solutions to the right of the dashed line of Figure 3.4. For bores of smaller amplitude, bottom friction can provide the necessary changes in R and/or S for steady flows, but at larger amplitudes wave-breaking and turbulence must occur.

3.4 Single-layer flow over topography with non-linearity and dispersion

We now return to the question of unidirectional flow over topography as in §2.3, but with the effects of wave dispersion included. Whereas hydraulic jumps produced by flow over topography were modelled as discontinuities in §2.3, in practice they may be undular if their amplitude is sufficiently small (Huang et al., 1982). The production of these upstream undular bores is a periodic phenomenon resulting from steady forcing. It is visually quite spectacular, with the succession of cnoidal-like waves being produced at the obstacle with no obvious cause of the periodicity. The author first saw this phenomenon in 1981 with the analogous system of a two-layer

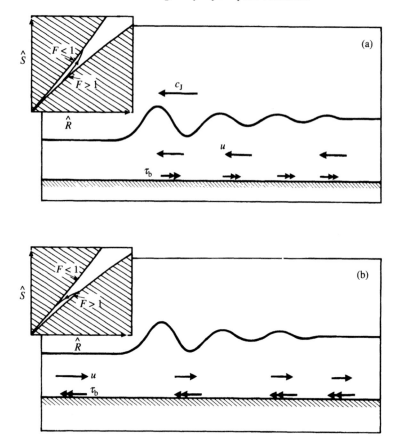

Figure 3.8 (a) A bore of steady form moving leftward into fluid at rest. Bottom frictional stress τ_b (denoted by double-headed arrows) causes energy ($\rho_0 QR$) and momentum flux ($\rho_0 S$) to increase on passing through the bore, so that the fluid follows the path on the $(\hat{R}, \hat{S})$ diagram as shown in the inset. (b) A bore at rest relative to the bottom. Here R and S both decrease as the fluid passes through the bore, as a result of bottom stress.

fluid (as described in Chapter 4), where it is even more spectacular because the waves are much larger, and move more slowly. Huang et al. described observations of flow caused by towing a model ship in a two-dimensional channel, and they reported upstream undular bores for F_0 in the range $0.75 < F_0 < 1.3$. For larger F_0 the bore was turbulent, as would be expected from the discussion of the previous section.

Theoretical studies of this phenomenon based on the Boussinesq equations have been described by Wu & Wu (1982), and a discussion of the mechanics of the process given by Wu (1987). Models of flows over one-dimensional obstacles of small height when F_0 is near unity were derived by Cole (1985) and Grimshaw &

Smyth (1986), who also consider more general situations as discussed in Chapter 5. From (2.10), the equation for topographically forced linear waves propagating against the flow is

$$\frac{\partial \eta}{\partial t} + (U - \sqrt{gd_0})\frac{\partial \eta}{\partial x} = \frac{U}{2}\frac{dh}{dx}, \tag{3.37}$$

and combining this with the KdV equation by simply adding the small advection and dispersion terms in a heuristic manner gives

$$\eta_t + (U - c_0)\eta_x - \frac{3c_0}{2d_0}\eta\eta_x - \frac{1}{6}c_0 d_0^2 \eta_{xxx} = \frac{U}{2}h_x, \tag{3.38}$$

where here $c_0 = \sqrt{gd_0}$. Equation (3.38) is a forced KdV equation (fKdV). It has been formally derived for forcing by a moving pressure pattern by Akylas (1984), and for forcing by topography in more general stratification by Grimshaw & Smyth (1986). Equation (3.38) is only formally applicable when a/d_0, $(d_0/L)^2$, $F_0 - 1$ and h_m/d_0 are all comparably small. This restricts it to the neighbourhood of $F_0 = 1$, $H_m = 0$, on the (F_0, H_m)-plane of Figure 2.11.

In solving (3.38), the natural starting point is to look for steady solutions that link an upstream state of uniform subcritical flow with a corresponding supercritical downstream state. For a given upstream state, this corresponding downstream state must be the same as that given by the hydraulic theory of §2.3.3 [to the order of the approximation involved in deriving (3.38)]. However, an examination of the integrated steady form of (3.38) (with a "top-hat" obstacle profile) shows that such a transition is not possible. Physically, this is because the restoring force due to wave dispersion limits the amplitude of the excursion of the free surface over the obstacle. Consequently, *if* the flow is steady, either or both of the upstream and downstream flows must be oscillatory.

Numerical studies by Grimshaw and Smyth, and analytical studies employing "modulation theory" by Smyth (1987) have shown that the character of the solutions of (3.38) depends on F_0 and H_m in the manner shown in Figure 3.9, when the uniform flow is commenced from a state of rest. The boundaries demarcating subcritical, supercritical and upstream-bore solutions are the same (to leading order) as those of Figure 2.11, and the same letters are used to denote them. The fKdV equation is not able to describe the non-uniqueness of the hydraulic solutions in the region between the lines AG and AE of Figure 2.11, so that these curves are identical in Figure 3.9, to this order of approximation.

Examples of numerical solutions for each of the four regions of Figure 3.9 are shown in Figure 3.10. The supercritical (a) and subcritical (d) solutions require little comment. The flows in the intervening region contain modulated cnoidal wavetrains both upstream and downstream, that replace the discontinuous hydraulic jumps of the hydrostatic theory of §2.3.3. Since the fKdV equation is inviscid, the upstream

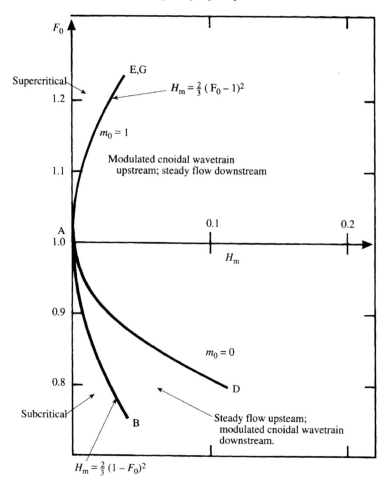

Figure 3.9 F_0–H_m diagram in the neighbourhood of $F_0 = 1$, showing the regions where solutions of (3.38) of different character are obtained (cf. Figure 2.11).

and downstream bores cannot assume a steady form and must be time-dependent wavetrains. In the upper region to the right of EAD, where the flow is represented by Figure 3.10b, the upstream wavetrain has solitary waves ($m = 1$) at its leading edge, and m decreases monotonically through the wavetrain to a value m_0 (with $0 \leq m_0 < 1$) at the obstacle.

As $t \to \infty$ and the length of the wavetrain increases, the part near the obstacle approaches a uniform cnoidal wavetrain of modulus m_0, whilst the distant "nose" is marked by a succession of increasingly isolated solitary waves. Near the upper boundary (AE), $m_0 \to 1$, and near the lower boundary (AD), $m_0 \to 0$. The rate of energy input into this wavetrain may be obtained from the QRS diagram,

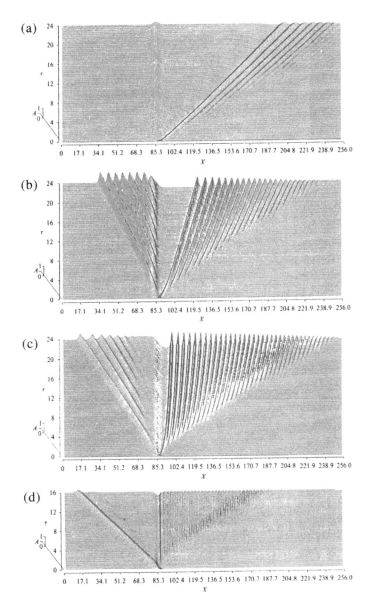

Figure 3.10 Typical solutions of the fKdV equation (3.38), for the impulsive commencement of flow over an obstacle. Specifically, $h(x) = \frac{2}{9}d_0\epsilon^4\mathrm{sech}^2(0.3X)$, where $X = \epsilon x/d_0$, $\tau = \epsilon^3 c_0 t/6d_0$, $F_0 = 1 + \epsilon^2\Delta/6$ and $\eta = \frac{2}{3}\epsilon^2 A d_0$. The term ϵ is an arbitrary but small number and Δ has the values given below. These flow types are found in the four regions of Figure 3.9, listed from the top downward. (a) Supercritical flow, $\Delta = 4$; (b) modulated upstream undular bore in region EAD of Figure 3.9, $\Delta = 0$; (c) as for (b), but with steady flow upstream in region BAD, $\Delta = -1.7$; (d) subcritical flow, $\Delta = -4$. (From Grimshaw & Smyth, 1986, reproduced with permission.)

Figures 3.4, 3.5, by determining the point corresponding to the periodic wavetrain with $m = m_0$. To the left of EAD, on the downstream side the flow is supercritical, and the modulated wavetrain is detached from the obstacle and lags behind it. It has solitary waves ($m = 1$) at the leading edge, and sinusoidal waves ($m = 0$) at the downstream trailing edge. In the lower region enclosed by BAD, the situation is to some extent the reverse of that in EAD. The upstream modulated wavetrain is now detached (Figure 3.10c), with m ranging from unity to zero within it, and the flow immediately upstream of the obstacle is steady. On the other hand, the downstream wavetrain is "attached" to the obstacle with $m = m_0 \leq 1$ there, decreasing to zero as one moves downstream through the bore. On AD, both bores are just attached, with $m_0 = 0$ upstream and $m_0 = 1$ downstream. As AB is approached, the amplitude of the upstream bore vanishes, and downstream $m_0 \rightarrow 0$, yielding linear lee waves. The modulated structure of these wavetrains is a natural consequence of the topographic forcing. On the upstream side this forcing produces (or tries to produce) a steady cnoidal wavetrain of modulus m_0 to the right of EAD, and a steady flow in BAD, and the unsteady modulated waves are the necessary junction between these states and the uniform flow.

If $h(x) < 0$, the numerical solutions of the fKdV are unsteady and irregular. This reflects the fact that the hydraulic solutions (2.55)–(2.58) are indeterminate if $F_0 = 1$, although they are everywhere sub- or supercritical if $F_0 \neq 1$.

Smyth (1988) has added viscous terms to (3.38) and found that the flow becomes steady near the obstacle in most cases. The amplitudes of the waves of an upstream bore remain constant with time as it propagates away from the obstacle, but they decay to zero spatially downstream from the leading wave. In spite of these complicating details, the relationship between the speed of the bore and its amplitude (as measured by the change in mean surface height associated with the wavy region) is given to good accuracy by the jump relationship (2.54), for both the inviscid and dissipative cases.

In general, steady (and also suitably unsteady) topographically forced flows may be interpreted using QRS theory. The total form drag F_D on an obstacle is given by the integral of the pressure over its surface (1.42), which from (3.20) is equal to $\rho \Delta S$, where ΔS is the change in S across the obstacle. The drag force can therefore be related to other flow properties via Figure 3.5, which also provides limits to its magnitude. Further, each of the various transitions shown in Figure 2.9 can be represented on the R, S diagram, and with the exception of the final time-dependent rarefaction, the whole sequence of transitions can be shown on the one diagram if changes of reference frame (and hence of Q) are accommodated. However, QRS theory is not directly applicable to time-dependent flows. Computations of the drag during the time-dependent flows when the upstream solitary waves are being continually forced (Lee et al., 1989) show a significant periodic variation, with the maximum drag occurring just before a wave detaches from the obstacle and moves upstream.

Numerical studies by Vanden-Broeck (1987) have implied that supercritical flows over obstacles may possess another steady solution in addition to the symmetric supercritical one given by (2.57). This additional solution corresponds to perturbation of a solitary wave by an obstacle, rather than perturbation of a uniform stream, and has only been reported for F_0 greater than about 1.2. It probably only applies to obstacles with restricted shapes that can fit comfortably under large-amplitude solitary waves.

If two obstacles are placed in a stream one behind the other, the number of possible flow states increases. Experiments by Pratt (1984) have identified four main flow regimes, as shown in Figure 3.11a. The occurrence of these regimes depends on the relative heights of the obstacles, as shown in Figure 3.11b. In flow regime I, where the first obstacle is higher than the second, the flow is controlled by the first obstacle and the flow is either wholly subcritical or supercritical over the second obstacle. In flow regimes II and III the two obstacle heights are comparable, and a non-linear wavetrain occurs in the region between the obstacles, with breaking waves in regime II and laminar waves in regime III. The observed amplitude of these waves in regime III increases with upstream Froude number, as would be expected.

This behaviour may be interpreted in terms of the $(\hat{R}, \hat{S})$ diagram as shown in Figure 3.11c. On passing over the first obstacle $\hat{S}$ is reduced, and the flow evolves from the subcritical state at P_1 to a cnoidal wavetrain at point P_2, which lies within the cusp. If the waves are breaking (and neglecting bottom friction), $\hat{R}$ decreases to point P_3 as the fluid passes through the wavetrain, and $\hat{S}$ then decreases to point P_4 on passing over the second obstacle. In regime IV, on the right of Figure 3.11b, the flow is controlled by the second obstacle which "floods" the smaller first obstacle, where the flow is wholly subcritical. If more than two obstacles are placed in the stream more variations on this theme would be produced, but no new phenomena would be expected. Similar flow properties are expected in a channel of variable cross-section (see §2.3.3) where there are two critical sections. Here, one may be due to a ridge and the second to a sidewall contraction, and a cnoidal-type wavetrain between them would be expected in suitable flow conditions.

The forced KdV equation continues to be a much-used tool to investigate one-dimensional non-linear wave motion in a variety of situations where the vertical structure of the flow is dominated by one mode (Grimshaw, 2010; Grimshaw & Helfrich, 2018).

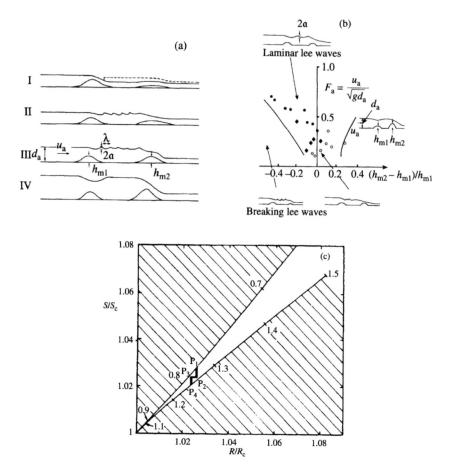

Figure 3.11 Single-layer flow over two obstacles. (a) Schematic diagrams of the four experimentally found flow regimes in the transcritical range, in order of increasing height of the downstream obstacle. (The upstream flow is specified by u_a, d_a, to be consistent with Figure 2.9a. I: $h_{m1} \gg h_{m2}$, flow supercritical between obstacles; II: $h_{m1} > h_{m2}$, breaking wavetrain between obstacles; III: $h_{m1} \approx h_{m2}$, laminar wavetrain between obstacles; IV: $h_{m1} < h_{m2}$, subcritical flow between obstacles. (b) Location of the flow regimes of (a), as functions of the upstream Froude number F_a and the height difference, $(h_{m2} - h_{m1})/h_{m1}$. Key: ○: $0.1 < 2a < 0.4$; ♦: $0.5 < 2a < 0.9$; ●: $1.0 < 2a < 1.5$. (c) The observed behaviour in regions II and III interpreted on the (R, S) diagram. ((a) and (b) from Pratt, 1984.)

3.5 Non-linear flow past three-dimensional obstacles

The study of the properties of flow past three-dimensional obstacles of finite height has a short history, but the phenomena involved are familiar to observers of the flow past stones in very shallow rivers.

For hydrostatic flow of a single layer we return to the governing equations (2.6)–(2.8). These are essentially the same as those for the two-dimensional (x, y) motion of a gas for which the ratio of specific heats $\gamma = 2$ (compared with 1.4 for air). However, the analogy breaks down when shock waves (i.e. hydraulic jumps) form; the physics of shock waves in gases is different from that of hydraulic jumps (§2.3.1), giving different relationships across the discontinuity.

The potential vorticity corresponding to that defined in §1.3.3 is $\zeta = \Omega/d$, where $\Omega = \partial v/\partial x - \partial u/\partial y$. For the inviscid equations, (2.6)–(2.8), we may then derive

$$\frac{D\zeta}{Dt} = 0, \tag{3.39}$$

which expresses the conservation of potential vorticity. However, this equation does not necessarily apply in regions where the inviscid equations are not valid, such as within hydraulic jumps, where energy dissipation occurs. The flow past obstacles of finite height gives rise to two-dimensional hydraulic jumps that are not perpendicular to the incident flow and may be curved, as seen in Figure 2.4, and it is convenient to first discuss the properties of these jumps.

3.5.1 Two-dimensional hydraulic jumps

We consider a stationary two-dimensional hydraulic jump (that is, two horizontal dimensions), shown schematically in Figure 3.12, with a uniform upstream fluid velocity U and depth d_u in the x-direction. At a representative point P, relative to the incident flow direction the jump has an inclination α, and the flow direction immediately downstream a corresponding inclination β. In general the jump may be curved, so that α and β vary along it. By analogy with the corresponding situation for shock waves in supersonic gas dynamics (see for example, Hayes & Probstein, 1966), if the curvature of the jump is not large we may expect the one-dimensional equations of §2.3.1 to apply in each section normal to the jump, with the velocity components along the jump being continuous across it. At point P, therefore, the velocity along the jump is $U \cos \alpha$, and denoting the normal velocities on the upstream and downstream sides by v_u and v_d and the depths by d_u and d_d respectively, (2.49) and (2.50) give

$$U d_u \sin \alpha = v_u d_u = v_d d_d, \tag{3.40}$$

$$d_u U^2 \sin^2 \alpha + \frac{1}{2} g d_u^2 = d_d v_d^2 + \frac{1}{2} g d_d^2. \tag{3.41}$$

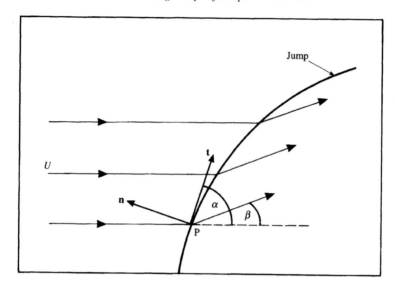

Figure 3.12 Schematic diagram of flow through a curved two-dimensional hydraulic jump. **t** and **n** denote the tangential and normal vectors to the jump at
P.

Defining the upstream Froude number F_u by $F_u = U/\sqrt{gd_u}$, these equations permit
the downstream properties to be expressed in terms of F_u and α, and give

$$\frac{d_d}{d_u} = \frac{1}{2}(G - 1), \qquad \text{where} \quad G = (1 + 8F_u^2 \sin^2 \alpha)^{1/2}, \qquad (3.42)$$

and

$$\tan \beta = \frac{\tan \alpha (G - 3)}{G - 1 + 2 \tan^2 \alpha}. \qquad (3.43)$$

Clearly we must have $F_u > 1$, and possible values for α lie in the range $1/F_u \leq \sin \alpha \leq 1$, with β vanishing at each end and having a maximum for α within this range.

Curved hydraulic jumps produce vorticity on their downstream side, as do curved
supersonic shock waves. The velocity immediately downstream of the jump is
$\mathbf{u}_d = U \cos \alpha \mathbf{t} - v_d \mathbf{n}$, where $\mathbf{t}$ and $\mathbf{n}$ are the tangential and normal unit vectors as
shown in Figure 3.12. Taking the curl of this velocity (by, for example, employing
polar coordinates centred on the *origin* of the radius of curvature at point P), gives

$$|\nabla \times \mathbf{u}_d| = \kappa U \cos \alpha \left(1 + \frac{1 + G}{4GF_u^2 \sin^2 \alpha}\right), \qquad (3.44)$$

where $\kappa = d\alpha/ds$ is the curvature of the jump at point P, s being distance along
it. For a jump curved as shown in Figure 3.12 the curvature is negative, so that for

$0 < \alpha < \pi/2$ the vorticity is negative, and for $\pi/2 < \alpha < \pi$ it is positive. This will apply to the upstream jumps shown in Figure 2.4a and c, for example. Hence, the dissipative processes within a curved jump can produce vorticity within the flow downstream.

Equations (3.40)–(3.43) may be used to obtain the properties of the reflection of a hydraulic jump (as described in §2.3.1) from a sidewall. The reflected jump makes a smaller (acute) angle with the wall than does the incident jump, although these angles tend to equality when the incident jump has small amplitude ($G \cong 3$) or is nearly perpendicular to the wall. In practice, however, such reflection is often not so simple. In gas dynamics, the reflection of shock waves from walls often takes the form of *Mach-stem reflection* (e.g. Hornung, 1986), where the incident shock joins the reflected shock at a point some distance from the wall; this point is connected to the wall by a third shock (the *Mach-stem*) that is normal to the wall and larger in amplitude than either of the others. Mach-stem reflection has also been observed and described for solitary waves (e.g. Miles, 1980; Tanaka, 1993) so that it is expected to play a role in the reflection of hydraulic jumps. But the details are complex, and aspects of Mach-stem reflection (including, in some cases, the reason why it must occur at all) remain obscure.

3.5.2 Hydrostatic flow past three-dimensional obstacles

We next consider the eventual steady flow (assuming that this exists) obtained past an isolated obstacle where the relative motion at speed U is commenced from a state of rest in fluid of depth d_0, so that the flow at large distances from the obstacle is initially uniform. For definiteness we consider obstacles that have a single maximum in height h_m, and have either circular horizontal cross-sections, or are elongated across the stream. Apart from the obstacle shape, the principal parameters of the flow are again $F_0 = U/\sqrt{gd_0}$ and $H_m = h_m/d_0$. A description of the properties of steady hydrostatic flow past two-dimensional obstacles of small height was given in §2.2.1, where it was shown that if F_0 is not close to unity, the flow for small H_m is wholly sub- or supercritical, as for one-dimensional flow. For $F_0 < 1$ the effect of the obstacle extends a considerable distance beyond it in all directions, but for $F_0 > 1$ disturbances are only found downstream, where they consist of waves emanating in beams at the Mach angle. Here we are primarily concerned with non-linear hydraulic effects that occur for larger H_m, and the linear flow solutions give an indication of the character of these larger-amplitude flows.

We may describe these flows with the same type of model as that used for one-dimensional flows in §2.3, with the governing equations (2.6)–(2.8). General analytical solutions to this non-linear system are not available and one must resort to numerical computation to obtain flow fields. However, it is possible to make a number of useful inferences about the flow properties from these equations, and we note some relevant conservation relations that apply, providing that the fluid has not passed through a dissipative region. The first is potential vorticity conservation (3.39), and uniform upstream conditions in steady flow imply that $\zeta = 0$ everywhere, so that $\partial u/\partial y = \partial v/\partial x$. Second, (1.29) implies that the total head R is uniform in the steady state, and for hydrostatic flow this yields

$$R(x, y) = \frac{1}{2}(u^2 + v^2) + g(d + h) = \frac{1}{2}U^2 + gd_0, \tag{3.45}$$

generalising (2.57) to two dimensions. For mass conservation, (2.8) gives $\nabla \cdot (d\mathbf{u}) = 0$. For obstacles that are symmetric about the direction of flow, in the vertical plane of symmetry this may be integrated to give

$$du = d_0 U - \int_{-\infty}^{x} \frac{\partial}{\partial y}(dv)\mathrm{d}x, \tag{3.46}$$

and corresponding expressions may be written down for other streamlines.

In conformity with the one-dimensional (x, z) flow, we expect that the linear solution gives a realistic qualitative picture of the flow for small obstacle heights up to the point where hydraulic transition and consequent hydraulic jumps occur. When this point is exceeded, for $F_0 < 1$ the upstream hydraulic jump corresponding to that of one-dimensional flow spreads out laterally and weakens as it propagates upstream with time, and disappears. The resulting upstream flow field has a tendency to diverge and spread laterally around the obstacle, decreasing the flux of fluid over it. This is reflected in the second term in (3.46). The extent of this divergence and its location is dependent on the obstacle shape, but if we assume that it is small at the ridge line (or crest) of the obstacle, we may write $ud = q(x, y)$, where q varies much more slowly with x than does either d or u. Ignoring any variation of q with x, we have locally on the centre-line

$$\left(\frac{u^2}{gd} - 1\right) \frac{\mathrm{d}d}{\mathrm{d}x} \cong \frac{\mathrm{d}h}{\mathrm{d}x}, \tag{3.47}$$

(cf. (2.59)) which gives the same hydraulic alternative (approximately) as for one-dimensional flows. The difference here is that $q < Q$ by an amount that depends on the divergence of the flow for $x < 0$, which depends on the obstacle shape. However, we may assume that as H_m increases, $q \leq Q$ when the hydraulic transition first occurs, so that the criterion for this transition lies at slightly larger H_m than that for one-dimensional flow.

Numerical solutions for $F_0 < 1$ with hydrostatic models have been described by Lamb & Britter (1984) for a cone-shaped obstacle, and by Schär & Smith (1993a) for an obstacle of the form $h = h_m/(1 + x^2/a^2 + y^2/a^2)^{3/2}$. Figure 3.13 shows steady-state flow fields and fluid depths for $F_0 = 0.5$, $H_m = 0.6$ (a, b, c) and

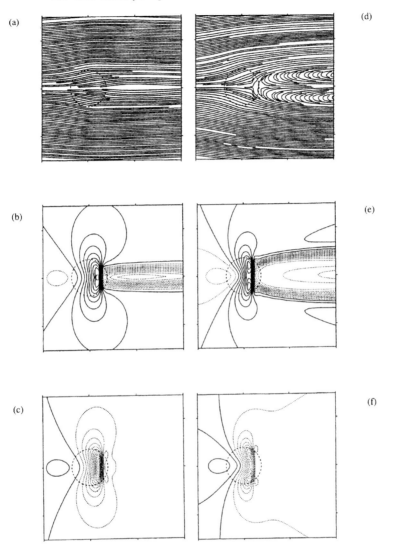

Figure 3.13 Steady-state solutions for hydrostatic flow past an axisymmetric obstacle of the form $z = h_m/(1 + r^2/a^2)^{3/2}$ where $r^2 = x^2 + y^2$, for $F_0 = 0.5$, $H_m = 0.6$ (regime IIa, Figure 3.14 for (a), (b) and (c); $F_0 = 0.5$, $H_m = 0.8$ (regime IIb, Figure 3.14 for (d), (e) and (f). Panels (a) and (d) show streamlines and the flow direction, (b) and (e) the fluid speed with contour spacing $0.05U$, and values less than $0.5U$ dashed; (c) and (f) show the height contours, at intervals of $0.05d_0$ with values less than $0.5d_0$ dashed. (Modified from Schär & Smith, 1993a, Fig. 5.)

$F_0 = 0.5$, $H_m = 0.8$ (d, e, f) from Schär & Smith. The upstream divergence is apparent in (a) and (d), and increases with H_m. Both cases are in the "hydraulic regime" with a hydraulic jump of finite length forming on the lee side. As described above, such jumps introduce vorticity into the flow downstream, and for $H_m = 0.8$ the flow divergence and jump strength have increased sufficiently in amplitude to give lee-side separation and reversed flow on the downstream centre-line. These simulations were numerically constrained to be symmetrical, and if this restriction is removed a periodic vortex street forms downstream (Schär & Smith, 1993b), where the vorticity in these vortices stems from the hydraulic jump.

The equations used were essentially inviscid except for the hydraulic jumps, so that viscous boundary-layer separation from the obstacle was excluded. However, in real situations, boundary-layer separation of flow around obstacles projecting through the layer is an additional process that can introduce vorticity into the wake region. Which of these two processes (hydraulic jumps or boundary-layer separation) is the more important for the behaviour of the wake in real situations may depend on the circumstances. For instance, one might expect boundary-layer separation to dominate for flow around a vertical-walled cylinder, and hydraulic jumps for hydrostatic flow past an obstacle with gentle gradients, but both may be important in each case.

A regime diagram for the various flow types is shown in Figure 3.14. In regions I and IV the flow is wholly sub- or supercritical, as for the linear solutions. In region II, the flow has become critical over part of the obstacle, and a hydraulic jump of finite length has formed over the lee side. The boundary shown between regimes I and II is that computed by Schär & Smith (1993a), which is close to the boundary obtained by Lamb & Britter (1984). The dashed line in region II separates the two flow types shown in Figure 3.13: to the right of it (in region IIb) the flow has a separated wake, and to the left it does not. Some aspects of these flows have been confirmed experimentally by Lamb and Britter using a two-layer representation of a single-layer flow.

In region IIb, the computed flow solutions have closed streamlines in the wake. In real flows at large Reynolds number, if such flows with closed streamlines are stable they eventually become dominated by viscosity, given sufficient time. In particular, if the circulation is driven by tangential stresses at the boundary and the depth d is approximately uniform, the vorticity within the eddy becomes uniform by diffusion (Batchelor, 1967, p. 538) in the time-scale L^2/v, where L is the radius of the eddy and v is the kinematic viscosity. If the eddy is circular, this implies rigid rotation. Large long-lived eddies are often observed in wakes behind islands in flows in suitably shallow water (e.g. Wolanski et al., 1984), although in practice their dynamics may be more complex than described here.

In region III the obstacle is high enough to project through the layer. For $F_0 < 1$, the boundary of this region may be obtained from (3.45) and the condition (from (2.59)) that $u^2 = gd$ at $h = h_m$. This implies that u vanishes with d as the depth of

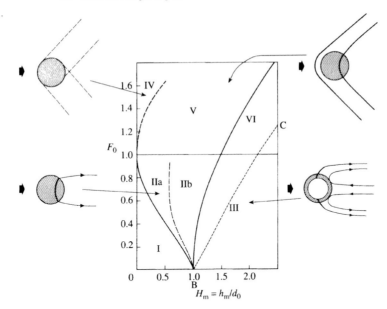

Figure 3.14 Flow regimes for single-layer hydrostatic flow past a circular obstacle, in terms of the upstream Froude number F_0 and $H = h_m/d_0$, the ratio of the maximum obstacle height to the upstream fluid depth. The sketches show the flow in plan view, with the shaded region denoting the obstacle, dashed lines denoting waves, heavy solid lines denoting hydraulic jumps, and light lines denoting streamlines. In region III the obstacle protrudes through the fluid layer, as indicated by the clear central region in the diagram. The dashed line BC is the line for blocking for one-dimensional flow denoted BC in Figure 2.11, included for comparison only. (Modified from Schär & Smith, 1993a.)

fluid over the obstacle decreases, and the boundary between regions II and III is

$$H_m = 1 + \frac{1}{2}F_0^2, \tag{3.48}$$

independently of the obstacle shape. If the obstacle is higher than this, (3.48) gives the maximum height that the fluid reaches on the obstacle surface.

When $F_0 > 1$, as H_m increases, there is no disturbance to the steady-state flow upstream of the obstacle until the point of hydraulic transition is reached. This occurs on the boundary of region IV which is shown schematically in Figure 3.14, as its location has yet to be calculated. As the analogy with gas dynamics would suggest and Figure 2.4 (p. 24) shows, in region V, a <-shaped stationary jump then forms with its vertex on the upstream side of the obstacle. Upstream of this jump the steady-state flow is unchanged from its form far upstream, and downstream of the jump the flow diverges and spreads laterally. This divergence causes the onset of critical flow to occur at a larger value of H_m than for two-dimensional flow, so that the boundary between regions IV and V lies to the right of its position in one-dimensional flow. The divergence also reduces the jump amplitude relative to its corresponding one-dimensional amplitude, on the central plane and elsewhere. Since the jump is stationary we may calculate its amplitude, and using the two-dimensional formula (2.54) locally with $c_J = U$ and $d_u = d_0$, we find that the depth d_d on the downstream side on the central plane is given by

$$F_0^2 = \frac{d_d}{2d_0}\left(1 + \frac{d_d}{d_0}\right). \tag{3.49}$$

Since we may assume that there is no divergence within the jump (i.e. the tangential velocities are equal), the centre-plane ($y = 0$) velocity u_d downstream of the jump is given by

$$u_d d_d = U d_0, \tag{3.50}$$

and the Bernoulli equation there is

$$\frac{1}{2}u^2 + g(d + h) = \frac{1}{2}U^2\left(\frac{d_0}{d_d}\right)^2 + g d_d. \tag{3.51}$$

If we again assume that $\partial q/\partial x \approx 0$ at $h = h_m$, we again obtain (2.59) there, giving a critical condition and hydraulic transition as for two-dimensional flow. As H_m increases, d and u decrease to zero at $h = h_m$, and (3.51) then gives the blocking height for $F_0 > 1$

$$H_m = \frac{1}{2}(1 + 8F_0^2)^{1/2} - \frac{1}{4} + [1 + (1 + 8F_0^2)^{1/2}]/16F_0^2. \tag{3.52}$$

This is a continuation of (3.48), and marks the boundary between regions V and VI as shown in Figure 3.14.

Figure 2.4 (p. 24) also implies that weaker hydraulic jumps occur on the downstream side as modifications of the trailing linear waves at the Mach angle, but their properties have yet to be explored. By analogy with transonic gas dynamics, we may expect that a myriad of complex phenomena involving "shocks" (hydraulic jumps) occur when $F_0 \approx 1$, depending on the shape of the obstacle. It would also appear that the double-state and hysteresis effect that occurs for supercritical one-dimensional flow (see §2.3.2) is still possible in this two-dimensional flow, for the same reasons. However, this effect is modified by the divergence over the obstacle and behind the jump, and is still subject to experimental confirmation.

The above discussion is applicable to obstacles of more-or-less arbitrary shape, including obstacles that are very long in the transverse direction, and the question arises as to the relevance of the one-dimensional flow solutions of §2.3.2 to flow over such obstacles. We first consider the case for $F_0 < 1$, with an obstacle that has length $2B$ and approximately uniform cross-section. After the motion has commenced from rest, the flow in the central region is essentially the same as the one-dimensional flow until time $t \sim B/\sqrt{gd_0}$, the time-scale for gravity waves to propagate from the ends of the obstacle to the central region. After this time the flow may be partly deflected laterally around the obstacle, and it approaches its steady two-dimensional state. This lateral divergence is felt upstream to a distance of order B. The resulting flow over the central part of the obstacle (or barrier) is similar to the one-dimensional flow for small H_m, where the value for the onset of hydraulic transition may be slightly larger.

However, although qualitatively similar the flow properties become progressively more different as H_m increases, to the point where the criteria for the onset of blocking are far apart [curves (3.48), (3.52)) versus (2.78); see Figure 3.14]. For $F_0 > 1$, the same general remarks apply, except that the upstream divergence is constrained to occur downstream of the jump. Here the flow is subcritical, so that transverse wave propagation may occur, and the jump moves further upstream as H_m increases.

In summary, the steady-state flows over a two-dimensional (x, z) barrier and the central region of a long but finite three-dimensional one may differ substantially if H_m is large enough, in contrast to the linearised flow solutions where they are substantially the same. A similar phenomenon that has some practical relevance occurs for stratified flows, as discussed in Chapter 13.

3.5.3 *The effects of bottom surface friction*

The above discussion assumes that the flow is essentially inviscid. However, in shallow water flows with gently sloping terrain, or in single-layer models of atmospheric flows over islands or mountains, bottom surface friction may be important. In such flows where $R_e = Ud/\nu \gg 1$, this factor is represented by the frictional drag term as given in (2.95) which in two dimensions has the form $C_D|\mathbf{u}|\mathbf{u}/d$, where C_D is a drag coefficient ($\sim 10^{-3}$). The relative importance of this term is measured by the "friction number" $C_D a/d$, where a is the horizontal length-scale (half-width) of the obstacle.

Numerical modelling studies (Grubišić et al., 1995) of single-layer flow with bottom friction have been used to address the question as to why atmospheric wakes behind islands are steady or oscillatory, as observations have shown that both occur. It is well known in two-dimensional fluid dynamics that wakes behind isolated obstacles are generally oscillatory if the Reynolds number is large, and this may be expected to be the case for oceanic wakes behind islands and seamounts. But in the atmosphere, where there is substantial airflow over the island, model results yield oscillating wakes if the friction number is small (~ 0.02), and steady wakes if it is not (~ 0.08). The implication is that, in the atmosphere, small islands will tend to have oscillatory wakes, and large islands relatively steady ones. Despite the simplicity of the model, observations of atmospheric wakes of various islands (including Hawaii) are generally consistent with these conclusions.

3.5.4 Supercritical hydrostatic flow past a varying sidewall

For flow in channels where the assumptions of §2.3.3 are not satisfied, because the channel is wide or the variations are too rapid, the formalism described there may still be adequate if the flow is subcritical. For supercritical flows, however, abrupt changes in the sidewalls cause different effects that may dominate the local flow, and these must be treated separately. We describe here the archetypal situation of a semi-infinite fluid flowing past a sidewall boundary or barrier that undergoes an abrupt change in direction through an angle β. The solutions for this case may then be used to infer what happens in more complex situations. The boundary has the form

$$y = Y_b(x) = 0, \qquad x < 0,$$
$$y = Y_b(x) = \tan\beta \cdot x, \qquad x > 0, \tag{3.53}$$

where $-\pi/2 < \beta < \pi/2$, and the fluid occupies the half-space $y > Y_b(x)$, with uniform velocity parallel to the boundary sufficiently far upstream. We consider steady flow with upstream Froude number F_0 given by $F_0^2 = U_0^2/gd_0$, where U_0, d_0 are the velocity and depth when $x \to -\infty$ as before.

If the flow is everywhere subcritical (i.e. the local Froude number $F = |\mathbf{u}|/\sqrt{gd}$ is everywhere less than unity), the fluid flows past the corner in a smooth fashion, with disturbance decreasing in amplitude with distance from the corner. Supercritical flow with $F_0 > 1$ is more interesting. For a "concave" corner where $\beta > 0$ is a finite angle, the flow undergoes an abrupt change from the uniform supercritical upstream state to a uniform downstream state across a hydraulic jump on the line (Figure 3.15a)

$$y = \tan\alpha \cdot x, \tag{3.54}$$

where from the equations of §3.5.1 we have

$$\frac{dd}{du} = \frac{1}{2}(G - 1), \quad \text{where} \quad G = (1 + 8F_0^2 \sin^2\alpha)^{1/2}, \tag{3.55}$$

where d_d and d_u are the (uniform) fluid depths on the upstream and downstream sides of the jump, respectively, and α is given implicitly in terms of F_0 and β by

$$\tan\alpha = \tan(\alpha - \beta) \cdot (G - 1)/2. \tag{3.56}$$

The position of the hydraulic jump depends on both the upstream Froude number and the angle of the sloping barrier. From (3.56) we deduce that as $\beta \to 0$, $\alpha \to 0$

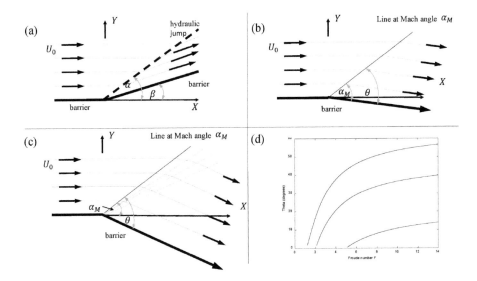

Figure 3.15 (a) Steady supercritical single-layer flow past a vertical barrier that has an abrupt change in the direction of a positive angle β. This produces a hydraulic jump at angle α, with deeper uniform flow and reduced value of F on the lee side. (b) The same as (a) but for a small negative angle β. Here there is an abrupt change in the flow at the Mach angle α_M to a direction parallel with the boundary, with a decrease in thickness d and an increase in F. (c) The same as (b), but for a larger negative angle β. The flow direction changes continuously over a range of Mach lines. (d) Solutions of equation (3.65) showing the change in Froude number with change of angle for supercritical flow around a corner, for initial Froude numbers $F_0 = 1.01$, 2 and 5.

and $0 < \beta < \alpha < \pi/2$. The flow downsteam of the hydraulic jump may be either sub- or supercritical (though most likely the latter), in the frame of reference of the concave corner. Hence the flow undergoes an abrupt change of direction from the upstream to the downstream state at this stationary hydraulic jump. Figure 7.9 of Henderson (1966) shows examples.

We next consider the corresponding case of flow with $F_0 > 1$ around a "convex" corner where $\beta < 0$. If β is small, say $\beta = -\Delta\theta$ where $0 < \Delta\theta \ll 1$, then the disturbance is linear and the flow undergoes a small change of direction (clockwise) at the Mach line (Figure 3.15b)

$$y = \tan \alpha_M \cdot x, \qquad \sin \alpha_M = 1/F_0, \qquad (3.57)$$

where α_M is the Mach angle (see §2.2.1) for this value of F_0. The flow at the Mach line consists of two orthogonal components: one directed along the line, and one directed normal to it. There is no variation of these flow components along this Mach line, and the flow is irrotational (no vorticity), so that the tangential velocity

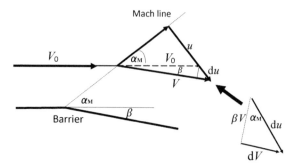

Figure 3.16 This diagram shows velocity vectors for the flow illustrated in Figure 3.15b. V_0 denotes the fluid velocity upstream of the Mach line, and u denotes the velocity component normal to the Mach line. Because of the change in direction of the barrier (through angle β), this normal component is increased by du so that the nett velocity changes from V_0 to vector V, parallel with the barrier change. The flow is irrotational, and hence the downstream velocity component parallel with the Mach line is unchanged, and similarly the stretched downstream normal component is uniform, along the Mach line. The triangle at lower right is an amplification of the small corner triangle indicated by the arrow.

component of the flow is unchanged downstream of the line. However, the normal velocity component on the downstream side is different, and is affected by changes in the boundary.

Figure 3.16 shows a plan of the velocity vectors for a small change $\beta = d\theta$ in the barrier angle. The fluid velocity approaching the Mach line has velocity V_0, and on the downstream side it has the velocity V. The latter consists of the velocity components along and normal to the Mach line, where the latter with initial velocity u is increased by the value du, so that initial velocity V_0 is increased and redirected to velocity V. This change is uniform along the Mach line. From the small triangle of Figure 3.16, one sees that

$$\tan \alpha_m = dV/\beta V, \tag{3.58}$$

giving a nett downstream flow parallel with the barrier with small incline β.

More generally, if β is not small, the change in flow direction is spread continuously over a range of angles θ (measured clockwise from the Mach angle α_M), constituting what is technically known as an expansion fan. This flow pattern is illustrated in Figure 3.15c. If the fluid speed (in whatever direction) is denoted V with depth d, the incoming fluid has speed $V = U_0$ and depth d_0, and

$$F = V/(gd)^{1/2}, \quad \text{where} \quad \sin \theta = 1/F, \tag{3.59}$$

the rate of change in total velocity V at the wavefront with respect to θ is then

$$\frac{dV}{d\theta} = V \tan \theta. \tag{3.60}$$

In general, for $\beta < 0$, no hydraulic jumps occur and the total head R (see §3.5.2) is constant everywhere so that (3.45) holds throughout the flow (where here $h = 0$). Differentiating (3.45) with respect to d we obtain throughout the flow

$$V \frac{dV}{dd} = -g, \tag{3.61}$$

and (3.60), (3.61) and (3.45) together give

$$\frac{dd}{d\theta} = -\frac{V^2}{g} \tan \theta. \tag{3.62}$$

With

$$F^2 = V^2/gd, \quad R = \frac{1}{2}V^2 + gd = \frac{1}{2}U^2 + gd_0, \tag{3.63}$$

(3.62) becomes

$$\frac{dd}{d\theta} = -\frac{(2(R - gd))^{3/2}}{g(2R - 3gd)^{1.2}}, \tag{3.64}$$

which integrates to von Kármán's solution (Ippen, 1951)

$$\theta = -\sqrt{3} \arctan \left(\frac{3}{F^2 - 1} \right)^{1/2} + \arctan \left(\frac{1}{F^2 - 1} \right)^{1/2} + \theta_0, \tag{3.65}$$

where

$$\theta_0 = \sqrt{3} \arctan \left(\frac{3}{F_0^2 - 1} \right)^{1/2} - \arctan \left(\frac{1}{F_0^2 - 1} \right)^{1/2}. \tag{3.66}$$

As θ increases (in a clockwise sense here) from zero at the Mach line, F increases monotonically from F_0, V increases and d decreases until the flow becomes parallel with the barrier at angle $-\beta$. The flow becomes thinner and faster in the range of angle where the direction changes. Figure 3.15d shows three curves where $F_0 = 1.01, 2$ and 5 respectively, where the ordinates $(\alpha_M - \theta)$ are β values. These show the change in Froude number for given changes in flow direction, with increasing velocity and decreasing depth.

These show that the adjustment in angle is largest initially, for all Froude numbers, and that the closer F_0 is to one, the larger is the change in angle that can be accommodated by this model. For the given flow assumptions, there is a limit to the magnitude of the turning angle, particularly for large F_0, and in the latter case other phenomena such as flow separation would be expected.

The corresponding phenomenon in supersonic gas dynamics is termed a "Prandtl–Meyer expansion fan", because the gas expands and its density decreases. In the hydraulic expansion fan here the fluid is thinning and accelerating. Equation (3.65) is also applicable to situations where the boundary direction varies continuously, with changes of either sign, provided that R is constant; that is, the wave fronts do not converge to form a hydraulic jump. Given this, the final downstream flow depends on the nett change in angle θ via (3.65), and not on the form of this change. Meteorological examples of this phenomenon are described in §5.2.3.

3.5.5 Non-hydrostatic effects with 3D obstacles: the KP equation

One may derive two-dimensional versions of the equations of §3.4 that incorporate both non-linearity and dispersion. Most of the studies of this type have examined the flow around three-dimensional obstacles in channels, so that the obstacles are not isolated, and this brings out some interesting phenomena of practical relevance. The observations of Huang et al. (1982) showed that the non-linear upstream disturbances from a three-dimensional obstacle (a ship) in a channel become one-dimensional (i.e. no y-dependence), as they propagate upstream from the obstacle. This effect may be modelled theoretically by two-dimensional (x, y) versions of the above Boussinesq and Kortweg–de Vries (KdV) equations. In particular, the extension of the latter by Kadomtsev & Petviashvili (1970) to give the KP equation has facilitated studies of non-linear wave propagation with spreading in two dimensions. In the notation of equation (3.12), this equation has the form

$$\left[\eta_t + c_0\eta_x + \frac{3}{2}\frac{c_0}{d_0}\eta\eta_x + \frac{1}{6}c_0d_0^2\eta_{xxx}\right]_x + \frac{1}{2}c_0\eta_{yy} = 0. \tag{3.67}$$

Its limitations are essentially the same as for KdV, and it describes wave motion that is primarily propagating in one (x) direction, but can also spread laterally. In particular, it accommodates lateral wave spreading and periodic structure in the y-direction. This equation has been used to explore the properties of three-dimensional waves of finite height in shallow water, sometimes produced as wakes. For forcing by topography in a uniform stream, in the same form as (3.38), it becomes

$$\eta_{xt} + (U - c_0)\eta_{xx} - \frac{3}{4}\frac{c_0}{d_0}(\eta^2)_{xx} - \frac{1}{6}c_0d_0^2\eta_{xxxx} - \frac{1}{2}c_0\eta_{yy} = \frac{U}{2}h(x, y)_{xx}. \tag{3.68}$$

An example of the properties of such flows is presented in Figure 3.17, which shows a solution of (3.67). This shows the development of flow from an initial state in which one half $(0 < Y < 3)$ of a linear channel of uniform depth contains a solitary wave of uniform height, and no motion at all on the other half $(3 < Y < 6)$. The side boundaries are vertical walls, and Figure 3.17 shows the development of

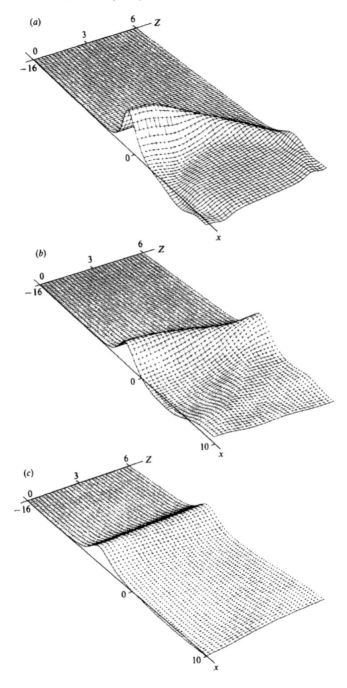

Figure 3.17 Evolution of non-linear free disturbance by the KP equation from an initial condition consisting of a soliton spanning only half of the channel (only half of the domain is displayed); (a) $T = 1$, (b) $T = 5$, (c) $T = 20$, where T is a measure of time. (From Katsis & Akylas, 1987, Figure 2, reproduced with permission.)

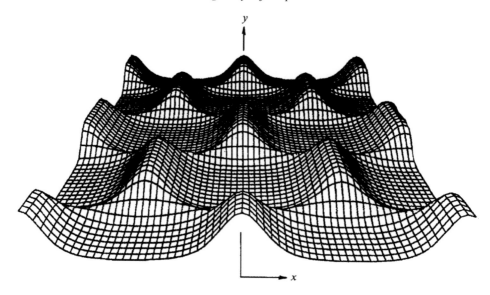

Figure 3.18 A solution of the KP equation that is periodic in two (x and y) dimensions. The wave field propagates in the x-direction and is periodic in both directions, but has sections in the y-direction where it is effectively one-directional, where the wave satisfies the one-directional KdV equation. (From Hammack et al., 1989, Figure 2, reproduced with permission.)

the flow at times $T = 1$, 5 and 20. The solitary wave is seen to propagate into the vacant part of the channel, and then progressively spread to a state that has uniform height in Y across the whole channel, of approximately half its initial value.

Another type of solution of the KP equation is represented by periodic patterns of two-dimensional cnoidal waves (Hammack et al., 1989), and an example is shown in Figure 3.18. Here the flow is doubly periodic and the waves are all moving in the x-direction, but the flow pattern appears to be one-dimensional in some sections. Similar flow patterns have sometimes been seen in waves approaching ocean beaches (Akylas, 1984).

In parallel with the forced KdV equation, the KP equation (in modified forms) continues to be used to explore non-linear wave motion in two horizontal directions that is dominated by a single vertical mode. Results compare well with those from more detailed models (Yuan et al., 2018).

4

Two-layer flow with jumps and topography

If they be two, they are two so
As stiff twin compasses are two,
Thy soul the fixed foot makes no show
To move, but doth, if the other do.
JOHN DONNE, *A Valediction: Forbidding Mourning.*

Here we discuss motion forced by obstacles in the flow of two contiguous homogeneous fluids of different densities that are aligned along one horizontal x-axis. We concentrate on those phenomena that are new compared with those of the single-layer system of Chapters 2 and 3, due to the differing densities and motion of the upper and lower layers. These phenomena involve the dynamics and mass conservation of the upper layer, and its interaction with the lower layer through the pressure at the interface between them. We also mostly restrict attention to flows that are aligned with one horizontal (x) axis, although multi-layer and continuously stratified flows through channels of non-uniform cross-section are included. The case of a pliant upper surface (see Chapter 1) is not addressed here, but see §11.3.4.

4.1 Basic equations

In this chapter we assume that the upper boundary is mostly rigid, so that there is only one interface. For some purposes the upper boundary is taken to be a free surface, but this is restricted to situations where it is dynamically passive, implying that fluid speeds are much less than the long free-surface gravity wave speed $\{\approx [g(d_1 + d_2)]^{1/2}\}$. We also assume that the interface is infinitely thin. The resulting equations still apply to situations where the interface has finite thickness (due, for example, to stratification and mixing), provided the motion has horizontal length-scales that are long compared with the thickness of the interface.

We take the layer depth, fluid velocity, pressure and density in the ith layer

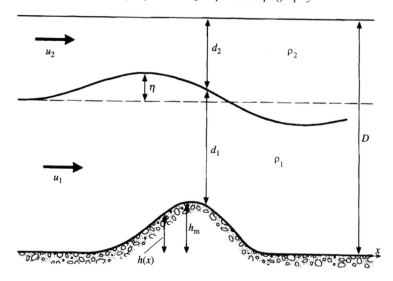

Figure 4.1 Definition sketch for the notation for two-layer flows.

to be $d_i, \mathbf{u}_i, p_i$ and ρ_i respectively, with $i = 1, 2$, and numbering upwards. If the undisturbed value of d_i far upstream is d_{i0}, then

$$d_{10} + d_{20} = D, \qquad d_1 = d_{10} + \eta(x,t) - h(x), \qquad (4.1)$$

where η is the displacement of the interface, D is the total depth far upstream, and h is the deviation of the lower boundary as before. A sketch showing notation is given in Figure 4.1. If we assume that the motion in each homogeneous layer is irrotational, then we may write

$$\mathbf{u}_i = \nabla \phi_i, \qquad \text{where} \quad \nabla^2 \phi_i = 0, \quad i = 1, 2. \qquad (4.2)$$

The equations of motion in each layer may then be expressed in the form

$$\nabla \left(\frac{\partial \phi_i}{\partial t} + \frac{1}{2} \mathbf{u}_i^2 + gz + \frac{p_i}{\rho_i} \right) = 0, \qquad i = 1, 2, \qquad (4.3)$$

and if the motion far upstream or downstream is undisturbed, this may be integrated to give

$$p_i = p_0 - \rho_i \left(g(z - d_{10}) + \frac{\partial \phi_i}{\partial t} + \frac{1}{2} (\nabla \phi_i)^2 \right), i = 1, 2, \qquad (4.4)$$

where p_0 is the undisturbed pressure at the interface. The usual conditions of continuity of pressure and displacement at the interface give

$$p_1 = p_2, \qquad (4.5)$$

$$w_1 = \phi_{1z} = \eta_t + \phi_{1x}\eta_x + \phi_{1z}\eta_z, \qquad w_2 = \phi_{2z} = \eta_t + \phi_{2x}\eta_x + \phi_{2z}\eta_z, \qquad (4.6)$$

at $z = d_{10} + \eta$, with the bottom and top boundary conditions for rigid boundaries

$$\nabla\phi_1.\nabla[z - h(x)] = 0, \quad \text{on } z = h(x),$$
$$\phi_{2z} = 0, \quad \text{on } z = D; \tag{4.7}$$

here, the x, z and t suffixes denote derivatives. If one compares the governing equations for the lower layer alone with those for a single layer (2.1)–(2.4), one sees that the only difference is the presence of the upper-layer pressure p_2 at the interface. Equations (4.4), (4.5) then give

$$\left(1 - \frac{\rho_2}{\rho_1}\right)g\eta + \frac{\partial\phi_1}{\partial t} + \frac{1}{2}(\nabla\phi_1)^2 = \frac{\rho_2}{\rho_1}\left[\frac{\partial\phi_2}{\partial t} + \frac{1}{2}(\nabla\phi_2)^2\right], \quad \text{on } z = d_{10} + \eta. \tag{4.8}$$

If the upper layer is deep so that $d_1/D \ll 1$, and the motion is hydrostatic (i.e. with horizontal scale L, where $d_1/L \ll 1$), mass conservation shows that the variations in $|\mathbf{u}_2|, \phi_2$ are $O(d_1/D)$ smaller than those of $|\mathbf{u}_1|, \phi_1$. Hence the terms on the right-hand side of (4.8) may be neglected, and the equations for the motion of the lower layer are the same as those for the single layer of Chapter 2 but with g replaced by $g' = (1 - \rho_2/\rho_1)g$. This is termed the "$1\frac{1}{2}$-layer model" approximation.

4.2 Linear waves

We consider small-amplitude perturbations to a flow where each layer has a uniform velocity U_i, with rigid horizontal boundaries above and below. Writing

$$\phi_i = U_i x - \frac{1}{2}U_i^2 t + \phi_i'(x, z, t), \quad i = 1, 2, \tag{4.9}$$

and substituting in (4.5) and retaining only the first-order terms in ϕ_1', ϕ_2' and η gives

$$\phi_{iz}' = \eta_t + U_i\eta_x, \quad i = 1, 2,$$
$$\rho_1(g\eta + \phi_{1t}' + U_1\phi_{1x}') = \rho_2(g\eta + \phi_{2t}' + U_2\phi_{2x}'). \tag{4.10}$$

We look for solutions satisfying the linearised form of (4.7) of the form

$$\phi_1 = A_1 \cosh kz e^{i(kx-\omega t)}, \quad \phi_2 = A_2 \cosh k(D-z)e^{i(kx-\omega t)}, \quad \eta = A_3 e^{i(kx-\omega t)}, \tag{4.11}$$

where A_1, A_2 and A_3 are constants. Equation (4.10) then gives the dispersion relation for waves with a rigid upper boundary

$$c = \frac{\omega}{k} = \frac{U_1 T_1 + U_2 T_2 \pm ((T_1 + T_2)\Delta\rho g - T_1 T_2(U_1 - U_2)^2)^{1/2}}{T_1 + T_2}, \tag{4.12}$$

where c is the phase speed of the waves, $\Delta\rho = \rho_1 - \rho_2$, and

$$T_i = \frac{\rho_i k}{\tanh kd_i}. \tag{4.13}$$

For any given k, (4.12) gives complex wave speeds implying instability if $(U_1 - U_2)^2$ is sufficiently large. This two-layer instability is quite real, but in many cases, in practice, it is inhibited by such factors as the finite thickness of the interface (in density, velocity or both) and surface tension (for immiscible fluids). Instability in stratified shear flows is discussed in more general terms in Chapter 9. For smaller values of shear and of k, (4.12) gives real (and realistic) expressions for wave speeds. In the hydraulic limit where $kd_i \to 0$, we have $T_i \to \rho_i/d_i$, and if $U_1 = U_2 = U$, then

$$c = U \pm \left(\frac{\Delta \rho g}{\frac{\rho_1}{d_1} + \frac{\rho_2}{d_2}} \right)^{1/2}. \tag{4.14}$$

Another special case that will be useful later is $U_1 = U_2 = U$ with $kd_2 \to \infty$, which gives

$$c = U \pm \left[\frac{\frac{\Delta\rho}{\rho_1} g \tanh k d_1}{k \left(1 + \frac{\rho_2}{\rho_1} \tanh k d_1 \right)} \right]^{1/2}. \tag{4.15}$$

This reduces to the single-layer expression (2.26) with $g\Delta\rho/\rho_1$ replacing g, if $\rho_2 = 0$, or $\tanh k d_1 \to 0$. The latter limit implies that $d_1/l \to 0$, where l is a wavelength, so that the flow is hydrostatic.

The generation of waves in two-layer systems by flow over obstacles with small height is described in §10.5.1, specifically equations (10.77)–(10.83). The remainder of this chapter is concerned with non-linear phenomena.

4.3 Equations for one-dimensional non-linear hydrostatic flow

We now assume that the horizontal length-scale L and time-scale of the motion are sufficiently large for the equation for vertical motion to be hydrostatic. A sufficient condition for this is $(D/L)^2 \ll 1$. The equations corresponding to (2.6)–(2.8) for a two-layer system are then

$$u_{it} + u_i u_{ix} = -\bar{p}_{ix}/\rho_i, \quad i = 1, 2$$
$$d_{it} + (d_i u_i)_x = 0, \tag{4.16}$$

where $\bar{p}_i$ denotes a vertical average of the pressure in the ith layer, and p_i is given by

$$p_1(x, z, t) = p_s + g\rho_1(h + d_1 - z) + g\rho_2 d_2,$$
$$p_2(x, z, t) = p_s + g\rho_2(h + d_1 + d_2 - z), \tag{4.17}$$

where p_s is the pressure at the top of the second layer. The equations of motion therefore become

$$\frac{\partial u_1}{\partial t} + \frac{\partial}{\partial x}\left[\frac{1}{2}u_1^2 + g(h + d_1 + \frac{p_2}{\rho_1}d_2) + \frac{p_s}{\rho_1}\right] = 0, \tag{4.18}$$

$$\frac{\partial u_2}{\partial t} + \frac{\partial}{\partial x}\left[\frac{1}{2}u_2^2 + g(h + d_1 + d_2) + \frac{p_s}{\rho_2}\right] = 0. \tag{4.19}$$

If the uppermost layer has a rigid upper surface, we have

$$h(x) + d_1 + d_2 = D, \tag{4.20}$$

and if it instead has a free upper surface, we have

$$p_s = \text{constant.} \tag{4.21}$$

The bracketed terms in (4.18), (4.19) constitute the Bernoulli functions (R_1, R_2) for each layer, and are conserved in steady non-dissipative flow. Taking the difference $\rho_1 R_1 - \rho_2 R_2$ removes p_s, giving a conserved function of u_i, d_i that may be used for analysis or as a diagnostic for the interfacial motion (e.g. Lawrence, 1993).

We now restrict ourselves to the case of a rigid upper surface, to focus attention entirely on the interfacial mode. Adding the two mass conservation equations in (4.16) then gives

$$d_1 u_1 + d_2 u_2 = Q, \tag{4.22}$$

where Q is the total volume flux. Q is taken to be independent of time and horizontal position, and may be taken to represent a steady form of a long wave on a free upper surface, termed an external or barotropic mode. In order to obtain two equations for two unknowns, we must eliminate p_s from (4.18), (4.19) by subtraction, yielding

$$\frac{\partial}{\partial t}\left(u_1 - \frac{p_2}{\rho_1}u_2\right) + \frac{\partial}{\partial x}\left[\frac{1}{2}\left(u_1^2 - \frac{p_2}{\rho_1}u_2^2\right) - \frac{g\Delta\rho}{\rho_1}d_2\right] = 0, \tag{4.23}$$

where $\Delta\rho = \rho_1 - \rho_2$, and similarly subtracting the conservation equations gives

$$\frac{\partial}{\partial t}(d_1 - d_2) + \frac{\partial}{\partial x}(d_1 u_1 - d_2 u_2) = 0. \tag{4.24}$$

We now define the new variables η and v, the natural or canonical variables for describing the internal or baroclinic mode, by

$$v = u_1 - \frac{p_2}{\rho_1}u_2, \qquad d_1 + h = d_{10} + \eta, \qquad d_2 = d_{20} - \eta, \tag{4.25}$$

where d_{10}, d_{20} represent values of d_1, d_2 respectively, far upstream, as in (4.1). Expressing u_1 and u_2 in terms of v and η from (4.22), (4.25) gives

$$u_1 = \frac{[\rho_2 Q/\rho_1 + (d_{20} - \eta)v]}{D - h - \delta}, \qquad u_2 = \frac{[Q - (d_{10} - h + \eta)v]}{D - h - \delta}, \tag{4.26}$$

where $\delta = (d_{10} - h + \eta)\Delta\rho/\rho_1$, and substituting these into (4.23), (4.24) gives the desired equations for η and v.

Writing $g' = g\Delta\rho/\rho_1$, we obtain

$$\frac{\partial\eta}{\partial t} + \frac{\partial}{\partial x}\left[\frac{(d_{20} - \eta)(d_{10} - h + \eta)v + (\eta - d_{20} + \frac{\rho_2}{\rho_1}(d_{10} - h + \eta))Q/2}{D - h - \delta}\right] = 0,$$

$$\frac{\partial v}{\partial t} + \frac{\partial}{\partial x}\bigg[g'\eta+$$
$$\frac{[(d_{20} - \eta)^2 - \frac{\rho_2}{\rho_1}(d_{10} - h - \eta)^2]v^2 + \frac{\rho_2}{\rho_1}[2Qv(D - h) - (1 - \frac{\rho_2}{\rho_1})Q^2]}{2(D - h - \delta)^2}\bigg] = 0.$$

$$(4.27)$$

These may be rewritten with a topographic forcing term containing $\partial h/\partial x$ on the right-hand side. Equations (4.27) are hyperbolic and may be expressed in characteristic form (see for example Whitham, 1974, p. 116).

At this point, in order to avoid undue complexity we make the Boussinesq approximation (page 7), which implies that in these equations we have

$$\delta = 0, \qquad v = u_1 - u_2. \qquad (4.28)$$

The following procedure with the full equations (4.27) and results can be found in Baines & Johnson (2016). The flow properties of the non-Boussinesq flows are generally similar to the Boussinesq case, though there are some differences in detail. Away from topography, where $h = 0$, these characteristic equations may be integrated exactly to give the "Riemann invariants", for which the Boussinesq form is

$$R_\pm = \arcsin\left(\frac{2\eta + d_{10} - d_{20}}{D}\right) \pm \arcsin\left(\frac{v}{(g'D)^{1/2}}\right) = \text{constant}, \qquad (4.29)$$

on the respective characteristics

$$\frac{dx}{dt} = c_\pm = \frac{Q}{D} + \frac{v(d_{20} - d_{10} - 2\eta)}{D} \pm \left[\left(g' - \frac{v^2}{D}\right)\frac{(d_{20} - \eta)(d_{10} + \eta)}{D}\right]^{1/2}. \qquad (4.30)$$

Here c_+ and c_- are the interfacial long-wave speeds, and are equal to those given by (4.12) with the Boussinesq approximation in the limit $k \to 0$. This system is similar to (2.45), (2.46) for a single layer, but more complex. We may express (4.30) in the dimensionless form

$$C_\pm = \frac{c_\pm - \bar{U}}{\sqrt{g'D}} = V(1 - 2r) \pm ((1 - V^2)r(1 - r))^{1/2}, \qquad (4.31)$$

where

$$\bar{U} = Q/D, \quad r = d_1/D, \quad V = v/\sqrt{g'D}. \qquad (4.32)$$

These wave speeds relative to the mean fluid velocity $\bar{U}$ are shown as contours of

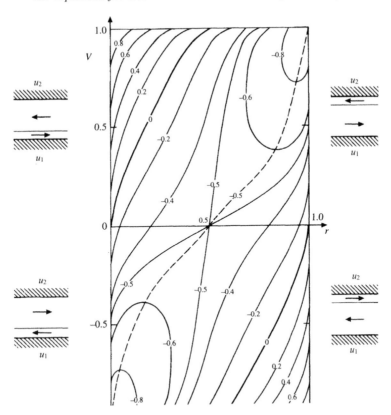

Figure 4.2 Long-wave speeds $C_\pm$ in a two-layer Boussinesq shear flow with a rigid lid as functions of r and V, relative to the mean motion, as given by (4.31). The diagram shows values of C_- for leftward-propagating disturbances. V denotes the velocity difference $(u_1 - u_2)/\sqrt{g'D}$, and r the fractional depth of the lower layer. In each case, u_1 is the velocity in the lower level, and u_2 that in the upper. The flow diagrams at the sides show the flow character in the neighbouring region of the (r, V) diagram. The dashed lines denote maxima in the wave speed for given V. Corresponding values for $C_+(r, V) = -C_-(r, -V)$.

constant C on the (r, V)-plane in Figure 4.2. Note that V is restricted to the range $|V| \ll 1$ for stability (i.e. real wave speeds), in this long-wave model.

Figure 4.2 displays a number of properties of wave propagation in sheared two-layer flows. These wave speeds are governed by two competing effects: wave propagation and advection. For $|V| \ll 1$, wave propagation dominates; the wave speeds in each direction are approximately equal, and greater than the fluid velocities of the layers (relative to the mean). For $|V|$ near unity, on the other hand, advection dominates; if r is also small (thin lower layer) or near unity (thin upper layer), the wave speed is close to the speed of the thin layer. For intermediate values of $|V|$, both advection and propagation have comparable effects on the wave speeds.

In short, wave speeds in the centre of the diagram are dominated by propagation, and those around its edges by advection. The quantities C_+ and C_- must lie in the range $-1 \leq C_+, C_- \leq 1$, but otherwise the wave speeds may take values which are outside, between, or even equal to the fluid speeds of the two layers. *Critical layer* phenomena as found in continuously stratified fluids (see Chapter 8) do not occur.

These equations may be applied to non-linear wave motion moving (leftwards, say) into undisturbed fluid, where $\eta = 0, v = v_0$. With the Riemann invariants given by (4.29), we have

$$R_+ = \arcsin(2r - 1) + \arcsin(v_0/\sqrt{g'D}), \qquad (4.33)$$

and R_- is constant on each characteristic $dx/dt = c_-$. These equations imply that η and v are constants on each characteristic $dx/dt = c_-$, in a fashion that is analogous to the equations for single-layer flows. From (4.30) and Figure 4.2 we may see that, under suitable circumstances, a monotonic change in η may steepen into a discontinuity or hydraulic jump, which must be accommodated into this framework as it was for single-layer flows. The character of these hydraulic jumps for two-layer flows requires special consideration, which we address in §§4.4 and 4.5.

4.4 Two-layer hydraulic jumps

Since the non-linear hydraulic model of §4.3 leads to the formation of discontinuities that take the form of hydraulic jumps, we need to be able to incorporate them into the model in order to complete the description of the dynamics. To do this, we must establish relationships between the flow conditions on the upstream and downstream sides of the jump. In general, two-layer hydraulic jumps are localised regions where the flow changes from one relatively uniform two-layer stream to another; energy dissipation, mixing and local wave phenomena may be involved. In many situations of interest here, mixing processes are much less severe than those in a gravity current, so that flow downstream of the jump may still be modelled by two layers. Figure 4.3 shows two examples of hydraulic jumps in miscible fluids where dissipation and mixing are evident. Figure 4.3a shows a strong visual similarity to a gravity current, except that the nose and associated ingesting of fluid into the head from below are absent because of the presence of the thin lower layer upstream.

Jumps occur in two common situations: in the first, a jump moves or propagates into an undisturbed region (as in Figure 4.3a), and in the second the jump is stationary, in the lee of an obstacle, for example (as in Figure 4.3b). For the propagating jump, the desired flow conditions are equivalent to a relation between the jump speed and its amplitude. These may be obtained if certain conditions may be assumed, and we first consider the simplest formulation of two-layer jumps, for

(a)

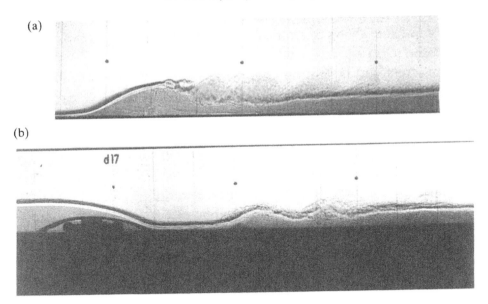

(b)

d 17

Figure 4.3 Examples of two-layer hydraulic jumps in miscible fluids. (a) A jump moving into fluid at rest with a shallow lower layer (depth initially 0.3 cm). The flow resembles a gravity current. (b) A downstream jump stationary relative to a moving obstacle. Mixing is evident in both layers. (Photographs courtesy of J. Simpson.)

which the following assumptions apply. The first three assumptions do not involve the detailed dynamics of the jump, and are as follows:

(1) that the flow is steady in a reference frame moving with the jump;
(2) that the top and bottom surfaces are horizontal through the jump, and frictional stresses there are negligible;
(3) that the layers maintain their identity through the jump, with negligible exchange of fluid between the layers.

This third assumption is not always satisfied in real jumps, and the effects of interfacial mixing are discussed below. However, with these three assumptions we may represent the hydraulic jump as a localised region separating two uniform but different two-layer flows, as depicted in Figure 4.4, which also serves to define the notation. We regard the variables on the left-hand (upstream) side with suffix "u" as known, and require expressions for the downstream variables (suffix "d") in terms of them. Note that this contains information about the jump speed, since we have taken axes relative to the jump. From continuity we have

$$d_{1d} + d_{2d} = D = d_{1u} + d_{2u},$$
$$d_{id}u_{id} = q_i = d_{iu}u_{iu}, \quad i = 1, 2,$$

(4.34)

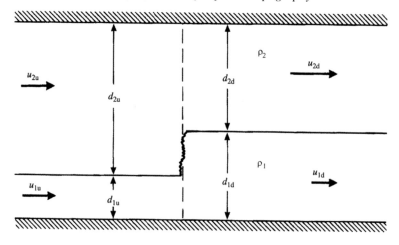

Figure 4.4 Notation sketch for a stationary two-layer hydraulic jump.

where q_i denotes the volume flux in the ith layer. From assumption (2) and the integral relations of §1.3 we also have that the momentum flux must be uniform, and (1.38) gives

$$S = \int_0^D p + \rho u^2 dz = \text{constant},$$ (4.35)

which implies

$$S = p_{su}D + \frac{1}{2}\Delta\rho g d_{1u}^2 + \rho_1 u_{1u}^2 d_{1u} + \rho_2 u_{2u}^2 d_{2u} + \frac{1}{2}\rho_2 g D^2$$

$$= p_{sd}D + \frac{1}{2}\Delta\rho g d_{1d}^2 + \rho_1 u_{1d}^2 d_{1d} + \rho_2 u_{2d}^2 d_{2d} + \frac{1}{2}\rho_2 g D^2.$$ (4.36)

This introduces an extra variable, $p_{su} - p_{sd}$, the difference between upstream and downstream pressures at the upper boundary, and we need another equation (in addition to the above four) in order to be able to solve for the downstream variables.

At this point, a number of different assumptions about internal dynamics of two-layer jumps have been proposed to determine this variable. The first suggestion (flow in the jump is hydrostatic) was by Yih & Guha (1955), followed by (no energy dissipation in the contracting layer) Chu & Baddour (1977) and Wood & Simpson (1984), and a third proposal (no energy dissipation in the expanding layer) by Klemp et al. (1997). These three dynamical assumptions gave mixed results with restricted applicability (though the KRS model applies well in the Boussinesq limit).

A more recent approach initiated by Borden & Meiburg (2013a) has resolved the issue for Boussinesq fluids by considering the vorticity budget across the jump. This procedure has been extended to fluids of arbitrary densities by Baines (2016), which provides a brief history of the preceding models. It follows that the nett production of vorticity within the jump is directly related to the differences in the pressure across the jump. The steady-state form of the vorticity equation in the frame of the jump is

$$\boldsymbol{u} \cdot \nabla \omega = \frac{1}{\rho^2} \nabla \rho \times \nabla p = -\nabla \times \left(\frac{1}{\rho} \nabla p \right), \tag{4.37}$$

where we assume that the flow is on a sufficiently large scale so that viscous effects may be neglected. Since the fluids are incompressible we have $\nabla \cdot \boldsymbol{u} = 0$, and if equation (4.37) is integrated over the region of the jump (area A) we have, for the left-hand side,

$$\int_A \boldsymbol{u} \cdot \nabla \omega \, dA = \int_A \nabla \cdot (\omega \boldsymbol{u}) \, dx dz = \int \omega \boldsymbol{u} \cdot d\boldsymbol{S}, \tag{4.38}$$

and S denotes the outward normal on the boundary surface of A upstream and downstream. The last term of (4.38) denotes the flux of vorticity across surface S. For jumps moving into fluid at rest, or fluid in uniform motion, there is no upstream vorticity.

On the downstream side the vorticity is concentrated in the vortex sheet between the two fluids, and the flux of it is given by the vortex sheet strength times the mean velocity of the sheet, which is (Borden & Meiburg, 2013a)

$$\int \omega \boldsymbol{u} \cdot d\boldsymbol{S} = -(u_{2d} - u_{1d}) \cdot (u_{2d} + u_{1d})/2 = -\frac{1}{2}(u_{2d}^2 - u_{1d}^2). \tag{4.39}$$

The area integral of the right-hand side of (4.37) may be reduced by Stokes's theorem to a line integral around the boundary of area A enclosing the jump

$$-\int_A \nabla \times \left(\frac{1}{\rho} \nabla p \right) dA = -\int \left(\frac{1}{\rho} \nabla p \right) \cdot d\boldsymbol{l}, \tag{4.40}$$

in the anti-clockwise sense. Since the density is uniform in each layer, the integral reduces to

$$-\int \left(\frac{1}{\rho} \nabla p \right) \cdot d\boldsymbol{l} = \left[\frac{1}{\rho_2} - \frac{1}{\rho_1} \right] [p_{iu} - p_{id}], \tag{4.41}$$

where p_{iu} and p_{id} denote the upstream and downstream pressures at the interface respectively (see Figure 4.4).

The upstream pressure at the interface is equal to $p_{iu} = p_{su} + g\rho_2(D - d_{1u})$, and the corresponding downstream pressure is $p_{id} = p_{sd} + g\rho_2(D - d_{1d})$. Putting these terms into equation (4.41) and equating it with (4.39) gives

$$\frac{1}{2}(u_{2d}^2 - u_{1d}^2) = \left[\frac{1}{\rho_2} - \frac{1}{\rho_1} \right] [p_{su} - p_{sd} + \rho_2 g(d_{1d} - d_{1u})]$$

$$= \frac{\Delta \rho}{\rho_1} \cdot \frac{p_{su} - p_{sd}}{\rho_2} + g'(d_{1d} - d_{1u}), \tag{4.42}$$

where $g' = g\frac{\Delta\rho}{\rho_1}$, and $\Delta\rho = \rho_1 - \rho_2$. These give

$$p_{su} - p_{sd} = \frac{1}{2}\frac{\rho_1\rho_2}{\Delta\rho}(u_{2d}^2 - u_{1d}^2) - g\rho_2(d_{1d} - d_{1u}).\qquad(4.43)$$

Eliminating $p_{su} - p_{sd}$ from equations (4.36) and (4.43), we obtain an expression for jump speed c_J in terms of the jump amplitude d_{1d}. It is convenient to express the jump speed in terms of the linear long-wave speed c_0 (given by (4.14)), so that

$$\frac{c_J^2}{c_0^2} =$$

$$\frac{r_d^2(1-r_d)^2\left[1 - r_u(1 - \frac{\rho_2}{\rho_1})\right]\left[1 + \left(\frac{\rho_1}{\rho_2} - 1\right)\frac{r_d+r_u}{2}\right]}{r_u(1-r_u)\left[\frac{r_d+r_u}{2} - r_u r_d + \left(\frac{\rho_1}{\rho_2} - 1\right)r_u r_d(1-r_d)^2 - \left(1 - \frac{\rho_2}{\rho_1}\right)r_d^2(1-r_u)(1-r_d)\right]},$$

$$\qquad(4.44)$$

where

$$c_0^2 \equiv \frac{\Delta\rho g}{\frac{\rho_1}{d_{1u}} + \frac{\rho_2}{d_{2u}}}, \qquad r_u = d_{1u}/D, \qquad r_d = d_{1d}/D.\qquad(4.45)$$

Following Borden & Meiburg, this model may be termed the "full vortex sheet (FVS)" model. In the Boussinesq limit in which $\rho_2 \to \rho_1$ it reduces to their equation (2.9)

$$c_J^2 = \frac{2r_d^2(r_d - 1)^2}{r_u(r_u + r_d - 2r_d r_u)}.\qquad(4.46)$$

It is also consistent with the following limits:

$$\text{if } r_d \to r_u : \quad c_J^2 \to c_0^2 \equiv \frac{\Delta\rho g}{\frac{\rho_1}{d_{1u}} + \frac{\rho_2}{d_{2u}}};\qquad(4.47)$$

$$\text{if } \rho_2 \to 0 : \quad c_J^2 \to \frac{gd_{1d}}{2}\left(1 + \frac{d_{1d}}{d_{1u}}\right),\qquad(4.48)$$

which is Rayleigh's relation for single-layer jumps (§2.3.1), and

$$\text{if } D \to \infty : \quad c_J^2 \to \frac{2g'd_{1d}^2}{d_{1u} + d_{1d}}.\qquad(4.49)$$

Upward jumps are possible for r_u values from zero up to slightly greater than 0.5 for ρ_2/ρ_1 near unity, and this upper limit increases as this ratio decreases. Figure 4.5 shows representative jump speeds (scaled with c_0) for $r_u = 0.035, 0.2$, for a range of density ratios from 0.1 to 0.99, the latter being very close to the Boussinesq limit. All of these curves have the same form – the jump speeds increase with increasing amplitude until a maximum is reached; for larger amplitudes (shown

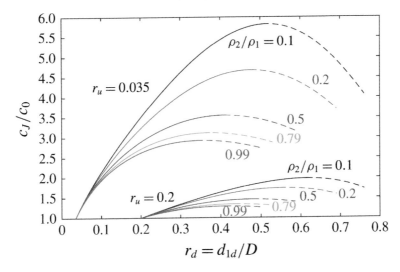

Figure 4.5 Hydraulic jump speeds c_J scaled with c_0 for the full range to the point of zero energy loss, for $r_u = d_{1u}/D = 0.035$ and 0.2. Amplitudes greater than those for maximum speed are shown dashed. (From Baines, 2016, reproduced with permission.)

dashed in Figure 4.5) the jump speed decreases, up to a termination point of maximum amplitude. Coresponding examples of hydraulic drops (where the upstream upper layer is thinner than the lower layer) are shown in Figure 4.6. A comparison between theoretical and observational jump speeds in immiscible fluids is shown in Figure 4.7.

Two other important properties of jumps are their energy dissipation and stability. The energy flux in a layered flow is given by $E_{fi} = \rho_i Q_i R_i$ for the ith layer (here $i = 1, 2$), where $Q_i = u_i d_i$ and R_i is the Bernoulli constant (see §1.3.1). The energy loss within the jump (i.e. the energy dissipation within it) is then given by E_{fi}upstream $- E_{fi}$downstream; this is generally larger in the contracting (upper) layer than in the expanding (lower) layer, unless the density ratio (ρ_2/ρ_1) is small. Plots of this energy loss with jump amplitude show that, for each layer, the energy dissipation increases with jump amplitude up to the point of maximum jump speed. For larger jump amplitudes (in the "dashed curves" region of Figure 4.5) it decreases, reaching zero at the termination point of the dashed curve. This termination point therefore represents a "conjugate state" that (in the frame of the jump) has the same energy flux in each layer as the original upstream flow (it has been proposed that combinations of these "dissipationless jumps" may constitute "flat solitary waves"; Lamb & Wan, 1998).

However, the hydraulic jumps in the dashed region of Figure 4.5 are believed to be of lesser significance than those with smaller amplitude (the solid curves), for

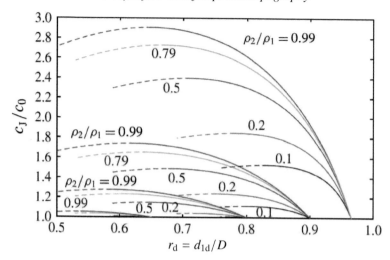

Figure 4.6 Hydraulic jump (hydraulic drop) speeds c_J scaled with c_0 as in Figure 4.5 for $r_u = 0.65, 0.8, 0.9$ and 0.965. Amplitudes greater than those for maximum speed are shown dashed. (From Baines, 2016, reproduced with permission.)

two reasons. The first is that such jumps are potentially unstable – a perturbation that increases (decreases) their amplitude will cause them to travel more slowly (faster), and in consequence the jump will tend to spread out and disperse, but led by a smaller jump of maximum speed. Second, it may be shown that for jumps with amplitudes greater than that of maximum speed, the flow on the downstream side is supercritical, relative to the jump (Baines, 2016). This means that downstream disturbances cannot propagate up to the jump and increase or affect it, so that such jumps cannot be built or maintained from downstream. Accordingly, only jumps from the solid (rising) parts of the curves of Figure 4.5 will be invoked in applications in following sections.

There is also the question of stability of the interface downstream of (and therefore also within) the jump. The flow is unstable to long waves if $v^2 > g'D$, where the characteristics (4.30) are complex. This condition is not reached for any of the curves shown in Figure 4.5, but it can occur for smaller r_u. However, in real situations, instability may occur for smaller shears and is governed by a local Richardson number criterion (see Chapter 9). This will cause increased (in addition to viscous) interfacial stresses, and in the case of miscible fluids, mixing. The latter will thicken the interface and may result in the effective transfer of fluid from one layer to another, or the creation of an intermediate layer.

It should be noted that, at the cost of additional complexity, this model of hydraulic jumps has been extended to two-layered flows in which the vorticity is spread over a

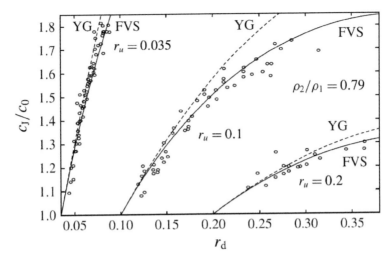

Figure 4.7 FVS model jump speeds for $r_u = d_{1u}/D = 0.035, 0.1$ and 0.2 compared with experimental observations from Baines (1984) made with water and kerosene ($\rho_2/\rho_1 = 0 : 79$). The results obtained with the Yih–Guha (YG) model are also shown, and the FVS model is clearly a better fit. (From Baines, 2016, reproduced with permission.)

finite range of heights, by Ogden & Helfrich (2016) and independently by Ungarish & Hogg (2018), which the latter term the "vortex wake" model. Application to other flows such as gravity currents is also possible. Another property of hydraulic jumps not addressed here is the nature and quantity of entrainment and mixing that may occur. An approach to this problem with hydraulic jumps in miscible fluids with one or two interfaces has been described by Milewski & Tabak (2015).

4.5 Hydrostatic flow over topography

We next examine the nature of two-layer flow over a two-dimensional obstacle in a fluid of total depth D where the flow develops from a known initial state, and the relative velocity $\overline{U} = Q/D$ of the mean flow is held constant with time. We concentrate on the case where the initial velocity of each layer is the same, so that $U = \overline{U}$. This would apply to situations where an obstacle is towed through fluid which is initially at rest, or to fluid which is set into relative motion by an externally imposed pressure gradient, or again, to the flow after an obstacle has been suddenly introduced into a moving stream. The equations are again restricted to the case of the Boussinesq approximation, but some results for non-Boussinesq flows are presented and the equations are given in Baines & Johnson (2016). As for the single-layer system, the initial conditions for each of these cases are slightly different. The system is presumed to be governed by the equations of §4.3, and at time $t = 0$ we take the layers to have the velocities and thicknesses given by

$$u_1 = u_2 = U, \quad d_1 = d_{10} + h, \quad d_2 = d_{20} - h, \quad \text{at } t = 0, \tag{4.50}$$

which correspond to the third case, the sudden introduction of the obstacle into a moving stream. The flow parameters may then be described by the dimensionless numbers

$$r = \frac{d_{10}}{D}, \quad H_m = \frac{h_m}{d_{10}}, \quad F_0^2 = \frac{U^2}{r(1-r)g'D}, \tag{4.51}$$

where h_m is the maximum value of h and F_0 is the Froude number of the undisturbed flow, defined as the flow speed divided by the linear wave speed, as before. For the most part we will describe the flow properties in terms of (F_0, H_m)-diagrams, for various values of r. The corresponding system with a free upper surface may be treated in a similar manner, but the details are more complex (Houghton & Isaacson, 1970).

If H_m is small, the commencement of motion (4.50) results in two transient disturbances, each having the form of the topography. One of these linear waves propagates against the stream and the other one with it, leaving a locally steady solution over the obstacle, in a manner similar to (2.13) for single-layer flows. We first look at these steady solutions where the upstream flow is undisturbed. Equations (4.27) in steady form may be manipulated to show that where $dh/dx = 0$, we must have either (with $v = u_1 - u_2$)

$$\frac{d}{dx}(\eta, v) = 0, \quad \text{or} \quad c_- = 0, \tag{4.52}$$

which is the "hydraulic alternative" for this two-layer system. As for single-layer flows, if $c_- < 0$, so that linear waves may propagate against the mean flow, we say that the flow is *subcritical*. If both $c_-, c_+ > 0$, the flow is *supercritical*. Provided $F_0 \neq 1$, if H_m is sufficiently small the first of these alternatives applies, and the flow is described by the integrals of (4.27):

$$\frac{(d_{20} - \eta)(d_{10} - h + \eta)v - ((d_{20} - d_{10} + h)/2 - \eta)Q}{D - h} = -\frac{Q(d_{20} - d_{10})}{2D},$$

$$\frac{v^2(d_{20} - d_{10} + h - 2\eta) + 2Qv}{2(D - h)} + g'\eta = 0, \tag{4.53}$$

where the constants on the right-hand side are determined by the upstream conditions. These give η and v as functions of h, and correspond to (2.56), (2.57) for a single layer. The solutions are everywhere subcritical for $F_0 < 1$, and supercritical if $F_0 > 1$.

The range of validity of these solutions is limited by the second possibility of

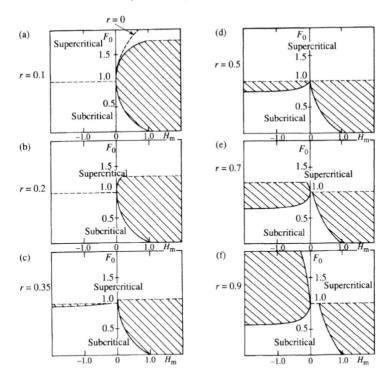

Figure 4.8 Boundary curves for wholly subcritical (when $F_0 < 1$) and supercritical ($F_0 > 1$) flow, for various r values, for uniform two-layer flow with a rigid lid, with the Boussinesq approximation. Curves where the flow is critical at the obstacle crest or trough are shown solid; dashed curves denote other boundaries of these flow regions. Regions where the upstream flow state is altered by the topography are shown shaded. In (a), the boundary curve for supercritical flow for an infinitely deep upper layer ($r = 0$) is shown dashed.

(4.52), critical flow at the obstacle crest, which gives the maximum obstacle height for which these solutions are possible, for given r and F_0. This may be obtained by differentiating (4.53) with respect to x and taking $dh/dx = 0$ at $h = h_m$, to obtain

$$r(1 - r)F_0^2 \left[\frac{r^2}{r_c^3} + \frac{(1-r)^2}{(1 - r_c - rH_m)^3} \right] = 1,$$

$$\frac{r(1 - r)F_0^2}{2} \left[\frac{r^2}{r_c^2} - \frac{(1-r)^2}{(1 - r_c - rH_m)^2} \right] + r_c + r(H_m - 1) = 0,$$

(4.54)

with $r_c = d_{1c}/D$, where d_{1c} is the lower layer depth at the obstacle crest. Eliminating r_c gives the boundary curves in (F_0, H_m)-space for the sub- and supercritical solutions, which are shown in Figure 4.8 for various values of r. Solutions in the non-shaded regions are given by (4.53); in the shaded regions the introduction of the obstacle alters the upstream flow conditions.

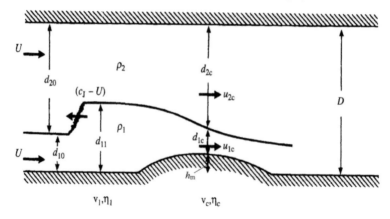

Figure 4.9 Definition sketch for two-layer flow over an obstacle, with an upstream jump.

Figures 4.8a–c show the bounding curves for three cases for $r < 0.5$. For $H_m > 0$, these are similar to that for the single layer ($r = 0$), except that the flow is supercritical for all H_m if F_0 is large enough, and this lower bound for supercritical flow approaches unity as $r \to 0.5$. For $H_m < 0$, we have two different solutions of (4.53) for $F_0 < 1$ and $F_0 > 1$, as for a single layer, with either solution possible if $F_0 = 1$. But if H_m is sufficiently large and negative, the sub- and supercritical regions are separated by a shaded region, visible in Figures 4.8c–f, which extends in to $H_m = 0$ as $r \to 0.5$. For $r = 0.5$ (Figure 4.8d), the behaviour for $H_m < 0$ is similar to that for $H_m > 0$; the subcritical solution applies below the curve, and the supercritical solution for $F_0 > 1$. For $r > 0.5$ (Figures 4.8e-f), the diagram for $H_m < 0$ is similar to that for $H_m > 0$ for $r < 0.5$ but in reversed form; for $H_m > 0$, the curves marking critical flow are confined to the range $F_0 < 1$, and do not approach the point $(H_m, F_0) = (0, 1)$.

In order to describe the flow to the right of the critical flow curves of Figure 4.8 (and we restrict consideration to $H_m > 0$), we must examine its temporal development more closely. If $d_1/D \ll 1$, the equations for the lower layer are approximately the same as those for a single layer (see §4.3). Hence we may expect the flow to have the structure shown in Figure 4.9, with an upstream-propagating hydraulic jump whose amplitude is controlled by a critical condition at the crest of the obstacle. The flow represented in Figure 4.9 has five unknowns: c_1, the speed of the jump relative to the obstacle, v_1 and η_1, the values of v and η at section "1" immediately downstream of the jump, and v_c and η_c (or r_c), the values of v and η (or r_c) at the obstacle crest. Once these five variables have been found the others readily follow. They are determined by the five equations:

(i) $c_1 = c_J - U$, where c_J is the velocity of the jump in fluid at rest, given by
 (4.44) or (4.46), whichever is appropriate;

(ii) conservation of mass through the jump in either of the two layers, which gives

$$u_{11} = \frac{d_{10}}{d_{11}}(U + c_J) - c_J; \qquad (4.55)$$

(iii), (iv) the integrated steady-state form of (4.27), connecting η_1, v_1 and η_c, v_c; and

(v) the critical condition $c_- = 0$ at the obstacle crest, which is

$$c_- = 0 = \frac{Q}{D - H_m} + \frac{v_c(d_{20} - d_{10} + h_m - 2\eta_c)}{D - h_m}$$
$$- \left[\left(g' - \frac{v_c^2}{D - h_m} \right) \frac{(d_{20} - \eta_c)(d_{10} - h_m + \eta_c)}{D - h_m} \right]^{1/2}. \qquad (4.56)$$

Alternatively, one may work with the layer velocities and depths downstream of the jump and at the obstacle crest, giving nine unknown variables, and use the mass conservation equation and Bernoulli equations (from (4.18)) for each layer.

The applicability of the model of Figure 4.9 is limited because hydraulic jumps have a maximum amplitude in two-layer systems, as seen in §4.4. If the topographic forcing is large enough to cause upstream disturbances with amplitudes larger than this maximum, the model must be modified. We can see how to do this by examining the governing equations. The conditions immediately downstream of a jump of infinitesimal magnitude are $r = r_u$, $v = 0$. If the jump amplitude and speed are increased, the downstream value of r increases and of v decreases, as determined by (4.45), (4.55), tracing a path in Figure 4.2. The jump may be regarded as being created by the accumulation of successive small disturbances propagating from the obstacle, and caused by successive small increases in the obstacle height up to the required value of H_m. This process continues while the linear wave speeds c_- increase, and successive waves are able to catch up with the jump and increase its magnitude. Eventually this process ceases, as Figure 4.2 shows that for larger disturbances (i.e. larger r and more negative v), the wave speeds progressively *decrease* with increasing nett disturbance amplitude and are unable to reach the jump, which has attained its maximum amplitude. As increasing H_m causes yet larger-amplitude disturbances, these become spread out behind the jump, as the larger-amplitude parts travel more slowly than the smaller-amplitude parts.

The appropriate model for this situation then has the form shown in Figure 4.10. The upstream flow consists of a hydraulic jump of maximum amplitude for the given upstream flow, followed by a *time-dependent* disturbance termed a *rarefaction*. This term is borrowed from gas dynamics, and is intended to imply that the disturbance

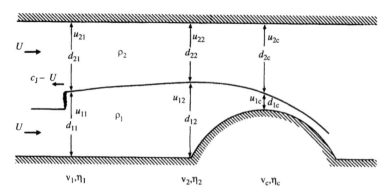

Figure 4.10 Definition sketch for two-layer flow over an obstacle, with an up-stream jump of maximum amplitude followed by a rarefaction. In comparison with Figure 4.9, note the increase in lower-layer depth upstream between sections "1" and "2".

is being rarefied, rather than the fluid, as it becomes progressively stretched out with time. If we regard the jump speed and the conditions downstream of it (as specified by v_1 and η_1) as being determined by the jump equations (4.44) and (4.55) for the maximum-amplitude jump, we have in effect four unknowns: v_2 and η_2 at section "2", immediately upstream of the obstacle, and v_c and η_c. At a sufficiently large time after the commencement of the motion, the flow over the obstacle may be presumed to be steady, so that the rarefaction only occurs *between* the jump and the obstacle. Then v_2, η_2 and v_c, η_c are related by the integrated steady-state form of (4.27). The other two equations required to determine these variables are given by, first, the fact that the Riemann invariant R_+ is constant through the rarefaction so that, from (4.29),

$$\arcsin\left(\frac{2\eta_1 + d_{10} - d_{20}}{D}\right) + \arcsin\left(\frac{v_1}{(g'D)^{1/2}}\right)$$
$$= \arcsin\left(\frac{2\eta_2 + d_{10} - d_{20}}{D}\right) + \arcsin\left(\frac{v_2}{(g'D)^{1/2}}\right), \qquad (4.57)$$

which links the conditions at section 1 with those of section 2 (cf. (2.47)). The second equation is the critical condition at the obstacle crest, (4.56).

The models represented by Figures 4.9 and 4.10 may be used to describe the flow properties as functions of F_0 and H_m. These are shown in Figures 4.11a, b and c for $r = r_u = 0.1, 0.35$ and 0.5, for $\rho_2/\rho_1 = 0.99$ – the near-Boussinesq case; and in Figure 4.12 for $r = 0.1$, $\rho_2/\rho_1 = 0.1$. In Figures 4.11a and 4.12 the curve BAE denotes the boundary of subcritical flow ($F_0 < 1$) and supercritical flow ($F_0 > 1$) as in Figure 4.8. To the right of this line the flow contains upstream jumps as in Figure 4.9, where the contours denote the amplitude of the jump. For sufficiently

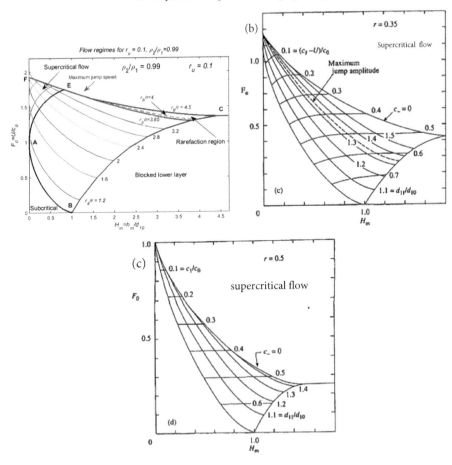

Figure 4.11 (F_0, H_m) diagrams for hydrostatic flow over obstacles (where $H_m = h_m/d_{10}$), showing the regions with the various different flow regimes, where the flow is wholly subcritical, wholly supercritical, or blocked, for $\rho_2/\rho_1 = 0.99$, the near-Boussinesq case. Lines in the central region show the amplitude d_{11}/d_{10} of upstream jumps, and (where appropriate) of rarefactions, using the jump formulation of Baines (2016), as described in Baines & Johnson (2016). (a) $r = r_u = d_{10}/D = 0.1$; (b) $r = 0.35$; (c) $r = 0.5$. Upstream disturbances in (c) are all rarefactions.

large Hm so that curve BC is reached, the lower layer is completely blocked. In Figures 4.11a,b the dashed line shows the amplitude of the jump of maximum speed, and above this line there is a rarefaction region where the flow has the form shown in Figure 4.10. In Figure 4.11c the upstream disturbance is all rarefaction. In Figure 4.11a, above the line EC the flow over the obstacle is again supercritical, but upstream disturbance for the appropriate value of F_0 remains. In the region AEF of Figure 4.11a, there are three possible flow states – supercritical flow, and two

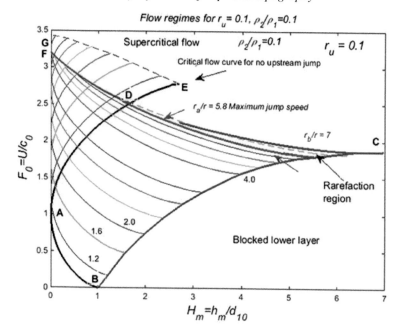

Figure 4.12 As for Figure 4.11, but for the parameters $r_u = 0.1$, $\rho_2/\rho_1 = 0.1$. The properties are essentially the same, except for the region DEGF, within which the flow has two possible flow states: supercritical flow, or critical flow at the obstacle crest with an upstream jump. (From Baines & Johnson, 2016.)

states with different amplitudes of upstream jump. This should be contrasted with Figure 4.12, which is the same except that the density of the upper layer is reduced ($\rho_2/\rho_1 = 0.1$); here the diagram is similar except that the region of multiple states (two only) extends above the line CDF to EG. In both cases, supercritical flow over the obstacle occurs above the line FC, and for F_0 less than the value at F, the flow will include upstream disturbances (Baines & Johnson, 2016).

Many of the above flow properties have been observed and described in Baines (1984). The experimental configuration is shown in Figure 4.13. The fluids used were kerosene (density 0.79 g/cm^3) and water, which are immiscible. The obstacle was mounted at the top, dipping into the free surface of the kerosene layer, rather than on the bottom, for practical reasons. The equations for this configuration are more complex than those for Figure 4.10 because the density difference is too large for the Boussinesq approximation to be applicable, and the displacement of the free upper surface must be allowed for. However, at the speeds of interest for the interfacial mode, the flow is subcritical with respect to the "external" or free-surface mode so that its effect is small, and the (F_0, H_m)-diagrams are similar to those of Figure 4.10. With one significant exception, the observed flow properties correspond very well

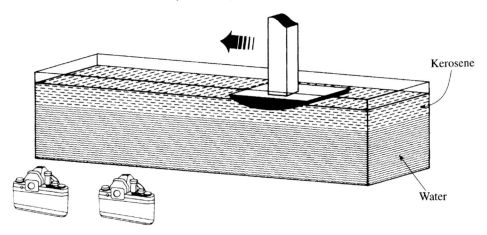

Figure 4.13 Experimental configuration for the observations of Figures 4.14–4.16. Note that the system is inverted relative to the normal configuration described in the text, with the obstacle (marked in black) at the top.

with the theoretical ones, including the locations of the boundaries of the various regions, and some examples of these flows are shown in Figures 4.14 and 4.15. Another observation is that flow separation may occur when the lower layer thickens on the lee side and the fluid moves against an adverse pressure gradient. This may introduce some asymmetry in the interface displacement about $h = h_m$ as discussed in §4.8. Further, multiple flow states have been observed experimentally in part of the region to the left of the AE curve of Figure 4.11, with water and kerosene and $r = 0.035$ (Baines, 1984), but not in all of it, and the difference can be related to surface friction in the experiments.

Figure 4.14 shows a number of flows with upstream bores. The kerosene layer is dark, and the inverted obstacle is white with a black centre section and is visible between the two supporting uprights. The figures are composites, made out of successive frames from a stationary camera, but approximate an instantaneous picture quite closely. For Figures 4.14a,b,c,e and f, we have $r = 0.035$, so that the initial kerosene layer thickness is only 3.5% of the total depth (though Figure 4.14d has $r = 0.2$). At small bore amplitudes as in Figure 4.14a, the bore is undular and laminar, with the amplitude of the waves decreasing to zero with distance behind the leading wave. This decrease is attributed to the effects of energy dissipation at the interface. As F_0 or H_m is made larger the waves may grow to very large amplitudes, and the flow may be turbulent on the lee sides and crests of the waves due to shear instability at the interface. Figure 4.14c shows that for sufficiently large F_0 and H_m the flow may be fully turbulent and resemble a gravity current, although in these immiscible experiments there would be no mixing (see also Long, 1974).

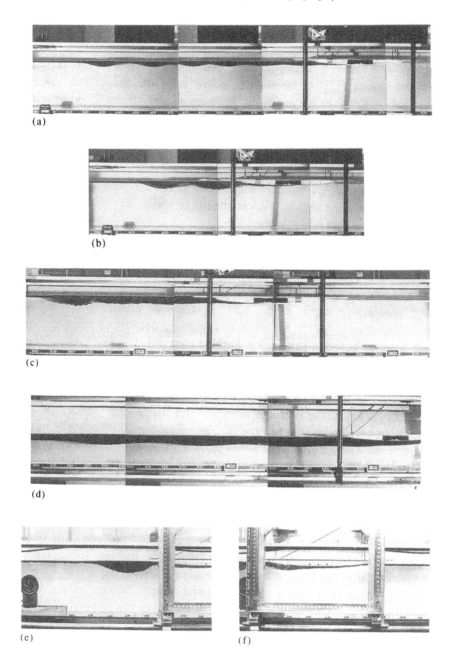

Figure 4.14 Examples of upstream two-layer bores produced by a moving obstacle, with the configuration of Figure 4.13. (a) $r = 0.035$, $F_0 = 1.12$, $H_m = 1.6$. (b) $r = 0.035$, $F_0 = 1.61$, $H_m = 1.6$. (c) $r = 0.035$, $F_0 = 1.61$, $H_m = 2.6$. (d) $r = 0.2$, $F_0 = 0.85$, $H_m = 0.317$. (e) $r = 0.035$, $F_0 = 1.82$, $H_m = 1.17$; (f) as for (e), but at a slightly later time. The camera is stationary.

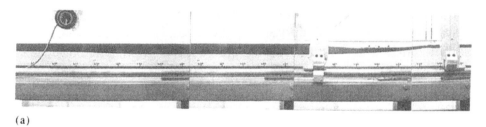

(a)

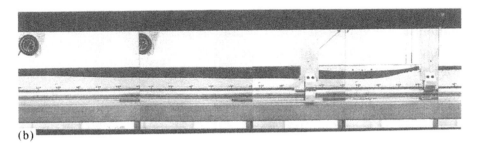

(b)

Figure 4.15 Examples of an upstream jump of maximum amplitude, followed by a rarefaction, with the same system as in Figures 4.13, 4.14. (a) $r = 0.35$, $F_0 = 0.64$, $H_m = 0.4$. (b) $r = 0.35$, $F_0 = 0.76$, $H_m = 0.69$. Note the different flow states over the obstacle. In (b), flow upstream of the obstacle is critical, giving supercritical flow at the crest and a downstream supercritical leap (see §4.8). Distance along the tank is marked in 10 cm intervals.

The process of generation of succeeding waves at the obstacle is quite spectacular, and is illustrated in Figures 4.14e,f. It is observed to occur by the thin layer of fluid at the leading face of the obstacle becoming thicker until a sufficiently large "lump" is formed, which is then able to propagate upstream of its own accord. The similarities with corresponding phenomena for single-layer flow are obvious, but are more striking here because the generated waves have larger amplitudes. The difference between Figures 4.14e,f shows that the second wave is growing, and that given enough time (or length of tank!) an enormous internal undular bore would result.

Figure 4.15 shows two flows with $r = 0.35$ where H_m is large enough to create a hydraulic jump of maximum amplitude, followed by a rarefaction. Here the hydraulic jump takes the form of a monotonic increase in the layer depth, which propagates without changing shape and occupies a length of about 30 cm, and the rarefaction is represented by a barely perceptible increase in layer depth over the expanding distance between this region and the obstacle. In Figure 4.15a, the flow is determined by a critical condition over the obstacle, as in Figure 4.14; but, in Figure 4.15b the flow is quite different. Here the flow is symmetric at the obstacle crest, which implies that the flow immediately upstream of the obstacle is critical (an "approach control"), and the upstream disturbance has reached its maximum amplitude for these conditions. The nature of the subsequent flow on the downstream side is discussed in §4.8.

4.6 Non-linear waves and internal bores

For two-layer motions that are not so long as to be hydrostatic, we may obtain an equation (or equations) for waves incorporating linear dispersion due to finite (but long) wavelength, and non-linear advection, where these two effects are small and comparable, as was the case for single-layer flows in §3.1. The nature of the dispersive terms may be inferred from the dispersion relation, which from (4.12) for fluid at rest with $\Delta\rho/\rho_1 \ll 1$, is

$$\omega^2 = g'k/(\coth kd_1 + \coth kd_2). \tag{4.58}$$

If k is small so that both $kd_1, kd_2 \ll 1$, then (4.58) has the form

$$\omega = c_0 k \left(1 - \frac{1}{6}k^2 d_1 d_2 + \cdots\right), \tag{4.59}$$

where c_0 is the non-dispersive wave speed given by (4.14). Equation (4.59) has the same form as (3.8), and therefore describes waves governed by the linearised KdV equation (3.9) where d_0^2 is replaced by $d_1 d_2$ (Benney, 1966). On the other hand, if d_2 is large so that $kd_2 \gg 1$ but $kd_1 \ll 1$, then (4.58) gives

$$\omega = c_0 k \left(1 - \frac{1}{2}|k|d_1 + O(kd_1)^2\right), \tag{4.60}$$

which corresponds to the equation (Benjamin, 1967; Ono, 1975)

$$\eta_t + c_0\eta_x + \frac{1}{2}c_0 d_1 \frac{\partial^2}{\partial x^2}\int_{-\infty}^{\infty} \frac{\eta(\xi,t)}{\xi - x}d\xi = 0, \tag{4.61}$$

where the horizontal bar denotes a Cauchy-principal-value integral. More generally, (4.58) with $kd_1 \ll 1$ but $kD = O(1)$ may be expressed as

$$c = c_0 \left[1 - \frac{1}{2}kd_1 \left(\coth kD - \frac{1}{kD}\right) + \cdots\right], \tag{4.62}$$

which reduces to (4.59) and (4.60) in the appropriate limits. This may be used to obtain a corresponding more general wave equation (Whitham, 1967), governing waves that satisfy (4.62).

The effects of non-linear advection may be obtained from the hydrostatic equations (4.27)–(4.29). For wave motion progressing to the right (say) into fluid at rest, the absence of waves moving to the left implies that the Riemann invariant R_- is constant, and this may be used to eliminate v (or η) to obtain an equation for η alone. For such waves we have $R_- = \arcsin(2r - 1)$, and for $|\eta/d_s| \ll 1$ from the first of (4.29) we obtain

$$\eta_t + c_0\eta_x + \frac{3c_0}{2d_s}\left[(1 - 2r)\eta - \frac{1 + 4r(1 - r)}{4d_s}\eta^2\right]\eta_x = 0, \qquad (4.63)$$

where $r = d_1/D$ and $d_s = r(1 - r)D$, retaining the first two terms in the expansion. The cubic non-linearity term is negligible except when r is close to $\frac{1}{2}$, where the quadratic term is small.

The above effects of wavenumber dispersion and non-linear advection (or amplitude dispersion) may be added to give combined equations for η. These equations may be formally derived by assuming that the two small dimensionless parameters a/d_s and $(d_s/L)^2$ have a given relationship, and by then following Whitham's 1974 procedure. Here a is the representative amplitude of the interface displacement, L is the horizontal length-scale, and d_s is a vertical length-scale such that the speed of linear interfacial waves c_0 is given by $c_0^2 = g'd_s$. Equation (4.61) with the quadratic non-linear term from (4.63) added is the *Benjamin–Ono equation*, which is formally applicable to flows with an effectively infinitely deep upper (or lower) layer. The more general "finite-depth" or "intermediate-long-wave" (ILW) equation based on (4.62) (Joseph, 1977; Kubota et al., 1978) includes the B–O and KdV equations in the appropriate limits, and analytical solutions for solitary waves have been found for all three.

Experiments to test the applicability of these various equations (Koop & Butler, 1981; Segur & Hammack, 1982) by comparing the solitary wave solutions with observed waves, have shown that the Korteweg–de Vries equation is more widely applicable than the others. Although the experiments were far from exhaustive, a parameter range for which the data are consistent with the Benjamin–Ono equation could not be found experimentally, and the range of applicability of the finite-depth or ILW equation was disappointingly small and for restricted wave amplitudes. Since these other equations are also more complex than the KdV, and do not seem to describe any new or different phenomena, we do not discuss them further here.

We therefore focus on the KdV equation as the basis for describing non-linear dispersive phenomena, but we will include the cubic non-linear term of (4.63) because it permits the description of some important effects. A formal derivation (e.g. Miles, 1979; Melville & Helfrich, 1987) yields the extended KdV equation (eKdV)

$$\eta_t + c_0\eta_x + \frac{3c_0}{2d_s}\left[(1 - 2r)\eta - 2\frac{(r^3 + (1 - r)^3)}{d_s}\eta^2\right]\eta_x + \frac{c_0d_s^2}{6r(1 - r)}\eta_{xxx} = 0, \quad (4.64)$$

where $r = d_{10}/D$, and $d_s = r(1 - r)D$ (the cubic non-linear term here differs from that in (4.63), but is equivalent to it when $2r \approx 1$). The term c_0 may be positive or negative. We regard (4.64) as the two-layer equivalent of (3.12), and treat it as a model equation which can give at least a qualitative description of the phenomena associated with two-layer bores. Note that in the limit $r \to 0$ (or 1), the wave-dispersion term becomes singular, whereas the other terms (to leading order) approach the single-layer forms. This reflects the inappropriateness of this equation for thin lower layers, where the term in (4.61) is applicable instead.

We may look for steady solutions of (4.64) that are functions of $x - ct$, and represent steady motions propagating with speed c. For these motions, (4.64) may be integrated twice. This introduces two constants of integration, but Benjamin (1966) has shown that one vanishes so that the η-term is absent, and hence

$$\frac{c_0 d_s^2}{3r(1-r)}\eta_x^2 = \mathcal{B}(\eta) \equiv B_0 + 2(c - c_0)\eta^2 - \frac{c_0}{d_s}(1 - 2r)\eta^3 + \frac{c_0}{d_s}[r^3 + (1-r)^3]\eta^4. \quad (4.65)$$

Here the constant term

$$B_0 = \frac{2d_s}{\rho_1 c^2}(\mathcal{F} - \mathcal{F}_0) = \frac{2d_s}{c_0}\int_0^D \left(\frac{u}{c}R - R_0\right)dz, \quad (4.66)$$

where $\mathcal{F}$ is the energy flux in this steady flow (see (1.44)) with R the total head, and $\mathcal{F}_0$ and R_0 are the energy flux and total head in undisturbed flow ($\eta = 0$) in the same frame of reference. For solitary wave solutions we must have $c > c_0$ and $B_0 = 0$, and for $0 < r < 1/2$, the quartic $\mathcal{B}(\eta)$ has the form shown in Figure 4.16a (cf. Figure 3.3 for the KdV equation). This shows that a solution starting from $\eta = 0$ must make a single excursion to $\eta = \eta_m$ and back, as for the corresponding solutions of (3.31). The corresponding solitary wave solution is

$$\eta = \frac{\eta_m}{\cosh^2 \theta - \mu \sinh^2 \theta}, \quad (4.67)$$

where μ is a parameter in the range $0 < \mu < 1$, and

$$\eta_m = a = \frac{\mu}{(1 + \mu)}\frac{(1 - 2r)}{(1 - 3r + 3r^2)}d_s, \quad (4.68)$$

$$\theta = \theta_0 + \left(x - c_0\left[1 + \frac{2}{3}\frac{(d_s/l)^2}{r(1 - r)}\right]t\right)/l, \quad (4.69)$$

where θ_0 is an arbitrary phase, $c_0^2 = g'd_s$ and

$$l^2 = D^2\frac{4}{3}\frac{(1 + \mu)^2}{\mu}\frac{r(1 - r)(1 - 3r + 3r^2)}{(1 - 2r)^2}. \quad (4.70)$$

We may take l as L, the horizontal length-scale. The wave speeds c are given by

$$c = c_0\left[1 + \frac{1}{2}(1 - 2r)\frac{\eta_m}{d_s} - \frac{1}{2}(1 - 3r + 3r^2)\left(\frac{\eta_m}{d_s}\right)^2\right]. \quad (4.71)$$

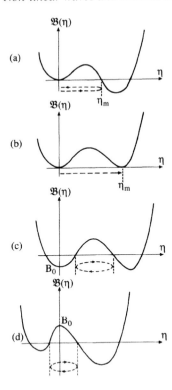

Figure 4.16 The quartic expression $\mathcal{B}$ on the right-hand side of (4.65). The dashed lines show the variation of η in the appropriate solution. (a) $B_0 = 0$, solitary wave (4.67). (b) $B_0 = 0$, monotonic inviscid bore solution (4.72). (c) $B_0 < 0$, non-linear periodic wavetrain. (d) $B_0 > 0$, non-linear subcritical wavetrain.

The form of these solitary waves depends only on the parameters r and μ. They are waves of elevation if $0 < r < 1/2$, and of depression if $1/2 < r < 1$. These waves are analogous to the single-layer solitary waves (3.14), modified by the addition of the cubic non-linear term, which has a small effect unless r is close to 1/2.

Equation (4.65) with $B_0 = 0$ also has the solution

$$\eta = \frac{\eta_m}{2}(1 + \tanh\theta), \qquad \mu = 1, \tag{4.72}$$

where θ is again given by (4.69), c by (4.71), and η_m and l by (4.68) and (4.70) respectively. There is only one such solution for each r, and this "inviscid bore" corresponds to the bore of maximum amplitude and zero (or very small) energy dissipation of hydraulic theory (§4.4), and observed in the two-layer experiments described in the previous section. It depends on the cubic non-linear term; the form of $\mathcal{B}(\eta)$ for this solution is shown in Figure 4.16b, where the extra bend in the curve permits the new form of behaviour (cf. Figure 3.3). A solution of (4.65) must move from $\eta = 0$ to $\eta = \eta_m$, and remain there. The bore amplitude and speed from (4.68) and (4.71) agree reasonably well with those obtained from the hydraulic models of §4.4 if $r(= r_u)$ is in the range $0.35 < r < 0.65$, so that r is close to $\frac{1}{2}$ and the bore amplitude is small; but they diverge rapidly outside this range.

The reversed or mirror-image form of solution (4.72) is also theoretically possible. If $\eta = \eta_m$, then

$$\eta = \frac{\eta_m}{2}(1 - \tanh\theta) \qquad (4.73)$$

is also a solution (Miles, 1981), as is suggested by Figure 4.16b, with η moving from η_m to zero. This permits a special form of "solitary wave", consisting of (4.72) followed at an arbitrary distance by (4.73), moving at the same speed. Such a wave has yet to be demonstrated experimentally, but it has been found numerically by Melville & Helfrich (1987) by using steady topographic forcing in (4.74) for a finite period of time.

If $B_0 < 0$, the solution of (4.65) must correspond to a flow with $\mathcal{F} < \mathcal{F}_0$, so that it has less energy flux than the uniform stream with no waves. This may be achieved by having a localised region of energy dissipation in the flow, such as breaking waves at the leading part of a bore. The curve $\mathcal{B}(\eta)$ is now lowered to a form shown in Figure 4.16c, and (4.65) has periodic solutions oscillating between the two zeros of $\mathcal{B}(\eta)$. These waves (which are cnoidal waves if the cubic non-linear terms are negligible) give the form of a steady undular bore, as in §3.3, for a single layer. As B_0 decreases to a minimum value, the wave amplitude decreases to zero giving a uniform stream.

4.7 Topographic forcing with non-linearity and dispersion

The eKdV equation may be extended further to include topographic forcing. If the topography has the form $z = h(x)$ on the lower boundary, and the fluid has the mean velocity U, the topography generates (mostly) waves propagating against the stream and these are governed by the forced extended KdV equation (feKdV) (Melville & Helfrich, 1987)

$$\eta_t + (U - c_0)\eta_x - \frac{3c_0}{2d_s}\left[(1 - 2r)\eta - 2\frac{(r^3 + (1-r)^3)}{d_s}\eta^2\right]\eta_x - \frac{c_0 d_s^2}{6r(1-r)}\eta_{xxx}$$

$$= \frac{(1-r)U}{2}\frac{dh}{dx}, \qquad (4.74)$$

where again $c_0 = [g'Dr(1-r)]^{1/2}$. If $2|\eta/d_s| \ll (1-2r)$, this equation is essentially the same as the forced KdV equation, (3.38), so that the solutions have the same properties and may be related to the descriptions in §3.4 by re-scaling the coefficients. The appropriate transformation to (3.38) (omitting the cubic non-linear

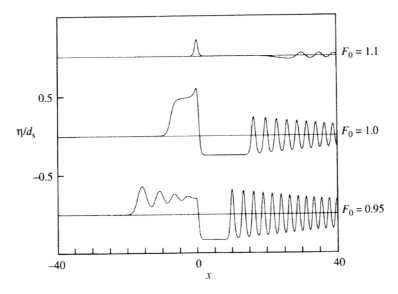

Figure 4.17 Numerical solution types of the feKdV equation (4.74) with $r = 0.35$ and an obstacle at $x = 0$. $F_0 = 1.1$: steady supercritical flow; $F_0 = 1.0$: upstream monotonic inviscid bore; $F_0 = 0.95$: upstream modulated wavetrain. x is scaled with the length of the obstacle. (From Melville & Helfrich, 1987, reproduced with permission.)

term) is

$$(x', t') = (r - r^2)^{1/2}(x, t), \qquad \eta' = (1 - 2r)\eta, \qquad h' = (1 - r)h, \qquad (4.75)$$

with c_0 and d_s (replacing d_0) remaining as for (4.74). The discussion and quantitative results of the previous chapter are then applicable here if x, t, η and h in (3.38) etc. are replaced by x', t', η' and h'.

The principal interest, therefore, is in the effect of the additional cubic non-linearity, in parameter ranges where it is significant. Figure 4.17 shows three flows with different values of $F_0 = U/c_0$, computed from (4.74) with $r = 0.35$. The first shows supercritical flow, the second a bore of maximum amplitude corresponding to the wave solution (4.72), followed (perhaps) by a short rarefaction, and the third a modulated undular bore of the type described by the fKdV equation. Figure 4.18 shows the regions of (F, H_m)-space where these solutions were obtained. These are seen to correspond reasonably well to the boundaries from hydraulic theory in Figure 4.11b.

For two-layer flow past a *three-dimensional obstacle* in a channel, when the lower layer is the thinner the flow has similar properties to those described for a single layer in §3.5 (Hanazaki, 1994). However, the extra parameters in two-layer flows

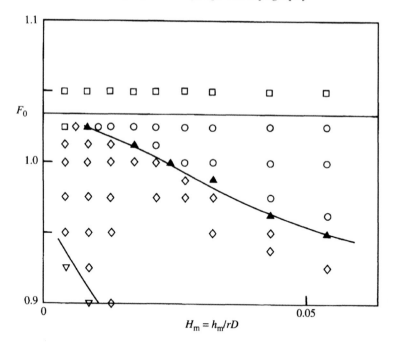

Figure 4.18 Regions of the (F_0, H_m)-plane for $r = 0.35$ where the solution types of Figure 4.17 were found numerically. □: steady supercritical flow; ○: monotonic upstream bore; ◇: undular upstream bore; ▽: steady subcritical flow; ▲ marks a transition between ○ and ◇. (From Melville & Helfrich, 1987, reproduced with permission.)

imply additional variations in behaviour of three-dimensional flows that have yet to be explored.

4.8 Downstream effects

In §4.5 we were predominantly concerned with flow properties on the upstream side of the obstacle, which were discussed without considering what might happen downstream. This is possible because, for the most part, these flows were controlled by a critical condition at the obstacle crest which effectively made the upstream flows independent of the downstream properties. When the flow is controlled in this way, the downstream behaviour is analogous to that of single-layer flows: the supercritical downstream state is terminated by a hydraulic jump, which may be attached to the lee side of the obstacle, or swept downstream if F_0 is large enough. This behaviour may be seen in the examples shown in Figures 4.14 and 4.15a. In such flows there is often substantial mixing associated with the jump, if the two layers are miscible. This mixing is not normally in the jump itself (as it is in single-layer flow), but occurs mostly in the supercritical region with large shear ahead of the jump (Lawrence, 1985), and stems from Kelvin–Helmholtz instability and the resulting "billows" (see Chapter 9). This may result in a much thicker interfacial region, and if this becomes thick enough it may be necessary to regard it as a separate third layer for dynamical purposes.

If the flow is subcritical and the obstacle is not too long, small-amplitude topography forces linear lee waves on the downstream side in the same manner as described in §2.2.2 for single-layer flows. As the topographic amplitude h_m increases, this lee wavetrain becomes non-linear. This behaviour may be described by (4.74) for moderately long obstacles, or for steady flows by (4.65) with B_0 generalised from (4.66) to (Benjamin, 1966)

$$B_0 = \frac{2d_s}{\rho_1 c^2}(\mathcal{F} - \mathcal{F}_0 + S_0 - S), \qquad (4.76)$$

where S is the *flow force* or *momentum flux* defined in (1.39) as S_{11}, and S_0 is its upstream value. On the downstream side, $S - S_0 = -F_D$, the drag force on the obstacle (from (1.42)). If we assume that there is no dissipation so that $\mathcal{F} = \mathcal{F}_0$, we have $B_0 > 0$, and with $c < c_0$ the quartic (4.65) then has the form shown in Figure 4.16d, giving a lee wavetrain in a wholly subcritical flow. Observations of non-linear lee wavetrains of this type, where the wave amplitude is substantially larger than that predicted by linear theory, have been described by Smith (1976) for both the laboratory and the atmosphere.

As described in §4.5 and indicated in Figure 4.11, if $F_0 < 1$ and the upstream lower-layer thickness becomes sufficiently large due to large H_m, the flow immediately upstream of the obstacle may become critical, because rarefactions of maximum amplitude have been sent upstream. Any further increase in the disturbance cannot propagate fast enough to move upstream, so that this flow state represents the largest deviation from the original flow that obstacles can produce in it, for given F_0. If the obstacle height is increased beyond this point, the upstream flow is unaltered and the flow over the obstacle will become supercritical (squeezing will increase the mean velocity) and satisfy the condition $\partial \eta/\partial x = 0$ at $h = h_m$, so that it is locally symmetric about the obstacle crest. In these circumstances, the flow on the downstream side must adjust to a deficit of fluid in the lower layer. Observations (Long, 1954; Baines, 1984; Lawrence, 1985, 1993) show that it does this by converting a supercritical stream with a deep lower layer over the lee side of the obstacle to another supercritical stream with a shallow lower layer, by a rapid descent of the interface. Examples of this behaviour are shown in Figures 4.15b and 4.19.

This remarkable phenomenon is known as a *supercritical leap*. In these transitions the flow appears to remain supercritical throughout (or nearly so), and it

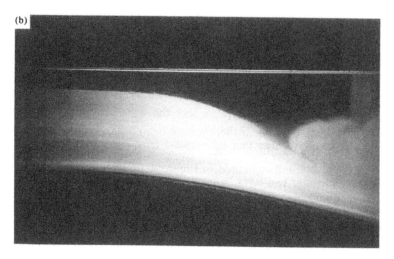

Figure 4.19 Two examples of lee-side supercritical leaps in steady flows with fixed flow rates of miscible fluids. In (a) the lee-side hydraulic jump is swept downstream, whereas in (b) it is stationary. Note the mixing in the strongly sheared regions. (Photographs supplied by G. Lawrence.)

is not described by a solution of the inviscid hydrostatic equations (4.18), (4.19). Lawrence (1993) has shown (in at least one example) that the supercritical leap involves dissipation, and that flow curvature is important. This transition is followed by a hydraulic jump that is either stationary or is advected downstream, and it approximately restores the flow to the original downstream conditions. This interesting dissipative phenomenon deserves further study.

5

Two-layer and stratified flow through contractions

I hope that you will remember that who seeketh two strings to one bow,
he may shoot strong but never straight.
QUEEN ELIZABETH I, *Letter to James VI.*

A note on terminology. When considering layered flow over topography, it is natural to number the layers from the bottom upwards, as a single bottom layer would be "layer one", whether there are significant overlying layers or not. For the case of layered flow through contractions, however, some previous authors have numbered layers from the top down (apparently following an influential paper by Wood, 1968). Here, to maintain consistency throughout this volume, the layers in equations have been numbered from the bottom upwards throughout.

We next consider hydrostatic two-layer flow through a channel whose cross-sectional area varies with distance along it, with vertical sidewalls, which may or may not contain some bottom topography. The principal feature is that the channel initially has a very large width constituting a reservoir, and that the fluid passes through a contraction or minimum width and subsequently broadens into a downstream reservoir. Here we only consider situations where the flow in each layer is in the same (positive x) direction, and defer discussion of exchange flows to the next chapter. Interest in this type of problem derives from processes such as extraction of water from power station cooling ponds, where the coolest water is most preferred. The channel dimensions in the (y, z)-plane are assumed to be small relative to that of the variations in the x-direction, so that flow properties are independent of y. The flow through a contraction is then envisaged to take place from one region that is effectively a large reservoir to another region with the same cross-section as the upstream reservoir. The upper boundary may be rigid at $z = D$, or may be a free surface.

The governing dynamical equations for flow in these two layers are the same as

(4.18)–(4.21), but (4.22) takes the form

$$b(x)(d_1 u_1 + d_2 u_2) = Q, \tag{5.1}$$

where Q now denotes a total volume flux. Throughout this chapter and the next we are mainly concerned with steady-state flow.

5.1 Two-layered flow through contractions with a free upper surface

With a free upper surface, the steady-state form of the equations (4.18), (4.19) and (4.21) reduces to the Bernoulli and continuity equations in the form

$$\frac{\rho_1 u_1^2}{2g} + \rho_1 d_1 + \rho_2 d_2 = \rho_1 D_1 + \rho_2 D_2, \tag{5.2}$$

$$\frac{u_2^2}{2g} + d_1 + d_2 = D_1 + D_2, \tag{5.3}$$

$$u_1 d_1 b = q_1, \qquad u_2 d_2 b = q_2, \tag{5.4}$$

where q_1 and q_2 denote the volume fluxes and D_1, D_2 the upstream thicknesses of the two layers. A diagram showing the notation is presented in Figure 5.1. If one follows Hugoniot's method (Binnie, 1972) and takes equations (5.2) and (5.3), and the derivatives of equations (5.4), one has four equations for the four derivatives of u_1, u_2, d_1 and d_2. These can be manipulated to produce a single equation for one of them, which we choose to be dd_1/dx, which takes the form

$$\left[\frac{\rho_1}{\rho_2} (1 - \frac{u_2^2}{g d_2})(1 - \frac{u_1^2}{g d_1}) - 1 \right] \frac{dd_1}{dx} = -\frac{1}{g b} \frac{db}{dx} \left[u_1^2 + \frac{\rho_1}{\rho_2}(1 - \frac{u_2^2}{g d_2}) u_1^2 \right]. \tag{5.5}$$

At the point of maximum contraction we have $db/dx = 0$, and if the gap there is sufficiently narrow, we may have $dd_1/dx \neq 0$, which implies that

$$(1 - F_1^2)(1 - F_2^2) = \frac{\rho_2}{\rho_1}, \tag{5.6}$$

where $F_i^2 = u_i^2 / g d_i$. From equations (5.2), (5.3) and (5.6) we may then obtain

$$\left(1 - 2\frac{(D_2 - d_2)}{d_2} - 2\frac{(D_1 - d_1)}{d_2}\right) \cdot \left(1 - 2\frac{\rho_2}{\rho_1}\frac{(D_2 - d_2)}{d_1} - 2\frac{(D_1 - d_1)}{d_1}\right) = \frac{\rho_2}{\rho_1}. \tag{5.7}$$

Examination of equation (5.7) shows that it has the solution

$$d_1 = \frac{2}{3} D_1, \qquad d_2 = \frac{2}{3} D_2 \tag{5.8}$$

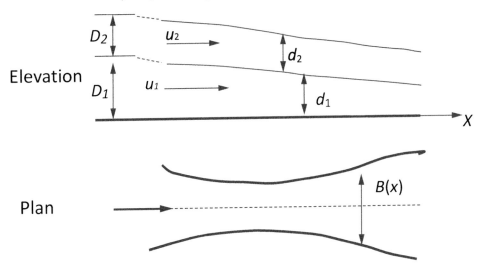

Figure 5.1 Notation diagram for two-layer flow through a contraction with a free upper surface over level terrain. The thickness of each layer at the point of maximum contraction is two thirds of its initial value.

at the point of maximum contraction, and that the nett transport in each layer is given by

$$q_1 = b\left(\frac{8g}{27}(D_1 + \frac{\rho_2}{\rho_1}D_2)\right)^{1/2} D_1, \qquad q_2 = b\left(\frac{8g}{27}(D_1 + D_2)\right)^{1/2} D_2. \qquad (5.9)$$

In other words, the flow is dominated by the free-surface mode.

If we specify Q, q_1 and q_2 instead of D_1 and D_2, the result of the above analysis again gives (5.9), which may be used to infer the values of D_1 and D_2. It should also be noted that these results are consistent with the expressions for the speed of long waves in two-layer systems with a free surface, as described in §4.2. These results were first obtained by Wood (1968) (using a different method), who pointed out that for these flows there was also a location upstream of the minimum gap where (5.6) was also satisfied (a position of "virtual control"). But it is clear from the above calculation that this is not important and is an incidental feature of these flows.

These expressions bear a remarkable similarity to the results for a single layer described in §2.3.3 – the free-surface mode dominates, and the internal wave mode plays no role. The flow undergoes a hydraulic transition from sub- to supercritical flow at the contraction, where the total depth of the flow is 2/3 of that in the upstream reservoir, as depicted in Figure 5.1 As the gap width b is progressively decreased further, the nett transport in each layer is reduced in proportion, ultimately to zero. The presence of supercritical flow on the downstream side implies the presence of a downstream hydraulic jump or equivalent form of transition to subcritical flow as the channel width broadens.

This solution may be extended to include the effect of a sill that is co-located with the maximum contraction. If the sill has the form $z = h(x)$ where h is the elevation above the bottom level, equations (5.2)–(5.3) become (see (4.18)–(4.19))

$$\frac{\rho_1 u_1^2}{2g} + \rho_1(h + d_1) + \rho_2 d_2 = \rho_1 D_1 + \rho_2 D_2, \tag{5.10}$$

$$\frac{u_2^2}{2g} + h + d_1 + d_2 = D_1 + D_2, \tag{5.11}$$

and equations (5.4) are unaltered. We may write the maximum height of the sill as $h_{max} = \gamma(D_1 + D_2)$, at the same location as the minimum contraction, so that $dh/dx = 0$ at the same location where $db/dx = 0$. A repeat of the above analysis then yields the result that, at the contraction,

$$\frac{d_1}{D_1} = \frac{2}{3}(1 - \gamma), \qquad \frac{d_2}{D_2} = \frac{1}{3}(2 + \gamma), \tag{5.12}$$

provided that the sill height γ is small. Introducing a sill has the effect of reducing the local thickness of the lower layer, and increasing the thickness of the upper layer.

5.2 Two-layered flow through contractions with a rigid upper boundary

If one imposes a rigid upper boundary, instead of a free surface, to the above flow the problem becomes more complex. This situation is very different, in that the total depth is known, the free-surface mode is absent, and the dynamics are dominated by the internal mode. For reference, a schematic picture is shown in Figure 5.2. Under these conditions the relevant equations for unidirectional steady-state flow through a contraction with vertical sidewalls and possible bottom topography $h(x)$ comprise the Bernoulli and continuity (mass conservation) equations for each layer, which here take the form

$$H_1 = \frac{1}{2}\rho_1 u_1^2 + \rho_1 g(d_1 + h) + g\rho_2 d_2 + p_s = \rho_1 g D_1 + g\rho_2 D_2 + p_s(r) \tag{5.13}$$

$$H_2 = \frac{1}{2}\rho_2 u_2^2 + \rho_2 g(d_1 + d_2 + h) + p_s = \rho_2 g(D_1 + D_2 + h) + p_s(r) \tag{5.14}$$

$$d_1 + d_2 + h = D, \tag{5.15}$$

where H_1 and H_2 are constants for steady flow, $h(x)$ denotes bottom topography, D the total depth with D_1 and D_2 the depths of the layers in the upstream reservoir, p_s denotes the unknown pressure at the upper boundary, and $p_s(r)$ denotes the value of p_s in the upstream reservoir. We also have the continuity equations (5.4): $u_1 d_1 b =$

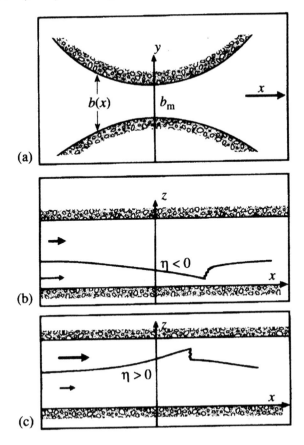

(a)

(b)

(c)

Figure 5.2 (a) Plan view of a contraction; (b) side view of flow with a critical condition at the minimum point with $\eta < 0$; (c) as for (b) but with $\eta > 0$. Solid arrows denote representative fluid velocities.

q_1, $u_2 d_2 b = q_2$. The pressure p_s must be eliminated by taking the difference: $H_1 - H_2$, so that we obtain

$$\frac{1}{2}u_1^2 + \frac{\Delta\rho}{\rho_1}g(d_1 - D_1 + h(x)) = \frac{1}{2}\frac{\rho_2}{\rho_1}u_2^2, \qquad (5.16)$$

where $\Delta\rho = \rho_1 - \rho_2$. The derivative of this equation with respect to x has the form

$$(F_1^2 + F_2^2 - 1)\frac{\mathrm{d}d_1}{\mathrm{d}x} = -\frac{1}{b}\frac{\mathrm{d}b}{\mathrm{d}x}(F_1^2 d_1 + F_2^2 d_2) + (1 - F_2^2)\frac{\mathrm{d}h}{\mathrm{d}x}, \qquad (5.17)$$

where F_1 and F_2 denote the internal Froude numbers for each layer, given by

$$F_1^2 = \frac{u_1^2}{\frac{\Delta\rho}{\rho_1}g d_1}, \qquad F_2^2 = \frac{u_2^2}{\frac{\Delta\rho}{\rho_2}g d_2}. \qquad (5.18)$$

Locations where

$$G^2 = F_1^2 + F_2^2 = 1, \tag{5.19}$$

denote critical conditions or "control points", notably at extreme values of b or h. At each location the flow rates of the layers are given by (5.4):

$$u_1 d_1 b = q_1, \qquad u_2 d_2 b = q_2, \qquad Q = q_1 + q_2. \tag{5.20}$$

It is convenient to scale the width b of the channel by $Q/D\sqrt{g'D}$, so we define $\hat{Q} \equiv Q/bD\sqrt{g'D}$, and at the minimum value of $b = b_\mathrm{m}$, $\hat{Q} = \hat{Q}_\mathrm{m}$.

At this point we make the Boussinesq approximation, which is customary because of the simplification that it provides, but also because it captures the main effects for all situations where the difference between the two densities is a small fraction of either. With this approximation (5.16) becomes

$$\frac{1}{2}u_1^2 + g'(\eta + h(x)) = \frac{1}{2}u_2^2, \qquad \eta = d_1 - D_1, \tag{5.21}$$

where $g' = \frac{\Delta\rho}{\rho_1 g}$, and in F_2, ρ_2 has the same value as ρ_1.

As for flow over obstacles, introduction of a contraction may cause changes to the upstream and downstream flow conditions in the final steady state. Flows that are initially supercritical upstream remain so in the contraction. Hence we concentrate on situations where the upstream channel is wide enough for the velocities there to be small and the flow subcritical. As was done in earlier sections for flow over obstacles and through contractions, we investigate steady-state solutions by examining the changes to the steady-state flows that occur when the minimum channel width b_m is progressively decreased in small increments. We assume that the upstream and downstream conditions are the same, but are not primarily concerned with how the downstream flow adjusts to downstream conditions. For these flows we identify two main types: first, those where the flux Q and the upstream layer depths d_{10}, d_{20} are fixed as b_m is varied, and second, those where Q, q_1 and q_2 are fixed, but the layer depths are not. This covers a broader range of conditions than those in §4.5, and we discuss these cases in turn.

5.2.1 Fixed Q, D_1, D_2, $h = 0$

We examine the situation of flow from one very large (infinitely wide) reservoir where the fluid is effectively stationary, to another in which the conditions are the same, through a channel of continuously varying width $b(x)$. We assume Q to be determined by external factors so that it remains constant, as is the level of the interface in the upstream reservoir, but the fluxes q_1 and q_2 are variable.

The fixed variables are g', D and $r = D_1/D$, and q_1 and q_2 are to be determined,

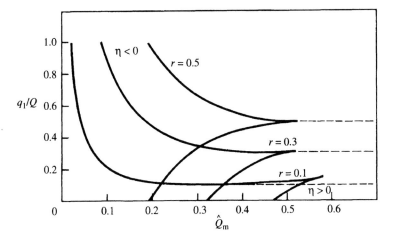

Figure 5.3 Curves denoting the conditions for critical flow ($G^2 = 1$) of two-layer fluid of uniform depth D through a contraction of minimum width b_m connecting two infinitely wide identical reservoirs. The parameters are $\hat{Q}_m = Q/(b_m D \sqrt{g'D})$, $r = D_1/D$, and q_1/Q. The dashed lines denote the flows $u_1 = u_2$. Curves for $r > 0.5$ may be obtained from those for $r < 0.5$ by replacing r by $1 - r$ and q_1 with q_2, so that $r = 0.7$ and 0.9 are mirror images of $r = 0.3$ and 0.1, respectively, about $q_1/Q = 0.5$.

but are subject to $q_1 + q_2 = Q$. Our objective here is to describe the changes to the flow (specifically, the changes in q_1/Q) that occur as $\hat{Q}_m$ is increased from (effectively) zero by decreasing channel width b_m, for given values of R and the initial value of q_1/Q, $(q_1/Q)_i$.

A special case is that where $(q_1/Q)_i = r$, so that the velocities are uniform and the flow is "barotropic". This corresponds to the experimental situation where the "contraction" is towed along a wide channel of two-layer fluid at rest, as in the towed obstacle experiments. As shown below, in this situation there is no interface displacement, and the velocity remains uniform with depth.

The more interesting case is where the velocities in each layer are different. If b_m is decreased ($\hat{Q}_m$ is increased) in a quasi-static fashion, the steady-state flow solutions remain subcritical until $G^2 = 1$ at $b = b_m$. Here equations (5.18)–(5.21) may be used to obtain curves where $G^2 = 1$ (here termed "critical curves") for given values of r, as a function of $(q_1/Q)_i$ and $\hat{Q}$, and these are shown in Figure 5.3 for $r = 0.1$, 0.3 and 0.5 (curves for $r > 0.5$ are the same as those for $r < 0.5$, with $(1 - r)$ replacing r and q_2 replacing q_1). Wholly subcritical flows through the contraction are described on this diagram by points moving with constant q_1/Q from $\hat{Q} = 0$ to values less than $\hat{Q}_m$, and back again as $\hat{Q}$ is decreased.

When $\hat{Q}_m$ reaches or exceeds the values on the curves $G^2 = 1$ of Figure 5.3, as in §4.5 the hydraulic alternative ensures that the flow at $b = b_m$ must be critical, and maintaining this condition can alter the value of q_1/Q. If $\hat{Q}_m$ suddenly exceeds the value on the curve in Figure 5.3 by a small amount, a hydraulic transition occurs at $b = b_m$, and this causes a wave of small amplitude to propagate upstream, with speed given by the appropriate point in Figure 4.2, causing a change to q_1/Q. If the increase in $\hat{Q}_m$ is not small, the upstream disturbance can be a hydraulic jump or a rarefaction, depending on the wave-speed behaviour shown in Figure 4.2. As this wave proceeds upstream where the channel widens, its amplitude decreases to zero, so that equations (5.20)–(5.21) still apply between point $b = b_m$ and the upstream reservoir. On the downstream side of $b = b_m$, the supercritical flow must be terminated by a stationary hydraulic jump in the expanding channel, at a point where its speed equilibrates with the flow speed. The possible flow types are shown in Figure 5.2b and c. Since the condition $G^2 = 1$ at $b = b_m$ still applies, the new condition must be represented on the curve (for given r value) on Figure 5.3, so that the change in $\hat{Q}_m$ gives the new value of q_1/Q. The flow on the upstream side may again by represented by a point moving from $\hat{Q} = 0$ to $\hat{Q}_m$ with constant q_1/Q.

In fact, all unidirectional flows that are critically controlled at the narrowest part of the channel are represented in Figure 5.3. We first consider $r = 0.5$ as an example. If $(q_1/Q)_i > 0.5$, most of the flow is in the lower layer, and $\eta < 0$ in the contraction. As $\hat{Q}_m$ increases, q_1/Q decreases (along the $r = 0.5$ curve) toward 0.5, and the magnitudes of η and v decrease, becoming zero where $q_1/Q = 0.5$ at $\hat{Q}_m = 0.5$. The two layers are now moving as one, with $\eta = v = 0$ everywhere. Further increase in $\hat{Q}_m$ causes increased velocity in the contraction without changing this state, and this is represented by the dashed line on line $r = 0.5$ in Figure 5.3. The flow is subcritical if $\hat{Q} < 0.5$ and supercritical if $\hat{Q} > 0.5$, but this distinction has no effect here, since there are no waves. "Closing the gap" has effectively squeezed the baroclinic or interfacial dynamics out of the flow. If $(q_1/Q)_i < 0.5$, the same behaviour is found with q_1/Q increasing with $\hat{Q}_m$ to 0.5, along the $r = 0.5$ curve.

If $r \neq 0.5$, the curves are asymmetric and the situation is somewhat different. These curves are similar for each value of r, and we take $r = 0.1$ as an example. If $(q_1/Q)_i > r = 0.1$, the behaviour is the same as described for $r = 0.5$: when the critical condition ($G^2 = 1$) for $r = 0.1$ is reached, further increase in $\hat{Q}_m$ (decrease in b_m) causes the value of q_1/Q to decrease along this critical curve. As q_1 decreases, η and v in the solution decrease in magnitude, and approach zero as q_1/Q approaches 0.1, at $\hat{Q}_m \approx 0.3$. When this state is reached, $\eta = v = 0$, and the two layers move together as one. Further increase in $\hat{Q}_m$ causes no further change in the internal dynamics, as for $r = 0.5$, and q_1/Q values are represented by the dashed line at $r = 0.1$.

If instead $q_1/Q < r = 0.1$, then η and v have opposite sign to that for $q_1/Q > r$. Critical flow in the contraction is reached at larger values of $\hat{Q}_m$ near 0.5, as Figure 5.3 shows. As $\hat{Q}_m$ is further increased, q_1/Q increases toward 0.1 along the critical line, and the magnitudes of η and v decrease to zero when the line $q_1/Q = 0$ is reached. It appears that experiments describing these flows have yet to be reported.

5.2.2 Fixed q_1, q_2, $h = 0$

Here the total discharges in each of the two layers are fixed, and the variable $r = D_1/D$ can take a range of values. The end conditions are taken at particular sections with given $\hat{Q}$, but the r-values (i.e. layer depths) are not specified except that the upstream flow is taken to be subcritical. These conditions are readily achieved in experimental studies (Armi, 1986; Lawrence, 1993). Here where $q_r = q_2/q_1$ is fixed, it is more appropriate to represent the flow in terms of the variables F_1 and F_2 in preference to u_1 and u_2, as in (5.17). The flow solutions may be usefully depicted on Froude number planes where the axes are F_1^2 and F_2^2, as shown in Figure 5.4.

Equations (5.17)–(5.22) still apply, and curves of constant $\hat{Q}$ (which mark the width b of the contracting channel – large $\hat{Q}$ means narrow channel) are given by

$$\frac{1}{F_1^{2/3}} + \frac{q_r^{2/3}}{F_2^{2/3}} = \left(\frac{1+q_r}{\hat{Q}}\right)^{2/3}, \quad q_r = q_2/q_1, \quad \hat{Q} \equiv Q/bD\sqrt{g'D}, \quad (5.22)$$

and these are denoted by the light solid lines in Figure 5.4. These curves are determined solely by conservation of volume of fluid, and are independent of the dynamical equation (5.16). Note that $\hat{Q}$ increases outward from the origin, so that, as fluid flows through the contraction, any solution must move from small $\hat{Q}$ values to larger ones, and then back.

After changing variables, equation (5.21) may be expressed in the form (Armi, 1986)

$$\frac{F_1^{-2/3}(1 + \frac{1}{2}F_1^2) - \frac{1}{2}q_r^{2/3}F_2^{4/3}}{F_1^{-2/3} + q_r^{2/3}F_2^{-2/3}} = r \equiv \frac{D_1}{D}, \quad (5.23)$$

provided that the flow is connected subcritically to a region with this value of r. The curves (5.23) may be plotted on the Froude number plane for any chosen value of q_r, and Figure 5.4 shows these (heavy solid) solution curves for the representative value of $q_r = 0.5$ (meaning that the transport in the lower layer is twice that of the upper), for various values of r. Curves representing flows that commence from a large upstream reservoir with a given r value begin at the bottom left-hand corner, and move upward as $\hat{Q}$ increases along the channel. The flow becomes critical if the curve reaches the straight line (5.19), and supercritical if it crosses it, but otherwise returns on the same path. There are two families of such curves: one family begins as subcritical flow near the origin and F_1^2-axis, crosses the critical line and becomes supercritical with large F_1, small F_2 – a thin, rapidly moving lower layer; the other

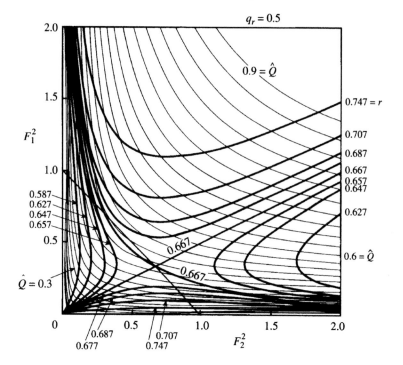

Figure 5.4 (F_1^2, F_2^2)-plane showing possible flow solutions (5.23) (heavy solid lines) for two-layer flows through a contraction with $q_r \equiv q_2/q_1 = 0.5$, labelled by their values of $r = D_1/D$. Line (5.19) denotes critical flow, and the light solid lines denote contours of constant $\hat{Q} = Q/(bD\sqrt{g'D})$, which is effectively a measure of channel width: larger values of $\hat{Q}$ imply narrower width. Here the lower layer has twice the transport of the upper layer. The corresponding figure for $q_r = 2$ where the upper layer has twice the transport of the lower, is given by exchanging axes: the abscissa is F_2^2 and the ordinate F_1^2, with the figure unchanged. (Adapted from Armi, 1986.)

begins near the origin and F_2-axis and becomes supercritical with large F_2, small F_1. There are also two branches of solution curves that are wholly supercritical, but are of no particular interest because they cannot be related to subcritical source regions. The value $r = 1/(1 + q_r)$ is a special case; here (5.23) is the straight line from the origin $F_1^2 = q_r F_2^2$, on which $\eta = 0$, and $u_1 = u_2$. This branches into three parts when it meets the critical line (5.19), all with the same r-value.

How does flow in a uniform channel with given q_r respond if it is progressively squeezed by a narrowing contraction? We describe this behaviour for the case of Figure 5.4, where $q_r = 0.5$ with $1/(1 + q_r) = 2/3$, and assume that the $\hat{Q}$ values at the upstream and downstream ends are small, but that r (the layer depths) can take a range of values. We begin with a value of $r > 2/3$ (a deep lower layer) at a point near the origin where $\hat{Q}_m$ is small. If the curve (5.23) for this r-value does not reach the critical line (5.19) because the gap is not small enough, the flow remains subcritical throughout, as noted above. If $\hat{Q}_m$ is increased (by decreasing b_m) to values above (5.19), the flow adjusts by sending a disturbance upstream to decrease the value of r, in order to keep the flow at $b = b_m$ critical, in a new steady state. The flow is now described by (5.23) with the new smaller r-value. The solution curve continues into the widening supercritical region (near the F_2^2-axis) and then adjusts with a hydraulic jump along a line of constant $\hat{Q}$, to the subcritical part of the curve. This gives subcritical flow at the downstream region (compatible with the downstream conditions), giving a flow of the form shown in Figure 5.2c. If $\hat{Q}_m$ is increased further, this process continues until r reaches the value $2/3$, where the subcritical part of the solution curve (5.23) has become a straight line. When $\hat{Q}_m$ for this state is exceeded, the flow on the downstream side adjusts to the state $u_1 - u_2 = 0$ everywhere, so that the baroclinic motion has been "squeezed out" of the flow. We may now increase $\hat{Q}_m$ to arbitrarily large values (narrower gaps) without changing this solution.

If $r < 2/3$ initially, then as b_m decreases the same picture applies for the solution curves near the F_2^2-axis of Figure 5.4. When the flow has become critical and $\hat{Q}_m$ is further increased, disturbances are sent upstream and r increases toward $2/3$. Here there is the difference that the limiting form of the curve at $r = 2/3$ has *two* points where the flow is critical, rather than one. The first point is a *virtual control*, upstream of which $\eta = v = 0$, and the second is the hydraulic transition at $\hat{Q} = \hat{Q}_m$ and larger F_2. Note that the flow is subcritical on both sides of the virtual control point (the intersection of the two straight lines). If $\hat{Q}_m$ is increased beyond this value, in these solutions the flow downstream of the virtual control must adjust to the "straight line" solution $\eta = u_1 - u_2 = 0$ with $r = 2/3$, as before. This is in spite of the fact that the contraction with the flow transition is narrower than that at the virtual control.

Flow on the downstream side of the contraction depends on the downstream value of r (i.e. conditions in the downstream reservoir). In practice, in the virtually controlled case, experiments (Armi, 1986) show that even if r is close to the upstream value the flow is more likely to remain supercritical (on either branch), with a subsequent hydraulic jump as the channel expands, rather than decelerate to subcritical flow through the virtual control point.

We may note that the solution properties when q_r is held constant and $\hat{Q}_m$ is increased (§5.2.2) are generally similar to those when r is held constant (§5.2.1). The solution curves in Figure 5.4 for given q_r and r are the same as those for fixed r in Figure 5.3, for the same values. In each case the final flow state is determined by the fixed value of r or q_r respectively.

This formalism with the Froude number planes with fixed q_1 and q_2 has been extended to include bottom topography in the form of a sill at the point of minimum contraction by Armi & Riemenschneider (2008). This is significant because maximum height of a bottom sill and maximum contraction tend to coincide in practical situations. The presence of the sill destroys the symmetry between the upper and lower layers in the pure-contraction case, and these differences are well displayed in the relevant Froude-number planes. The procedure can be applied to both unidirectional and exchange flows.

5.2.3 *The limit* $d_2 \to \infty$

If the depth of the upper layer becomes infinite, the velocity of the upper layer may still have some effect on the motion of the lower layer. However, with the hydrostatic approximation the converse does not apply, and a value for $u_2(x)$ may be specified independently. If $d_1 = D_1 + \eta$, and $u_i = u_{i0}$, for $i = 1, 2$, in the upstream reservoir, which may or may not be infinitely wide, (5.22) becomes

$$\frac{1}{2}u_1^2 + g'\eta = \frac{1}{2}(u_{10}^2 - u_{20}^2) + \frac{1}{2}u_2(x)^2, \qquad (5.24)$$

and the critical condition (5.23) is

$$\frac{u_1^2}{g'(D_1 + \eta)} = 1. \qquad (5.25)$$

If the downstream conditions are such that the flow does become critical at the minimum gap, these give at $b = b_{\mathrm{m}}$,

$$\eta = -\frac{1}{3}D_1\left[1 - \frac{u_{10}^2 - u_{20}^2 + u_2(x)^2}{g'D_1}\right], \qquad (5.26)$$

$$u_1^2 = \frac{2}{3}gD_1 + \frac{1}{3}[u_{10}^2 - u_{20}^2 + u_2(x)^2]. \qquad (5.27)$$

We may identify two special cases. First, u_2 is constant throughout, so that the upper layer is not affected by the contraction; these equations are then the same as those for a single layer (2.88), with g' replacing g. Second, $u_2 = u_1$ throughout, as may occur if both layers experience the same contraction. In this case $\eta = 0$ throughout, and $u_1^2 = g'D_1$ at $b = b_{\mathrm{m}}$. Grace (1991) has reported observations of a stationary hydraulic jump downstream of flow of cold foggy air through a topographic gap over water (Backstairs Passage), where the first of these cases applies, and a $1\frac{1}{2}$-layer model seems to be adequate to describe the phenomenon.

When the flow is supercritical, the assumptions of uniform conditions of flow across a channel of varying breadth may not be valid if the channel is wide or the changes are too rapid. In particular, the wave disturbances produced by an isolated side-boundary described in §3.5.4 should also be found in systems where the $1\frac{1}{2}$-layer model is applicable.

Such phenomena have in fact been observed in the atmosphere off the coast of northern California (Winant et al., 1988). Suitable conditions occur over the sea adjacent to a coastline with relatively steep topography (the sidewall), where there is a lower layer of cold air driven (at supercritical Froude number) along the coastline by a large-scale pressure field, and surmounted by a warmer deep upper layer. Density profiles of this type may occur for many reasons, but in the northern California case they arise because of a relatively cold sea with subsiding upper level air in a synoptic high-pressure system. In this locality, with supercritical southward motion of the lower layer, an expansion fan (yielding flow with larger F – see §3.5.4) has been observed at a convex region (Point Arena), together with a hydraulic jump further downstream emanating from a concave region (Stewarts Point), giving subcritical flow on its downstream side. The same phenomena are probably common in other locations where the atmospheric conditions are similar, and Winant et al. speculate that they also occur off the western coasts of South America, North and South Africa, and off Somalia, among others. These concepts can also be helpful in interpreting an observed flow structure when the boundary variations are not so simple.

5.3 Non-linearity with dispersion in contractions

In channels where the cross-sectional variation is small and gradual but not hydrostatic, and the flow is close to critical, the flow may be described by an feKdV equation of an analogous form to (4.74) (Tomasson & Melville, 1991; Clarke & Grimshaw, 1994). As with all models of this type, the formal validity of this equation depends on suitable relative magnitudes of these three small parameters (dispersive effects, lateral variation and departure from criticality). When the cubic non-linear term is not important, the principal change relative to the hydrostatic model is that the hydraulic jumps upstream and downstream are replaced by undular bores in the form of modulated wavetrains, as described in §3.4.

In the special case where the channel walls are vertical as in §5.2, with the Boussinesq approximation the fkdV equation does not apply to the flows where $\eta = 0$, v $= 0$ (i.e. $u_1 = u_2$) everywhere. However, if the Boussinesq approximation is not made, a modified form of this equation may be derived, where the term $(U - c_0)\eta_x$ is replaced by one of the form $[(U - c_0)\eta]_x$ (Clarke & Grimshaw, 1994). The resulting perturbations from a contraction with $u_1 = u_2$ initially now have a signal in η, v governed by this equation. The solution has a similar form to those for flows governed by (4.74), but the perturbation is smaller in magnitude by a factor $(\Delta\rho/\rho)^{1/2}$, and develops on a correspondingly longer time-scale.

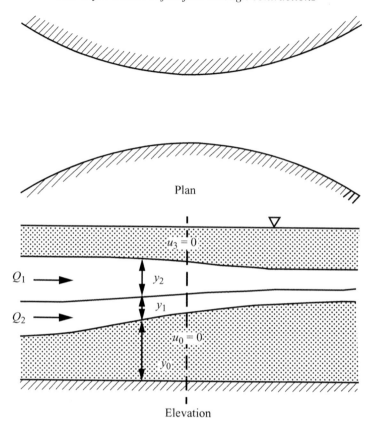

Figure 5.5 Two layers flowing through a contraction. (From Williams & Armi, 1991, reproduced with permission.)

5.4 Multi-layered flow through contractions

In reservoirs and cooling ponds it is sometimes necessary to withdraw water (for external purposes) from some particular level, or from a chosen range of densities. This sets up a flow field in a restricted range of densities that is known as a withdrawal layer, with the difference from the above flows being that there is now stationary fluid above and/or below this layer. In general the resulting flow may be complex, but if the sink is separated from the main body of the (large) reservoir by a contraction (such as that shown in Figure 5.2a), the resulting steady-state flow may be calculated in a similar manner to that described above. The simplest case of interest consists of two flowing layers, as shown in Figure 5.5. All fluid in the large upstream reservoir is assumed to be effectively stationary. The flow is assumed to be steady and hydrostatic – no variation with time is considered, horizontal length-scales are long, and the Boussinesq approximation is not made.

The system has four layers of comparable thickness, numbered from 0 to 3, with the uppermost (layer 3) and lowest (layer 0) stationary. The symbols u_i, d_i, D_i and ρ_i denote respectively the velocity, thickness, upstream reservoir thickness and the density of the ith layer. The volume flow rate in each of the flowing layers is determined by the strength of withdrawal, and is assumed to be uniformly constant:

$$q_i = u_i d_i b, \qquad i = 1, 2 \tag{5.28}$$

where $b(x)$ is again the channel width. The lower stagnant layer may be removed, and the upper stagnant layer may be absent initially, but if both are entirely absent the flow is the same as that described in §5.2. The Bernoulli equations for the three lowest layers ($i = 0, 1, 2$) may be written as

$$
\begin{aligned}
(\rho_2 - \rho_3)&g d_2 + (\rho_1 - \rho_3)g d_1 + (\rho_0 - \rho_3)g d_0 \\
&= (\rho_2 - \rho_3)g D_2 + (\rho_1 - \rho_3)g D_1 + (\rho_0 - \rho_3)g D_0, \\
\tfrac{1}{2}\rho_1 u_1^2 &+ (\rho_2 - \rho_3)g d_2 + (\rho_1 - \rho_3)g(d_0 + d_1) \\
&= (\rho_2 - \rho_3)g D_2 + (\rho_1 - \rho_3)g(D_0 + D_1), \\
\tfrac{1}{2}\rho_2 u_2^2 &+ (\rho_2 - \rho_3)g(d_0 + d_1 + d_2) \\
&= (\rho_2 - \rho_3)g(D_0 + D_1 + D_2).
\end{aligned}
\tag{5.29}
$$

The first of these equations may be used to eliminate d_0, so that we obtain four equations for the variables u_1, u_2, d_1 and d_2, in terms of q_1, q_2, the densities and the conditions in the reservoir, $D_0, \ldots, D_3$. These in turn give equations for the variables d_1, d_2 in the form

$$
\begin{aligned}
\frac{1}{2}\frac{\rho_1}{\beta_{12}g}\left(\frac{q_1}{b d_1}\right)^2 + A d_1 + B d_2 &= A D_1 + B D_2, \\
\frac{1}{2}\frac{\rho_2}{\beta_{12}g}\left(\frac{q_2}{b d_2}\right)^2 + B d_1 + C d_2 &= B D_1 + C D_2,
\end{aligned}
\tag{5.30}
$$

where $\beta_{mn} = \rho_m - \rho_n$, and we have

$$
A = \frac{\beta_{13}}{\beta_{12}}\left(1 - \frac{\beta_{13}}{\beta_{03}}\right), \quad
B = \frac{\beta_{23}}{\beta_{12}}\left(1 - \frac{\beta_{13}}{\beta_{03}}\right), \quad
C = \frac{\beta_{23}}{\beta_{12}}\left(1 - \frac{\beta_{23}}{\beta_{03}}\right),
\tag{5.31}
$$

which are all constants depending on the densities (if the density differences at the three interfaces are the same, we have $A = 2/3$, $B = 1/3$, $C = 2/3$). Taking derivatives, we then obtain

$$
\frac{d d_1}{dx} = \frac{1}{b}\frac{db}{dx}\frac{\Delta_2}{\Delta_1}, \qquad
\frac{d d_2}{dx} = \frac{1}{b}\frac{db}{dx}\frac{\Delta_3}{\Delta_1},
\tag{5.32}
$$

where

$$
\begin{aligned}
\Delta_1 &= (A - F_1^2)(C - F_1^2) - B^2, \qquad \Delta_2 = (C - F_2^2)F_1^2 d_1 - B F_2^2 d_2, \\
\Delta_3 &= (A - F_1^2)F_2^2 d_2 - B F_1^2 d_1,
\end{aligned}
\tag{5.33}
$$

and F_1, F_2 are Froude numbers defined as

$$F_1^2 = \frac{\rho_1 u_1^2}{\beta_{12} g d_1}, \qquad F_2^2 = \frac{\rho_2 u_2^2}{\beta_{12} g d_2}. \tag{5.34}$$

As in single-layer flow through contractions (§2.3.3), at the point of minimum width where $db/dx = 0$, we must have $\Delta_1 = 0$, and where Δ_1 vanishes the numerators must also vanish. For single-layer flows, this specifies a critical flow condition. In the present system with three interfaces (four layers), there are three internal wave modes (see §8.1.1 – for the first mode the three interfaces move together; the second is the "sausage" mode, and in the third the three interface displacements alternate). With mean motion only in the central two of the four layers (and the four layers have comparable thickness), only the second and third modes can become stationary, and this is expressed in the condition $\Delta_1 = 0$, which specifies the critical conditions for these two modes. The critical mode at $b = b_{\rm m}$ is the second mode; mode 3 is critical at some point upstream of the contraction minimum, and is determined by the condition that Δ_1, Δ_2 and Δ_3 are all zero.

After some algebra, in the same manner as in §5.1, one can show that at both of these critical points the layer thicknesses d_1 and d_2 are in the same ratio as in the reservoir: $d_1/d_2 = D_1/D_2$, and that

$$d_1/D_1 = d_2/D_2 = 2/3, \quad \text{at } b = b_{\rm m}. \tag{5.35}$$

Further, since the fluxes Q_1 and Q_2 are the same for all sections, the velocities u_1 and u_2 must have the same ratio at these points. There is a solution of the above equations that gives the same ratios (for the u_i, d_i) for all positions between the reservoir and contraction minimum, and Q_1 and Q_2 are given in terms of the reservoir conditions and $b_{\rm m}$ in the form (Wood, 1968; Binnie, 1972; Williams & Armi, 1991)

$$
\begin{aligned}
q_1 &= \left(\frac{2}{3}\right)^{3/2} \left[\frac{\beta_{12} g}{\rho_1} \left(\frac{\rho_2 + (\rho_1 + \rho_2) D_1/D_2}{\rho_0 + \rho_1 + \rho_2}\right)\right]^{1/2} b_{\rm m} D_1 D_2^{1/2}, \\
q_2 &= \left(\frac{2}{3}\right)^{3/2} \left[\frac{\beta_{23} g}{\rho_2} \left(\frac{\rho_1 + \rho_0(1 + D_1/D_2)}{\rho_0 + \rho_1 + \rho_2}\right)\right]^{1/2} b_{\rm m} D_2^{3/2}.
\end{aligned}
\tag{5.36}
$$

This flow has been termed *self-similar* (Wood, 1968; Benjamin, 1981) because the vertical flow profile remains similar to itself (the thicknesses of the layers are always in the same ratio) at all positions, and is a manifestation of the same phenomenon as in §5.1.

This form of analysis may also be applied to "plunging flows" through a contraction, in which the lower stagnant layer is absent, and the upper stagnant layer is only present in part of the flow, downstream of a separation point, S. Where the stagnant layer is present there are two interfaces, and hence two internal wave modes, 1 and 2, but upstream of the stagnant layer there is only mode 1. Both moving layers become thinner and the fluid accelerates as it enters the contraction, and the stagnant upper layer increases in thickness. Here, mode 2 is critical at the upstream "virtual control", and mode 1 is critical at the minimum width of the contraction. If the total flow rate is progressively increased, in the succession of resulting steady flows the separation point S moves progressively downstream, past the virtual control point, and then past the main control point at the maximum contraction. The corresponding steady flows may be obtained by the same form of analysis as that described above, noting that the number of control points changes from 2 to 1 to 0 (Williams & Armi, 1991). It may also be applied to situations where separation points occur in the lower layer. Experiments with controlled two-layer flow inputs through a contraction that realise flows consistent with these equations have been described by Williams & Armi (1991) for a variety of flow conditions.

5.5 Continuously stratified flow through contractions

The above layered self-similar flow solution can be generalised to arbitrary (stable) density profiles, as follows. We assume the same geometry as shown in Figure 5.6, but with an arbitrary stable density profile in the reservoir for $0 < z < z_0$, with stationary homogeneous fluid (of density ρ_s) at higher levels. The coordinate x again denotes the direction of flow of fluid with density $\rho > \rho_s$ toward and through a contraction. The hydrostatic pressure in the reservoir relative to the level z_0 is then

$$p(z) = g \int_z^{z_0} \rho(z')dz'. \tag{5.37}$$

At each point within the contraction, the elevation of the stream surface emanating from level z in the reservoir is denoted $y(x, z)$, and the local pressure is given by

$$p(y) = g \left((z_0 - y(x, z_0))\rho_s + \int_y^{y(x,z_0)} \rho(x, y')dy' \right), \tag{5.38}$$

where all fluid above the level $y(z_0)$ is assumed to be stationary with density ρ_s. The Bernoulli equation for the flow on each stream surface is

$$\frac{1}{2}\rho u^2 + \rho g y(x, z) + p(y(x, z)) = H(z) = \rho g z + p(z), \tag{5.39}$$

which by integrating by parts may be expressed as

$$\rho u(x, y(x, z))^2 = 2g \left[(z_0 - y(x, z_0))\Delta\rho - \int_z^{z_0} (z' - y(x, z')) \frac{d\rho(z')}{dz'} dz' \right], \tag{5.40}$$

where $\Delta\rho = \rho(z_0) - \rho_s$. For steady flow the volume flux $q(x, z)$ on each stream surface must be constant, so that the flux within a layer of thickness $dy(x, z)$ must be

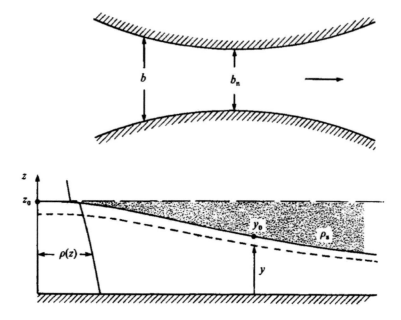

Figure 5.6 Stratified fluid flowing through a horizontal contraction. (From Armi & Williams, 1993, reproduced with permission.)

the same as the corresponding layer of thickness dz, so that $q(z)dz = u(x, z)dy(x, z)$, and

$$q(z) = u(x, z)b(x)\frac{dy(x, z)}{dz}. \qquad (5.41)$$

Hence we may write

$$f(z) = \frac{\rho(z)q(z)^2}{gb_m^2}$$

$$= 2\frac{b(x)^2}{b_m^2}\left(\frac{dy}{dz}\right)^2\left[(z_0 - y(x, z_0))\Delta\rho - \int_z^{z_0}(z' - y(x, z'))\frac{d\rho(z')}{dz'}dz'\right]. \qquad (5.42)$$

The x-dependence in this equation is contained in $b(x)$ and $y(x, z)$. Writing $\lambda(x) = (b(x)/b_m)^2$, the self-similar solution has the form

$$y = k(\lambda)z, \qquad (5.43)$$

and substituting this into (5.42) gives

$$f(z) = 2\lambda k^2(1 - k)\left[z_0\Delta\rho - \int_z^{z_0}z'\frac{d\rho}{dz'}dz'\right]. \qquad (5.44)$$

This implies that $\lambda(k^2 - k^3)$ must be a constant (since the other terms in the equation are independent of x), which we need to determine. Regarding λ as a function of k, we must have a minimum value of λ of unity where $d\lambda/dk = 0$. With this condition we obtain $k = 2/3$ when $\lambda = 1$, so that the constant is determined to be $4/27$, and hence (Wood, 1968; Benjamin, 1981; Armi & Williams, 1993)

$$q(z)^2 = \frac{8}{27} \frac{g b_m^2}{\rho(z)} \left[z_0 \Delta \rho - \int_z^{z_0} z' \frac{d\rho}{dz'} dz' \right].$$

(5.45)

This self-similar solution is a generalisation of the two-layer flow solution above – the flow properties are determined by the density profile in the reservoir, and the minimum width of the contraction. As with the two-layer flow, it is the only steady solution available for such flows. It also corresponds with the solution for homogeneous "single-layer" flow through contractions described in §2.3.3. This correspondence may be extended further to apply to self-similar stratified flows from reservoirs of stationary fluid with the general flow of homogeneous layers that satisfy the shallow water equations (Yih, 1969), of which flow through contractions is just one example.

One of the low order wave modes (generally mode 1 or 2) is critical at $b = b_m$, which is a control section for the flow. Which mode this is depends on whether the flow is concentrated near the upper or lower boundary (giving mode 1), or in the centre (mode 2), which in turn depends on the nature of the stratification in the reservoir. All the higher order modes must be critical at a succession of locations between the reservoir and the minimum cross-section. These are sometimes described as virtual controls, but they do not control the flow, though in the two-moving-layer case described above, they do enable the calculation of the flow properties. Also, these critical flow sections (both virtual and at $b = b_m$) are for different wave modes in each flow, which contrasts with those for two layers in §5.2, where they are all for the single mode.

The flow rate profile $q(z)$ in the above flows is determined by the density profile in the reservoir, and b_m. If a progressively larger flow rate is forced, by stronger withdrawal by sinks downstream, the increased flow must come from a wider range of densities. Flow changes through a continuous set of self-similar flows would result, until the flow becomes supercritical at $b = b_m$ for all internal wave modes. At this point the flow has become uniform (barotropic) upstream of the contraction, and the flow pattern downstream of it depends on the distribution of the sinks. Experimental results on a variety of such flows have been described by Armi & Williams (1993).

6

Exchange flows

All government, indeed every human benefit and enjoyment,
every virtue, and every prudent act,
is founded on compromise and barter.

EDMUND BURKE,
A Vindication of Natural Society, Vol. II.

The term "exchange flows" denotes situations where fluid flows in both directions between two regions or reservoirs, through some sort of constriction. The fluid is being "exchanged" in a general sense, and this exchange is usually driven by a difference in density. This behaviour is very common in nature. Flows through doorways and windows are examples, and are described in §6.3. On a larger scale, it is often found where an estuary, lake or lagoon opens to the sea. In a "normal" estuary, fresh water from rivers flows out to sea over heavier (colder, saltier) water below. This may cause mixing and turbulent entrainment of some of the underlying fluid, which increases the nett volume of the outflow. This entrained fluid is replaced by salty (or colder) water flowing inshore below it, giving a two-way exchange.

Alternatively we may have the "inverse estuary" situation. This occurs in regions where there is little or no river flow but rather substantial evaporation in the shallow inshore region. This produces salty water that sinks and flows out to sea at low levels, being replaced by sea water flowing in at near-surface levels. In some places the river inflow or the evaporation is concentrated in a lagoon that is connected to the sea via a channel or constriction of some sort, which governs the exchange between the two regions.

Another example is two large bodies of water such as lakes connected by a narrow channel, where each lake has different evaporation or inflow properties. The idealised inviscid two-layer hydraulic model represents the simplest prototype for the essential dynamics, and this model is often quite realistic if two relatively homogeneous bodies of water are involved. The Strait of Gibraltar is a prominent location where this type of exchange flow exists, and to which this hydraulic model has been applied (Armi & Farmer, 1988).

A specific feature of exchange flows is the distinction between two types of flow, termed respectively *maximal* exchange, where the two-way flow is controlled by two critical flow conditions that are independent of the reservoirs or sources of fluid, and *submaximal* exchange, where only one critical condition is present, and conditions in (at least) one of the reservoirs influences the nett exchange rate. In flow with maximal exchange, the flow between the two critical sections is subcritical, and flow immediately outside this region is supercritical with possible hydraulic jumps.

6.1 Two-layer exchange flow in a uniform channel over topography

This situation is the same as that discussed in §4.5 except that the flow in one layer is now reversed. The problem of determining the resulting flow may be tackled by the same methods. In particular, we may define a Froude number F_0 based on the flat-bottom flow, and for small obstacle height h_m, the flow will be subcritical for $F_0 < 1$ and supercritical for $F_0 > 1$. Flow regimes where the lower layer is partially or totally blocked may be calculated for larger h_m, together with associated "upstream" jumps and rarefactions, provided that the flow is stable. These "upstream" disturbances may lie on either side of the obstacle, depending on the initial flow state. Some flow solutions have been described in terms of the (F_1^2, F_2^2)-plane by Farmer & Armi (1986).

6.2 Two-layer exchange flow through contractions

The description of this case builds on the development in Chapter 5, and is examined in some detail because of the insight it provides into the hydraulics of layered flows, and the elegant nature of the final results. We consider exchange flow between two large reservoirs through a contraction of uniform depth D with vertical sidewalls, where the lower-layer motion is from right to left (q_1, $u_1 < 0$), and the upper-layer motion from left to right (q_2, $u_2 > 0$), as depicted in Figure 6.1. As fluid flows through the contraction the speeds there increase, but if it is wide enough the flow there remains subcritical. The upper layer has depth d_{2l} on the left-hand side, and d_{2r} on the right. Here d_{2l} and d_{2r} are different and with the above choice of directions, we have $d_{2l} \geq d_{2r}$. The equations (5.13)–(5.21) are still applicable and the possibilities that $G^2 = 1$ at the minimum value of $b(x)$, and also where $F_1^2 \left(1 + \frac{d_1}{d_2}\right) = 1$, still hold.

The total flux Q and conditions in the two large reservoirs are regarded as being determined by external factors, and are not affected by conditions at the contraction. Since the total depth D is presumed to be constant, we may use $r_l = d_{1l}/D$ and $r_r = d_{1r}/D$ as the defining parameters for the reservoirs, although d_{2l} and d_{2r} are more appropriate if D varies and becomes infinite (downwards) in a reservoir, for example.

There are four possible categories of such exchange flows, of increasing complexity, which we will deal with in turn. In the first two of these the interface levels in the two reservoirs are equal, so that $r_l = r_r$; Category I then has $Q = 0$, and Category II has $Q \neq 0$. Categories III and IV have unequal reservoir levels so that $r_l < r_r$ with $Q = 0$ in Category III and $Q \neq 0$ in Category IV. The conceptual device of examining how the flow changes as b_m is decreased is also useful here.

6.2.1 Category I: $r_l = r_r$, $Q = 0$, $h(x) = 0$

As in Chapter 5, we begin by considering the nature of the changes to the flow between two reservoirs with a very wide contraction. From (5.13)–(5.21), the flow is everywhere subcritical with little displacement of the interface, with equal fluxes q_1 and q_2. We then progressively decrease the minimum gap width, b_m. From (5.20), (5.21) we may determine the value of b_m where the flow becomes critical, and this is given in Figure 6.1 as a function of $\hat{q}_{1m} = |q_1|/b_m D\sqrt{g'D}$ and r in the reservoirs ($\hat{q}_i, \hat{q}_{im}$, with $i = 1,2$, are defined in the same manner as $\hat{Q}, \hat{Q}_m$). For values of $\hat{q}_1$ below this curve, the flow is everywhere subcritical.

However, once the curve is reached, further reduction in b_m must cause $|q_1|$ (and hence $|q_2|$) to decrease in proportion so that the flow remains critical at $b = b_m$, and to conform to (5.20)–(5.21), q_{1m} remains equal to the value on the curve for given r. This is achieved by sending a wave "upstream", to the right if $r < 0.5$, to the left if $r > 0.5$, and on the "downstream" side the flow becomes supercritical with a stationary hydraulic jump, as in §5.2. If $r = 0.5$, hydraulic jumps do not form and the interface remains horizontal. Hence, if the minimum gap width b_m is decreased beyond the minimum value for critical flow, the local conditions (u_i, d_i) at $b = b_m$ remain constant and critical. This causes the fluxes q_1 and q_2 to decrease in proportion to b_m, for all r. This behaviour is similar to that of a single layer flowing through a contraction, as described in §2.3.3.

We note in passing that if q_1 and q_2 are fixed and r_l and r_r are allowed to vary (as for fixed q_1, q_2 in Chapter 5), the behaviour is different. When the critical curve of Figure 6.1 has been reached, as b_m is decreased the R-value on the "upstream" side migrates along it to $r = 0.5$, where it remains. The flow now has $u_1 = -u_2$, $\eta = 0$ everywhere, and this state is maintained as b_m is decreased further, provided the flow is stable.

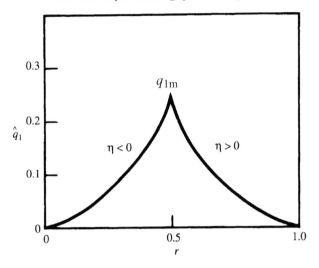

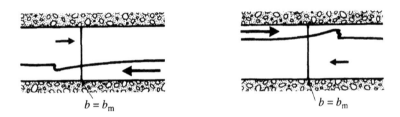

Figure 6.1 Exchange flow (category I) through a contraction between identical reservoirs ($r = r_1 = r_r$), with zero nett transport ($Q = 0$). Either $\hat{q}_{im} = |q_1|/(bD\sqrt{g'D}$ must lie on the curve (where the flow is critical at $b = b_m$, specifying a maximum value of q_1 or q_2 for given b_m), or below it, in which case the flow is everywhere subcritical. As in Chapter 5, $d_1 = d_{10} + \eta$.

6.2.2 *Category II:* $r_1 = r_r$, $Q \neq 0$, $h(x) = 0$

We follow the same procedure as above, and consider the changes that occur as b_m is progressively decreased. Here we have

$$Q = q_1 + q_2, \tag{6.1}$$

where Q may be positive or negative and is imposed by external factors, and q_1 and q_2 are initially negative and positive respectively. The terms Q, D and r are constants in this squeezing process, but q_1 and q_2 may vary. We scale the contraction width b by $|Q|/D\sqrt{g'D}$, defining $\hat{Q} = |Q|/bD\sqrt{g'D}$, $\hat{Q}_m = |Q|/b_mD\sqrt{g'D}$ as before, and the curves for critical flow (obtained as above), expressed in terms of q_2/q_1, are shown in Figure 6.2 for various values of r. The ratio q_1/q_2 is plotted in the upper

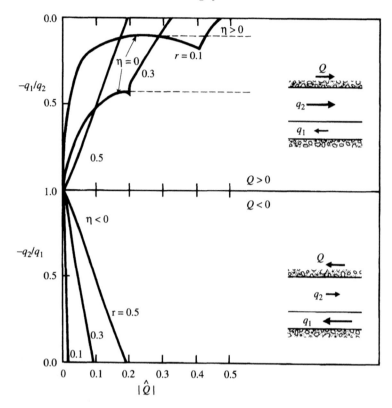

Figure 6.2 Exchange flow (category II) through a contraction between identical reservoirs ($r = r_1 = r_r$), with non-zero nett transport ($Q \neq 0$): the conditions for critical flow. For a state with given initial q_1 and q_2, the flow is everywhere subcritical if $\hat{Q}_m$ is less than the value on the curve for the appropriate value of r. For larger $\hat{Q}_m$, q_1/q_2 must adjust to keep the flow critical at $b = b_m$. The dashed lines denote the flows $u_1 = -u_2$. The curves for $r = 0.7, 0.9$ (not shown) are mirror images of those for $r = 0.3, 0.1$, respectively, about the centre-line $q_1/q_2 = -1$.

half of the diagram (where $Q > 0$), and q_2/q_1 in the lower half (where $Q < 0$). If $\hat{Q}_m$ lies to the left of the curve (for given r), the flow is everywhere subcritical, and it becomes critical at $b = b_m$ when the curve is reached.

If $\hat{Q}_m$ is increased further, the flow must vary q_2/q_1 to keep the flow critical at $b = b_m$. When these curves on Figure 6.2 reach the upper and lower limits on the diagram where q_1 or q_2 are zero, they are contiguous with the corresponding curves for the same r value in Figure 5.3. These two junctions between the two figures create a continuous "circular" ordinate for the joint figure, the top of Figure 6.2 joining on to the bottom of Figure 5.3 and the bottom of Figure 6.2 joining with the top of Figure 5.3 (allowing for a change in flow direction). The implications of this are as follows.

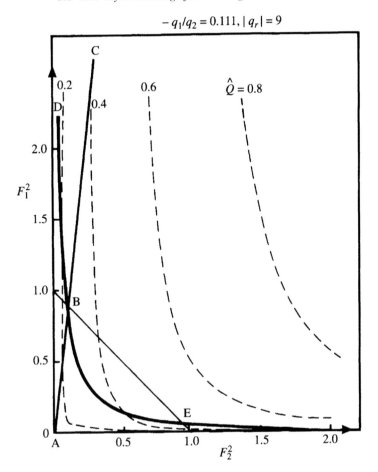

Figure 6.3 (F_1^2, F_2^2)-diagram for exchange flow with $q_r = -q_2/q_1 = 9$, showing the solution curves for $r = 1/(1 + q_r) = 0.1$ (heavy solid curves) and curves of constant $\hat{Q}$ (dashed).

We first discuss the case with $r = 0.5$. Here η is negative for $Q < 0$ and positive for $Q > 0$, for wholly subcritical flows. As $\hat{Q}_m$ increases to and past the point where the flow becomes critical, the flow develops the same sub/supercritical transitions with downstream hydraulic jumps as shown in Figure 6.1, for η of the same sign. With increasing $\hat{Q}_m$, the flow adjusts q_1 and q_2 to keep the flow critical at $b = b_m$, with q_2 decreasing to zero if $Q < 0$, and q_1 decreasing to zero if $Q > 0$. Further increase in the value of $\hat{Q}_m$ causes q_1 or q_2 to change sign and the flow to "transfer" to Figure 5.2, where the discussion of §5.2 then applies. The flow therefore ultimately tends to the state of two co-flowing layers with equal velocity and thickness, which is reached when $\hat{Q}_m = 0.5$.

When $r \neq 0.5$ the curves are more complex, but they are topologically equivalent to each other for all r, and we may take $r = 0.1$ as a representative example. For $Q < 0$ (and $r = 0.1$), the behaviour is essentially the same as for $r = 0.5$: once the critical curve in the lower part of Figure 6.2 is reached, further increase in $\hat{Q}_m$ causes $|q_2/q_1|$ (and hence $|q_2|$) to decrease to zero and then become positive. The flow is represented by the corresponding curve on the upper part of Figure 5.2, and the subsequent behaviour is described in §5.2. The same applies, *mutatis mutandis*, if $Q > 0$ and $(-q_1/q_2) < 0.111$; here $\eta > 0$, the upstream wave propagates to the left as in Figure 5.1c, and the flow transfers to Figure 5.2 as for $r = 0.5$.

For $(-q_1/q_2) > 0.111$, however, the situation is different. Here $\eta < 0$, and the upstream wave and the downstream hydraulic jump propagate to the right as for $Q < 0$ (with the form depicted on the left side of Figure 6.1). As $\hat{Q}_m$ increases to 0.24, the amplitude of the interface displacement decreases, and the flow evolves to the state at $(-q_1/q_2) = 0.111$ where $u_2 = -u_1$ and $\eta = 0$ on the "upstream" side to the right of the critical section at $b = b_m$. This is represented by the curve ABD in Figure 6.3, the (F_1^2, F_2^2)-diagram for this value of q_1/q_2, showing the solution curves for $r = 0.1$. In addition to ABD there are two other solution curves: ABC and ABE. The first, ABC, is the flow state $u_2 = -u_1$ everywhere, and is represented by the line $F_1^2 = -(q_2/q_1)F_2^2$. This is a special flow, in that the requirement for critical flow at $b = b_m$ disappears. From (5.14) we have

$$u_2 = -u_1 \quad \text{everywhere} \Rightarrow \eta = 0, \quad \text{and} \quad \frac{1}{v}\frac{\partial v}{\partial x} = -\frac{1}{b}\frac{\partial b}{\partial x}, \qquad (6.2)$$

so that

$$vb = \text{constant}, \qquad (6.3)$$

in this solution. The curve ABE corresponds to the critical solution for $(-q_1/q_2) < 0.111$. If $\hat{Q}_m$ is increased further (> 0.24), the first of these two solutions is the most likely to occur from the process of gradual increases of $\hat{Q}_m$, although the second is possible if the flow is suitably disturbed. The first solution is possible for (arbitrarily) larger values of $\hat{Q}_m$ and is represented by the dashed line in Figure 6.2, but in practice it will be limited by shear-flow instability, since v increases without limit as b_m decreases. We may speculate that this may provide a disturbance that causes the flow to "jump" to the other solution with $\eta > 0$ when $\hat{Q}_m$ is large enough, to eventually reach the state where v = 0, as before.

It follows that if the conditions within two effectively infinite reservoirs are the same, when some arbitrary flow with $Q \neq 0$ is established between them and the contraction is then squeezed by reducing b_m, the flow tends to a state where either $u_1 = u_2$ or $u_1 = -u_2$, depending on r and the initial value of q_1/q_2. If the latter occurs, further squeezing may eventually also result in the state $u_1 = u_2$. Note that this result does *not* occur when $Q = 0$ (category I), which is an exceptional special case in which critical flow at $b = b_m$ is maintained as b_m decreases.

6.2.3 *Category III:* $r_1 < r_r$, $Q = 0$, $h(x) = 0$

In situations where the levels of the interface in the two reservoirs are unequal, steady flows that are everywhere subcritical are not possible. If a subcritical flow were to be taken as an initial condition, the resulting pressure gradient would accelerate the flow to the point where the reservoir conditions became equal (if this were possible), or the flow would be controlled or limited by some condition. It follows that we cannot approach the understanding of these flows by starting with a very broad contraction with a simple steady subcritical flow, as was done above, and then following its development as the minimum width b_m decreases. Instead, we begin by considering the possible conditions at the point of maximum contraction. From (5.17) and (4.25) we see that at $b = b_m$, we must have either $G^2 = 1$ or $d/dx(\eta, v) = 0$. With a fundamental asymmetry in the imposed conditions and $Q = 0$, the latter condition does not occur. Hence, in the case of reservoirs joined by a contraction, the controlling condition is achieved by a critical flow condition, which limits the rate of exchange of the fluids.

We may identify three different types of flow, termed type L for "lower subcritical", type U for "upper subcritical", and type M for maximal exchange respectively. Type L flows are those where the narrowest section at $b = b_m$ is subcritically connected to the right-hand reservoir (with thinner *lower* layer, as in the left-hand diagram of Figure 6.1); in type U flows it is subcritically connected to the left-hand reservoir (with thinner *upper* layer, as in the right-hand diagram of Figure 6.1), and for type M it is subcritically connected to neither. Solutions for this configuration of the flow are best described in terms of the interface levels in the reservoirs, specified in terms of r_1 and r_r. As for layers that are flowing in the same direction, solution curves for exchange flows that are connected subcritically to a reservoir (i.e. types L and U) may be expressed in the form (5.23) (the direction is immaterial because the velocity terms are squared) where here $q_r = -q_2/q_1$ (defined so that q_r is always positive), and $r = r_1$ or r_r depending on circumstances. Here $r = r_1$ for type L, and $r = r_r$ for type U.

Since q_r is fixed (and equal to 1), these solution curves may be shown on a single (F_1^2, F_2^2)-diagram as in Figure 6.4, which shows a number of possible flows. All of these solutions are critical at $b = b_m$, and hence must cross or touch the critical line $F_1^2 + F_2^2 = 1$. If $r = r_r \ll 1$ the solution curve (for type L) is close to the F_1^2-axis, then $\eta < 0$ in the contraction, and a stationary hydraulic jump exists on the left-hand side, as shown in (a), (b) and (c) of Figure 6.4. There is only one value of the flux q_1 for given b_m. The location and intensity of the hydraulic jump is determined by the requirement that, across the jump, the flow must make the transition along a curve of constant $\hat{q}_1$, given by

$$F_1^{-2/3} + F_2^{-2/3} = q_1^{-2/3}, \qquad (6.4)$$

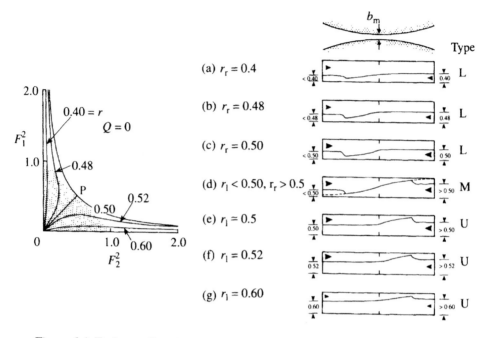

Figure 6.4 Exchange flow (category III) through a contraction between unequal reservoirs with zero nett transport. *Left*: the Froude number plane of solutions. *Right*: sideviews (a)–(g) showing some representative interface profiles and flow types for given values of r_1 and r_r with $r_1 < r_r$. (a), (b) and (c) are type L; (e), (f) and (g) are type U; and (d) is type M. (Adapted from Armi & Farmer, 1986.)

to another solution curve (5.23) that has $r = r_1$. As r_r increases, this type L is obtained up to $r_r = 0.5$, and the magnitude of $\hat{q}_{1m}$ increases up to this point. Similarly if r_1 is close to 1, the same situation applies in reverse: solution curves for type U are possible for $r_1 > 0.5$, and are illustrated in (e), (f) and (g) of Figure 6.4. They have $\eta > 0$, lie close to the F_2^2-axis, and are subcritical on the left-hand side with $r = r_1$. The hydraulic jump now lies on the right. The ranges of values of r_1 and r_r where solutions of types L and U are found are shown in Figure 6.5. The value of $\hat{q}_{1m}$ for type L for $r_1 < r_r$ is given by the curve in Figure 6.1 with $r_r = r$, and the same behaviour applies for the other regimes L and U, *mutatis mutandis*.

The third type of solution, type M, is found in the range $0 \le r_1 \le 0.5 \le r_r \le 1$. This flow is also critical at $b = b_m$, but is supercritical in different ways on each side of this section. Waves cannot propagate *toward* the critical section from either the left or the right, but can propagate *away* from it on both sides. The solution is illustrated in Figure 6.4d, and it generally has a hydraulic jump on each side to return the flow (along curves of constant $\hat{q}_1$) to solutions (5.23) with $r = r_1$ and r_r that can connect to the reservoir conditions. It is represented in the Froude

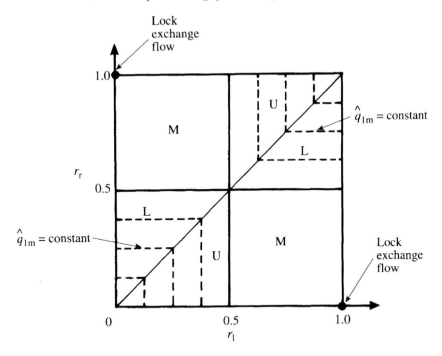

Figure 6.5 Category III as for Figure 6.4, showing the ranges of r_1 and r_r where flow types L, U and M are obtained for $Q = 0$. The straight line $r_1 = r_r$ denotes category I.

number plane by the curve through point P that is tangent to the critical line. In the special case known as "lock exchange flow", where $r_1 = 0$, $r_r = 1$, so that the fluid is exchanged between two homogeneous reservoirs, there are no hydraulic jumps and this curve describes the whole solution. This type M solution has a constant value of $\hat{q}_{1m} = 0.25$ throughout the region where it is found on the (r_1, r_r)-plane (Figure 6.5), and this is the maximum value of $\hat{q}_1$ possible for any r_1, r_r when $Q = 0$. This implies that q_1 and q_2 are the largest possible for given values of b_m, D, ρ_1 and ρ_2 (and $Q = 0$), and hence these flows are termed *maximal exchange* solutions. For the same parameters, flow types L and U have smaller magnitudes of q_1 and q_2, and are termed *submaximal*.

6.2.4 Category IV: $r_1 < r_r$, $Q > 0$, $h(x) = 0$

The situations discussed in the preceding sections lead, with a sense of climax, to this most general case of exchange flow between two unequal reservoirs with nett flux between them. Relative to category III we now have an additional parameter, the total mass flux Q. The externally imposed flow conditions are now specified by the three independent dimensionless parameters r_1, r_r and $\hat{Q}_m = Q/b_m D\sqrt{g'D}$, and we will express the flow properties in terms of them. We concentrate on the situation where $r_1 < r_r$ and $Q > 0$; other situations may then be inferred from symmetry in the vertical (up/down), since this is a Boussinesq fluid, and symmetry in the horizontal (left/right). The flux Q is related to $q_r(= -q_2/q_1)$ by

$$Q = q_2(1 - 1/q_r) = q_1(1 - q_r),\tag{6.5}$$

and also

$$\hat{q}_{2m} \equiv q_2/(b_m D\sqrt{g'D}) = \hat{Q}_m q_r/(q_r - 1).\tag{6.6}$$

On the Froude number plane where $q_r = $ constant, the curves of constant $\hat{q}_{2m}$ (and $\hat{Q}_m$) are given by

$$\frac{1}{(q_r F_1)^{2/3}} + \frac{1}{F_2^{2/3}} = \frac{1}{\hat{q}_2^{2/3}}.\tag{6.7}$$

We again have the three types of solutions, namely L, U and M, defined in the same way, and we first look for solutions of types L and U. These are governed by (5.23) with $r = r_r$, r_1 respectively, and the critical condition (5.21) at $b = b_m$. The values of q_1 and q_2 may then be inferred from the curves of Figure 6.2 for given r, and the flows have the character shown in the insets in Figure 6.1 (type L on the left, type U on the right). When $\hat{Q}_m = 0$, type L is found where $0 \le r_1 \le r_r \le 0.5$; as $\hat{Q}_m$ increases in this region, q_1 decreases and q_2 increases, to the point where $q_r = 1/r_r - 1$, and $u_1 = -u_2$ at $b = b_m$. This marks the onset of type M flow. Solution of the above equations shows that this occurs where

$$\hat{Q}_m = (1 - 2r_r)[r_r(1 - r_r)]^{1/2},\tag{6.8}$$

and is independent of r_1. The curve (6.8) is shown in Figure 6.6a, and a three-dimensional perspective of the surface is given in Figure 6.6c.

From the properties of types L and U, we may see that type L has $u_1 + u_2 < 0$ everywhere, type U has $u_1 + u_2 > 0$ everywhere, and type M has $u_1 + u_2 = 0$ somewhere in the flow. These conditions help to distinguish the boundaries of these flow regions in $(\hat{Q}_m, r_1, r_r)$-space. If $\hat{Q}_m$ is increased beyond the values of (6.8), the point where $u_1 + u_2 = 0$ moves "upstream" from the narrowest point of the channel (to the left in the present configuration), to give a structure of the form shown in Figure 6.7a, where $q_r = 2$ and $\hat{Q}_m \approx 0.167$. This flow is doubly supercritical, as for $\hat{Q}_m = 0$ (shown in Figure 6.4d), but for $\hat{Q}_m > 0$ the two supercritical flow regions are separated by a subcritical region connecting the critical flow point at $b = b_m$, and the virtual control point where $u_1 + u_2 = 0$ (Wood, 1970). Small-amplitude disturbances produced outside this central subcritical region cannot propagate into it.

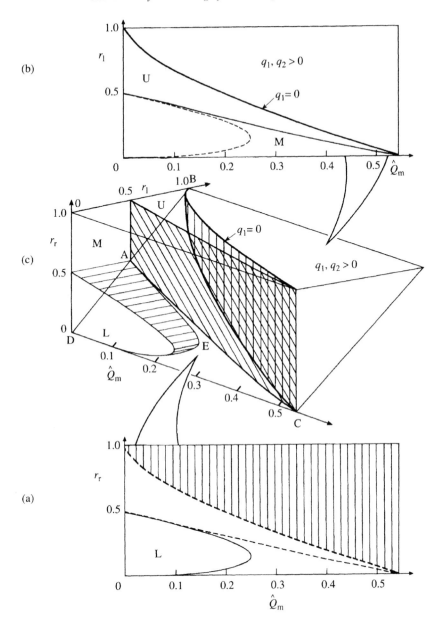

Figure 6.6 Exchange flow between reservoirs with $r_i < r_r$. Ranges of parameters r_1, r_r and Q_m where flow regimes L (lower subcritical), U (upper subcritical) and M (maximal), are obtained, are shown. (a) Boundary curves in the plane $r_1 = 0$; the dashed lines denote the horizontal projections of the lines AC and BC in panel (c). (b) Boundary curves in the plane $r_1 = 1$; the dashed line denotes the projection of the curve AED (which lies in the plane $r_r = 1$). (c) A three-dimensional perspective of the flow regimes L, U and M in $(r_1, r_r, \hat{Q}_m)$-space. The letters A, B, C, D and E refer to the points indicated.

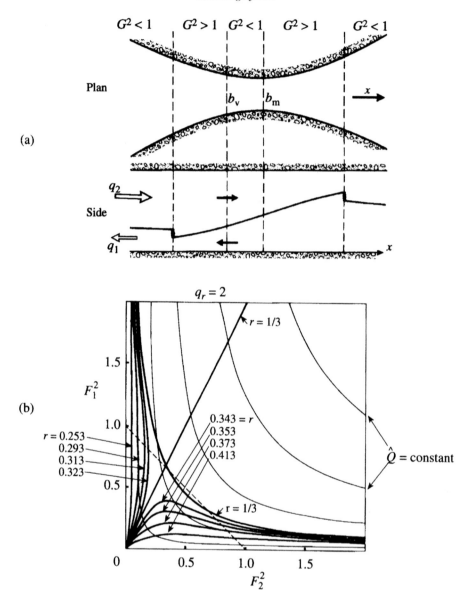

Figure 6.7 A schematic example of a flow solution with $\hat{Q} > 0$, $q_r = 2$, with the (F_1^2, F_2^2)-plane of possible solutions depending on r_l, r_r values. Open arrows denote fluxes, solid arrows denote velocities. (b) Here, r is constant for flow solutions between the hydraulic jumps, but because of the jumps it cannot be equated with r_l or r_r.

The conditions at the two critical points may be used to obtain the flow variables, including the fluxes in each layer. If u_i, d_i at $b = b_m$ and at the virtual control

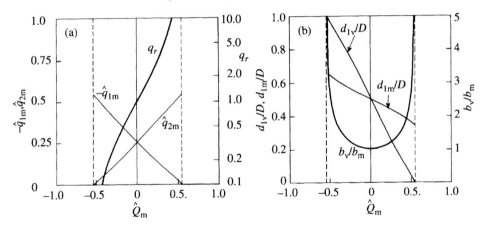

Figure 6.8 (a) Type M (maximal exchange) flow solutions between unequal reservoirs; $\hat{q}_{1m}$, $\hat{q}_{2m}$ and q_r, as functions of $\hat{Q}_m$. These relations hold within the region marked M in Figure 6.6c. (b) As for (a), but showing b_v/b_m, d_{1v}/D and d_{1m}/D.

$(b = b_v)$ are denoted respectively by u_{im}, d_{im} and u_{iv}, d_{iv} we have (Armi & Farmer, 1986)

$$d_{1m} + d_{2m} = d_{1v} + d_{2v} = D, \tag{6.9}$$

$$q_i = u_{im}d_{im}b_m = u_{iv}d_{iv}b_v, \qquad i = 1,2, \tag{6.10}$$

$$\frac{u_{1m}^2}{g'd_{1m}} + \frac{u_{2m}^2}{g'd_{2m}} = 1, \qquad \frac{u_{1v}^2}{g'd_{1v}} + \frac{u_{2v}^2}{g'd_{2v}} = 1, \tag{6.11}$$

$$u_{2v} = -u_{1v}, \tag{6.12}$$

$$Q = q_1 + q_2, \tag{6.13}$$

$$\frac{1}{2}(u_{2m}^2 - u_{1m}^2) + g'd_{2m} = g'd_{2v}. \tag{6.14}$$

Equation (6.14) follows from (6.12) and momentum conservation between the two critical flow points. These equations contain nine unknowns (u_{im}, d_{iv} and b_v), and if d_i is scaled with D, u_i with $\sqrt{g'D}$ and b_v with b_m, solving these equations enables the variables to be expressed as functions of $\hat{Q}_m$ alone. This means that the nature of the solution in the central section (near $b = b_m$) is independent of the precise values of r_l and r_r and depends only on $\hat{Q}_m$, although r_l and r_r determine the limits of this maximal exchange flow. Properties of the solutions of these equations are shown in Figure 6.8, up to the maximum value of $\hat{Q}_m = (2/3)^{3/2}$. Figure 6.8a shows q_r, and the fluxes q_{1m} and q_{2m}; the latter vary approximately linearly with $\hat{Q}_m$. Figure 6.8b shows b_v and the layer thicknesses at the critical points. The variation in b_v and d_{1v} with $\hat{Q}_m$ shows the considerable lateral movement of the virtual control point.

This central region must be terminated on each side by hydraulic jumps as before, in order to connect to subcritical reservoirs with given r_l and r_r. As $\hat{Q}_m$ increases, the virtual control point moves leftwards and the left-hand hydraulic jump moves rightwards and its amplitude decreases (the right-hand hydraulic jump, on the other hand, moves rightward and its amplitude *increases*). Where these two meet, the hydraulic jump amplitude has decreased to zero, and the left-hand reservoir is connected subcritically to the virtual control point. Here the F_{iv}^2 must satisfy (5.23) with $r = r_l$ and this together with (6.11)–(6.13) yields that for this condition we have

$$\hat{Q}_m = (1 - 2r_i)[r_i(1 - r_i)]^{1/2}b_v b_m. \tag{6.15}$$

This curve is shown in Figure 6.6b, and the surface in 3D perspective is drawn in Figure 6.6c.

If $\hat{Q}_m$ exceeds the value given by (6.15) the flow is of type U, as shown in Figure 6.6b,c. As $\hat{Q}_m$ increases further, u_1 and u_2 both increase to the point where $u_1 = 0$ (and $F_1 = 0$) everywhere. The equations for type U show that this occurs where

$$\hat{Q}_m = \left[\frac{2}{3}(1 - r_r)\right]^{3/2}. \tag{6.16}$$

If $\hat{Q}_m$ exceeds this value and $r_l > 0$, then u_1 becomes positive, and the flow is no longer an exchange flow. The equations of Chapter 5 will then apply, with the minor difference that the reservoir conditions are unequal. In the special case where $r_l = 0$ so that the depth of the denser layer in the left-hand reservoir is zero, when $\hat{Q}_m > (2/3)^{3/2}$ the steady-state solutions show that the lower layer remains at rest and a "front" forms where the interface intersects the bottom. This front moves to the right as $\hat{Q}_m$ increases, as described in some detail by Armi & Farmer (1986) for the lock exchange case ($r_l = 0$, $r_r = 1$).

We may note from all of the above that if $Q \neq 0$, $r_l \neq 0$, the result of progressively reducing the minimum gap width b_m is to reduce the exchange between the reservoirs to zero at a finite value of b_m. With further reduction the flow becomes unidirectional and then uniform, with two co-flowing layers with $\eta = v = 0$, $r = r_l$ everywhere upstream of a downstream hydraulic jump. The latter enables the flow to adjust to the conditions of the downstream reservoir, and apart from this the internal or interfacial motion has been "squeezed out" of the system.

There are numerous complications that can arise in practical circumstances, and the above idealised model provides only an introductory framework for the basic dynamics. However, the concept of maximal exchange, where the fluxes in the two layers are determined by the dynamics in the contraction and are largely independent of the properties in the reservoirs (within limits), may be extended to include other factors such as friction and time-dependence (Armi & Farmer, 1987). The principal requirement for this is that the central flow remain isolated from the reservoirs on each side by a supercritical region.

In many practical instances, such as a lagoon opening to the sea, one or both reservoirs may be finite in size. The above results for infinite reservoirs may be applied to finite reservoirs where the properties change with time by allowing r_l (and/or r_r as appropriate) to vary slowly, in a manner proportional to the flow rate, so that the solution varies with time in a quasi-static manner. Another complication is that a channel may have more than one contraction or ridge, and the Strait of Gibraltar is a prime example. This may be treated by an extension of the methods described above, and the reader is referred to Farmer & Armi (1986) for further details. Whether the actual flow through the Strait of Gibraltar is maximal or submaximal has been a matter of some debate (Garrett et al., 1990).

Dalziel (1992) considered exchange flows through contractions that have triangular and parabolic cross-sections (the (y, z)-plane), narrower at the bottom, with and without sills (bottom topography). There are obvious differences, but generally the same processes and phenomena apply. And the effects of turbulent mixing at the interface of two-layer exchange flows (with a partially mixed central layer) have been explored by Hogg et al. (2001a). A parallel study by the same authors (Hogg et al., 2001b) has examined the properties of (artifiicially generated) internal waves propagating through or within these (mostly two-layer) flows, and the effects that they may have on critical conditions and hydraulic control.

6.3 Exchange flows through doorways and windows

The foregoing analysis has application to buoyancy-driven or forced flow of stratified air inside buildings. A common situation is that where a doorway or window separates two rooms or large spaces containing air of different temperature, so that an exchange flow is set up between them when the door or window is opened. This geometry does not satisfy the usual (slowly varying) requirements for hydrostatic flow, but nonetheless, the hydrostatic equations are found to give a realistic description of the observed phenomena (Lane-Serff, 1989). The flow usually separates on passing through such an opening, and the channel boundaries may be taken on the separation streamlines or surfaces. Hence we may assume that the equations of §§ 5.2, 6.2 provide a satisfactory description of these flows. Equations (5.10), (5.11) then imply that $G^2 = 1$ at the door or window, so that the flow is critical there with

$$\frac{u_1^2}{g'\overline{d_1}} + \frac{u_2^2}{g'\overline{d_2}} = 1, \qquad (6.17)$$

where u_i and $\overline{d_i}$ represent the mean velocity and depth respectively in the ith layer.

Most *windows* are sufficiently far from other boundaries, such as walls, floors and ceilings, for them to be treated as isolated openings in an infinite plane wall, for present purposes. If a window is opened between two large volumes containing homogeneous air of different temperatures (say cold air inside, warm air outside), the exchange flow with $Q = q_1 + q_2 = 0$ (a common case in practice) is complex (a descending and spreading three-dimensional gravity current on one side, a rising plume on the other), but at the opening it is symmetric about the vertical, and the interface is at the mid-level of the opening. Equation (6.17) then gives that

$$u_1 = |u_2| = \frac{1}{2}(g'D_w)^{1/2}, \tag{6.18}$$

where $D_w = 2\overline{d}_1 = 2\overline{d}_2$ is the vertical size of the window. The flow through the window in each direction is therefore controlled by the window size and the temperature difference, and is maximal.

For *doorways* that extend from the floor to the ceiling, the analysis of §6.2 is applicable, with $r_l = 0$ and $r_r = 1$, or vice versa. For pure exchange flows with $Q = 0$, (6.18) is again applicable, with $D_w = D$. The case of forced flows (due, for example, to air conditioning) with $Q \neq 0$, is of Category IV in §6.2.4. This theory predicts that the exchange is blocked so that the flow is unidirectional if $\hat{Q}_m > (2/3)^{3/2}$, but its applicability to this extent is uncertain.

The situation with the more usual doorway, where the door height D_d is less than the ceiling height D, is more complex. Here the thickness d_2 of the upper layer in the doorway relative to the lower is reduced, so that for $Q = 0$, $\overline{d}_2/D_d$ lies in the range

$$0.375 \leq \overline{d}_2/D_d \leq 0.5, \tag{6.19}$$

where the lower limit applies to the case of no lateral contractions (a doorway as wide as the room), and the upper limit is for a pure contraction (Dalziel, 1991).

6.4 Multi-layer and continuously stratified exchange flows

6.4.1 Three-layer flows

A fluid with three layers of different densities can support two internal wave modes (rather than just one), and this leads to a greater variety of the types of exchange flows that may occur. In particular, in general two types of critical sections may occur (where a wave may be stationary, propagating in either direction), one for each mode. These are commonly called "control sections", but they may or may not in fact exert any control over the flow. Lane-Serff et al. (2000) have described laboratory experiments of a variety of three-layer exchange flows through a contraction, which

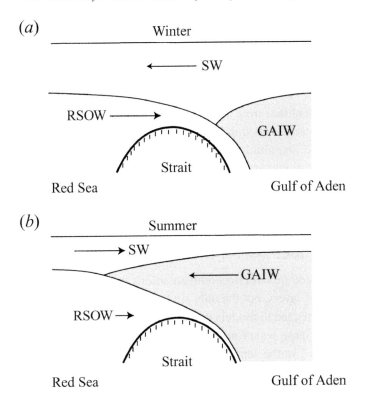

Figure 6.9 Schematic showing the flow through the Bab al Mandab strait at the mouth of the Red Sea, during (a) the winter monsoon, (b) the summer monsoon. SW denotes surface water, GAIW Gulf of Aden intermediate water, and RSOW Red Sea outflow water. There is significant seasonal change in all three components, including a large change in the transport and direction of the surface water. (From Lane-Serff et al., 2000, reproduced with permission.)

illustrate the range of possible flow types that may occur, and found good agreement with a three-layer hydraulic model.

A focus for comparison with oceanic data is the flow through the Bab al Mandab Strait, between the Gulf of Aden and the Red Sea. Here field observations show that the flow has three identifiable layers, or water masses, with the bottom layer of dense (salty) water flowing out of the Red Sea, as shown in Figure 6.9. The uppermost layer can flow in either direction, and the middle layer may be stationary and only exist on one side. This situation also includes a sill at the narrowest point, which the functional procedure can readily accommodate. The seasonal cycle in throughflow is strongly influenced by the variability of the local volume of the Gulf of Aden intermediate water (GAIW), which is caused by seasonal variation in upwelling in the Gulf of Aden, due to the annual monsoon cycle. A three-layer model of the exchange flow by Smeed (2000), with vertical sidewalls, produced qualitative results that were consistent with observations. In fact, the breadth within the strait varies with depth, and three-layer modelling studies by Siddall et al. (2002) containing this variation gave good representation of the observations, not only at the two extremes of Figure 6.9, but through the whole annual cycle. Four different flow types were identified over the annual cycle, two of which were only critical for mode 2, and these results implied that the injection of the GAIW into the Strait greatly reduces the magnitude of the exchange in summer by as much as 30%. Subsequent observational studies of Bab al Mandab (Pratt et al., 2000; Smeed, 2004) have shown that tides and the annual cycle make the flow more complex than the above simple model would imply.

6.4.2 *Multi-layered and continuous flows*

The functional procedure with Bernoulli equations for each layer is applicable to an arbitrary number of layers, but the only studies to date that are more complex than three layers are restricted to models that address the problem of stratified exchange flow between two large reservoirs, and assume that the flow has the property of being "self-similar", in the sense of Wood (1968). This means that the heights of streamlines or stream surfaces are everywhere in the same ratio. This methodology has also been used to address the ventilation of Scandinavian coastal embayments due to seasonal variation of offshore waters (Stigebrandt, 1990).

If a layered exchange flow is assumed to have one layer that is at rest, the moving layers above it behave independently from the oppositely directed layers below (Engqvist, 1996), so that the flow resembles two multi-layer stratified flows as described by Wood (1968), one above the other. The governing equations are as in §6.2: the continuity and hydrostatic Bernoulli equations for each layer, the whole driven by a surface pressure gradient, as illustrated in Figure 6.10 for seven layers, with the flow controlled by critical flow conditions at or near the point of maximum contraction. The stagnant layer effectively separates the dynamics into two regions: above and below. Critical flow occurs for each region at the narrowest part of the contraction, but they are decoupled from each other (see also §8.1.1). At the narrowest point the depth of the moving flow from each region is 2/3 of its upstream value. For the flows in both regions, hydraulic jumps occur on the downstream side as the flow adjusts to conditions in the downstream reservoir. Figure 6.11 shows six flow types, in which the difference between them is due to the magnitude and direction of the surface pressure gradient – right to left at the top and upper region, left to right in the lower region. There is only the one control section for the flow in each region.

These solutions have been extended to continuous stratification by Hogg &

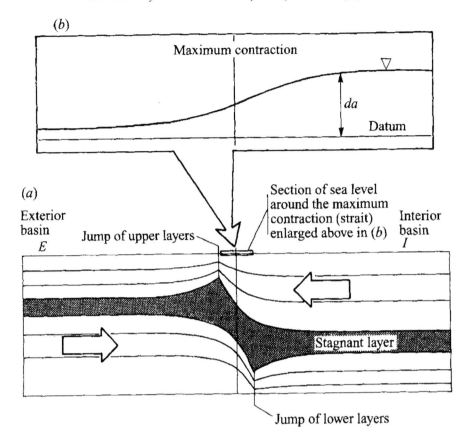

Figure 6.10 (a) A diagram showing a seven-layer exchange flow between two large reservoirs with the fourth layer stagnant. The sharp points on the downstream sides denote hydraulic jumps. (b) Sea level variation in the vicinity of the central contraction, showing a right-to-left pressure gradient. (From Engqvist, 1996, reproduced with permission.)

Killworth (2004), using the same equations in forms that are continuous in the vertical direction. The exchange flow is again through a contraction between two large reservoirs, and may contain nett barotropic transport as well as two-way exchange. The assumption is made that the flow is self-similar in the same manner as above which, with suitably simple upstream conditions on each side, enables an analytic solution. The assumption of self-similarity was tested by comparing the results with numerical simulations from a time-dependent numerical model that made no such assumptions.

Two different self-similar flow types were identified, as shown in Figure 6.12. The first type, C1, is essentially the continuously stratified version of Engqvist's solution, with a stagnant layer of everywhere finite thickness between the two counter-moving

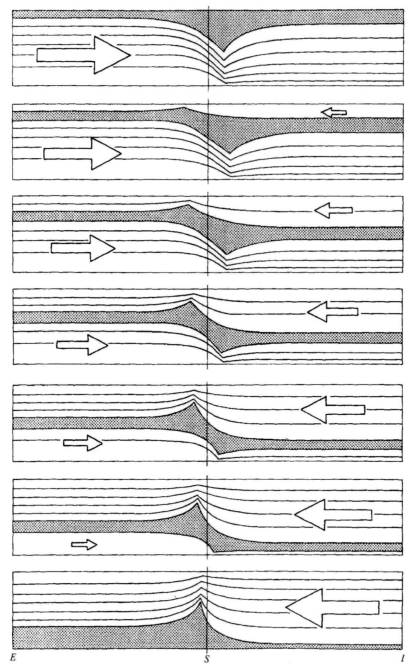

Figure 6.11 Possible flow solutions between two large reservoirs with seven discrete layers. The shaded layer is stagnant, and arrows denote flow direction. Each
flow corresponds to a different surface level pressure gradient. The vertical height
reduction at the strait contraction is 2/3 of each initial flow. (From Engqvist, 1996,
reproduced with permission.)

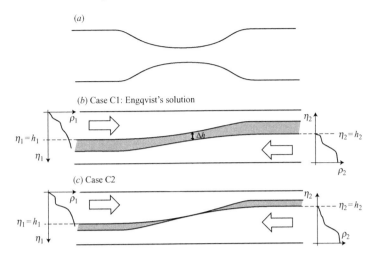

Figure 6.12 Schematic of the stratified exchange flow. (a) Plan view of the contracting channel. (b) Case C1: Engqvist's solution for two decoupled stratified layers separated by a homogeneous layer with a minimum distance of Δh. (c) Case C2: a flow solution where the two flowing layers touch at a single point. In both cases the upstream stratification and height coordinates are specified for each reservoir. η_1 and η_2 denote the elevation in the two reservoirs with density profiles, with (mostly) $\rho_1(\eta) < \rho_2(\eta)$. (From Hogg & Killworth, 2004, reproduced with permission.)

stratified layers. In the second type, C2, the stagnant layer thickness has been reduced to zero at a single point, where the upper and lower layers make contact with each other. The flow solutions are determined by the self-similar assumption and the density profiles in each reservoir, indicated on the left and right of Figure 6.12b,c. The case C2 clearly marks the limit of the Engqvist-type solutions.

Some examples of the self-similar flow solutions compared with the numerical solutions in steady state are shown in Figure 6.13. This shows four examples of the analytic self-similar solution on the left, and the steady-state result from the time-dependent integration on the right. The relevant parameters are $r_\rho = \delta_V / \delta_H$ – where δ_V is the vertical density difference and δ_H is the horizontal density difference – and q_0, which is a measure of the nett fluid flux through the channel. Both reservoirs are linearly stratified with the same density gradient (with the same value of δ_V), but the mean density of each reservoir is offset by a small amount.

Figure 6.13i shows a comparison between the self-similar solution (left) and the steady-state numerical result on the right, for $r_\rho = 4$ (large vertical density gradient, small horizontal gradient), and no nett throughflow ($q_0 = 0$). This is an "Engqvist-type" solution, and there is clearly good overall agreement, with an error of order 3%. Figure 6.13ii shows the same comparison but with a mean throughflow from

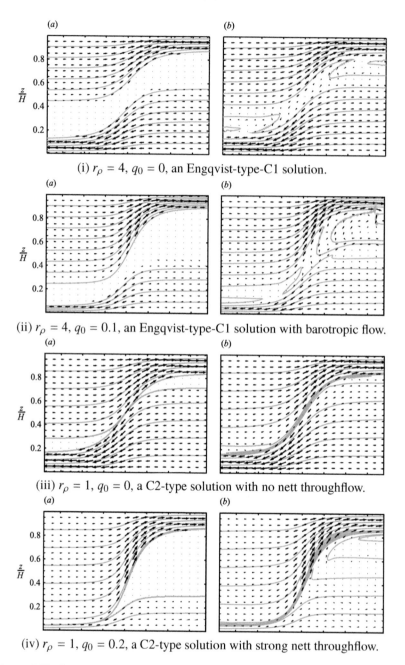

(i) $r_\rho = 4$, $q_0 = 0$, an Engqvist-type-C1 solution.

(ii) $r_\rho = 4$, $q_0 = 0.1$, an Engqvist-type-C1 solution with barotropic flow.

(iii) $r_\rho = 1$, $q_0 = 0$, a C2-type solution with no nett throughflow.

(iv) $r_\rho = 1$, $q_0 = 0.2$, a C2-type solution with strong nett throughflow.

Figure 6.13 A comparison between self-similar analytic solutions (left) and the steady-state result of time-dependent integrations (right) of uniformly density-stratified exchange flows through the contraction shown in Figure 6.12a. The contours show isopycnals, and arrows denote velocity vectors. (From Hogg & Killworth, 2004, reproduced with permission.)

left to right with $q_0 = 0.1$. Here there are some differences between the self-similar solution (left) and the steady-state of the time-dependent integration (right), but there is good general overall agreement. Both of these flows are Case C1.

Figure 6.13iii shows an example of Case C2, with $r_\rho = 1$ and $q_0 = 0$. In the self-similar solution (left), the stagnant region has a finite thickness everywhere except at $x = 0$, but in the numerical solution stagnant regions are only seen near the ends. The difference can be attributed to the effect of viscosity in the numerical model. Figure 6.13iv shows the same Case C2 flow as 6.13iii but with a strong throughflow with $q_0 = 0.2$. In the analytical solution the layers touch at $x = -0.36$ (1/4 width from left-hand edge). Here some differences are starting to emerge, notably on the right, though the main features are still similar.

Overall, in the range of parameters where self-similar flows are possible, they appear to give a good description of these stratified exchange flows, at least where viscous effects are negligible. The above analysis on exchange flows is inviscid, but when fluid in contact is flowing in opposite directions, the effects of friction may be significant. The problem has been addressed by Winters & Seim (2000), Zaremba et al. (2003) and Gu & Lawrence (2005). There is no space for the details here, but results show that friction can change the hydraulic control locations, and both interfacial and bottom friction can play important roles in determining the exchange flow rate.

7

Gravity currents, downslope and anabatic flows, and stratified hydraulic jumps

Gravity is only the bark of wisdom's tree, but it preserves it.

CONFUCIUS, *Analects*

This chapter consists of four sections. The first describes the horizontal flow of dense fluid (notionally termed *gravity currents* or *density currents*) into a uniform or density-stratified environment. The second section describes the various flow types of gravity currents down slopes, where again the environment may be uniform or density-stratified. The third section describes models of hydraulic jumps that may occur in stratified environments, with potential application to deep oceanic flows, and the fourth describes the nature of buoyant flows upslope.

7.1 Gravity currents over horizontal terrain in uniform environments

Gravity currents are common phenomena in nature, and occur when a mass of heavy fluid is released or discharged into a large volume of less dense fluid above a solid surface, so that the flow is driven by the difference in densities. Examples are: sea breezes, flow from thunderstorm downdraughts, turbidity currents on the ocean floor, snow avalanches, and ground level flow of cold air into a heated room after a door has been opened. For a description of their many and various manifestations the reader is referred to Simpson (1997), and for the history of dynamical details, to Linden (2012). This section provides an introductory description of the properties of gravity currents propagating into homogeneous fluids.

7.1.1 Unidirectional gravity currents

Figure 7.1a shows a typical example of a unidirectional gravity current in the laboratory, caused by salt water released into fresh water at a constant rate at one end of a long tank. It shows the characteristic leading head with mixing on the downstream side. The low-level heavy fluid in the current flows towards the head, where it rises and is mixed with the surrounding fluid to form a sheared stratified layer, which lies above the dense inflow. Mixing in the upper part of and behind the head is due to Kelvin–Helmholtz shear instability (see §9.2.1). The leading part of the head is an elevated overhanging nose (Figure 7.1b), caused by the frictional drag of the lower boundary. Upstream fluid below the level of this nose is ingested into the head, and its upward motion due to buoyancy within the head increases the turbulence level and causes the flow to be three-dimensional (Simpson, 1997, Chapter 11; Härtel et al., 2000). This is manifested in the characteristic evolving "lobe and cleft" structure of the leading face, visible in duststorms, for example (Simpson, 1997). Despite this evolving pattern the two-dimensional mean flow is steady, provided that the supply of fluid at a given distance behind the head is constant.

Figure 7.1b shows typical velocity and density profiles behind the head in a frame of reference moving with it. The speed of the head is U, and u_1, d_1 the speed (taken as positive) and depth of the layer flowing towards the head. If we assume that D is very large, the externally imposed variables may be taken to be $g' = g\Delta\rho/\rho_1$ (where $\Delta\rho = \rho_1 - \rho_2$), and the inflow rate of heavy fluid relative to the ground, Q. This gives $Q = (U + u_1)d_1$. Other quantities such as d_1 and $d_1 + d_3$, the height of the head, are determined by the dynamics of the system. From scale analysis alone we may expect that (cf. (3.34))

$$d_1 \sim (Q^2/g')^{1/3}, \quad U + u_1 \sim (Qg')^{1/3}. \tag{7.1}$$

The effect of the additional variable D is shown in Figure 7.2. The observed head speed U is approximately equal to $1.2(g'd_1)^{1/2}$ for $d_1/D \ll 1$, with a steady decrease with increasing d_1/D (Figure 7.2a; Huppert & Simpson, 1980). The dashed line shows theoretical values from two inviscid models with no bottom stress and no mixing, giving similar results, based on conservation of mass and momentum flux (Benjamin, 1968) or vorticity balance (see §3.5) (Borden & Meiburg, 2013b).

Figure 7.2b shows the observed inflow velocity relative to the head, u_1 (Figure 7.1b), where $q_1 = u_1d_1$ is the flux of heavy fluid being mixed in the head with the lighter surrounding fluid to form the stable overlying layer. The term u_1/U is approximately constant (≈ 0.15) with d_1/D, with perhaps a very slight increase. Hence the fluid velocity behind the head is 15% greater than the head speed. These observations show that the dependence on d_1/D is weak, and support and justify the scaling (7.1). The mixing process is seen to have a substantial effect on the overall dynamics, but it is not dominant. Of the other characteristic properties, the total height of the head is typically $3d_1$ if $d_1/D \ll 1$, but is less than this otherwise, and the nose height is typically $0.1d_1$–$0.2d_1$. These conclusions are based on laboratory

(a)

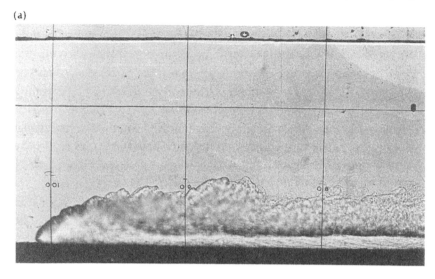

(b)

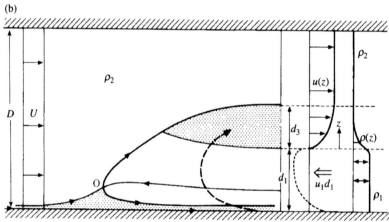

Figure 7.1 (a) Shadowgraph of a gravity current head, moving from right to left; $d_1/D = 0.06$, $\Delta\rho/\rho = 0.008$, and head height is $4.5d_1$. (Photograph courtesy of J. Simpson.) (b) A two-dimensional model of the flow near a gravity current head, in axes moving with the head. O is a stagnation point at the nose. Representative velocity and density profiles behind the nose are shown on the right. (Modified from Simpson & Britter, 1979.)

observations, but should still be applicable to environmental flows at very large Reynolds numbers (Simpson & Britter, 1979).

If a gravity current encounters an obstacle with a height comparable to the depth of the current, part of the flow continues over the obstacle to constitute a reduced gravity current on the downstream side, and part is reflected to constitute a hydraulic

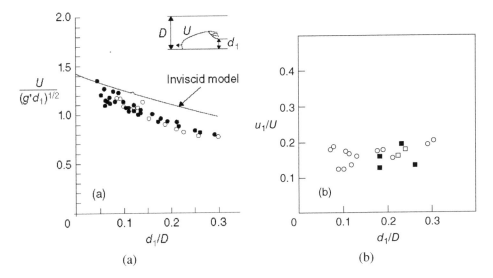

(a)

(b)

Figure 7.2 (a) Observed speeds U of the head, scaled with $(g'd_1)^{1/2}$, as a function of d_1/D. ($\circ$): steady-state moving floor experiments; ($\bullet$): lock exchange experiments. The dashed line is from the inviscid, non-mixing model of Benjamin (1968). (b) Observed fluid speed u_1 towards the head, relative to head speed U, as a function of d_1/D. (Modified from Simpson & Britter (1979); different symbols refer to different measurement techniques.)

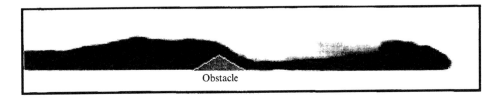

Figure 7.3 A typical image of flow of a rightward-moving gravity current over an obstacle, with a reflected hydraulic jump upstream and reduced flow on the downstream side. (From Lane-Serff et al., 1995, reproduced with permission.)

jump propagating against the current on the upstream side. An illustration is shown in Figure 7.3.

If the flow rate of dense fluid released into a channel of finite depth is very large, a variety of different phenomena is possible. Theoretical studies have shown that a range of different steady-state flows can be obtained if the flow is supercritical (i.e. faster than the wave speed on the interface of the intruding fluid) (Hogg et al., 2016). In particular, it is possible that the flow may become "choked" at some locations, meaning that the local flow rate cannot supply enough fluid to keep up with the head. Alternatively, near the source the depth of the released fluid could become large enough to fill the whole channel. Experiments realising these phenomena have yet to be reported.

7.1.2 Slumping gravity currents

If a large (but finite) body of stationary dense fluid is released into a fluid of lesser density, either at one end of a long channel or in an axisymmetric form, it collapses by passing through three successive stages (Huppert & Simpson, 1980). The first is the *slumping stage*, during which the dense fluid collapses and spreads as a coherent body, until its depth is less than 10% of the total depth of fluid. During this stage the thickness of this body initially decreases with distance from the starting point, to the head. However, its height soon becomes approximately uniform over its total length, in the time-scale for waves on its upper interface to travel the length. The height and speed of the head and rectangle then decrease as the fluid body expands further, and this spreading speed is significantly reduced by counterflow of the intruded fluid.

In the second stage it continues to spread as a collapsing rectangle (or cylinder) but now with the same speed as a gravity current, as described above – the *gravity current stage*. Here the spreading speed continues to decrease with the height of the rectangle. In the third *viscous stage*, the depth/thickness of the dense fluid has reduced to the point where the spreading rate is controlled by a dynamical balance between buoyancy and viscous forces. The second stage may be absent if viscous forces become significant while the flow is still in the slumping stage.

7.2 Gravity currents in density-stratified environments

The behaviour of gravity currents over level terrain moving in density-stratified environments depends very much on the vertical profile of the stratification. Gravity currents take the form of deformable obstacles moving into a stratified environment, and the analysis of flow over solid obstacles (in Chapters 10 and 11) constitutes a helpful guide to the gravity current case. Some experimental studies have been described by Rottman & Simpson (1983, 1989), Maxworthy et al. (2002), and in Chapter 13 of Simpson (1997). Without addressing all the details, this section describes the main effects of stratified environments by considering the flow of gravity currents in the two extremes: two-layer and linear density-stratified environments, and these are discussed in turn.

7.2.1 The propagation of gravity currents into two-layer stratified environments

Numerical experiments by White & Helfrich (2012), building on previous work by Holyer & Huppert (1980) and White & Helfrich (2008), have identified five distinct flow regimes, or "Types" for steady gravity currents entering two-layer flows. These experiments involved a two-dimensional two-layer fluid initially at rest, with a (effectively) steady inflowing gravity current at $x = 0$, for a finite time, with free-slip boundary conditions. The undisturbed fluid has a total depth D, with lower and upper layers having densities ρ_1, ρ_2, and depths d_1, d_2, respectively. Long waves on the stationary interface have speeds c_0 given by

$$c_0^2 = \frac{(\rho_1 - \rho_2)g}{\rho_1/d_1 + \rho_2/d_2}, \tag{7.2}$$

as described in §4.2. The inflowing gravity current has (effectively) constant depth h and velocity U_0 at $x = 0$, giving an effective Froude number $F = U_0/c_0$.

Examples of the five flow types are shown in Figure 7.4. Type I (with lower layer much thicker than the gravity current) denotes subcritical flow, and is shown in Figure 7.4a. There is a (almost imperceptible) depression in the interface over the leading nose of the gravity current, with (imperceptible) transient waves sent upstream. In Type II (Figure 7.4b) the gravity current generates an upstream undular bore, which contains fluid from both the intrusion and the lower layer. In Type III (Figure 7.4c) with a deeper lower layer, the upstream motion contains a combination of an upstream bore and a rarefaction wave, which move progressively further ahead of the main intrusion with time. Type IV has no parallel in the topographic case, and involves supercritical flow but with a body of mixed fluid constituting a "solitary wave" at the leading edge of the intrusion (Figure 7.4d). Lastly, Type V (Figure 7.4e) consists of wholly supercritical flow, with approximately uniform height behind the head. The analysis is two-dimensional, which (unduly) enhances the Kelvin–Helmholtz instabilities on the interface behind the head, in cases (c) and (d) but most particularly (e).

The various parameter ranges for the above flow types may be displayed on a Froude-number/gravity current thickness diagram for any given two-layer stratification, and three examples are shown in Figure 7.5. These may be (loosely) compared with Figure 4.11, of two-layer flow over topography, and give a remarkable overview of the range of possible flow phenomena under these conditions.

7.2.2 The propagation of gravity currents into linearly stratified environments

We consider heavy gravity currents of density ρ_c released from a finite body of fluid behind a lock of height h into uniformly stratified fluid of total depth D and buoyancy frequency N, with bottom and top densities ρ_b and ρ_0 respectively. The

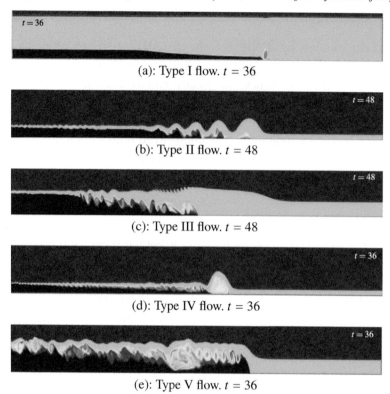

(a): Type I flow. $t = 36$

(b): Type II flow. $t = 48$

(c): Type III flow. $t = 48$

(d): Type IV flow. $t = 36$

(e): Type V flow. $t = 36$

Figure 7.4 Sections at particular times of different gravity currents propagating into initially stationary two-layer fluids, showing various flow regimes. (a) Type I: subcritical flow. $d_1/D = 0.45$, where D is the total depth, and d_1 the undisturbed depth of the lower layer; $S = 0.75$, where $S = (\rho_1 - \rho_2)/(\rho_c - \rho_2)$, where ρ_c is the density of the inflow and ρ_1 and ρ_2 are the densities of the lower and upper layers respectively. The lower layer is light coloured and the upper layer is dark (most of the upper layer is not included in the picture). There is a small downward displacement in the interface above the gravity current front. (b) Type II: upstream undular bore: $d_1/D = 0.1$, $S = 0.9$; (c) Type III: upstream bore with rarefaction: $d_1/D = 0.3$, $S = 0.9$; (d) Type IV: supercritical flow with trapped wave at front: $d_1/D = 0.1$, $S = 0.5$; (e) Type V: supercritical gravity current: $d_1/D = 0.3$, $S = 0.5$. (From White & Helfrich, 2012, reproduced with permission.)

two important parameters in this system are the Froude number, F_r, and the density ratio, R, defined by

$$F_r = V/(ND/\pi), \quad \text{and} \quad R = \frac{\rho_c - \rho_0}{\rho_b - \rho_0}, \tag{7.3}$$

where V is the velocity of the fluid. Experimental and theoretical modelling (Maxworthy et al., 2002; Ungarish & Huppert, 2002) of various runs showed different

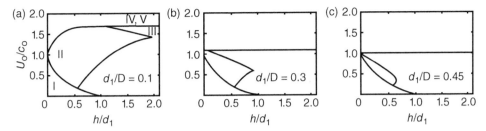

Figure 7.5 Diagrams for the flow regimes of Figure 7.4 for different two-layer stratification, in terms of the Froude number (U_0/c_0) of the gravity current and the ratio of current thickness to lower layer thickness at inflow. (a) $d_1/D = 0.1$; (b) $d_1/D = 0.3$; (c) $d_1/D = 0.45$. The regimes I, II, III, IV and V shown in (a) carry over to (b) and (c). (Adapted from White & Helfrich, 2012.)

behaviour depending on whether the initial Froude number F_0 was subcritical ($F_0 < 1$) or supercritical ($F_0 > 1$).

For subcritical flows, the speed of the head of the current is effectively constant out to a finite but significant distance from the source (provided h is not small or R close to unity), and the Froude number F_r is dependent on the parameter R in the form

$$F_r = F_0 = C + K \log_{10}(R),\qquad(7.4)$$

where C and K are constants that depend on h/D. Waves developed on the current behind the head as it progressed. After this constant-speed period the flow was significantly affected by bottom friction, and the velocity decreased to near zero but with some oscillations at the leading edge.

For supercritical flows ($F_0 > 1$) the picture is significantly different. Here the speed of the head is larger, and progressively decreases with increasing distance. Conspicuous mixing tends to occur behind the head, and no significant following wavetrain develops. The velocity of the head decays monotonically.

The range of parameters of these experiments was not large, and did not reach into the regime of strong flows described in §7.1.1 (Hogg et al., 2016). Some generalisations of these results, for inviscid steady-state flows with varying stratification, have been given by White & Helfrich (2008).

7.3 Gravity currents down slopes

7.3.1 Homogeneous environments

When dense fluid is released into homogeneous fluid at a steady rate at the top of sloping terrain, the nature of the resulting gravity current depends on the slope of the terrain. When the slope angle θ is small ($\theta \le 0.5°$), the head speed decreases with distance from the source, but at larger slopes the buoyancy force balances the frictional drag at the surface and the head travels at a constant speed (Britter & Linden, 1980). This head speed U_h is given by

$$U_h = 1.5(\pm 0.2)(g_0' Q_0)^{1/3}, \quad \text{for} \quad 5° \le \theta \le 90°, \tag{7.5}$$

where $g_0' = g \Delta \rho / \rho$ and Q_0 denote the buoyancy and volume flux at the source. The head speed is approximately 60% of the fluid speed, so that fluid accumulates in the head from behind, and also by direct entrainment from the environment, a process which also occurs in the following flow behind the head (Ellison & Turner, 1959). The volume of the head increases with increasing slope (as well as downslope distance), and the flow has the pattern of a turbulent plume at a slope of 90°, where the dominant process on both the head and the following flow is entrainment (Morton et al., 1956).

   Downslope flows in the atmosphere may also be caused by radiative cooling on a sloping surface, causing "katabatic winds". This phenomenon is presumably quite common after sunset over sloping terrain when mean winds are light, but field studies of it are few (Manins & Sawford, 1979). Rather than bottom friction, mixing with the overlying (relatively stagnant) air is the dominant form of drag on this downflow, as determined by the local Richardson number criterion (see Chapter 9).

7.3.2 Downslope flows into stratified environments

This section is mostly concerned with steady-state flow patterns that result from the release of dense fluid (as a gravity current) from a continuous source at the top of a slope into a density-stratified environment, rather than with transient features (such as the structure of the head). If dense fluid is released into an environment that consists of horizontal layers of homogeneous fluid separated by density interfaces, in general terms the descending fluid penetrates each interface if its density at that level is greater than that of fluid below the interface (Wells & Wettlaufer, 2007; Monaghan, 2007); otherwise, it spreads horizontally at the interface level. If the inflowing fluid has a range of densities, this may result in fluid spreading at several interfaces (Cortes et al., 2015).

   In environmental situations (such as downslope flows in the ocean) it is more common for the host fluid to be continuously stratified, and this introduces a variety of new phenomena. Here we discuss the bulk properties of the flow, and the various phenomena that are associated with them. If we consider a two-dimensional vertical slice through the flow, the principal variables are Q, the volume flow rate downslope, G, the buoyancy of the inflow, and N, the buoyancy frequency of the environment, defined by

$$G = g'(z) = \frac{g\Delta\rho(z)}{\rho_0}, \qquad N^2 = -\frac{g}{\rho_0}\frac{d\rho_0(z)}{dz}, \qquad (7.6)$$

where z is the vertical coordinate, g gravity, $\rho_0(z)$ the density of the environment and $\Delta\rho(z)$ is the (positive) difference between the densities of the downflow and environment at level z. Important dimensionless parameters are the buoyancy number B and Reynolds and Richardson numbers R_e and R_i, defined by

$$B = \frac{QN^3}{G^2}, \qquad R_e = \frac{Q}{\nu} \qquad \text{and} \qquad R_i = \frac{G\bar{d}^3 \cos\theta}{Q^2}, \qquad (7.7)$$

where $\bar{d}$ denotes the mean normal thickness of the downflow, and B relates the properties of the downflow to the stratification, and accordingly varies with position downslope. At the top of the slope, where $z = 0$, it takes the value $B(0) = B_0$. The principal variables governing the steady flow are then B_0, the slope angle θ and (it turns out) the Prandtl number, $Pr = \nu/\kappa$, where ν and κ denote viscosity and diffusivity of density respectively.

From laboratory and numerical studies, there are three main types, or flow regimes, of these downslope flows: the gravity current regime, when bottom slopes are small, the plume regime (when they may be large), and a third regime in which the downflow splits into two components that separate from the slope, through a transition that may resemble a hydraulic jump. If the bottom slope is constant, the regime normally does not change with depth until the flow separates from the bottom surface. Animations showing the development in the laboratory of all three flow regimes may be seen as supplementary material to Baines (2005) on the web site of the *Journal of Fluid Mechanics*.

The gravity current regime

Extending the term from horizontal flows, the gravity current regime applies in regions where the bottom slope is sufficiently small. An example of a flow in this regime is shown in Figure 7.6, which shows a sequence of profiles of a gravity current descending a slope of 12°, at subsequent times. The downflowing dense fluid has been dyed with fluorescene and illuminated by a thin laser beam, scanned along a vertical section. The flow consists of a dense bottom current, with a region above it of dense fluid that has become detached and partially mixed with the overlying stratified environment. In the bottom frame the dense fluid has reached its equilibrium level, and is spreading in a broad outflow around this level. Fluid that has become detached and mixed above it is also slowly spreading into the environment at all levels. The process causing the mixing and detachment of the fluid from the downflow can be identified as resulting from Holmboe instability (see

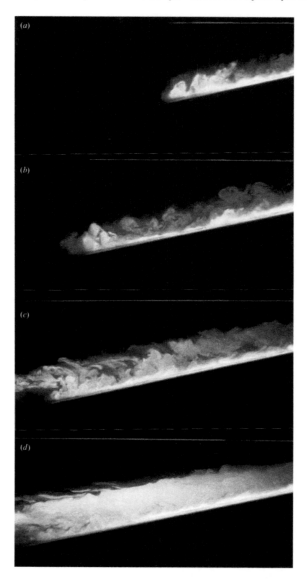

Figure 7.6 A time sequence of the development of the flow for a typical experiment at $\theta = 12°$, with the inflowing dense fluid dyed with fluorescene ($B_0 = 0.0058$, $R_e = 273$). These pictures show instantaneous vertical sections near the centre-line of the tank, illuminated by a thin vertical sheet of light from a scanned laser beam. Frames (a) and (b) show the initial gravity current head, and (d) is at a much later time than the first three. Frames (b–d) show the steady dense downflow with its main outflow at the bottom, and the growth of mixed, detrained fluid over the range of depths between the level of the source and this main outflow. The accumulation of detrained fluid is seen in the increasing intensity of the dye in the region above the main downflow, which slowly moves to the left. (From Baines, 2001, reproduced with permission.)

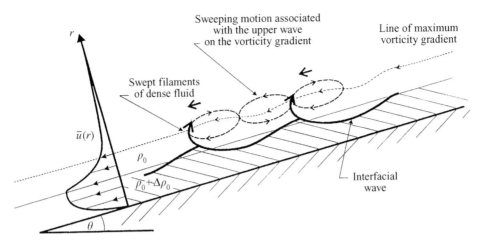

Figure 7.7 A schematic diagram showing the mean velocity profile in region II for a 12° slope, and the process of Holmboe instability, which is due to the mutual interaction between a gravity wave on the interface and a vorticity wave on the vorticity gradient above it (see Chapter 9). The main dynamical forces maintain the mean velocity and density profiles, so that the instability process keeps recurring, and wisps of detrained fluid result. Here, *r* denotes the coordinate normal to the slope. (From Baines, 2001, reproduced with permission.)

Chapter 9), as depicted in Figure 7.7. This generates eddies just above the downflow that sweep wisps of dense fluid from it that mix into the environment. The remaining fluid in the downflow spreads at its neutral level, which lies above the level of the initial density, implying some entrainment into the downflow during descent.

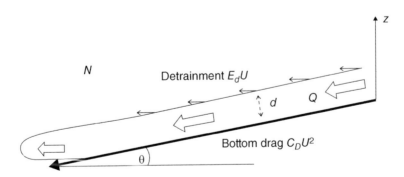

Figure 7.8 Definition sketch for a detraining downslope gravity current into a density-stratified environment.

The main features of this flow may be captured by a simple model of three equations, and a sketch depicting the notation is shown in Figure 7.8. If one assumes that the flow is in steady-state, with *s* the downslope distance, the equations for mass and buoyancy become, with variables Q, G and d:

$$\frac{dQ}{ds} = (E_e - E_d)\frac{Q}{d}, \tag{7.8}$$

$$\frac{dG}{ds} = -N^2 \sin\theta - E_e G/\bar{d}, \tag{7.9}$$

$$\frac{\partial}{\partial s}\left(\frac{Q^2}{\bar{d}}\right) = G\bar{d}\sin\theta - C_D\frac{Q^2}{\bar{d}^2} - \frac{\cos\theta}{2}\frac{\partial}{\partial s}\left(\bar{d}^2 G\right), \tag{7.10}$$

where C_D is the drag coefficient on the underlying surface. Here E_e and E_d denote entrainment and detrainment coefficients which must be determined from experiments. The best simple fit to the data (Baines, 2001), particularly for small ($\leq 12°$) slope angles gives

$$E_d = E_e + 0.2B^{0.4}\sin\theta. \tag{7.11}$$

The values of entrainment coefficients are better known, and have the form (Fernando, 1991; Strang & Fernando, 2001; Baines, 2001):

$$E_e = \begin{cases} C_1(\theta)/0.1, & 0 < R_i < 0.1, \\ C_1(\theta)/R_i, & R_i > 0.1, \end{cases} \tag{7.12}$$

where $C_1(\theta)$ is a function of the slope angle, with a value of 1.2×10^{-4} for $\theta > 5°$, increasing to 10^{-3} as $\theta \to 0$.

These flows are obtained if the bottom slope and the buoyancy number B_0 are sufficiently small (as discussed below), but if this is not the case the flow is in the plume regime.

The plume regime

The dynamics of downflows in this regime are essentially a two-dimensional version of those described by Morton et al. (1956), who introduced the concept of the entrainment coefficient E. The regime is contiguous with the vertical slope case, and the flows may be appropriately described as "plumes", as distinct from gravity currents. Some representative flows are pictured in Figure 7.9, and a notation sketch is shown in Figure 7.10. The downflow is essentially turbulent with an upper boundary that is vaguely defined, with conspicuous billows. These billows are a manifestation of Kelvin–Helmholtz instability (see Chapter 9), and entrain environmental fluid into the downflow through most of its length. At the bottom of the downflow, the current overshoots its level of neutral density and springs back in the form of a large circulating eddy, which then feeds into a broad region of outflow centred on the mean neutral-density level (Baines, 2002; Baines, 2005). Some of this sprung-back fluid is re-entrained into the boundary current and re-circulated in the "eddy" constituting the springback.

Apart from this intrusive outflow region, the main effect on the environment

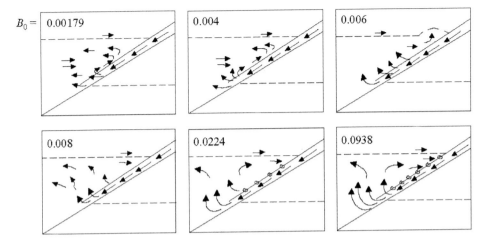

Figure 7.9 Sketches of the observed flow for six representative values of B_0 for a slope angle of 30° slope with a smooth surface. These diagrams are indicative of the observed flow patterns, and are not intended to be quantitative. They have been constructed from examination of video tapes of the dye movements as viewed from the side. In the region immediately above the bottom current, the flow is turbulent, and the arrows represent mean rather than steady motion. The upper dashed line denotes the upper limit of dyed fluid outside the initial downflow, which may be due to detrainment or springback. The lower dashed line denotes the initial equilibrium level of the dense fluid in the tank. As noted in the text, the first two runs are of a very unusual type, with two outflow regions separated by an entrainment region. (From Baines, 2005, reproduced with permission.)

is one of entrainment causing nett flow toward the slope at all levels, including the overshoot region immediately below the intrusion. The relevant steady-state equations are:

$$\frac{dQ}{ds} = E\frac{Q}{d},\tag{7.13}$$

$$\frac{dG}{ds} = -N^2 \sin\theta - EG/\overline{d},\tag{7.14}$$

$$\frac{d}{ds}\left(\frac{Q^2}{\overline{d}}\right) = S_2 G\overline{d}\sin\theta - C_D\frac{Q^2}{\overline{d}^2}.\tag{7.15}$$

Here E is the entrainment coefficient for plumes, which dates back to Morton et al. (1956), and appropriate values are (Turner, 1986)

$$E = \begin{cases} 0.08, & R_i < 0, \\ \frac{0.08 - 0.1R_i}{1 + 5R_i}, & 0 < R_i < 0.8, \\ 0, & R_i > 0.8. \end{cases}\tag{7.16}$$

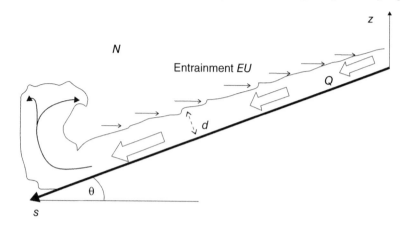

Figure 7.10 Definition sketch for an entraining downslope gravity current into a density-stratified environment.

In these flows R_i is generally somewhat less than unity, so that the entrainment dominates in (7.14) and $\bar{d}$ increases downslope.

The boundary between these two regimes occurs because of the breakdown of the gravity current regime, which occurs as the thickness of the downflow becomes small. Equations (7.9) and (7.10) may be combined to yield

$$(1 - R_i \cos\theta)\frac{d\bar{d}}{ds} = 2E_e + C_D - E_d - R_i \left(\sin\theta + \frac{E_e}{2}\cos\theta + \frac{\sin\theta}{2}(R_i B^2 \cos^2\theta)^{1/3} \right).$$

(7.17)

As $\bar{d}$ becomes small R_i decreases, and in the gravity current regime it is necessary that the right-hand side of (7.16) not become negative, or the current thickness would decrease to zero. Hence, taking the right-hand side of (7.16) to be zero with both $\bar{d}$ and R_i zero, together with (7.10), a limiting criterion for the gravity current model is

$$C_D + E_e - 0.2B^{0.4}\sin\theta \geq 0.$$

(7.18)

The term E_e depends on θ and R_i, but may be assumed to have a maximum value of about 10^{-3}. When the left-hand side of (7.17) becomes negative, the smallness of R_i and the consequent increase in shear will cause greater instability, turbulence, mixing and entrainment, so that the plume model will become applicable. Experimental points are shown in Figure 7.11 for a smooth sloping bottom surface, and clearly demonstrate the transition region between gravity currents and plumes. Some experimental runs with a much rougher (doormat) surface had the effect of moving the boundary to the right (larger values of B_0), as expected (Baines, 2005).

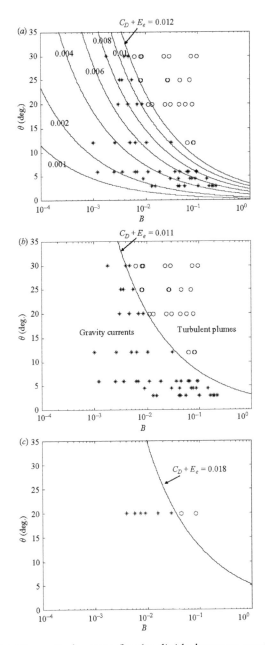

Figure 7.11 (a) Theoretical curves for the divide between gravity currents and plumes on the (B, θ)-plane as given by (7.17), for various values of $C_D + E_e$. All of the experimental points from smooth surfaces are also shown, with asterisks denoting gravity currents and circles denoting plumes. (b) As for (a), but showing the single value $C_D + E_e = 0.011$. (c) As for (b), but showing the single value $C_D + E_e = 0.018$, together with the points for the doormat surface at $20°$. (From Baines, 2005, reproduced with permission.)

Figure 7.12 An example of the double outflow regime on a slope of 30°, with $B_0 = 0.004$. The thickness of the downflow is small, but it is still turbulent and has the character of an entraining plume, which produces the main outflow at the bottom, as for Figure 7.3. Here, however, fluid appears to rise from this region above the downflow to a second outflow region above. In crossing the intermediate region, it is drawn toward the downflow by the entrainment, with conspicuous upward bubbling occurring just above it. (From Baines, 2008. © La Société canadienne de météorologie et d'océanographie, reprinted by permission of Informa UK Limited, trading as Taylor & Francis Group, www.tandfonline.com on behalf of La Société Canadienne de Météorologie et d'Océanographie.)

The hydraulic jump regime

In the gravity current regime for steep, 30°, slopes and small, ≤ 0.004, values of B_0, the flow pattern can differ from the above cases in that it presents two separated regions of outflow at different levels (see Figure 7.12), associated with a form of hydraulic jump in the downflow (Baines, 2005 – see supplementary material). Here the thickness of the downflow abruptly increases and the velocity decreases at a particular location, so that supercritical flow (Froude number $F_r > 1$) transitions to subcritical ($F_r < 1$). Similar flows have also been observed in some numerical simulations that show a complex flow pattern (Marques et al., 2017), but in spite of the fact that the Froude number often exceeds unity at the top of the slope, the phenomenon seems to be rare in steady flows over uniform slopes.

Oceanic observations

Downslope flows are common in the atmosphere, but systematic observations are few and the flows are generally more complex than those described above, so that models based on steady flow down uniform slopes are mostly not applicable. However, in the ocean downslope flows have been observed in a number of locations where these assumptions seem generally appropriate (Baines, 2008).

One such location is the outflow from the Red Sea, where dense saline water is formed by evaporation and flows downslope to the Indian Ocean (Peters et al., 2005). There are two main outflow channels with initial bottom slope of about $0.2°$ which steepens to about $2.7°$, and current thickness of about 100 m and velocity of $1 \, ms^{-1}$. From these parameters the lab experiments imply that these flows are detraining in the gravity current regime, which is consistent with observations.

A second location is the overflow of cold dense water from the continental shelf of the Antarctic Ross Sea (Gordon et al., 2004). Under the influence of the Coriolis force, the effective bottom slope is $7°$, and with a thickness of 200 m and a velocity of about $0.5 \, ms^{-1}$ the flow is in the entraining plume regime. These flows are complicated by the presence of tides, but this is consistent with observations that show that the volume flux increases by a factor of two at depths of 2000 m.

Another region with major impact on the deep ocean circulation is the Denmark Strait Overflow. This consists of relatively steady deep westward flow over the sill between Greenland and Iceland. Near the sill the bottom slope is small, and the flow is expected to have the mixing properties of a detraining gravity current, with detrainment increasing with bottom slope (Girton & Sanford, 2003). At a north–south section 20 km west of the sill, where the slope and velocity have increased, in the northern (shallower) part of the section the flow has the criteria for an entraining plume, but at deeper water where the downslope velocity is less, it remains a detraining gravity current.

Evaporation from the Mediterranean Sea causes its surface waters to be denser (saltier) than those in the Black Sea, and also denser than water in the Atlantic Ocean near Gibraltar. As a result there is exchange flow at both locations: deep dense water outflow through the Bosphorus to the Black Sea, and also through the lower part of the Strait of Gibraltar to the Atlantic. On the Black Sea continental shelf, available data (Ünülata et al., 1990) imply that the flow is a detraining gravity current. But on the continental shelf this fluid descends at a slope angle of $10°$, which implies that it is in the entraining plume regime. In contrast, the outflow through the Straits of Gibraltar to the Atlantic passes over more complex topography which implies that it behaves as a detraining gravity current in some locations, an entraining plume in others, with possible hydraulic jumps associated with some local topographic features.

7.4 Hydraulic jumps in stratified flow

Stratified hydraulic jumps seem to be fairly common events in the atmosphere in certain locations, but detailed descriptions of them are few. One example that has been described in some detail is the "morning glory" of northern Australia (e.g. Smith et al., 1982), which is a travelling undular bore that forms over the Cape York peninsula on a daily basis. Its structure is made visible by a contrast between cloudy and clear air, which is due to the local water vapour profile. Similar phenomena are probably common in other locations, but are invisible because the air is drier. Flow over mountains can produce jumps on the downstream side (a numerical example appears in frames (e) and (f) of Figure 10.17), and they have been reported downstream of downslope windstorms in the lee of the Rocky mountains (Karyampudi et al., 1991). On a smaller scale, hydraulic jumps appear to be common in low-level nocturnal flows. Here radiative cooling can produce a layer of cold air near the ground; over sloping terrain, this can produce or strengthen downslope currents that must adjust to subcritical conditions further downstream through a jump. An example is the common "gully wind" phenomenon over the city of Adelaide (W. Grace, private communication). Perhaps the best-known example of this kind is the "Loewe phenomenon", associated with the katabatic winds near the coast of Antarctica. For most of the year over most of the continent, radiative cooling causes a cold downslope supercritical katabatic wind with a typical depth of 200–300 m, that flows toward the coast where it reaches speeds of up to $60\,\mathrm{ms}^{-1}$. There, because of the change in slope and the different conditions prevailing over the sea, this flow reverts to a subcritical state through a jump with spectacular turbulent cloud structure, that has often been observed at coastal stations (Lied, 1964; Pettre & Andre, 1991).

The essence of these jumps may be roughly approximated by one- and two-layer models, but stratified jumps are turbulent if their amplitude is sufficiently large, and the associated mixing presents additional problems in modelling them. Detailed observations of their structure in both atmosphere and ocean are rare. The variety of problems involved, particularly in the context of the ocean, have been described by Thorpe (2010), and these have been refined to a practical model (at least in the oceanographic context) by Thorpe & Li (2014). The model has then been applied to observations at a number of oceanographic locations by Thorpe et al. (2018). The overall structure is reasonably simple, and is as follows.

The model is depicted in Figure 7.13. The base ground level is horizontal, and the model is only concerned with flow conditions on the upstream and downstream sides of the jump. On both sides, the vertical profiles of both the fluid velocity and density anomaly are constant up to a given height, above which they decrease linearly to zero. The overlying fluid is stationary, with uniform density, and the downstream density of the lower layer is the same as upstream. If subscript 1 denotes upstream of the jump and subscript 2 denotes downstream of it, the velocity and density profiles of the moving layers may be written as

$$u_i(z) = U_i f_i(z/h_i), \qquad \rho_i(z) = \rho_0 \left[1 - \Delta + 2\Delta f_i(z/h_i)\right], \qquad (7.19)$$

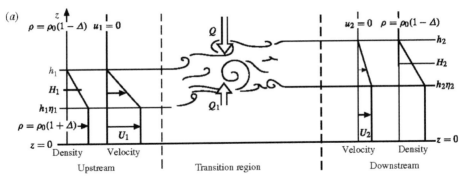

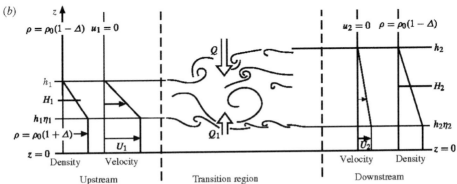

Figure 7.13 Sketches showing the model representation of a turbulent hydraulic jump or transition in a stratified shear flow over a plane boundary at $z = 0$: (a) a mode 1 transition and (b) a transition of mode 2. Here Q and Q_1 represent the fluxes of volume of density $\rho_0(1 - \Delta)$ from above and of density $\rho_0(1 + \Delta)$ from below into the transition zone. (From Thorpe & Li, 2014, reproduced with permission.)

where the vertical profiles both upstream and downstream are given by

$$f_i(z/h_i) = \begin{cases} 0, & z \geq h_i \\ (1 - z/h_i)/(1 - \eta_i), & \eta_i \leq z/h_i \leq 1 \\ 1 & 0 \leq z/h_i \leq \eta_i \leq 1. \end{cases} \tag{7.20}$$

Note that the thickness of the sum of the two layers increases with passage through the jump, but case (b) implies much greater mixing within the jump than in case (a), so that the thickness of the lower layer decreases. The gradient Richardson number in the central (mixed) interfacial layer ($\eta_i h_i < z < h_i$) is

$$R_{i_i} = 2g\Delta h_i(1 - \eta_i)/U_i^2, \qquad i = 1, 2, \tag{7.21}$$

and the model specifies that on the downstream side, $R_{i_2} = 1/3$. This ensures that the downstream flow is stable, a necessary condition for a realistic hydraulic jump.

There is one additional consideration when comparing theory with observations, namely that the upstream flow may be potentially unstable to conventional shear (Kelvin–Helmholz) instability (see §9.2.1). Whether or not the presence of this feature can be detected in the flow may depend on the quality of the observations.

The Froude number of the upstream flow is (perversely) defined as

$$F_r = \frac{U_1^2}{g\Delta h_1}. \tag{7.22}$$

Figure 7.12 shows properties of these model-jumps in terms of the upstream conditions η_1 and F_r. Internal waves can propagate upstream in the shaded region of Figure 7.14a, so that jumps are not possible there; upstream conditions for real jumps must lie to the right of the solid line. Possible solutions for hydraulic jumps are obtained from the equations conserving volume, mass and momentum fluxes between sections 1 and 2, as described by Thorpe & Li (2014). These equations yield one solution of (mostly) the mode 2 type for points in region C of Figure 7.14b, two solutions (one of mode 2 and one of mode 1) in region B, and no solutions in region A. Note that the model is formulated over horizontal terrain, but is also applicable to downslope flows over sloping terrain.

This model has been applied by Thorpe et al. (2018) to three oceanic flows. The first is the deep flow through the Samoan Passage, in which cold saline deep water from the South Pacific flows to the North, through a gap of width of order 40 km. There are several locations where the flow pattern suggests hydraulic jump structure, and the model has been applied and compared with two of them near 8° South. The model gives a point (or points) in region C of Figure 7.14b, and a section of the potential density field for this location is shown in Figure 7.15. The flow is situated over a downslope, and the flow pattern clearly indicates classic hydraulic structure. Comparisons between model and flow at other locations in the Red Sea and Mediterranean outflows are less distinct, but suggest a hydraulic jump in the latter but not in the former (Red Sea outflow). All the cases observed were of mode-2 type, where the dense bottom layer contracts and the mixing layer expands. Overall, the results suggest that this relatively simple model can be very useful in identifying and evaluating these complex fluid structures.

7.5 Anabatic flows

In contrast to katabatic (downslope) flows, anabatic flows are directed upslope, driven by buoyancy in the fluid. The phenomenon is not uncommon in the atmosphere, and can arise from sunlight incident on sloping terrain, which warms a layer of overlying air. If the heated region is small, one would expect that this would give rise to a local plume, but if the heated region is extensive, upslope anabatic flow

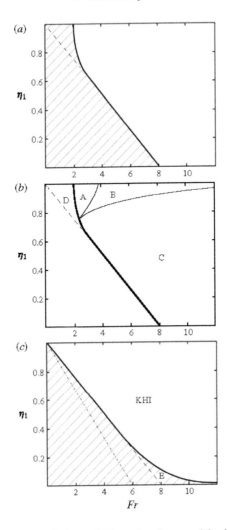

Figure 7.14 A summary of the stability of a flow and hydraulic jumps in the (η_1, F_r)-plane. (a) Internal waves can propagate upstream in the hatched region, and consequently no stationary hydraulic jumps are formed there. One has $R_i = 1/4$ on the line joining ($\eta_1 = 0$, $F_r = 8$) to ($\eta_1 = 1$, $F_r = 0$), with smaller values of R_i to its right. (b) The region $R_i < 1/4$ is divided as follows: A, in which no jumps may occur; B, in which jumps of modes 1 and 2 are possible; and C, in which only one jump, generally of mode 2, is possible. Flows in B and C are supercritical and the remaining area of the (η_1, F_r)-plane is subcritical. In D, $R_i < 1/4$ and the flow is unstable to KHI (see §9.2.1) but, because waves can propagate upstream (as shown in panel (a)), no stationary jumps can occur. (c) The hatched region is where the flow is stable to KHI. Its boundary (thick line) is the stability boundary separating stable flow (to the left) from unstable flow (to the right). One has $R_i < 1/4$ in the stable region E at small η_1 to the right of the stability boundary where (as shown in panel (b)) hydraulic jumps may occur. The dot-dashed line corresponds to $R_i = 1/3$. (From Thorpe et al., 2018, reproduced with permission.)

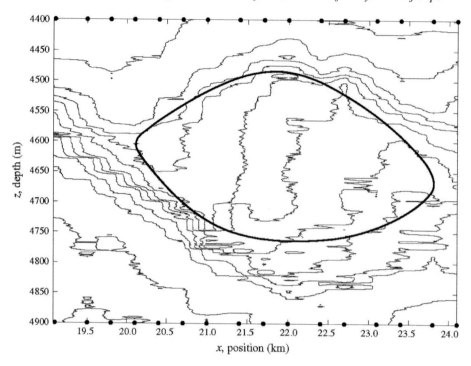

Figure 7.15 Contours of potential density between $x = 19.1$ km and $x = 24.1$ km and depths ranging from 4400–4900 m in the Samoan Passage. The mean horizontal locations of vertical profiles made by towed ship-lowered instrument package (Alford et al., 2013) are indicated by dots on the x-axis. The flow becomes supercritical at $x \approx 20$. The approximate position of the mixing region associated with the hydraulic jump is indicated by the oval-shaped curve. (From Thorpe et al., 2018, reproduced with permission.)

can result in air that can reach a maximum thickness of 1–2 km at about mid-day. A representative velocity U_m of this upslope flow is given by

$$U_m \approx 10(\alpha F_s H)^{1/3}, \tag{7.23}$$

where α is the slope of the terrain, F_s is the buoyancy flux due to solar radiation (e.g. Turner, 1973), and H is the thickness of the associated convective boundary layer (Hunt et al., 2003).

The effect of various different slopes on these flows has been examined experimentally by Hocut et al. (2015), who found that on slopes steeper than about 20° the upslope flow separated and fed into a rising plume. If the terrain slope β was in the range $10° < \beta < 20°$, the length of the upslope flow was smaller than for more gentle slopes, and it terminated with a rising plume that entrained additional environmental fluid. If slope $\beta < 10°$, the extent of the flow along the slope was substantially reduced, with rising flow occurring over a large proportion of the available sloping terrain. Field observations made on Granite Mountain during the *MATERHORN* project (see §14.2.7) were consistent with these results.

8

Waves in stratified fluids

The breaking of a wave cannot explain the whole sea.

VLADIMIR NABOKOV, *The Real Life of Sebastian Knight*

We now discuss the propagation of disturbances with small amplitudes through stratified fluids. For the most part this implies linear disturbances, but we also examine the behaviour of wave motion that becomes non-linear in certain regions known as critical layers. Linear wave theory is very useful for interpreting some of the non-linear phenomena produced by topography, as we saw for one- and two-layer systems in Chapters 2–5, and will see again in Chapters 9–13. In this chapter we mostly confine attention to waves in two dimensions (x, z), for simplicity. Generalisations to the third (y) dimension are discussed at the end in §8.11.

8.1 Waves in multi-layered models

A natural progression from the preceding chapters leads us to consider multi-layer models. Systems with three or more well-defined and (relatively) homogeneous layers are not uncommon in nature. For example, when mixing occurs at the interface of two-layer flows (and this is quite common for exchange flows), an intermediate layer of mixed fluid is produced, and this may have a significant effect on the dynamics. A model consisting of a number of homogeneous layers of fluids with different densities is also sometimes used to approximate a continuously stratified fluid where no layers are apparent. Hence it is important to understand the behaviour of such systems, and how this differs from that of continuously stratified fluids.

8.1.1 Layers with uniform density and velocity

We consider small disturbances on a basic state consisting of n layers as shown in Figure 8.1 but with a horizontal lower boundary, and surmounted by an infinitely

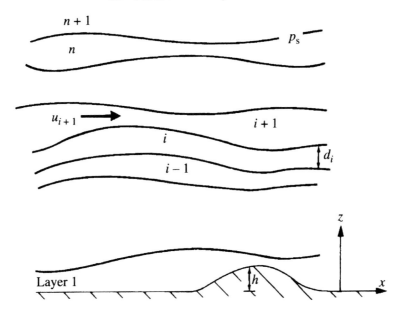

Figure 8.1 The configuration and notation for a system of n homogeneous layers, surmounted by an infinitely deep layer. We denote by u_i and d_i the velocity and thickness of the ith layer, and p_s denotes the pressure at the upper boundary of the nth layer.

deep $(n + 1)$th layer. The ith layer has density ρ_i, pressure p_i, velocity $\mathbf{u}_i$, thickness d_i, with undisturbed uniform horizontal speed U_i and vertical displacement η_i. The interface at the top of the ith layer is then at

$$z = \sum_{j=1}^{i} d_j = Z_i + \eta_i, \qquad \text{where} \quad Z_i = \sum_{j=1}^{i} \hat{d}_j. \tag{8.1}$$

The motion is governed by potential flow within the layers, with boundary conditions (1.25) at the interfaces. Where the mean properties of a fluid are discontinuous, boundary conditions across the junction require some care. To obtain a linear solution, we require values of p_i, p_{i+1} at a fixed level, namely $z = Z_i$. We may obtain such analytically continued values of p_i and p_{i+1} by expanding these functions in Taylor series about their values at the interface $z = Z_i + \eta_i$. This gives

$$p_j(x, Z_i, t) = p_j(x, Z_i + \eta_i, t) - \eta_i \frac{\partial p_j}{\partial z}(x, Z_i + \eta_i, t) + O(\eta^2), \quad j = i, i + 1. \tag{8.2}$$

Since (1.25) gives

$$p_i(x, Z_i + \eta_i, t) = p_{i+1}(x, Z_i + \eta_i, t), \tag{8.3}$$

we may use the vertical component of the equation of motion

$$\frac{Dw_i}{Dt} = -\frac{1}{\rho_i}\frac{\partial p_i}{\partial z} - g, \tag{8.4}$$

to eliminate $\partial p/\partial z$ from (8.2), and obtain

$$p_i(Z_i) - p_{i+1}(Z_i) = \Delta_i \rho g \eta_i, +O(\eta^2), \qquad i = 1,\dots,n, \tag{8.5}$$

where $\Delta_i \rho = \rho_i - \rho_{i+1}$. If we look for disturbances that have the form

$$\eta_i = a_i e^{ik(x-ct)}, \qquad i = 1,\dots,n, \tag{8.6}$$

where the a_i are constants and c and k are horizontal wave speed and wavenumber, then (1.25) gives

$$w_j = ik(U_j - c)\eta_i + O(\eta^2), \quad j = i, i+1, \qquad \text{at} \quad z = Z_i, \tag{8.7}$$

and with potential flow within each layer, (8.4)–(8.7) give

$$
\begin{bmatrix}
\Delta_1\rho - T_1 - T_2 & \dfrac{T_2}{\cosh k\bar{d}_2} & 0 & & 0 \\
\dfrac{T_2}{\cosh k\bar{d}_2} & \Delta_2\rho - T_2 - T_3 & 0 & & 0 \\
 & & \ddots & & \\
0 & 0 & \dfrac{T_i}{\cosh k\bar{d}_i} & & 0 \\
0 & 0 & \Delta_i\rho - T_i - T_{i+1} & & 0 \\
0 & 0 & \dfrac{T_{i+1}}{\cosh k\bar{d}_{i+1}} & & 0 \\
 & & & \ddots & \dfrac{T_n}{\cosh k\bar{d}_n} \\
0 & 0 & 0 & \dfrac{T_n}{\cosh k\bar{d}_n} & \Delta_n\rho - T_n - T_{n+1}
\end{bmatrix}
\begin{bmatrix}
a_1 \\ a_2 \\ \\ a_{i-1} \\ a_i \\ a_{i+1} \\ \\ a_{n-1} \\ a_n
\end{bmatrix}
= [0], \tag{8.8}
$$

where

$$T_i = \frac{\rho_i k(U_i - c)^2}{g \tanh k\bar{d}_i}, i = 1,2,\dots n; \qquad T_{n+1} = \frac{\rho_{n+1} k(U_{n+1} - c)^2}{g}. \tag{8.9}$$

The determinantal equation of the tri-diagonal matrix in (8.8),

$$\det[\text{matrix in (8.8)}] = 0, \tag{8.10}$$

gives the linear wave speeds c, and (8.8) then gives the structure of the internal wave modes or eigenfunctions, the column vectors $[a_i]$. There are $2n$ solutions. In general, the eigenfunctions for these solutions are all different, but they may coincide in pairs in special circumstances, such as when the U_i are all zero. If we have a rigid boundary at $z = D = Z_n$ instead of an interface, the same equations (8.8) are obtained except that a_n is zero and the nth row and column are deleted, giving $2(n-1)$ solutions. In the long-wave limit as $k\bar{d}_i \to 0$ for all i, we have

$$\cosh k\bar{d}_i \to 1, \qquad T_i \to \frac{\rho_i(U_i - c)^2}{gd_i}, \qquad i = 1, 2, \ldots, n, \quad T_{n+1} \to 0. \quad (8.11)$$

In general, if the velocity differences $|U_{i+1} - U_i|$ are sufficiently large, (8.10) has complex (conjugate) roots for any given k, implying that the flow is unstable. This applies for even a single interface, which is unstable for sufficiently large k for any $|U_{i+1} - U_i| \neq 0$. The reasons for this are discussed in §9.2.

Apart from questions of stability, the flow parameters do not impose restrictions on the values that c may have, relative to the flow velocities (cf. the case of continuous stratification discussed below). Figure 8.2 shows a simple example of the effect of increasing shear on wave speeds (relative to the central layer) in a three-layered system, with equal layer thicknesses d, density increments $\Delta\rho$ and velocity differences ΔU. For $0.44 < \Delta U/(g'd)^{1/2} < 1.5$, one or both wave speeds lie within the range of fluid speeds. In particular, it is quite possible that c may equal the velocity of one particular layer, U_i say, as this example shows. When this occurs we have $T_i = 0$, and the above determinantal equation (8.10) becomes

$$
\det \begin{bmatrix}
\Delta_1\rho - T_1 - T_2 & \dfrac{T_2}{\cosh k\bar{d}_2} & & \\
\dfrac{T_2}{\cosh k\bar{d}_2} & \Delta_2\rho - T_2 - T_3 & & \\
& & \ddots & \\
& & & \Delta_{i-1}\rho - T_{i-1}
\end{bmatrix}
$$

$$
\times \det \begin{bmatrix}
\Delta_i\rho - T_{i+1} & \dfrac{T_{i+1}}{\cosh k\bar{d}_{i+1}} & & \\
\dfrac{T_{i+1}}{\cosh k\bar{d}_{i+1}} & \Delta_{i+1}\rho - T_{i+1} - T_{i+2} & & \\
& & \ddots & \\
& & & \Delta_n\rho - T_n - T_{n+1}
\end{bmatrix} = 0,
$$

$$(8.12)$$

where again, the only non-zero terms are on the three central diagonals. Hence, for a propagating wave mode with $c = U_i$, one and only one of these determinants need vanish. If the first determinant vanishes but the second does not, the structure of the eigenfunction will be such that the wave propagates on the interfaces $\eta_1, \eta_2, \ldots, \eta_{i-1}$, whilst we must have $\eta_i = \eta_{i+1} = \cdots = \eta_n = 0$, and vice versa if the second determinant vanishes but the first does not. Therefore $c = U_i$ implies that the wave energy in this mode is totally confined to one side of the ith layer.

If we regard a wave mode propagating horizontally on many layers as the sum of an upward and a downward propagating wave, then where $c = U_i$, the ith layer acts as a *total reflector* of the wave motion on each side of it. The structure of this

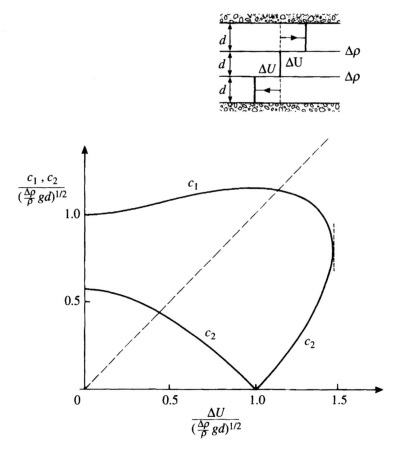

Figure 8.2 Long-wave speeds for the fast and slow modes in a sheared three-layer system with equal layer thicknesses d, and velocity and density differences ΔU, $\Delta \rho$ as a function of the magnitude of the shear. The wave and fluid speeds are symmetrical about the central layer. The system is unstable to long waves when $(\Delta U)^2/g'd > 1 + 2/\sqrt{3}$.

mode (or any disturbance travelling at the same speed) on one side of this ith layer is not affected by the thickness of the latter, nor by the nature of the mean flow on the other side of it. In fact, in the long-wave limit this ith layer acts precisely like an infinitely deep inert layer, as may be seen by comparing the long-wave forms of the matrices in (8.8) and (8.12). If $U_i - c$ is small but not zero, T_i is $O(U_i - c)^2$, and (8.12) contains an extra term of $O(U_i - c)^2$. In the associated mode, the $\eta_i, \ldots, \eta_n$ (or $\eta_1, \ldots, \eta_{i-1}$, as appropriate) will also be $O(U_i - c)^2$, and the transmission of the wave energy through the ith layer will be non-zero but small. In principle, for these results to apply, there is no limit to the number of layers or how thin they may be. We return to this subject in our discussion of "critical layers" in §8.9.

8.1.2 Layers with uniform density and vorticity

The preceding layered model may be generalised to situations where the velocity *gradient* is uniform in each layer, so that the mean velocity is continuous everywhere including the junctions of the layers. This is a step closer to reality if one wishes to approximate a continuous velocity profile with a layered one. The mean velocity is represented in the form

$$U(z) = U(Z_{i-1}) + (z - Z_{i-1})U_i', \qquad Z_{i-1} < z < Z_i, \qquad (8.13)$$

where $U_i' = dU/dz$. Since the vorticity in each layer is conserved and uniform, the equations of motion for perturbations within a layer are the same as if this motion were irrotational. For a small disturbance of the form of (8.6), then (1.25) implies that (8.5) and (8.7) are still applicable at pliant boundaries, where w is now continuous everywhere since $U(z)$ is. These equations then give, corresponding to (8.8),

$$
\begin{bmatrix}
L_1 & \mathfrak{I}_2 & & 0 & 0 & 0 & & 0 & 0 \\
\mathfrak{I}_2 & L_2 & & 0 & 0 & 0 & & 0 & 0 \\
& & \ddots & & & & & & \\
0 & 0 & & L_{i-1} & \mathfrak{I}_i & 0 & & 0 & 0 \\
0 & 0 & & \mathfrak{I}_i & L_i & \mathfrak{I}_{i+1} & & 0 & 0 \\
0 & 0 & & 0 & \mathfrak{I}_{i+1} & L_{i+1} & & 0 & 0 \\
& & & & & & \ddots & & \\
0 & 0 & & 0 & 0 & 0 & & L_{n-1} & \mathfrak{I}_n \\
0 & 0 & & 0 & 0 & 0 & & \mathfrak{I}_n & L_n
\end{bmatrix}
\begin{bmatrix}
a_1 \\
a_2 \\
\\
a_{i-1} \\
a_i \\
a_{i+1} \\
\vdots \\
a_{n-1} \\
a_n
\end{bmatrix}
= [0], \quad (8.14)
$$

where

$$
L_i = \Delta\rho_i + (\rho_i U_i' - \rho_{i+1} U_{i+1}')[U(Z_i) - c]g^{-1}
$$
$$
- \left(\frac{\rho_i}{\tanh k\bar{d}_i} + \frac{\rho_{i+1}}{\tanh k\bar{d}_{i+1}} \right) \frac{k[U(Z_i) - c]^2}{g}, \qquad (8.15)
$$
$$
\mathfrak{I}_i = \frac{\rho_i k}{g \sinh k\bar{d}_i}[U(Z_i) - c][U(Z_{i-1}) - c], \qquad i = 1, \dots, n,
$$

where $\bar{d}_{n+1}$ is infinite. If instead the upper boundary is rigid at $z = D = Z_n$, $a_n = 0$ and the last row and column of (8.14) are deleted, as for (8.8). The determinant of the matrix in (8.14) gives the wave speeds as for (8.10), and (8.14) reduces to (8.8) in the limit $\bar{d}_i \to 0$, $U_{i+1}' \to 0$ for all i even or odd.

Equation (8.14) has a generally similar appearance to (8.8) apart from the term involving the differences in $\rho_i U_i'$ added to $\Delta_i\rho$. There are $2n$ solutions for c, which may have values equal to the fluid speeds within the layers, or between or outside them. If $c = U(Z_i)$, and $\Delta_i \rho \neq 0$ it follows that $a_i = 0$, and the structures of the eigenfunctions above and below this level are independent, a situation corresponding to (8.12). If $\Delta_i \rho = 0$, all terms in the ith row have the factor $[U(Z_i) - c]$, but the vanishing of this term is a spurious solution (introduced to accommodate the density variation) and it must be factored out to obtain the genuine roots for c.

If one considers a single isolated interface, for example the ith with $\overline{d}_i, \overline{d}_{i+1} \rightarrow \infty$, and make the Boussinesq approximation for simplicity, we have

$$c = U(Z_i) - \frac{\Delta \zeta}{4k} \pm \left[\left(\frac{\Delta \zeta}{4k} \right)^2 + \frac{g'}{2k} \right]^{1/2}, \qquad (8.16)$$

where $\Delta \zeta = U'_i - U'_{i+1}$ is the discontinuity in vorticity and $g' = g \Delta \rho / \overline{\rho}$. If $\Delta \zeta = 0$ this gives two gravity waves, one propagating in each direction, and if instead $g' = 0$, there is one wave with

$$c = U(Z_i) - \frac{\Delta \zeta}{2k}, \qquad (8.17)$$

if the spurious root $c = U(Z_i)$ is removed. Equation (8.17) represents a wave propagating to the left if $\Delta \zeta > 0$. The propagation is due to the kinematic effect of the vorticity perturbation at the interface, and the mechanism is illustrated in Figure 8.3. A sinusoidal displacement causes a corresponding sinusoidal perturbation of the vorticity as shown, which has the form of a sinusoidal vortex sheet. This implies a velocity field as shown by the arrows and hence leftward propagation. This mechanism is common in fluids where the mean vorticity (or potential vorticity) is not uniform, and is the basis for Rossby waves in rotating fluids, for example.

Some of the roots for c may be complex implying instability, but the situation for a single isolated interface is always stable (provided U is continuous) as (8.16) shows. The mechanism causing instability is discussed in §9.2.

8.2 Continuously stratified fluids: equations

We next consider wave propagation in fluids where the density varies continuously. If the fluid has velocity and density profiles $U(z)$ and $\rho_0(z)$ respectively in the undisturbed state, the linearised form of (1.1)–(1.4) which govern small disturbances

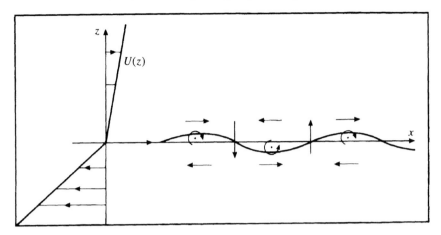

Figure 8.3 The kinematics of wave propagation on a vorticity interface, with vorticity discontinuity $\delta\zeta > 0$. A sinusoidal displacement of the interface gives a sinusoidal vortex sheet perturbation as shown by the circular arrows. This implies an associated velocity field as shown by the straight arrows, as may be seen by adding the advective effects of the two neighbouring vorticity regions at points where $\eta = 0$, for example. Hence the displacement and velocity are always out of phase by 1/4 wavelength, and the perturbation travels as a periodic wave to the left.

is

$$\rho_0 \left[\left(\frac{\partial}{\partial t} + U\frac{\partial}{\partial x} \right) u' + \frac{dU}{dz}w' \right] = -\frac{\partial p'}{\partial x}, \tag{8.18}$$

$$\rho_0 \left(\frac{\partial}{\partial t} + U\frac{\partial}{\partial x} \right) w' = -\frac{\partial p'}{\partial z} - \rho'g, \tag{8.19}$$

$$\left(\frac{\partial}{\partial t} + U\frac{\partial}{\partial x} \right) \rho' + \frac{d\rho_0}{dz}w' = 0, \tag{8.20}$$

$$\frac{\partial u'}{\partial x} + \frac{\partial w'}{\partial z} = 0, \tag{8.21}$$

where u', w', p' and ρ' denote perturbation quantities from the respective mean values $U, 0, p_0$ and ρ_0, and viscosity and diffusion have been omitted. Equation (8.21) implies that we may define a perturbation stream function ψ by

$$u' = -\frac{\partial\psi}{\partial z}, \qquad w' = \frac{\partial\psi}{\partial x}. \tag{8.22}$$

If we make the Boussinesq approximation for convenience and eliminate all other variables in favour of ψ, we obtain

$$\left(\frac{\partial}{\partial t} + U\frac{\partial}{\partial x} \right)^2 \left(\frac{\partial^2}{\partial x^2} + \frac{\partial^2}{\partial z^2} \right) \psi + N^2\frac{\partial^2\psi}{\partial x^2} - \frac{d^2U}{dz^2}\left(\frac{\partial}{\partial t} + U\frac{\partial}{\partial x} \right)\frac{\partial\psi}{\partial x} = 0. \tag{8.23}$$

We look for disturbances of the form

$$\psi = \hat{\psi}(z)e^{ik(x-ct)}, \qquad \eta = \hat{\eta}(z)e^{ik(x-ct)}, \qquad (8.24)$$

where $\hat{\psi} = (U - c)\hat{\eta}(z)$, and obtain the equation for $\hat{\psi}$:

$$[U(z) - c]^2 \left[\frac{d^2\hat{\psi}}{dz^2} + \left\{ \frac{N^2(z)}{[U(z) - c]^2} - \frac{\frac{d^2U}{dz^2}}{(U - c)} - k^2 \right\} \hat{\psi} \right] = 0. \qquad (8.25)$$

If $U(z) - c \neq 0$ in the field of flow, this implies that

$$L(\hat{\psi}) \equiv \frac{d^2\hat{\psi}}{dz^2} + \left\{ \frac{N^2(z)}{[U(z) - c]^2} - \frac{\frac{d^2U}{dz^2}}{(U - c)} - k^2 \right\} \hat{\psi} = 0. \qquad (8.26)$$

With suitable upper and lower boundary conditions, this equation has eigenfunction solutions (waves) with discrete eigenvalues (wave speeds) c. For internal wave solutions the bracketed term in (8.26) must be positive. If this term is positive below a given level and negative above it, the character of the solution must change from oscillatory (i.e. sinusoidal-like) to non-oscillatory (exponential-like) as z increases.

Unlike the equations in §8.1 governing disturbances in layered flows, (8.25) has *singularities* where $c = U(z)$ at some level, with corresponding singular solutions. In the same way in which the equation $x^2 f(x) = 0$ has the solution $f(x) = 0$, but also has the solutions $f(x) = \delta(x)$ and $f(x) = \delta'(x)$ where $\delta(x)$ denotes the Dirac delta function, (8.25) has the "solution" (8.26), but also has the solutions

$$L(\hat{\psi}) = A\delta(U - c) \qquad \text{and} \qquad L(\hat{\psi}) = BN(z)\delta'(U - c), \qquad (8.27)$$

where A and B are constants. These equations have singular solutions for any value of c in the whole range of U, for all k. These solutions are required in addition to those of (8.26) in order to provide a complete description of the behaviour of motion governed by (8.23) (Case, 1960). The regions of the fluid close to the singularities are termed *critical layers* or *critical levels*, and this is the principal additional feature contained in (8.27). The character of the solutions in these regions is described in detail in §8.9. In practice, the behaviour of the flow in critical layers often involves additional physical factors, such as viscosity and non-linearity. On the other hand, the non-singular solutions of (8.26) describe the behaviour of internal waves without critical layers, and we consider the nature of these solutions in the next few sections.

Equation (8.26) shows that the vorticity gradient term $U'' = d^2U/dz^2$ provides a restoring force similar to that of the density gradient term but is "one-signed", promoting wave propagation in one direction only. In terms of the displacement $\hat{\eta}$,

(8.25) may be written

$$\frac{d}{dz}\left((U-c)^2\frac{d\hat{\eta}}{dz}\right) + [N^2 - (U-c)^2 k^2]\hat{\eta} = 0. \tag{8.28}$$

If we assume that $U - c \neq 0$ everywhere, then defining

$$\xi = \int_{z_0}^{z} \frac{dz}{(U-c)^2}, \tag{8.29}$$

(8.28) becomes

$$\frac{d^2\hat{\eta}}{d\xi^2} + (U-c)^2[N^2 - (U-c)^2 k^2]\hat{\eta} = 0. \tag{8.30}$$

Equations (8.28), (8.30) have the interesting property that the vorticity gradient term of (8.25) has been removed. This has implications for the case of homogeneous fluids ($N = 0$), where (8.30) has only non-oscillatory solutions. If $U'' \neq 0$ between rigid boundaries where $\hat{\eta} = 0$ at $z = z_1, z_2$ say, (8.30) has no discrete wave mode solutions at all (e.g. Burkill, 1956); the discrete spectrum of wave mode solutions of (8.14) with c within the range of U has become continuous and the "modes" are singular. However, the kinematic wave propagation mechanism due to advection of vorticity perturbations (embodied in (8.17) and Figure 8.3) still applies: for an arbitrary initial disturbance, the subsequent motion is made up of singular disturbances governed by (8.27). This motion moves generally leftwards if $U'' < 0$; it may grow initially, but it eventually decays with time, and its energy is lost to the mean flow (though numerical studies have shown that, for homogeneous fluids, this transient growth may be large enough to promote instability and transition to turbulence in practical situations: Trefethen et al., 1993). If either or both of the upper and lower boundaries is pliant (see §1.2) rather than rigid, a finite number of discrete wave modes is possible (with c outside the range of U), in addition to the singular motion.

8.3 Waves in finite-depth systems

We next consider solutions to (8.26) in finite-depth systems in which the flow is stable (see Chapter 9) everywhere and singularities (i.e. critical levels) are avoided. For simplicity we will mostly consider systems with two rigid boundaries – at $z = 0, D$, since our main purpose is to provide an outline of the dynamics. The boundary conditions are therefore that $\psi = 0$ on $z = 0, D$. The parallel results for a pliant upper surface may be readily derived from the same equations with the boundary conditions (1.25). In general terms, when $U(z)$ and $\rho(z)$ may be discontinuous at a material surface, the boundary conditions on $\hat{\psi}$ may be obtained by integrating (8.25) across the boundary, with the given profiles of $U(z)$ and $N(z)$.

8.3.1 Waves in fluid with uniform velocity and buoyancy (frequency)

If N and U are independent of z, with a rigid lower boundary we have the solutions

$$\hat{\psi} = \sin mz, \qquad c = U \pm \frac{N}{(k^2 + m^2)^{1/2}}, \tag{8.31}$$

and for a rigid upper boundary the eigenvalues are

$$m = \frac{j\pi}{D}, \qquad j = 1, 2, 3, \ldots, \tag{8.32}$$

so that

$$c_j = U \pm \frac{N}{[k^2 + (j\pi/D)^2]^{1/2}}. \tag{8.33}$$

In contrast, for a pliant upper boundary at $z = D + \eta$, at and above which the velocity and density are continuous, the eigenvalues are given by

$$mD \cot mD + kD = 0. \tag{8.34}$$

For long, hydrostatic waves, where $kD \to 0$, we have

$$m = \left(j + \frac{1}{2}\right)\frac{\pi}{D}, \qquad j = 0, 1, 2, 3, \ldots. \tag{8.35}$$

For these modes the phase at the upper boundary differs by 1/4-wave-length from those with a rigid boundary. With frequency ω given by $\omega = kc$, the velocity of energy propagation associated with the wave, the group velocity c_g, is given by

$$c_g = \frac{d\omega}{dk} = U \pm \frac{Nm^2}{(k^2 + m^2)^{3/2}}. \tag{8.36}$$

Clearly, $c_g \to c$ as $k \to 0$, and the waves become hydrostatic. With a rigid upper boundary, long waves of mode j may propagate energy upstream if $K/j > 1$, where $K = ND/\pi U$. Here $1/K$ is the *Froude number* for the first mode, defined in the same way as in the preceding chapters.

8.3.2 Waves in fluid with uniform vorticity and buoyancy (frequency)

If N is constant but U varies linearly with z, and we take axes moving with the mean velocity of the fluid, we may write

$$U = U_{\mathrm{m}}\left(\frac{2z}{D} - 1\right), \tag{8.37}$$

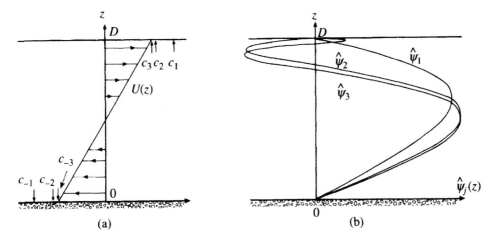

Figure 8.4 (a) The velocity profile with constant shear described by (8.37). The speeds (8.43) of the three fastest modes, scaled by U_m, are marked. (b) Stream functions for the first three eigenfunctions with $c_j > 0$ for constant N and the velocity profile of (a), for the long-wave limit, given by (8.43).

where U_m is the maximum velocity (see Figure 8.4a). Here we may define an important parameter – the local or gradient *Richardson number*

$$R_i \equiv \frac{N^2}{(dU/dz)^2},$$

(8.38)

which in general is a function of position. Here (8.37) gives $R_i = (ND/2U_m)^2$, which is uniform throughout the fluid. Equation (8.26) with rigid boundaries then has the following solutions. We define

$$Y = \frac{kD}{2}\left(\frac{c}{U_m} + 1 - \frac{2z}{D}\right),$$

(8.39)

and

$$
\begin{aligned}
Y &= Y_1 \equiv \frac{kD}{2U_m}(c + U_m), &\text{at } z = 0, \\
Y &= Y_2 \equiv \frac{kD}{2U_m}(c - U_m), &\text{at } z = D.
\end{aligned}
$$

(8.40)

The solution may then be written

$$\hat{\psi} = Y^{1/2}\left(\frac{I_{i\mu}(Y)}{I_{i\mu}(Y_1)} - \frac{I_{-i\mu}(Y)}{I_{-i\mu}(Y_1)}\right),$$

(8.41)

where the $I_{i\mu}$ denote modified Bessel functions (Watson, 1966), and $\mu = (R_i - 1/4)^{1/2}$. The dispersion relation is

$$I_{i\mu}(Y_1)I_{-i\mu}(Y_2) - I_{i\mu}(Y_2)I_{-i\mu}(Y_1) = 0,$$

(8.42)

specifying the eigenvalues for c. Provided $R_i > 1/4$, these eigenvalues all lie outside the range of fluid speeds, and have limit points at the maximum and minimum fluid velocities, $\pm U_m$. In the long-wave limit $kD \to 0$, these functions become

$$\hat{\psi} = \left(1 + \frac{U_m}{c_j}\left(1 - \frac{2z}{D}\right)\right)^{1/2} \sin\left\{\mu \log\left[\frac{1 + \frac{U_m}{c_j}\left(1 - \frac{2z}{D}\right)}{1 + \frac{U_m}{c_j}}\right]\right\}, \tag{8.43}$$

with the eigenvalues (cf. Figure 8.2)

$$c_j = U_m \coth(j\pi/2\mu),), \quad j = \pm 1, \pm 2, \ldots . \tag{8.44}$$

The structure of the lowest three modes propagating in the positive x direction is shown in Figure 8.4b for $\mu = 1$. Note that the local vertical wavelength decreases as $U - c$ decreases, giving smaller vertical velocities and larger horizontal ones. As $U_m \to 0$, these eigenvalues and eigenfunctions tend to those of the unsheared case (8.31). If $0 \le R_i \le 1/4$, the motion is similar to that for $N = 0$: there are no wave modes, but the system is stable (Case, 1960; Miles, 1961).

8.3.3 Waves in fluids with arbitrary profiles of $U(z)$ and $N(z)$

Many of the properties for uniform R_i (8.37)–(8.44) also apply to *arbitrary* stable stratified shear flows, such as that shown in Figure 8.5a, with general boundary conditions of the form $a\hat{\psi} + bd\hat{\psi}/dz = 0$, at $z = 0, D$. In particular, provided that $R_i > 1/4$ everywhere in the range $0 < z < D$, there is a denumerably infinite set of normal modes and corresponding eigenvalues where the latter satisfy (Bell, 1974; Ince, 1926, Chapter 10)

$$0 < U_{\max} < \cdots < c_3 < c_2 < c_1 < U_{\max} + \frac{a_1 N_{\max} D}{(\pi^2 + k^2 D^2)^{1/2}},$$

$$U_{\min} - \frac{a_2 N_{\max} D}{(\pi^2 + k^2 D^2)^{1/2}} < c_{-1} < c_{-2} < c_{-3} < \cdots < U_{\min}, \tag{8.45}$$

where $U_{\max}$ and $U_{\min}$ are the respective maximum and minimum of U in $0 < z < D$, $N_{\max}$ is the maximum value of N, a_1 and a_2 depend on the boundary conditions, and

$$\begin{aligned} c_j &\to U_{\max}+ \\ c_{-j} &\to U_{\min}- \end{aligned} \quad \text{as} \quad j \to \infty. \tag{8.46}$$

For rigid boundaries at $z = 0, D$, we know $a_1 = a_2 = 1$. Further, for $j > 0$, we have $dc_j/dk < 0$ for all k, so that the maximum velocity of each mode occurs at $k = 0$. The group velocity for mode j,

$$c_{gj} = c_j + k\frac{dc_j}{dk}, \tag{8.47}$$

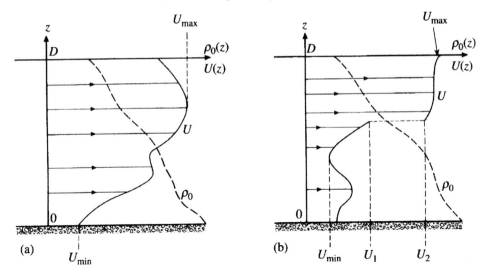

Figure 8.5 (a) A representative "arbitrary" stratified shear flow. (b) A stratified shear flow with a gap in the velocity profile, where no fluid has velocity $U(z)$ in the range $U_1 < U(z) < U_2$. Wave modes with speeds in this range are now possible, with the ends as limit points.

is never more than c_j, and becomes equal to it as $k \to 0$. Corresponding results hold for the modes with $j < 0$. If $N = 0$ and $dU/dz = (U_2 - U_1)/(z_2 - z_1)$ which is constant in an interior range $z_1 < z < z_2$, then internal wave modes exist with wave speeds in this range, with U_1, U_2 as limit points. This behaviour persists in the limit $z_1 \to z_2$, where the velocity profile is discontinuous as in Figure 8.5b.

In these stratified shear flows, the Richardson number (defined by (8.38)) has major significance. It may be shown with the use of Kneser's oscillation theorem from the theory of differential equations applied to (8.30) (Bell, 1974; Swanson, 1968) that stratified shear flows support internal waves of all frequencies less than N where $R_i > 1/4$, but not where $R_i < 1/4$. The quantity $R_i^{1/2}$ is the ratio of the time-scales of two factors, namely the gradients of buoyancy and fluid velocity. The buoyancy gradient N promotes wave propagation, and the velocity gradient suppresses it by continually rotating the material lines of fluid particles toward the horizontal, where they become indefinitely extended. The phase lines of internal waves propagating through the same fluid are subject to the same tendency, particularly if the period of the wave is much greater than the time-scale of the shear. Whether or not this process is strong enough to suppress local wave propagation is measured by the criterion $R_i < 1/4$. This fundamental property has important consequences for the stability of stratified shear flows, as discussed in Chapter 9. It should be noted, however, that some limited horizontal wave propagation may still be possible when $R_i < 1/4$, and even when $R_i = 0$, when one-way propagation on the vorticity gradient U'' may occur (as in Poiseuille flow).

In general, the set of eigenfunctions obtained for any particular flow can be made orthogonal. However, such a set will almost certainly not be complete, in the sense of being able to describe the evolution of some arbitrary initial disturbance, or to fit prescribed conditions at some specified value of x. This is because these conditions generate components that have wave speeds c lying within the range of fluid speeds, resulting in critical layer phenomena and temporally evanescent terms as mentioned above. The only known exceptions to this are cases where U is constant or N and U'' vanish in layers, as in §8.1. For the latter, if the set of n modes propagating against (or with) the flow is made orthogonal, the set will also be complete, since they are required to describe the motion of n layers.

Waves with $k = 0$ are of special interest. These waves have constant phase (i.e. no variation with x), and are termed "columnar disturbance modes" or more briefly, "columnar modes". They may be generated by, for example, the sudden imposition of a $u(z)$ (and hence $\psi(z)$) profile at $x = 0$ that does not change with time, and has been chosen so that it may be completely represented by the free internal wave modes described above. If this disturbance is introduced suddenly at $t = 0$, then an infinite set of columnar modes is generated, and these propagate upstream (if fast enough) and downstream at their respective long-wave speeds. The leading regions or fronts of each of these modes (where the amplitude increases from zero to near the final value) may be discontinuous initially, but as time increases their widths expand as $(Nt)^{1/3}$, due to wave dispersion (McEwan & Baines, 1974).

The propagation of the front of a columnar mode through a region results in nett changes to the mean velocity and density profiles in the region. To describe these, we let the initial velocity and density profiles be $U_0(z)$, $\rho_0(z)$ with N-profile $N_0(z)$. After the passage of the front of a columnar mode with stream function $\epsilon\hat{\psi}_j(z)$, where ϵ is a small amplitude parameter, the resulting velocity profile is

$$U(z) = U_0(z) - \epsilon\frac{d\hat{\psi}_j}{dz}, \tag{8.48}$$

and the corresponding change in the density field may be obtained by integrating (8.20) to give

$$\rho = \rho_0(z) - \epsilon\left[\frac{\frac{d\rho_0}{dz}\hat{\psi}_j(z)}{U_0 - c_j}\right], \quad N^2(z) = N_0^2(z) - \epsilon\frac{d}{dz}\left[\frac{N_0^2\hat{\psi}_j(z)}{U_0 - c_j}\right], \tag{8.49}$$

where c_j is the speed of the columnar mode. In general, the structure and wave speeds of modes propagating on this new stratified shear flow will be slightly different from their previous values.

As with ordinary ($k \neq 0$) modes, a columnar mode may be reflected from a vertical barrier such as an endwall of a tank, although in practice this can only occur when $U_0 = 0$. When this reflection occurs, the reflected velocity perturbations cancel out the incident ones (to leading order in ϵ), leaving zero nett motion. However, the associated density perturbations of the reflected wave *add* to those of the incident, and this density change is the nett result (locally) of the passage of the incident and reflected waves. This may be seen from (8.49) with $U_0 = 0$, where $\hat{\psi}_j$ and c_j for the reflected wave have reversed sign from those of the incident wave.

8.4 Waves in infinitely deep stratified fluids

The upper radiation condition (§1.2) applies when the fluid is stratified to $z = \infty$ so that waves may propagate upward out of the region of interest without downward reflection. We may represent this condition by placing an infinitely deep layer of fluid with suitably constant N and U on top of the lower fluid region that is of primary interest. In this upper semi-infinite region a general linear disturbance may be expressed in the form

$$\psi(x, z, t) = \text{Re} \int_0^\infty A(k)e^{i(kx+nz-\omega t)}dk, \tag{8.50}$$

for the stream function ψ, where "Re" specifies that the real part is taken, and where $A(k)$ is some suitable complex function. From (8.26) n and ω are related to k by the dispersion relation

$$(\omega - Uk)^2 = \frac{N^2 k^2}{k^2 + n^2}. \tag{8.51}$$

The upper radiation condition states that, since there are no sources of energy at infinity, the motion must not contain any Fourier wave components that carry energy downward (Lighthill, 1965). This implies that the group velocity

$$\mathbf{c}_g \equiv \nabla_k \omega = \left(\frac{\partial\omega}{\partial k}, \frac{\partial\omega}{\partial n}\right), \tag{8.52}$$

must have an upward vertical component: that is, $\frac{\partial\omega}{\partial n} > 0$ for all Fourier components. For a stationary fluid ($U = 0$), each plane wave component $e^{i(kx+nz-\omega t)}$ has the phase and group velocities

$$\mathbf{c} = \frac{\omega}{k^2 + n^2}(k, n), \quad \mathbf{c}_g = \frac{\omega n}{k(k^2 + n^2)}(n, -k). \tag{8.53}$$

These are orthogonal, so that the group velocity vector is aligned with the wave crests as shown in Figure 8.6. Since the energy flux associated with the waves propagates with the group velocity, this is also directed along the wave crests; the horizontal component of $\mathbf{c}_g$ is in the same direction as that of the phase velocity $\mathbf{c}$,

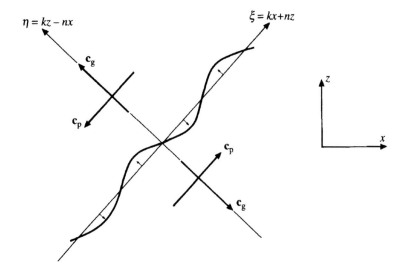

Figure 8.6 Phase and group velocities for a plane internal wave in a fluid at rest. The sinusoidal line depicts the velocity (or displacement) perturbation in a plane wave of the form $e^{i(kx+nz-\omega t)}$, with k and n both positive. The flow is uniform along the lines of constant phase $kx + nz$, which is the alignment of the wave crests and troughs. If $\omega > 0$ (resp. < 0), the phase velocity is upward (resp. downward), and the associated phase and group velocities are given below (resp. above) the curve. If k or n is negative, the corresponding diagram is obtained from a mirror image of this one about a vertical line.

but the vertical component is *oppositely* directed. A wave field (or packet) that is growing and extending upwards with time, for example, has downward-propagating phase lines.

We may note two special cases. First, the limit $n \to 0$ leads to waves with $\omega = N$ with the crests aligned vertically. These waves have no horizontal fluid velocity (i.e. $u = 0$ everywhere) and also have *zero* group velocity, and hence cannot transport energy through the fluid. The motion has degenerated to a set of independently oscillating vertical columns of fluid. Secondly, the limit $k \to 0$ implies $\omega \to 0$ and $w \to 0$, so that all the motion is horizontal. These Fourier components are the infinite-depth analogues of the columnar disturbance modes described in §8.3. In a sense, the motion may be regarded as independently moving horizontal slabs. However, in this limit the group velocity $\mathbf{c}_g = (N/n, 0)$, which is not zero, implying that energy is being propagated horizontally.

A useful concept when discussing wave propagation in non-homogeneous fluids is that of a "wave packet"; this denotes a wave field that has approximately uniform frequency and wavenumber over a given region (in which the mean flow is also reasonably homogeneous), so that it propagates at the local group velocity and remains together as a distinct entity. A wave packet with $n = 0$ will remain stationary relative to the fluid, but one with $k = 0$ will propagate horizontally at speed N/n.

We next consider the situation where N and U are constant throughout the whole fluid for $z > 0$, so that (8.50), (8.51) hold in this region. Whereas the spectrum of vertical wavenumbers for a finite-depth system is discrete, here it is continuous. It is useful to plot (8.51) in (k, n)-space for a particular frequency ω, giving the "wavenumber surface" (the term surface being used to include a possible third dimension) or "Lighthill diagram". As was done in §2.2.2 for surface waves, this diagram enables us to infer the character of the wave field produced by a source of given frequency. The most interesting and relevant case is $\omega = 0$, corresponding to a steady pattern of waves caused by steady forcing of some kind, shown in Figure 8.7.

This surface of possible wave numbers consists of a circle of radius N/U, and the n-axis, $k = 0$, twice. The arrows denote vectors perpendicular to the wavenumber surface (in the direction of ω increasing), and from (8.52), (8.53) it may be seen that these point in the direction of the group velocity in *physical* space (with x and z axes aligned with k and n) for waves with that particular wavenumber vector $\mathbf{k} = (k, n)$. For waves generated in the vicinity of the origin in physical (x, z)-space, therefore, wave energy in the resulting steady-state flow pattern continuously propagates away at the group velocity in various directions. At radial distances $r = (x^2 + z^2)^{1/2}$ somewhat greater than the size of the source (a mountain, say), the wave field in any particular direction consists of waves with wavenumbers whose corresponding arrows point in that direction. Hence wavenumbers on the circle are found on the downstream or lee side (the "lee waves"), and the energy associated with the wavenumber (k, n) propagates in the direction of that wavenumber. This fact results from adding the horizontal advection due to the wind to the group velocity (8.52) of waves moving relative to the fluid. The waves with wavenumbers on the $k = 0$ axis constitute the columnar disturbances, which propagate purely horizontally. For $|n| > N/U$ these waves are only found downstream, but for $|n| < N/U$ they may propagate either upstream or downstream.

If the hydrostatic approximation is made, (8.23) becomes

$$\left(\frac{\partial}{\partial t} + U\frac{\partial}{\partial x}\right)^2 \frac{\partial^2 \psi}{\partial z^2} + N^2 \frac{\partial^2 \psi}{\partial x^2} - \frac{d^2 U}{dz^2}\left(\frac{\partial}{\partial t} + U\frac{\partial}{\partial x}\right)\frac{\partial \psi}{\partial x} = 0. \tag{8.54}$$

With U constant, the wavenumber surface corresponding to Figure 8.7 is shown in Figure 8.8; the surface for the columnar motions is not affected since they are essentially hydrostatic phenomena, but the circle of lee wavenumbers has degenerated to horizontal lines at $n = \mp N/U$. Such waves are only found directly above the source. Consequently, the hydrostatic approximation is only applicable for waves with $k \ll N/U$, as implied by (1.20).

If the source of wave energy is specified, and the motion is linear, the amplitude

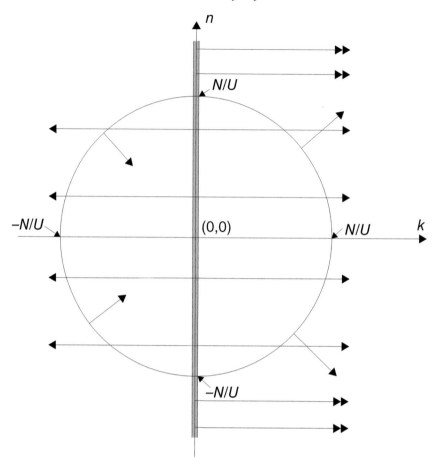

Figure 8.7 The wavenumber surface for steady flows ($\omega = 0$) in a fluid with uniform mean velocity U and buoyancy frequency N, from (8.51). The surface consists of a circle of radius N/U, and the n-axis (twice). The arrows, normal to the surface, denote the direction of group velocity of the waves in physical space, with k corresponding to x, and n to z.

of the disturbance in the far-field (where r is large) may be calculated by the method of stationary phase (see Lighthill, 1978, for a detailed description). This may readily be done for the lee waves when the waves are generated by flow over topography, but it does not apply to the columnar disturbances generated in this way since their source is usually non-linear (see Chapter 10). If N and U are not constant, columnar disturbances will still be present, with their structure determined by (8.26) with $k = 0$; they are not affected by the upper radiation condition.

An internal wave that propagates vertically into a region in which N and U vary with depth will be affected in two main ways by these variations. These effects are

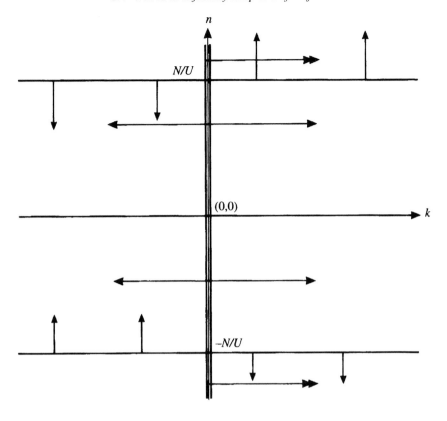

Figure 8.8 As for Figure 8.7, but for hydrostatic flow ($k \ll n$).

manifested in the solutions of (8.26) and appear in the changing structure of $\hat{\psi}$ with height z. Firstly, if N varies with z but U is constant, an upward-propagating wave will be partially reflected downward by the density inhomogeneities. This effect is typical of general wave propagation through an inhomogeneous medium, in which the waves are reflected or scattered by medium variations. Here, propagation is not possible (and reflection becomes total) if N decreases to the point where $N < |\omega - Uk|$. If U varies with z instead of (or in addition to) N, downward reflection will also occur for the same reasons. If N and U both vary together in the same region, the reflections from the shear usually reinforce those from the density variations (Mied & Dugan, 1975). If U becomes sufficiently large, we again have total reflection when $|Uk - \omega|$ exceeds N (see Figure 8.11a, p. 231). However, when U varies, the second effect becomes manifest, namely that the wave may exchange energy with the background mean flow. This means that the energy of the wave field will change as a wave (packet) propagates vertically through the fluid, quite apart from the downward scattering phenomenon from inhomogeneities in U and N, which redistributes the wave energy without altering the total. The discussion of this phenomenon is deferred until §8.7.

The reflection of upward-propagating internal waves from a given change in $U(z)$ or $\rho_0(z)$ is strongest if these inhomogeneities of the background medium are concentrated or abrupt. This means that the "refractive index", the {bracketed term} in (8.26),

$$m^2 = \frac{N^2}{(U-c)^2} - \frac{U_{zz}}{(U-c)} - k^2, \tag{8.55}$$

changes in a nearly discontinuous manner. Further, this downward reflection will be total if m^2 vanishes at some height and is negative everywhere above it. Moreover, m^2 must be positive for wave propagation, and if it vanishes at some level $z = z_1$ it may usually be expressed locally in the form

$$m^2 = B(z_1 - z) + O(z_1 - z)^2, \tag{8.56}$$

where B is a constant. Equation (8.26) then has Airy function (Abramowitz & Stegun, 1968, p. 466) solutions in the vicinity of z_1, with $\hat{y}$ oscillatory below it and exponentially decaying above. These motions are termed "trapped", in the sense that the wave energy is essentially trapped below the level $z = z_1$; such waves propagate purely horizontally below this level, as finite-depth modes. If m^2 again becomes positive at a higher level $z = z_2$ above z_1, waves may again propagate in this higher region. The wave energy may leak from the lower region across (z_1, z_2) to this upper one, giving a "leaky" mode, and this situation is described in §8.5.

We may define the *intrinsic frequency*

$$\omega' = \omega - Uk, \tag{8.57}$$

which is the frequency of the wave as seen by an observer moving with the local mean flow. If it so happens that $d^2U/dz^2 = 0$ at the "turning height" $z = z_1$, we have

$$m^2(z_1) = k^2 \left(\frac{N^2(z_1)}{\omega'^2} - 1 \right) = 0, \tag{8.58}$$

so that the intrinsic frequency is equal to the local buoyancy frequency. In general, we expect trapped modes to occur if m^2 decreases with height for fixed k. Examples of such modes in nature are common, such as high-frequency (< 10 minute period) waves on low-level stable layers in the atmosphere, and, with inverted geometry, on the oceanic thermocline. In general, a system where $m^2 \rightarrow m_\infty^2 > 0$ as $z \rightarrow \infty$, will support both a continuous spectrum of waves (in the vertical) for small k and a discrete one for larger k.

A plane wave propagating in fluid at rest (with given frequency $\omega < N$) that is incident on a sloping (non-horizontal) plane solid boundary, is reflected as another plane wave with phase lines and velocity at the same angle to the horizontal (and same frequency) as the incident wave, though the wavelength and group velocities are different. If the reflecting surface is not a plane but is curved or has bumps in it, some of the incident wave energy is reflected back in the opposite direction to that of the incident wave (Baines, 1971a). If the reflecting surface is ∩-shaped, so that only the lower part of the wave is reflected, the other (upper) part continues unaffected apart from the neighbourhood of the dividing tangential phase line. In steady-state conditions there is no effective diffraction of this wave energy around the obstacle, apart from a stationary wave component that adjoins the tangential line (Baines, 1971b). In both situations, the reflected waves are governed by an integral equation.

8.5 Trapped and leaky modes

In trapped modes, wave energy is confined to a horizontal wave guide with a discrete spectrum of vertical wavenumbers. Each mode also has a distinct dispersion relation between wave speed and horizontal wavenumber, like the finite-depth modes of §8.3. Solutions for trapped modes for a variety of situations are described in Gossard & Hooke (1975), Chapter 11. In the atmosphere, for example, most modes are leaky to some extent, in the sense that wave energy may propagate (leak) into the upper atmosphere, causing the mode to lose energy. Since this situation is so common in practice, it is useful to describe here some prototype situations and their properties in more detail. These are useful in discussing forced motions. We concentrate here on the properties of leaky modes in steady flows, and their relationship to purely trapped ones. The results may be generalised to non-steady propagating modes by a change of axes.

Steady forcing of a stratified fluid, such as flow over topography, results in steady disturbances, and these satisfy (8.24), (8.25) with $c = 0$. We may write these equations in the form (using w instead of ψ for convenience later)

$$w' = \hat{w}(z).e^{ikx}, \tag{8.59}$$

$$\frac{d^2\hat{w}}{dz^2} + (l^2 - k^2)\hat{w} = 0, \tag{8.60}$$

where "Scorer's l^2" is defined by (Scorer, 1949)

$$l^2 = \frac{N(z)^2}{U(z)^2} - \frac{U_{zz}}{U}. \tag{8.61}$$

If l is constant, (8.60) has the solution

$$\hat{w} = e^{inz}, \tag{8.62}$$

where

$$n^2 = l^2 - k^2. \tag{8.63}$$

If a rigid boundary is placed at $z = 0$, the solution

$$w' = e^{ikx}.(e^{inz} - e^{-inz}), \tag{8.64}$$

with n given by (8.63) with arbitrarily chosen k, represents the reflection of a downward-propagating wave at the boundary. This wave is exponentially decaying if k is complex.

Trapped modes are only possible if there are inhomogeneities in the fluid. The simplest prototype situation is that where the vertical profiles of $N(z)$ and $U(z)$ combine to give

$$l = \begin{cases} l_1, & 0 < z < z_1, \\ l_2, & z > z_1. \end{cases} \tag{8.65}$$

If $l_1 > l_2$, this situation may support trapped modes, provided that there are solutions with k real and $l_2 < k < l_1$. Equation (8.65) may be satisfied by a variety of N and U profiles, and for simplicity we assume that ρ and U are continuous, so that the boundary conditions at $z = z_1$ are that w' and dw'/dz are continuous. With a rigid lower boundary at $z = 0$, the structure of a trapped mode is

$$\hat{w} = \begin{cases} \sin m_1 z, & 0 < z < z_1, \\ \sin m_1 z. e^{-m_2(z-z_1)}, & z > z_1, \end{cases} \tag{8.66}$$

where $m_1^2 = l_1^2 - k^2$, $m_2^2 = k^2 - l_2^2$, and

$$m_2 \tan m_1 z_1 + m_1 = 0. \tag{8.67}$$

This equation specifies possible values for k. Trapped modes correspond to real roots for k, and there can only be a finite number of them. This number depends on $(l_1^2 - l_2^2)^{1/2} z_1$, and may be zero if $(l_1^2 - l_2^2)^{1/2} z_1 < \pi/2$. These modes are effectively trapped below the level $z = z_1$, with the disturbance having a vertically decaying, standing-wave character above this level. There is also an infinite number of complex solutions for k, with $k = k_r + ik_i$. These are now "leaky modes", with upward-propagating energy, where the complex wavenumber has discrete values. Note that there is also a continuous spectrum of solutions to (8.60) with (8.65), with k real for $k < l_2$.

A more relevant and complex situation for leaky modes is a three-layer system, where

$$l = \begin{cases} l_1, & 0 < z < z_1, \\ l_2, & z_1 < z < z_2, \\ l_3, & z > z_2, \end{cases} \tag{8.68}$$

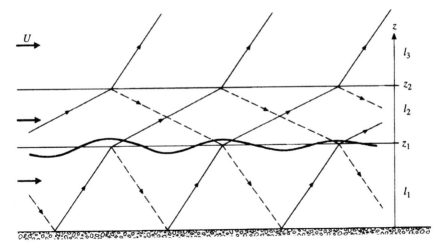

Figure 8.9 Energy flux in a stationary leaky mode with three-layer structure, with $l_1 > l_2 < l_3$, drawn for $l_1 = l_3$. Solid straight lines denote upward energy flux, dashed lines downward energy flux. The mode decays from left to right as $e^{-k_i x}$ in all three layers, and a representative interface displacement at $z = z_1$ is shown. In each of the two lower layers, the motion may be represented by an upward- and a downward-propagating plane wave, and the energy fluxes of these are shown. The nett effect in these two layers is that energy flows rightward and upward. In all three layers, the constant phase lines (not shown) slope upward to the left.

where $l_1 > l_2 < l_3$, and ρ and U are again supposed continuous at z_1, z_2, as before. Solutions to (8.60) then have the form

$$
w = \begin{cases}
\sin n_1 z, & 0 < z < z_1, \\
\frac{1}{2}\cos n_1 z_1 \left(\tan n_1 z_1 + \frac{in_1}{n_2}\right) \times \\
\quad \left[e^{-in_2(z-z_1)} + \frac{n_2+n_3}{n_2-n_3}e^{in_2(z-z_1)}e^{-2in_2(z_2-z_1)}\right], & z_1 < z < z_2, \\
w(z_2)e^{in_3(z-z_2)}, & z > z_2,
\end{cases}
\tag{8.69}
$$

where

$$
n_j^2 = l_j^2 - k^2, \qquad j = 1, 2, 3, \tag{8.70}
$$

with the real part of n_j taken as positive, and k is given by

$$
\left(\tan n_1 z_1 - \frac{in_1}{n_2}\right)(n_2 - n_3) = e^{-2in_2(z_2-z_1)}\left(\tan n_1 z_1 + \frac{in_1}{n_2}\right)(n_2 + n_3). \tag{8.71}
$$

The situations that are of interest here are those where the wave would be trapped if $z_2 \to \infty$, or $l_3 \to l_2$, so that the system would reduce to (8.65) with the solution (8.66), with a solution of (8.67) with k real and positive. If l_3 is greater than this value of k, the corresponding solution (8.69)–(8.71) has k complex, with

$$k_r, k_i > 0, \qquad \text{Re } n_j > 0, \qquad \text{Im } n_j < 0, \qquad j = 1, 2, 3. \qquad (8.72)$$

The mode decays in the downstream direction as $e^{-k_i x}$, with energy leaking upwards through the region $z_1 < z < z_2$ to the uppermost region, where it propagates freely. This solution contains two complex waves in the central region: the first has upward phase propagation, downward energy propagation, and decreases with increasing z, whereas the second, with much lower amplitude at z_1, has downward phase propagation, upward energy propagation, and increases upward. The increase in amplitude with z for $z > z_2$ is physically consistent, as it is for (8.64). Figure 8.9 shows schematically the energy flux in this mode associated with these constituent components, and these modes may be used to interpret observed waves induced by flow over topography as described in succeeding chapters.

8.6 The effects of molecular viscosity and diffusion on internal waves

Molecular viscosity and diffusion act to slowly decrease the amplitude of internal waves. In many geophysical situations their nett effect is small on time-scales of several periods, although it may be significant in the neighbourhood of critical layers. We may obtain a simple description of their effects as follows.

We consider a vertically propagating wave in a stratified fluid where N is constant and $U = 0$. The full equation for ψ for linear disturbances incorporating viscosity ν and diffusivity of density κ (see (1.1), (1.3) and (8.23)) is then

$$\left(\frac{\partial}{\partial t} - \kappa \nabla^2 \right) \left(\frac{\partial}{\partial t} - \nu \nabla^2 \right) \nabla^2 \psi + N^2 \frac{\partial^2 \psi}{\partial x^2} = 0. \qquad (8.73)$$

We look for solutions of the form

$$\psi \sim e^{i(\alpha z - \beta t)} . e^{i(kx + nz - \omega' t)}, \qquad (8.74)$$

where ω', k and n are related by the inviscid dispersion relation (8.51) with $U = 0$, and we assume that κ, ν are sufficiently small so that

$$\kappa, \nu \ll \omega' / M^2, \quad \text{where } M^2 = k^2 + n^2. \qquad (8.75)$$

Substituting (8.74) into (8.73) gives

$$\beta + \frac{\omega' n}{k^2 + n^2} \alpha = -i \frac{(\kappa + \nu)}{2} (k^2 + n^2)[1 + O(\kappa M^2 / \omega', \nu M^2 \omega')], \qquad (8.76)$$

and hence

$$\psi \sim e^{i\alpha(z + \omega' nt / M^2)} . e^{-\frac{1}{2}(\kappa + \nu) M^2 t} . e^{i(kx + nz - \omega' t)}, \qquad (8.77)$$

or alternatively,

$$\psi \sim e^{-i\beta(M^2/\omega'n)(z+\omega'nt/M^2)} \cdot e^{-\frac{1}{2}(\kappa+\nu)M^4z/\omega'n} \cdot e^{i(kx+nz-\omega't)}. \tag{8.78}$$

From (8.53), the vertical component of the group velocity $\mathbf{c}_g$ is $-\omega'n/M^2$, so that the first factor in these expressions describes the vertical propagation of the wave energy, or wave packet as a whole. In particular, it specifies the direction in which the "packet" is moving. The second factor describes the effect of viscosity and diffusivity on the wave as it propagates, and the third represents the plane wave. Equation (8.77) shows that a wave (or wave packet) moving vertically with the group velocity will be damped exponentially in time at the rate $\frac{1}{2}(\kappa+\nu)M^2$. Equation (8.78) by contrast, shows that a vertically propagating wave is damped spatially at the rate $\frac{1}{2}(\kappa+\nu)M^4/|\omega'n|$.

8.7 Energy and momentum transport in a non-uniformly moving fluid

Waves propagating through a stratified fluid transport energy and momentum in the process. If we again assume that the wave amplitudes are small, so that linearised equations are applicable, multiply (8.18) by u', (8.19) by w' and (8.20) by ρ' and add, we obtain

$$\left(\frac{\partial}{\partial t} + U\frac{\partial}{\partial x}\right)E + \frac{\partial}{\partial x}(p'u') + \frac{\partial}{\partial z}(p'w') = -\rho\frac{dU}{dz}u'w', \tag{8.79}$$

where

$$E = \frac{1}{2}\rho_0(u'^2 + w'^2 + N^2\eta^2) \tag{8.80}$$

is the total wave energy per unit volume (i.e. the wave energy density) relative to the local mean flow, η is the vertical displacement of a streamline, and we have used $\rho' = -\eta(d\rho_0/dz)$, which follows from the power series expansion of $\rho_0(x, z + \eta)$ in η. If there is no mean shear, we may identify $p'\mathbf{u}'$ as the energy flux in the local reference frame, and (8.79) constitutes an equation of conservation of wave energy. However, if $dU/dz \neq 0$, so that the mean fluid motion is not uniform, (8.79) has an extra term on the right-hand side which represents a source or sink of wave energy, arising from the interaction of the wave with the shear in the mean flow. Hence the wave energy is *not* conserved, but is exchanged with the mean motion. If we consider internal wave fields that are steady in time, $U(z)E + p'u'$ and $p'w'$ in (8.79) may be regarded as the horizontal and vertical fluxes of wave energy respectively in a fixed frame of reference, so that the left-hand side is the divergence of this flux. If viscosity ν and diffusivity of density κ are included as in (1.1), (1.6), an additional energy dissipation term $\mathcal{D}$ given by

$$\mathcal{D} = \rho_0(\nu\mathbf{u}' \cdot \nabla^2\mathbf{u}' + \kappa N^2\eta\nabla^2\eta) \tag{8.81}$$

should be added to the right-hand side of (8.79).

For waves with a well-defined single frequency ω in the frame of reference in which the mean flow is $U(z)$, the variation of wave energy in a sheared flow may be conveniently described in terms of another local quantity, the *wave action* $\mathcal{A}$, defined by

$$\mathcal{A} = E/\omega' = E/(\omega - Uk). \tag{8.82}$$

This is the local wave energy density E divided by the local frequency ω', where both are defined relative to axes moving with the mean flow. The wave action is therefore an intrinsic property of the waves, independent of the coordinates used. Equation (8.79) with the dissipation term $\mathcal{D}$ included may then be manipulated to give

$$\frac{\partial \mathcal{A}}{\partial t} + \nabla \cdot (\mathbf{c}_g \mathcal{A}) = -(\nu + \kappa)(k^2 + n^2)\mathcal{A}, \tag{8.83}$$

(Bretherton, 1966; Grimshaw, 1974) where $\mathbf{c}_g$ is the group velocity relative to fixed axes, given by (8.52). Equation (8.83) shows that, in the absence of dissipation, $\mathcal{A}$ is conserved in a frame of reference moving with the group velocity. Hence, in this frame moving with the wave packet, E varies in proportion with ω'.

We next consider the effect of these waves on the background mean flow. For present purposes, the mean will be defined to be one of two things: a horizontal average, taken over a wavelength for a periodic wave, or the total horizontal integral for a non-periodic wave field that decreases to zero as $|x| \to \infty$ (the latter would be applicable to a leaky lee-wave field produced by flow over a localised region of topography, for example). Both of these means will be denoted by an overbar, and it is generally only necessary to specify which is implied when specific cases are considered. Waves propagating into a new region of fluid will cause the mean flow, defined in this way, to change. However, if the waves have small amplitude a, the corresponding alterations to the mean flow are $O(a^2)$, so that we may still regard the mean flow as constant in the equations that describe the waves to $O(a)$.

The effect of waves on the mean motion may be seen by integrating (1.1)–(1.4) with respect to x (neglecting viscosity and diffusion). We write

$$\mathbf{u} = \bar{\mathbf{u}} + \mathbf{u}', \qquad p = \bar{p} + p', \qquad \rho = \bar{\rho} + \rho', \tag{8.84}$$

where the mean values of the dashed quantities are zero, and where we suppose that the flow in the absence of waves has the form

$$u = [U(z), 0], \quad p = p_0, \quad \rho = \rho_0, \quad \text{with} \quad dp_0/dz = -\rho_0 g. \tag{8.85}$$

Integrating (1.4) gives

$$\frac{\partial \bar{w}}{\partial z} = 0, \tag{8.86}$$

which implies that $\bar{w} = 0$ in situations with a horizontal lower boundary. From (1.1), (1.3) we then obtain, with the Boussinesq approximation,

$$\frac{\partial \bar{u}}{\partial t} = -\frac{\partial \overline{u'w'}}{\partial z}, \tag{8.87}$$

$$\frac{\partial \bar{p}}{\partial z} = -\bar{\rho}g - \frac{\partial \overline{w'^2}}{\partial z}, \tag{8.88}$$

$$\frac{\partial \bar{\rho}}{\partial t} = -\frac{\partial \overline{\rho'w'}}{\partial z}, \tag{8.89}$$

where the averages relate to periodic or isolated wave fields (in x) as specified above, and in the latter case $\bar{u} \rightarrow U(z)$ as $|x| \rightarrow \infty$. For linear internal waves we have

$$\rho' = -\eta d\rho_0/dz, \quad \text{and} \quad w' = \partial\eta/\partial t + U\partial\eta/\partial x, \tag{8.90}$$

where η is the vertical displacement of a particle from its initial undisturbed state, so that

$$\overline{\rho'w'} = -\frac{1}{2}\frac{d\rho_0}{dz}\frac{\partial}{\partial t}\overline{\eta^2}. \tag{8.91}$$

Equation (8.89) may then be integrated to give

$$\bar{\rho} = \rho_0 + \frac{1}{2}\frac{d\rho_0}{dz}\frac{\partial}{\partial z}\overline{\eta^2}. \tag{8.92}$$

Hence from (8.86), (8.92) the waves introduce small changes to $\bar{p}$ and $\bar{\rho}$ (from p_0 and ρ_0), but these are generally of secondary importance. Equation (8.87), by contrast, shows that the mean flow is accelerated by the divergence of $\overline{u'w'}$, so that the latter may be identified as the vertical flux of horizontal momentum.

Equation (8.87) shows that if $\partial\overline{u'w'}/\partial z = 0$, the mean flow is steady. If on the other hand we *assume* that the overall flow is steady and integrate (8.79) with respect to x, we obtain

$$\frac{d\overline{p'w'}}{dz} = -\rho_0\frac{dU}{dz}\overline{u'w'}. \tag{8.93}$$

We may also multiply (8.18) by $\rho_0 U u' + p'$ and integrate to obtain

$$\frac{dU}{dz}(\overline{p'w'} + \rho U\overline{u'w'}) = 0. \tag{8.94}$$

For these equations to be interesting we require $dU/dz \neq 0$, and (8.93), (8.94) together give (Eliassen & Palm, 1961)

$$U\frac{d\overline{u'w'}}{dz} = 0. \tag{8.95}$$

Hence, provided that $U \neq 0$ so that critical layers are avoided, for steady flows the vertical flux of horizontal momentum $\overline{u'w'}$ is independent of height z, and the vertical flux of wave energy $\overline{p'w'}$ is proportional to $U(z)$. If $\overline{p'w'}$ is positive, $U\overline{u'w'}$ is negative. As a wave propagates vertically in this steady state, therefore, if U is positive and increases with height then so will the upward flux of wave energy. If U decreases toward zero, on the other hand, so will this energy flux.

We may imagine a situation where an internal wave of constant amplitude (either horizontally periodic or of finite horizontal extent) is introduced from below into a fluid where $U(z) > 0$, and begins to propagate upward and against the flow (Figure 8.10a). A non-zero gradient of $\overline{u'w'}$ must be present in the leading part of this wavy region where the wave has not yet reached its maximum amplitude, and this must result in a deceleration of the mean flow. However, when the wave reaches its steady-state amplitude, $\overline{u'w'}$ becomes uniform with height, the deceleration stops, and the mean flow becomes steady. If $U(z)$ is not constant, this steady state involves a steady exchange of wave energy with the mean flow. The reverse process occurs when the source of the wave field is "switched off", and the mean flow is accelerated back to its original value. These changes to the mean flow may be calculated by first solving equations (8.18)–(8.21) for the internal wave field. The associated changes to the mean flow with time are then given by integrating (8.87), with (8.88) and (8.89).

The above equations are for an inviscid fluid. In practice, the effect of dissipative processes on the waves may be very significant, and if viscosity and diffusion are included, (8.79) contains the extra term $\mathcal{D}$ (see (8.81)) on the right-hand side, where $\overline{u'w'}$ is the rate of dissipation of energy due to molecular processes. Equation (8.93) will then contain the horizontal average of this term, $\overline{\mathcal{D}}$, and corresponding dissipative terms appear on the right-hand sides of (8.94) and (8.95). Equation (8.87) becomes

$$\frac{\partial \overline{u}}{\partial t} = -\frac{\partial \overline{u'w'}}{\partial z} + \nu \nabla^2 \overline{u}, \tag{8.96}$$

where the last term denotes viscous stresses in the mean flow, and is generally small. On the other hand, when viscous dissipation of the internal waves is taken into account, the wave momentum flux divergence term is not zero in the steady state and its effect may be quite substantial (Figure 8.10b). If the wave field is maintained for a long time, this term may cause a steady secular change to the mean flow while the wave field persists, and this is *not* reversed when the wave production is "switched off". Dissipation of the internal waves, therefore, can make a significant and lasting contribution to the mean motion, and we investigate this further in the next section.

If internal wave fields persist for long enough, the changes to the mean flow may become large enough to have a significant effect on the wave propagation. The best

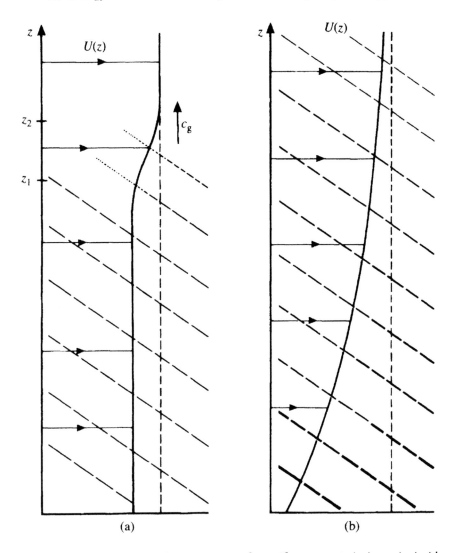

Figure 8.10 (a) The vertical propagation of a uniform wavetrain in an inviscid non-diffusive fluid, introduced into a mean flow U at or below $z = 0$. The front of the wavetrain moves upward at the group velocity, and at this time it is located between z_1 and z_2. Below this front the wavetrain is steady. The mean flow is decelerated in the wave front by a finite amount (given by the second term on the right-hand side of (8.108)), and remains constant below it. The diagonal lines denote the phase lines of the wavetrain. (b) As for (a) but with viscosity included. The wave front has passed through the top of the diagram and the wavetrain is now quasi-steady, with its amplitude decreasing upwards due to viscous dissipation. The mean flow is being slowly decelerated due to the dissipation of the wave and the associated divergence of the momentum flux (the third term on the right-hand side of (8.108)), causing a decrease in U. This is most rapid at the bottom where the dissipation is largest.

example of this is the experimental demonstration by Plumb & McEwan (1978) of the analogue of the quasi-biennial oscillation of the equatorial atmosphere (Plumb, 1977), where the waves and mean flow interact in a slow secular oscillatory fashion.

It should be noted that in rotating systems where motion is permitted in the y-direction, the situation is more complicated than is described here, and mean vertical circulations may arise. An overview of the topic, with examples, has been provided by McIntyre (1980) and Andrews (1980).

8.8 The "slowly varying" or WKB approximation

So far in this chapter we have considered linear disturbances of small amplitude, which propagate vertically through fluid where N and U vary with height. These variations may cause the waves to be reflected downward (partially or wholly), according to the solutions of (8.23)–(8.25). If these variations are sufficiently gradual, these downward reflections are minimal, and may be neglected. In this situation the wave properties may be described by use of the "slowly varying" approximation. This approximation is often known as the WKB, or WKBJ approximation, after Wentzel, Kramers, Brillouin and Jeffreys, who independently derived it in 1925, although Gill (1982) has pointed out that it was derived much earlier by Liouville and Green in the 1830s. Here we refer to it by the descriptive name, the "slowly varying" approximation. It enables the derivation of equations that govern the interaction between the wave field and the mean flow, under conditions where the waves may adjust gradually to changes in the mean flow as they propagate. Dissipation due to viscosity and diffusion is included, since these are important, overall, as described above. The equations were first derived in this context by Bretherton (1966), and the viscous terms were added by Grimshaw (1974), to which the reader is referred for full mathematical details.

We assume that the mean flow and the mean wave properties only vary with z and t, and that

$$\frac{L}{\rho}\frac{d\rho}{dz} \ll 1, \qquad \frac{L}{U}\frac{dU}{dz} \ll 1, \qquad \frac{\nu + \kappa}{NL^2} \ll 1, \tag{8.97}$$

where $L = 2\pi/n$ is a representative vertical wavelength of the waves, and ν and κ are the viscosity and diffusivity, respectively. Since $n \approx N/U$ in many cases, (8.97) also requires that the Richardson number, (8.38), be large. Locally, therefore, waves may propagate as plane waves of the form $e^{i(kx+nz-\omega't)}$, relative to axes moving with the local mean flow U, which do not "feel" the large-scale variation of U and N. If ω is the frequency of the wave relative to fixed axes, we have

$$\omega' = \omega - Uk, \qquad \omega'^2 = \frac{N^2 k^2}{k^2 + n^2}, \tag{8.98}$$

as for (8.51). We may express the stream function ψ for the whole wave field in a form which is consistent with this property, namely

$$\psi = \psi_c(z,t)e^{i\phi(x,z,t)}, \tag{8.99}$$

where ψ_c is the amplitude and ϕ a phase function, which varies much more rapidly with z and t than does ψ_c, so that the derivatives of ψ are dominated by those of ϕ. The wave number $\mathbf{k}$ and frequency ω are then defined by

$$\mathbf{k} = \nabla\phi, \qquad \omega = -\frac{\partial\phi}{\partial t}, \tag{8.100}$$

where $\mathbf{k}$ and ω are also functions of z and t, varying on the scale of variation of the mean flow. From (8.100) it follows that

$$\nabla \times \mathbf{k} = 0, \qquad \frac{\partial\mathbf{k}}{\partial t} + \nabla\omega = 0, \tag{8.101}$$

where the latter equation expresses the conservation of wave crests. In the present context, these simplify to

$$k = \text{constant}, \qquad \frac{\partial n}{\partial t} + \frac{\partial\omega}{\partial z} = 0. \tag{8.102}$$

Equations (8.98) and (8.102) constitute the ray equations (analogous to geometrical optics) for this situation. They may be solved, if $N(z)$ and $U(z)$ are known, to obtain ω and n as functions of z and t. We may then obtain the wave amplitudes from equations for the wave action, $\mathcal{A}$, of §8.7. In the present situation equation (8.83) takes the form

$$\frac{\partial\mathcal{A}}{\partial t} + \frac{\partial}{\partial z}(c_{gz}\mathcal{A}) = -(\nu + \kappa)(k^2 + n^2)\mathcal{A}, \tag{8.103}$$

where $c_{gz} = \partial\omega/\partial n$, the vertical component of the group velocity, so that $\mathcal{A}$ and E, and hence the wave amplitude, may be calculated as functions of z and t.

In the "slowly varying" approximation, we neglect the effect of the variation in U and N and the dissipative terms on the local properties of the waves, and only include them when considering the variation in the wave properties over the longer distances where these changes become manifest. From (8.18), therefore, we may write

$$\overline{u'w'} = \frac{k}{\omega - Uk}\frac{\overline{p'w'}}{\rho_0}, \tag{8.104}$$

generalising (8.94), and since $\overline{p'u'}$ is the energy flux relative to the local mean flow, we have

$$\overline{p'w'} = c_{gz}E = c_{gz}\omega'\mathcal{A}. \tag{8.105}$$

Hence, (8.87) for the mean motion may be written

$$\frac{\partial \bar{u}}{\partial t} + \frac{k}{\rho_0} \frac{\partial}{\partial z}(c_{gz}\mathcal{A}) = 0. \tag{8.106}$$

Since $\mathcal{A}$ has been determined from the above equations, (8.106) permits the determination of the resulting changes in $\bar{u}$ from the initial profile $U(z)$, after the wave field has been propagating (possibly in an unsteady manner) for an arbitrary time. Furthermore, although the waves are assumed to have locally linear properties, this change in $\bar{u}$ is not restricted to being small, although it must still satisfy (8.97). From (8.103), (8.106) we may obtain

$$\frac{\partial}{\partial t}\left(\bar{u} - \frac{k}{\rho_0}\mathcal{A}\right) = \frac{k}{\rho_0}(\nu + \kappa)(k^2 + n^2)\mathcal{A}, \tag{8.107}$$

so that

$$\bar{u}(z,t) = U(z,0) + \frac{k}{\rho_0}[\mathcal{A}(z,t) - \mathcal{A}(z,0)]$$

$$+ \frac{k}{\rho_0}(\nu + \kappa)\int_0^t [k^2 + n(z,t')^2]\mathcal{A}(z,t')dt'. \tag{8.108}$$

Figure 8.10b shows an example of the effects described by this equation. The presence of the waves adds a component $k\mathcal{A}/\rho_0$ to the mean velocity, relative to the situation when the waves are absent. If the wave field is steady ($\omega = 0$) so that the waves are propagating against the flow, with negligible dissipation, this component is part of the constant mean flow; for upward propagating steady waves with $U > 0$, $k\mathcal{A}$ is negative, so that the presence of the waves decreases the flow speed. The secular viscous term acts in the same direction. Equations (8.98), (8.102), (8.103) and (8.108)), with U replaced by $\bar{u}$, constitute a set of evolution equations for $n(z,t)$, $\mathcal{A}(z,t)$ and $u(z,t)$, which may be solved numerically for given initial and upper and lower boundary conditions. For further details on the interaction between waves and mean flows see Craik (1985).

How significant is this contribution of the waves to the mean flow? We may take $E \approx \rho_0 N^2 \eta^2$, so that for steady flow

$$k\mathcal{A}/\rho_0 = -N^2\eta^2/U. \tag{8.109}$$

With $N = 10^{-2}\mathrm{s}^{-1}$, $\eta=100\,\mathrm{m}$, we have $k\mathcal{A}/\rho_0 \approx -1/U\mathrm{ms}^{-1}$ with U in ms^{-1}, which is very small in an atmospheric context unless U is very small. For a large-amplitude wave with $\eta=500\,\mathrm{m}$ (i.e. a crest to trough height of 1 km) we have $k\mathcal{A}/\rho_0=-25/U$, so that the wave contribution is significant ($>1\,\mathrm{ms}^{-1}$) if $U<25\,\mathrm{ms}^{-1}$, but not large. With small viscosity and diffusivity, the dissipative term is small unless n is large. Hence the wave-mean-flow interaction is generally small unless $(c-U)/N\eta$ is small.

This condition implies that the waves are close to a critical level, and we discuss this bizarre phenomenon in the next section.

8.9 Critical layers

Critical layers and critical levels occur quite frequently in nature, and they have a profound effect on wave propagation. As outlined above, a critical level for a wave with horizontal phase speed c is a level or height where $U(z) = c$; the expression *critical layer* is used to denote the region of fluid close to the critical level where the dynamics are strongly influenced by it. We may distinguish two main types of critical layer: those where $R_i < 1/4$, and those where $R_i > 1/4$. In the first case the motion near the critical level is non-oscillatory, and it includes the homogeneous situation where $R_i = 0$. The mechanics of this situation have been discussed at length elsewhere (e.g. Craik, 1985). Here our primary concern is the interaction between critical layers and internal waves, and we assume that $R_i > 1/4$ throughout this section.

There are two main ways in which internal waves may encounter critical levels. Firstly, waves propagating vertically may enter a region where $U(z) - c \rightarrow 0$, and we concentrate on this situation in this section. Secondly, a wave "trapped" in a horizontal wave guide may propagate horizontally into a region where U or N vary with x, so that $U - c \rightarrow 0$ at some level within the flow; alternatively, waves may be generated locally by coherent forcing in a layer within which $U - c = 0$. A common example occurs when flow over topography forces an internal wave that breaks, resulting in a localised region of stationary fluid. This region may be regarded as a localised critical layer induced by the wave (see Chapter 11).

We consider the behaviour of a plane wave of small amplitude propagating vertically towards a critical level, as governed by linear inviscid equations. We assume that a wave component has the form

$$\psi = e^{-inz}.e^{ik(x-ct)}, \qquad k, \ n, \ c > 0, \tag{8.110}$$

in a region of uniform N and U, and that this wave propagates vertically into a region where U monotonically increases to give $U = c$ at $z = z_c$ (see Figure 8.11b). If we assume that the flow is steady, the behaviour of such a linear disturbance is governed by (8.25), which becomes singular at $z = z_c$. In order to determine how the wave behaves near this level, and what the changes in the solution are across it, we need to consider how the wave field grows with time to reach a steady state, or alternatively, to introduce other physical factors that give new terms in the equation that remove the singularity, such as non-linearity or viscosity. Once the character of the solution with these new factors has been determined, the magnitude of the latter may be reduced to zero to give the corresponding linear inviscid steady solution.

We first consider the inviscid time-dependent problem, and the simplest way of doing this is to permit the wave speed c to be complex, with $c = c_r + ic_i$. Solutions with $c_i > 0$ correspond to wave motion forced from below (for $n > 0$) with forcing growing at a rate kc_i. This is physically plausible if kc_i is small, and the singularity in (8.25) is circumvented (an equivalent but slightly different procedure is to introduce *Rayleigh friction*, which is an artificial drag force proportional to the local fluid velocity). If we assume that $c_i |d^2U/dz^2|/(dU/dz)^2 \ll 1$, standard mathematical techniques may be used to show that the solution of (8.25) in the neighbourhood of the critical level $z = z_c$ has the form (Booker & Bretherton, 1967)

$$\hat{\psi}(z) = A_1 \left(z - z_c - \frac{ic_i}{dU/dz} \right)^{1/2+i\mu} + A_2 \left(z - z_c - \frac{ic_i}{dU/dz} \right)^{1/2-i\mu}, \tag{8.111}$$

where A_1 and A_2 are arbitrary constants and

$$\mu = (R_i - \frac{1}{4})^{1/2} > 0, \tag{8.112}$$

where R_i is the Richardson number defined by (8.38). In the limit $c_i \to 0$, (8.111) may be written

$$\hat{\psi}(z) = \begin{cases} |z - z_c|^{1/2}(A_1 e^{i\mu \log|z-z_c|} + A_2 e^{-i\mu \log|z-z_c|}), & z > z_c, \\ -i|z - z_c|^{1/2}(A_1 e^{\mu\pi} e^{i\mu \log|z-z_c|} + A_2 e^{-\mu\pi} e^{-i\mu \log|z-z_c|}), & z < z_c, \end{cases} \tag{8.113}$$

which is the steady inviscid solution.

By analogy with the form of plane waves (e^{-inz}), we see that the first part of this solution ($e^{i\mu \log|z-z_c|}$) denotes downward phase propagation and hence upward energy propagation, particularly if $\mu \gg 1$ where the waves are locally sinusoidal. Similarly, the second part of the solution denotes downward energy propagation. Both parts are discontinuous at $z = z_c$ by a factor $e^{-\mu\pi \pm i\pi/2}$, and as R_i and μ increase, the fraction of the wave transmitted through the critical layer becomes negligible. With u' and w' given by (8.22), the horizontal mean of the Reynolds stress is given by

$$\overline{u'w'} = -\frac{ik}{4}(\hat{\psi}^* \hat{\psi}_z - \hat{\psi} \hat{\psi}_z^*), \tag{8.114}$$

where the asterisk denotes complex conjugate, and substituting (8.111) we deduce

$$\lim_{c_i \to 0} \overline{u'w'} = \begin{cases} \dfrac{\mu k}{2}(|A_1|^2 - |A_2|^2), & z > z_c, \\ -\dfrac{\mu k}{2}(|A_1|^2 e^{2\mu\pi} - |A_2|^2 e^{-2\mu\pi}), & z < z_c. \end{cases} \tag{8.115}$$

Each part of $\overline{u'w'}$ is discontinuous in magnitude across $z = z_c$ by a factor $e^{-2\mu\pi}$ and

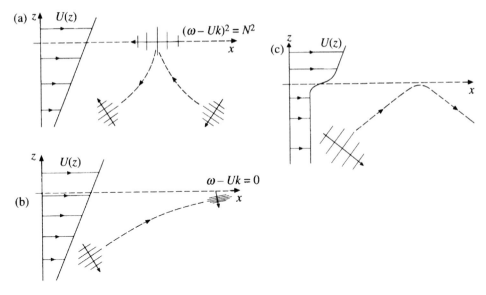

Figure 8.11 (a) Schematic representation of the reflection of a plane internal wave at a turning level where $\omega' = N$ in a uniform shear flow. A wave packet follows the curved dashed line. (b) As for (a), but showing a wave gradually encountering a critical level, where $\omega' = 0$. (c) As for (b), but showing a wave encountering a critical level abruptly, so that most of the wave energy is reflected.

changes sign, but is constant elsewhere as expected from (8.95). The vertical wave energy flux

$$\overline{p'w'} = -\rho_0 \overline{(U - c)u'w'} \qquad (8.116)$$

is positive (i.e. upward) for the first part of the solution, and negative (i.e. downward) for the second part, as expected from (8.113). This solution shows that a wave incident on a linear inviscid critical layer does not by itself produce a reflected wave. Instead, the incident wave produces a transmitted wave on the other side of the critical layer with a greatly reduced amplitude, but still propagating in the same direction, and the associated momentum flux is toward (or away from) the critical layer on *both* sides, and is discontinuous there.

A physical interpretation of the behaviour represented by this solution is as follows. A plane internal wave (or wave packet) propagating upward (say) in a uniform shear flow experiences a rotation of its phase lines, since these are advected as though they are fixed in the fluid (Hartman, 1975). This results in a curved trajectory for the wave packet (described for slowly varying cases by (8.98), (8.102), (8.103)), and depending on their initial orientation the phase lines will either become vertical, with intrinsic (local) frequency $\omega' = N$, or progressively more horizontal, with $\omega' = 0$ as the wave approaches a critical layer. When the former occurs, the waves have reached a turning height, and are reflected back downward with ω' decreasing, as in Figure 8.11a. When $\omega' \to 0$, on the other hand, the wave propagation speed relative to the fluid similarly decreases, and the resulting motion becomes primarily advected by the mean flow (Figure 8.11b; Phillips, 1966; Hartman, 1975). (Note w' and ψ decay with time as $[(dU/dz)t]^{-3/2}$, and u', ρ' and η as $[(dU/dz)t]^{-1/2}$, as the wave approaches the critical level.) The lost wave energy is transferred to the mean flow. For a continuously generated incident wave, the various constituent decaying parts superimpose to give a horizontal velocity that grows as $[(dU/dz)t]^{1/2}$, at $z = z_c$, leading to $u' \to |z - z_c|^{-1/2}$, as $t \to \infty$.

The above inviscid picture is incomplete in the steady state because of the singularity at $z = z_c$, which implies that momentum is being exchanged between the wave and the mean flow at this level, and nowhere else. Since nature abhors singularities, non-linear effects or viscosity and diffusion must be important in practice, although their relative importance will depend on the situation. Further, the flow is implicitly unsteady because the mean flow must change. If the incident wave amplitude is sufficiently small, we may expect that viscosity and/or diffusion will be important near the critical level, and that non-linear effects will be negligible until a long time has elapsed. Numerical studies of this steady linear flow with viscosity and diffusion have shown that the transmitted momentum flux across the critical level is again smaller than the incident by the factor $e^{-2\mu\pi}$, and that the reflected energy is still negligible (Hazel, 1967).

If the wave amplitude is sufficiently small so that the equations are linear, therefore, the theoretical behaviour of critical layers in viscous fluids seems fairly clear. However, this description is rarely adequate in practice. If an internal wave has a gradual approach to a critical level (as distinct from one where it encounters it abruptly due to a sudden change in U, for example), the increasing viscous dissipation due to the increase in n and decrease in wave speed may be sufficient to ensure that the wave never reaches level z_c at all. If the amplitude of the incident wave is increased, we may expect non-linear effects to become more significant, and the question arises: is there a criterion that specifies when the linear viscous/diffusive model ceases to be valid? Results from laboratory experiments (Koop, 1981; Koop & McGee, 1986; Thorpe, 1981) and numerical computations (Fritts & Geller, 1976; Fritts, 1982; Winters & D'Asaro, 1989) for the case where N and dU/dz are approximately constant have shown that, as the amplitude of the incident wave is increased, non-linear advection eventually dominates viscosity and diffusion, and the wave steepness grows until the density field becomes statically unstable in the vicinity of the critical level. However, linear wave theory gives a good description of the flow up to the point of instability (Fritts, 1982), and the vertical wavenumber increases as the critical level is approached.

Hence we may employ the "slowly varying" approximation to obtain a criterion for the onset of instability as a wave approaches its critical level (even though, in this approximation, the wave never actually reaches this level). We consider a plane wave approaching a critical level where N and dU/dz are constant, with a vertical wave action flux $c_{gz}\mathcal{A} = \rho_0 n_0 \overline{w_0'^2}/k^2$, where n_0 and w_0' are the values of n and w' at some arbitrary level $z = 0$, well below the critical level. With the slowly varying formalism, as the wave approaches the critical level its amplitude first increases to a maximum, and then decreases due to viscosity and diffusion. If the maximum amplitude is sufficiently large, the wave field will be unstable. To obtain this maximum amplitude we integrate (8.98), (8.102) and (8.103), assuming that the mean and wavy motion are both steady, so that $\overline{u}$ is constant in time. We write $(\overline{u}) = U(z)$, and assuming that dU/dz is constant, (8.98) implies that

$$n(z)^2 = N^2/U(z)^2 - k^2. \tag{8.117}$$

We assume that instability occurs when $d\rho/dz \geq 0$ in the wave field, so that the fluid becomes statically unstable and overturns. This implies that $u = U + u' \leq 0$ somewhere. From (8.99), (8.100), this is equivalent to the condition

$$|n\psi_a| \geq U(z). \tag{8.118}$$

Integrating (8.103), assuming no overturning, gives

$$\log(n\psi_a^2/n_0\psi_{a0}^2) = -(\nu + \kappa)\int_0^z \frac{N^4}{knU^5}dz. \tag{8.119}$$

If we also assume that at the maximum amplitude we have $n \gg k, n_0$, from (8.118), (8.119) the necessary criterion for the density field to be statically stable everywhere is

$$n_0 k^2 \psi_a^2 \text{ or } n_0 \overline{w_0'^2} < \frac{e}{6}(\nu + \kappa)^2 Nk\left(\frac{dU}{dz}\right)^{-1} = \frac{e}{6}(\nu + \kappa)^2 kR_i^{1/2}. \tag{8.120}$$

Note that although the internal wave amplitude is reduced by viscosity and diffusion, much of the internal wave energy is not dissipated but is converted to kinetic energy of the mean flow, as in (8.108).

Equation (8.120) applies in the simplest case of constant N and wind shear. It is necessary to evaluate it separately for other profiles. Some tests of this criterion have been made by Koop & McGee (1986), and the results are shown in Figure 8.12. Figure 8.12a,b show the observed and computed density fields (on the same basis as in §8.8 when the static stability criterion is satisfied, and Figure 8.12c,d show the same fields when it is not, and the results are consistent. Equation (8.118) is also marginally satisfied in the experiments of Plumb & McEwan (1978), in which waves approaching the critical level appeared to break without violent overturning, and substantial changes occurred in the mean flow over a long period of time.

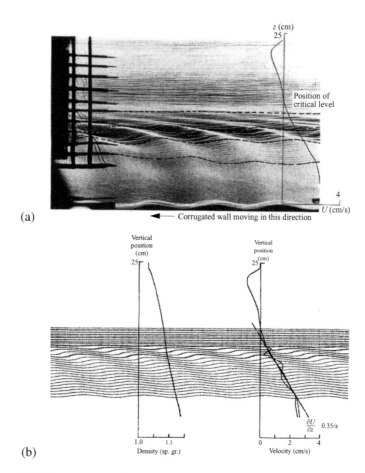

(a)

(b)

Figure 8.12 Observations of critical layers in two shear flows. (a) Shadowgraph image of a wave field generated by a corrugated lower boundary (wavelength 7.5 cm) towed at speed 2.5 cm/s from right to left, moving with the boundary. The measured velocity profile in this frame is shown on the right. Dashed lines have been superimposed on the photograph to accentuate the isopycnal displacements, and the uppermost shows the position of the critical level. (b) shows the displacement field for the situation in (a) in the same frame, calculated using the slowly varying wave action model described in §8.8. In the density profile, sp. gr. denotes specific gravity. In the velocity profile, the straight line denotes the undisturbed mean, and the dashed line the computed variations. (c) (opposite) As for (a), but with a (longer) 15 cm wavelength lower boundary towed at 3.88 cm/sec. Small-scale turbulence is now apparent in the wave field below the critical level. (d) The theoretical displacement field for (c), computed as for (b), which is now statically unstable. (From Koop & McGee, 1986, reproduced with permission.)

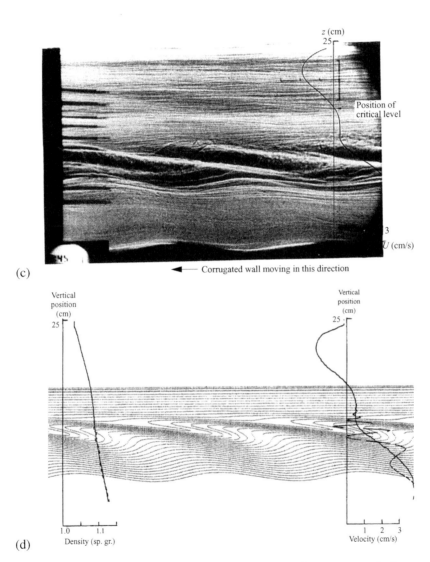

(c)

(d)

Figure 8.12 continued. The configuration and notation for a system of n homogeneous layers, surmounted by an infinitely deep layer. Here u_i and d_i denote the velocity and thickness of the ith layer, and p_s denotes the pressure at the upper boundary of the nth layer.

If the amplitude of the incident wave action flux is increased beyond the values of (8.120), greater overturning and resultant mixing are expected at the critical level, or just below it, yielding patches of homogeneous fluid elongated in the horizontal. The dominant process is simply static instability ($R_i < 0$) rather than shear-induced Kelvin–Helmholtz instability ($R_i < 1/4$), apparently because the two criteria are satisfied almost coincidentally in practice, and the former acts more quickly. The development of this overturning may be dependent on three-dimensional disturbances (Winters & Riley, 1992), that have structure perpendicular to the mean shear. To date, laboratory and numerical experiments have not gone far beyond this point, and the details are no doubt complicated. However, in general the nett result of a sustained incident wave of sufficiently large amplitude is expected to be a well-mixed and near-homogeneous region near the critical level.

It is also quite possible, and indeed quite common, for a vertically propagating internal wave to encounter a critical level *abruptly*, so that (8.120) and the associated dissipative processes are largely circumvented. Whether or not this occurs depends on several factors such as the shape of the $U(z)$ profile and the vertical wavelength and group velocity. A schematic example is shown in Figure 8.11c. Either way, the incident wave arrives in the vicinity of the critical layer at sufficient amplitude to cause overturning and mixing. Parts of the wavetrain arriving at later times will therefore encounter a layer of mixed fluid at or near the critical level.

The effect of a neutrally (or weakly) stratified layer where $U = c$ on a propagating internal gravity wave was discussed in §8.1, in the description of layered models. Regardless of the number of layers or their thickness, the critical layer of finite vertical extent acted as a perfect reflector of wave energy, with its upper and lower boundaries reflecting as pliant surfaces to waves incident from above and below the layer, respectively. This simple picture implies that processes that result in overturning and mixing at the critical level, *regardless of their cause*, will make the critical level act as a pliant (or free) surface reflector. Further, this situation should persist for as long as the incident wave impinges on the layer. Numerical simulations of mountain-induced waves encountering critical levels (Bacmeister & Pierrehumbert, 1988, described in Chapter 11), which involve a localised packet of waves with various wavenumbers but all having the same critical level, and where viscosity is not dominant, show substantial reflection from the region, because the critical layer locally becomes a region of near-uniform u and ρ. In the atmosphere the left-hand side of (8.120) is frequently much larger than the right-hand side, but for internal waves in the ocean they are more nearly comparable.

There are some studies that suggest that the above picture of the characteristics of internal waves incident on critical layers is not complete. If one assumes a steady state, it is possible to obtain near-inviscid solutions for internal wave modes with critical layers where non-linear terms are important near the critical level, and a substantial flux of wave energy occurs across the layer (e.g. Kelly & Maslowe, 1970; Maslowe & Redekopp, 1980). These solutions contain "cat's eyes" of recirculating homogeneous fluid, and assume that the wave amplitude is sufficiently large to overwhelm viscosity and diffusion. Such flows have not yet been realised in laboratory or time-dependent numerical experiments, and given that this critical layer structure is peculiar to each wave, it seems doubtful that they would be relevant to realistic situations, but the question has yet to be settled.

When a horizontally propagating wave mode encounters a critical level due to U varying with x, the above remarks again apply. Since a horizontally propagating mode may be decomposed into an upward- and a downward-propagating wave (as in §8.4), the critical level appears in the midst of the wave field, with waves impinging on it from both sides simultaneously. In the linear viscously dominated region, waves will be damped on both sides of the critical level, so that the mode will decay quite rapidly. And if non-linear wave-breaking occurs, reflection from the "mixed" region will also occur on both sides, with much less wave damping than in the viscous case.

In summary, it would appear that the overall properties of internal waves incident on critical layers where U and N vary gradually and $R_i > 1/4$, may be described fairly simply, in terms of two main types of behaviour.

(i) If the incident wave action flux is sufficiently small and the approach to the critical level is sufficiently gradual, dissipative processes may be large enough to limit the wave amplitude so that it never reaches the critical level, and to prevent overturning.

(ii) At the other extreme, where the wave action flux is larger or the approach to the critical level is abrupt, so that substantial overturning occurs near the critical level, the nett result is expected to be a horizontal, homogeneous mixed layer. This will result in almost total reflection, based on the simple linear wave theory of §8.1 and numerical studies of mountain-induced waves.

The bounds on the first regime have been determined, but those on the second are not precisely known. There must be a transition region between the two, where the mixed layer is not present everywhere, and both absorption and reflection occur. However, it is important to note that in both these theoretical extremes, and in relevant numerical and laboratory experiments, there is no significant transmission of wave energy *through* the critical layer.

The above remarks apply specifically to an internal wave field that is "switched on" to become a more-or-less periodic wave or a steady state, and then encounters or forms a critical layer. It assumes that all (Fourier) components have the same horizontal phase speed, and hence the same critical level. If instead the waves have a spread of wave speeds, the critical "level" will be spread over a corresponding range of heights. Further, time-dependence of either the wave or mean flow may vary the conclusions as stated here. For instance, a transient wave packet has a spread of wave speeds, and hence results in a broader critical layer, and a time-varying background flow may cause the critical layer to disappear altogether (Fritts, 1982).

8.10 Wave-overturning and saturation

When several different internal wavetrains coexist in the same region of fluid, they may interact with each other through non-linear terms (e.g. Phillips, 1977). The velocity field of one advects each of the others. The theory for these interactions is complex unless the waves are of fairly small amplitude, and there is no clear distinction between a field of non-linearly interacting internal waves and "turbulence" in a stratified fluid. However, the waves may combine to cause the fluid to overturn in small localised patches (Orlanski & Bryan, 1969; McEwan, 1973) that cause irreversible mixing, and this is more common if the wave amplitudes are larger. In the atmosphere, where $\rho_0(z)$ decreases exponentially with height (see Chapter 1), vertically propagating waves increase in amplitude as the packets rise because energy conservation implies that η and u increase as $1/[\rho_0(z)]^{1/2}$ as $\rho_0(z)$ decreases. This implies that an ensemble of different wave packets will tend to break in a patchy fashion, losing their wave energy in the process. In accordance with §8.7, this implies that the associated momentum flux of the waves is reduced in these regions, and this results in a local stress on the mean flow.

Assessing the overall effect of a complex field of interacting internal waves on the larger-scale flow is a non-trivial task. However, a concept that has proved useful for this purpose and has some measure of support from observations is the "saturation hypothesis" (Lindzen, 1981). If we suppose that motion at low levels in the atmosphere generates wavetrains that propagate vertically and have substantially the same properties and orientation, the wave amplitude will grow with height and the waves will begin to break at a particular level. The "saturation hypothesis" assumes that in breaking, the waves are not totally destroyed, and the breaking is only partial; as the waves continue to propagate vertically past the initial breaking point, their amplitude is controlled by the criterion that they remain marginally on the point of overturning, by the criterion of static instability. The envisaged process is analogous to that of spilling breakers on the surface of the ocean approaching a beach, where the waves progressively lose amplitude due to the partial breaking. For the internal waves, this implies that the wavetrain loses energy, and there is an associated momentum flux divergence in the region. If the properties of the wavetrain at low levels are known, their properties at upper levels, and their momentum flux as a function of height, may be calculated on the basis of this hypothesis.

8.11 Wave propagation in three dimensions

So far in this chapter we have only considered internal wave motion in the (x, z)-plane. Much of this is applicable in three dimensions. However, there are some aspects that require additional treatment, and as background for the discussion in Chapters 12–15 we next consider the properties of internal wave propagation in the three-dimensional world.

With a coordinate system (x, y, z), we assume that the background mean flow has the form $[U(y, z), V(x, z), 0]$, which is the most general form for a non-divergent horizontal mean flow field. The horizontal shear $(\partial U/\partial y, \partial V/\partial x)$ is large enough to be significant in the atmosphere if its magnitude is comparable with the Coriolis frequency $(2\Omega. \sin(latitude))$, where Ω is the Earth's angular velocity). This is not normally the case, except close to the centre of fairly deep cyclones and kindred phenomena. Since we are omitting rotational effects in this discourse we also omit horizontal shear effects, and assume that $\mathbf{U} = [U(z), V(z)]$.

The equations governing internal gravity wave propagation (corresponding to (8.18)–(8.21)) are then

$$\rho_0 \left[\left(\frac{\partial}{\partial t} + U\frac{\partial}{\partial x} + V\frac{\partial}{\partial y} \right) u' + \frac{\partial U}{\partial z}w' \right] = -\frac{\partial p'}{\partial x}, \tag{8.121}$$

$$\rho_0 \left[\left(\frac{\partial}{\partial t} + U\frac{\partial}{\partial x} + V\frac{\partial}{\partial y} \right) v' + \frac{\partial V}{\partial z}w' \right] = -\frac{\partial p'}{\partial y}, \tag{8.122}$$

$$\rho_0 \left(\frac{\partial}{\partial t} + U\frac{\partial}{\partial x} + V\frac{\partial}{\partial y} \right) w' = -\frac{\partial p'}{\partial z} - \rho'g, \tag{8.123}$$

$$\left(\frac{\partial}{\partial t} + U\frac{\partial}{\partial x} + V\frac{\partial}{\partial y} \right) \rho' + \frac{d\rho_0}{dz}w' = 0, \tag{8.124}$$

$$\frac{\partial u'}{\partial x} + \frac{\partial v'}{\partial y} + \frac{\partial w'}{\partial z} = 0. \tag{8.125}$$

Since we cannot define a stream function for general three-dimensional motion, we must work with some other variable such as w', ρ', p' or η, where

$$\eta = -\rho'/(d\rho_0/dz), \quad w' = \left(\frac{\partial}{\partial t} + U\frac{\partial}{\partial x} + V\frac{\partial}{\partial y} \right)\eta', \tag{8.126}$$

and we choose w' here. Eliminating u', v', p' and ρ' from these equations then gives

$$\left(\frac{\partial}{\partial t} + U\frac{\partial}{\partial x} + V\frac{\partial}{\partial y} \right)^2 \left(\frac{\partial^2}{\partial x^2} + \frac{\partial^2}{\partial y^2} + \frac{\partial^2}{\partial z^2} \right) w' + N^2 \left(\frac{\partial^2}{\partial x^2} + \frac{\partial^2}{\partial y^2} \right) w'$$
$$- \left(\frac{\partial}{\partial t} + U\frac{\partial}{\partial x} + V\frac{\partial}{\partial y} \right) \left(\frac{d^2U}{dz^2}\frac{\partial}{\partial x} + \frac{d^2V}{dz^2}\frac{\partial}{\partial y} \right) w' = 0, \tag{8.127}$$

as the main governing equation corresponding to (8.23), for both w' and η. With U, V and N functions of z only, a wave generated at low levels (for example) of the form

$$w' = \hat{w}(z).e^{i(kx+my-\omega t)}, \tag{8.128}$$

will propagate with k, m and ω remaining constant. Substituting into (8.127) then gives for $\hat{w}$

$$\frac{d^2\hat{w}}{dz^2} + \left[\frac{N^2(k^2 + m^2)}{(kU + mV - \omega)^2} - (k^2 + m^2) - \frac{k\frac{d^2U}{dz^2} + m\frac{d^2V}{dz^2}}{(kU + mV - \omega)} \right] \hat{w} = 0. \tag{8.129}$$

If U, V and N are constant, the [bracketed term] may be represented by n^2, the square of the vertical wavenumber, so that $\hat{w}$ has the form $\hat{w} = e^{inz}$, and the dispersion relation between wave frequency ω and the wavenumber components (k, m, n) is

$$(\omega - kU - mV)^2 = \frac{N^2(k^2 + m^2)}{k^2 + m^2 + n^2}. \tag{8.130}$$

The phase velocity c of the wave has the direction $\mathbf{k} = (k, m, n)$ and is given by

$$\mathbf{c} = \frac{\omega}{|\mathbf{k}|^2}\mathbf{k} = \left[\mathbf{k} \cdot \mathbf{U} \pm N \left(\frac{k^2 + m^2}{k^2 + m^2 + n^2} \right)^{1/2} \right] \frac{\mathbf{k}}{k^2 + m^2 + n^2}. \tag{8.131}$$

This should be distinguished from the velocity $\mathbf{c}_p$ of a line of constant phase in any particular direction $\mathbf{r}$ say, which is given by

$$\mathbf{c}_p = \frac{\omega}{\mathbf{k} \cdot \mathbf{r}}\mathbf{r}. \tag{8.132}$$

Now $\mathbf{c}_p$ is equal to $\mathbf{c}$ if $\mathbf{k}$ and $\mathbf{r}$ have the same direction, but otherwise its magnitude is greater, and it may become infinite if $\mathbf{r}$ becomes perpendicular to $\mathbf{k}$. The group velocity is given in the usual way by

$$\mathbf{c}_g = \left(\frac{\partial\omega}{\partial k}, \frac{\partial\omega}{\partial m}, \frac{\partial\omega}{\partial n} \right)$$

$$= (U, V, 0) \pm \frac{Nn}{(k^2 + m^2)^{1/2}|\mathbf{k}|^3}(kn, mn, -k^2 - m^2), \tag{8.133}$$

which is perpendicular to $\mathbf{c}$ if the horizontal advection is omitted. The vertical momentum flux at each point is now $(u'w', v'w')$, and this may be averaged in a variety of ways.

Steady flow fields where $\omega = 0$ may be produced by steady forcing, from uniform flow over topography, for example. As in §8.4, we may plot the wavenumber surface (Lighthill diagram) for (8.130) for $\omega = 0$, with $V = 0$ so that the wind is aligned with the x-axis, and the result is shown in Figure 8.13a. This should be contrasted

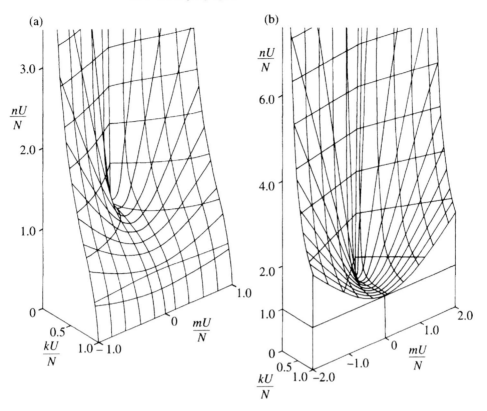

Figure 8.13 (a) Three-dimensional wavenumber surface (Lighthill diagram) for steady flow with uniform N and U (in the positive x-direction) for waves forced by an isolated source, obtained from (8.130) with $V = 0$. The surface is symmetric about the $(k = 0, n = 0)$-planes, and only the quarter with $k, n > 0$ is shown. Normals to this surface with positive x-components denote the direction of the group velocity of the waves in physical space, as in Figure 8.7. (b) The same as for (a) but with the hydrostatic assumption $(k^2 + m^2 \ll n^2)$. This is the three-dimensional form of Figure 8.7.

with Figure 8.6. The diagram shows the wavenumbers of motion that may be forced by steady flow over an isolated obstacle. Normals to the surface denote the group velocity vectors for the respective wavenumbers, showing the direction in physical space in which such waves may be found if the k, m and n axes are interpreted as x-, y- and z-axes. All waves have a component of group velocity directed downstream. In contrast to the Kelvin ship wake of Chapter 2 where the waves are confined to lie within a wedge with a definite angle, here waves are possible in every direction on the downstream side, so that the theoretical wedge angle is $\pi/2$. Note also that the purely horizontally propagating modes (with $k = 0$) of the two-dimensional case are totally absent here for $n < N/U$; they are still present for $n > N/U$, with distinct directions depending on n.

If we make the hydrostatic approximation in (8.127), the term $\nabla^2 w'$ is effectively replaced by $\partial^2 w'/\partial z^2$, which implies that $k^2 + m^2 \ll n^2$ in (8.130). The wavenumber surface for this hydrostatic system is shown in Figure 8.13b. This is effectively an expansion of the form of Figure 8.13a in the neighbourhood of $(0,0,n)$. Note that no hydrostatic waves are produced with $n < N/U$.

When $\mathbf{U} = (U, V)$ varies with height, from (8.129) critical levels arise where $\omega = \mathbf{k} \cdot \mathbf{U}$, and for steady flows for which $\omega = 0$, where $\mathbf{k} \cdot \mathbf{U} = 0$. In fact, as (8.129) shows, the vertical structure of the wave is solely dependent on the flow properties projected in the direction of horizontal propagation. If the horizontal wind has constant magnitude but changes direction with height by rotating around, for example, only the component in the direction of $\mathbf{k}$ affects the vertical wave propagation. Equation (8.129) effectively reduces to the one-dimensional form, (8.26), for each component, and the associated properties described in §§8.4 to 8.7 follow suit. However, waves produced by an isolated source and having a spread of horizontal wavenumbers (k, m) with a distribution such as shown in Figure 8.13a or b have a corresponding variation of properties in the vertical. In particular, critical levels may occur over a range of different heights, depending on their horizontal direction of propagation (k, m).

The above equations may be used to address the question of the stability of a stratified mean flow profile $[U(z), V(z)]$ that varies in both direction and speed with height. In the normal mode approach, a general disturbance may be assumed to have the form (8.128), and its stability properties are governed by (8.129) with suitable boundary conditions. We may then choose axes so that $m = 0$, in which case (8.129) reduces to (8.26), regardless of $V(z)$. Hence the three-dimensional flow will be stable if every two-dimensional vertical cross-section through it is stable, but not otherwise.

9

The stability of stratified flows

Neither a borrower nor a lender be;
for loan oft loses both itself and friend,
and borrowing dulls the edge of husbandry.

W. SHAKESPEARE, *Hamlet*

The flows described in this book are assumed to be predominantly laminar, but fluid flows in general can be potentially unstable. This can lead to transition to turbulent flow, which generally has significantly different properties from laminar flows. The stability of laminar flows is a much studied topic. The literature is quite extensive, and results are described in books by Lin (1955), Chandrasekhar (1961) and Drazin & Reid (1981, 2004). In many situations the mathematical details can be quite complex. This chapter addresses mechanisms for instability, in stratified shear flows in particular, but avoids the complexity and concentrates on how the instability processes work.

In fluid dynamics, there are two main types of instability. The first occurs when dense fluid lies above lighter fluid, in an environment where gravity acts downward, which can result in convection (a process termed Rayleigh–Taylor instability). Whether or not convection and consequent turbulence occurs depends on the geometry, the fluid viscosity and the diffusion of heat. The second type of instability occurs in fluids that flow horizontally but the fluid velocity is sheared, or non-uniform. This process is termed Kelvin–Helmholtz instability, and the mechanics of why these flows can be unstable and potentially turbulent are much less obvious. The purpose here is to provide an outline of when such flows (particularly stratified flows) are likely to become unstable, and what the nature of this process is.

9.1 Stability of stratified shear flow: a general criterion

There is a simple sufficient condition for the stability of general steady stratified shear flows, generally known as the *Miles–Howard theorem*. Its derivation is quite short, and is as follows.

A random small initial two-dimensional disturbance in a stratified fluid will excite a whole spectrum of modes with horizontal wavenumbers k. These may be taken to have the form (8.24), and the evolution with time and the vertical structure of such components will be governed by (8.25). Such components will grow with time if c is complex, implying instability (if $\hat{\psi}$ and c are complex, the complex conjugate $\hat{\psi}*$ and $c*$ will also be a solution). We suppose that the fluid lies between levels z_1 and z_2, where z_1 and z_2 may be at horizontal boundaries or at infinity; ψ is presumed to vanish at each of z_1 and z_2. We define

$$\phi = (U - c)^{1/2}\hat{\psi}(z), \tag{9.1}$$

where c and $\hat{\psi}$ may be complex with $c = c_r + ic_i$. Substituting for $\hat{\psi}$ in (8.25), multiplying by ϕ^*, the complex conjugate of ϕ, and integrating over the depth of the fluid we obtain

$$\int_{z_1}^{z_2} \left\{ (U - c)\left(\left|\frac{d\phi}{dz}\right|^2 + k^2|\phi|^2 \right) + \frac{1}{2}\frac{d^2U}{dz^2}|\phi|^2 \right.$$
$$\left. + \left[\frac{1}{4}\left(\frac{dU}{dz}\right)^2 - N^2 \right] \frac{|\phi|^2}{(U - c)} \right\} dz = 0. \tag{9.2}$$

The imaginary part of this equation is

$$c_i \int_{z_1}^{z_2} \left(\left|\frac{d\phi}{dz}\right|^2 + k^2|\phi|^2 + \left[N^2 - \frac{1}{4}\left(\frac{dU}{dz}\right)^2 \right] \frac{|\phi|^2}{|U - c|^2} \right) dz = 0. \tag{9.3}$$

If the local or gradient *Richardson number*

$$R_i \equiv \frac{N^2}{(dU/dz)^2} > \frac{1}{4}, \tag{9.4}$$

at all depths, the only way (9.3) can be satisfied is by $c_i = 0$, for all k. Hence if (9.4) is satisfied at all levels, the fluid must be stable to all small disturbances (Miles, 1961; Howard, 1961). If $R_i < 1/4$ somewhere, the flow may be stable or unstable, and in many cases it is unstable. (The equations used in this short proof are based on the Boussinesq approximation, but as Miles and Howard showed, if this approximation is not made the conclusions are the same, with N defined as in (1.5).)

When $N = 0$, additional necessary criteria for inviscid instability apply, notably that we must have $d^2U/dz^2 = 0$ at some level (this is *Rayleigh's criterion*). When considering topographic effects, it is convenient to imagine that the topography is introduced into a known stable flow in which $R_i > 1/4$ everywhere, and then to study the additional effects caused by this topography. Instability of the flow field may result from such effects. We therefore have a clear-cut necessary criterion for instability, but this gives little indication of the mechanics of the instability process when it occurs, or the nature of the resulting growing disturbance. This is addressed in the next section.

9.2 The process and products of the instability of shear flows

Steady horizontal flows in which $R_i < 1/4$ somewhere may be unstable. The significance of the Richardson number is that it is a ratio of (the squares of) the time-scales of two competing processes: the mean shear and the buoyancy. When $R_i = 1/4$, in one buoyancy period $(2\pi/N)$ the shear advects a line of fluid particles from nearly horizontal with a slope of $-1/(2\pi)$ through the vertical to a slope of $+1/(2\pi)$. This substantial effect is sufficient to prevent wave propagation, as the example of uniform shear in §8.3 and the discussion in §8.9 show, so that in an unstable flow there is a range of depths where propagation is not permitted. The physical causes of shear flow instability are less obvious than those for other fluid dynamical instabilities such as convection, and the essence of the process is best illustrated by a simple prototype, given in §9.2.1. This is followed by two further examples with more complex geometry in §§9.2.2–9.2.3. Some additional examples are described by Caulfield (1994) and Carpenter et al. (2010, 2011).

9.2.1 The stability of an isolated shear layer

The mechanism described here has been shown to apply to a large class of velocity profiles (Baines & Mitsudera, 1994). We take the system

$$U(z) = \begin{cases} U_0, & z > d, \\ U_0 z/d, & -d < z < d, \\ -U_0, & z < -d, \end{cases} \qquad (9.5)$$

with homogeneous fluid ($N = 0$ everywhere), as shown in Figure 9.1a. This velocity profile is an idealised form of a typical shear layer. It consists of two vorticity interfaces where the vorticity discontinuities are of equal magnitude and opposite sign. Each interface along supports a single free wave mode as described in §8.1 and Figure 8.3, but they propagate in opposite directions. For disturbances of the form (8.24), the solution to (8.25) (with $N = 0$) is (Rayleigh, 1896, Section 368)

$$\psi = \hat{\psi}(z)e^{ik(x-ct)}, \quad \text{where} \quad \hat{\psi}(z) = e^{-k|z-d|-i\theta} - e^{-k|z+d|+i\theta}, \tag{9.6}$$

$$\left(\frac{c}{U_0}\right)^2 = \frac{(1-2kd)^2 - e^{-4kd}}{(2kd)^2}, \tag{9.7}$$

where $\hat{\psi}(z)$ depends on c. This disturbance consists of a wave on each interface that has the same form as that of the isolated free mode, but the presence of the other interface implies that these have a specified relative phase θ for given c, and the latter is given by (9.7) rather than (8.17). If $0 < kd < k_s d \approx 0.64$, c is imaginary with $c = ic_i$ (Figure 9.1b), and the flow is unstable, growing exponentially with time. Then θ is given by

$$\tan 2\theta = -\frac{2kdc_i/U_0}{1-2kd}, \tag{9.8}$$

and lies in the range $0 < \theta < \pi/2$. As Figure 9.1b shows, the maximum growth rate lies near the middle of the unstable range of kd.

Figure 9.1a shows the structure of a typical growing disturbance, which is the sum of two *stationary* (i.e. non-moving) waves. If the vertical velocity from the lower wave is in phase with the displacement of the upper wave, it will increase the amplitude of the latter by advection. From symmetry the velocity field of the upper wave will increase the amplitude of the lower wave in the same way, resulting in exponential growth in the amplitude of *both* waves. The total vertical velocity field at each interface has a component that is in phase with the local displacement, causing growth (or decay), plus a component that is 1/4-wavelength out of phase with it that controls the speed of propagation. For the unstable cases, the latter renders the pattern stationary so that advective growth can occur, and this causes the finite bandwidth of the instability around the condition for exact resonance, $kd = 0.5$.

This kinematic mechanism depends on the mutual advection of two otherwise free waves that are *propagating in opposite directions*. Because the vertical velocity of each free wave leads its displacement by 1/4-wavelength in its direction of propagation, the two waves are able to force each other symmetrically and in sympathy. As Figure 9.1a shows, this means that in the resulting growing disturbances the phase for the vertical displacement leans forward with increasing z, whereas that for the vertical velocity leans backward.

If one calculates the perturbation energy of each single free wave mode, one finds that the wave on the lower interface has positive energy and that on the upper, negative energy, in the sense that the presence of the wave adds to or reduces the total energy of the fluid motion (in some suitable frame of reference). This concept of negative energy waves may be used to characterise the instability process

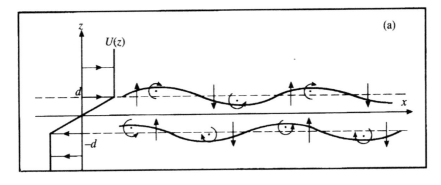

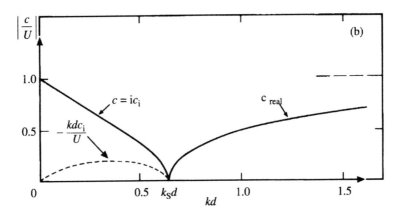

Figure 9.1 (a) The mechanism for instability for two vorticity interfaces in a homogeneous fluid, with the velocity profile as shown on the left. The diagram shows the interface displacements and associated vorticity perturbations (circular arrows) for a typical unstable mode. Maxima in vertical velocity at each interface are denoted by the arrows. (b) Wave speeds (real and imaginary, solid line) and growth rates (dashed line) from (8.20).

(Cairns, 1979; Craik, 1985), and provides an energetic picture that complements the kinematic mechanism described here.

In the limit $d \to 0$, the profile (9.5) tends to a single vortex sheet. Here the mechanism of instability may instead be described in kinematic terms by the self-advection of this sheet (Batchelor, 1967). This mechanism also applies if the vortex sheet coincides with a density interface; the process is then termed *Kelvin–Helmholtz* instability, where an initial disturbance rolls up to give the familiar "billow" structure. Laboratory demonstrations of the development of these instabilities of sheared flow in a tilted tank with two-layer and continuous stratification have been depicted by Thorpe (1968) and Turner (1973).

What happens if uniform density stratification (with uniform N) is introduced to this picture? Stratification above or below the shear layer has no effect on stability, but stratification with $|z| < d$ may, because it can affect interaction between the waves on the two vorticity interfaces. If N is small so that $R_i < 1/4$, the flow is still unstable in the same manner as above, because the motion between the two interfaces is not wave-like (see §8.3). However, if N is large enough so that $R_i > 1/4$, the motion between the two interfaces has the form of propagating waves, and each wave must have a critical level (where the flow velocity equals the phase speed) in this region. When encountering this level the amplitude of the wave across it is reduced by the factor $e^{-\mu\pi}$, where $\mu = (R_i - 1/4)^{1/2}$, and also with a phase change of $\pi/2$, with negligible reflection (see §8.9). This reduction in amplitude is sufficient to destroy the mutual interaction between the two modes that causes instability (Baines & Mitsudera, 1994; Booker & Bretherton, 1967).

9.2.2 Holmboe instability

Situations where the density varies on a much smaller vertical scale than the velocity may be represented by a system where a density interface is embedded within a shear layer. This system produces *Holmboe instability* (Holmboe, 1962), which has the following character. The situation examined by Holmboe had three interfaces: (9.5) with a density interface at $z = 0$, but it suffices to consider the simpler system of one vorticity interface and one density interface, namely

$$U(z) = \begin{cases} U_0, \\ U_0 z/d, \\ U_0 z/d, \end{cases} \qquad \bar{\rho}(z) = \begin{cases} \rho_0, & z > d, \\ \rho_0, & 0 < z < d, \\ \rho_0 + \Delta\rho, & z < 0, \end{cases} \qquad (9.9)$$

as shown in Figure 9.2a. Disturbances of this system corresponding to (9.6)–(9.8) have the form

$$\hat{\psi}(z) = Ae^{-k|z-d|} + Be^{-k|z|}, \qquad (9.10)$$

where $B/A = e^{kd}[2kd(1 - c/U) - 1]$, and c satisfies

$$(c^2 - c_1^2)(c - c_2) + c_1^2(U_0 - c_2)e^{-kd} = 0, \qquad (9.11)$$

with $c_1^2 = g'/2k$ and $c_2 = U_0(1 - 1/2kd)$. Thus c_1 and c_2 are the speeds of the free waves on the two interfaces as given by (8.16)–(8.17). From the above discussion of two vorticity interfaces, we may expect instability when a rightward-propagating gravity wave on the lower interface "resonates" with the vorticity wave on the upper. This implies $c_1 = c_2$, which gives

$$J \equiv \frac{g'd}{2U_0^2} = kd\left(1 - \frac{1}{2kd}\right)^2. \qquad (9.12)$$

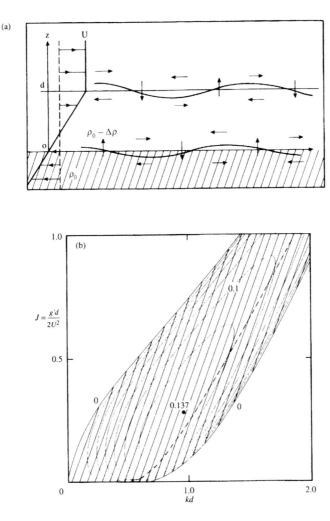

Figure 9.2 The mechanism of Holmboe-type instability, for the simplest prototype case of a density discontinuity $\Delta\rho$ at $z = 0$, and a vorticity discontinuity U' at $z = d$. The velocity profile is shown on the left and the lower layer is shaded. The figure shows the interface displacements and the perturbation fluid velocities (denoted by arrows) in the frame of reference moving with a growing mode (with its speed denoted by the dashed vertical line in the velocity profile), in which they are stationary apart from changes due to growth. A rightward-propagating gravity wave on the density interface (below) resonates with the leftward-propagating wave on the vorticity interface (above), at wavelengths where the speeds of these free waves are approximately equal. At the phase configuration shown, the vertical velocity field of the upper (lower) wave increases the amplitude of the lower (upper) wave. The resulting growth rate of each wave is (approximately) proportional to the difference in phase between its own displacement and velocity fields, *relative* to that of a freely propagating interfacial wave. (b) The region of instability in the (J, kd)-plane, with contours of growth rates c_i/U_0. The dashed line denotes (9.12).

This is plotted as the dashed line in Figure 9.2b, which also shows the curves of constant growth rate in the unstable region of the (J, kd)-plane. Figure 9.2a shows the interface displacements and the vertical and horizontal perturbation velocities near the interfaces for a typical growing mode. The same kinematic advection mechanism as for §9.2.1 again applies, with the phase of the displacement field leaning forward with height, and that of the velocity field leaning backward. For the three-interface system of (9.5) with a density interface at $z = 0$, the growth rates are almost the same as for the two-interface system of Figure 9.2, except where J is small and the mechanism of Figure 9.1 is dominant.

9.2.3 Disturbances in a radiating system

We next look at the complementary situation where the velocity varies much more rapidly with height than the density, by examining another idealised profile. We return to the velocity profile (9.5), and let the buoyancy frequency N have the form

$$N(z) = \begin{cases} 0, & |z| < d, \\ N_0, & |z| > d, \end{cases} \tag{9.13}$$

where the radiation condition of §8.4 applies to motion in both semi-infinite stratified fluids above and below the shear layer. The stream function for disturbances may be written in the form

$$\hat{\psi} = \begin{cases} A e^{i n_1 z}, & z < -d, \\ B e^{kz} + C e^{-kz}, & |z| < d, \\ D e^{-i n_3 z}, & z > d, \end{cases} \tag{9.14}$$

where A, B, C and D are constants, and

$$n_1^2 = \frac{N_0^2}{(U_0 + c)^2} - k^2, \qquad n_3^2 = \frac{N_0^2}{(U_0 - c)^2} - k^2. \tag{9.15}$$

The terms n_1 and n_3 must either be real and satisfy the radiation condition, or have negative imaginary parts giving exponentially evanescent $\hat{\psi}$ for $|z| > d$. The significant properties of this system are obtained by satisfying the boundary conditions at $z = \pm d$ to obtain c. They are shown in Figure 9.3 in terms of $N_0 d / U_0$ and kd, and these may be understood as follows.

Each interface in isolation (with the central region regarded as semi-infinite) supports wave modes of the form

$$\hat{\psi}(z) = \begin{cases} e^{-kR(z-d)}, & z > d, \\ e^{k(z-d)}, & z < d, \end{cases} \tag{9.16}$$

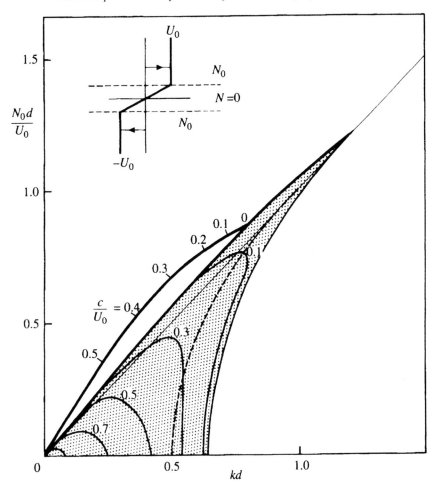

Figure 9.3 Properties of a shear layer with uniform stratification above and below (shown in inset). The system is unstable, with disturbances in the stippled region growing with $c_r = 0$ and contours of c_i/U_0 shown. The dashed line shows the condition for resonance of two free modes, as for Figure 9.2. On the two heavy solid lines (one on the boundary of the unstable region), the system admits steady solutions where waves are radiated to infinity above and below the layer. c varies along the uppermost of these with the values of c/U_0 as shown, and $c = 0$ on the lower. Infinite over-reflection (see text) occurs for an incident wave with N_0d/U_0, kd and c coinciding with the values on these two curves.

for the upper wave, where

$$R = \frac{1 - (N_0d/U_0)^2}{1 + (N_0d/U_0)^2},$$

(9.17)

and similarly for the lower, with wave speeds

$$c = \pm U_0 \left\{ 1 - \frac{1}{2kd} \left[1 + \left(\frac{N_0 d}{U_0} \right)^2 \right] \right\}, \tag{9.18}$$

the "+" sign applying to the upper layer. From the preceding examples in this section we may expect the flow to be unstable at or near the "resonant" condition for these modes to be stationary relative to each other, and equating the two speeds in (9.18) gives

$$N_0 d / U_0 = (2kd - 1)^{1/2}. \tag{9.19}$$

This curve is plotted as the dashed line in Figure 9.3, and since R must be positive the curve stops at $N_0 d / U_0 = 1$. The unstable region of Figure 9.3 is approximately centred on (9.19). Here the growing disturbances consist of trapped modes with $c_r = 0$, where $c = c_r + i c_i$. Contours of constant c_i / U_0 are shown, and $c = 0$ on the boundaries of the region. The behaviour of these growing modes is essentially the same as that in the preceding examples of Figures 9.1, 9.2, with the interface displacements having a forward (i.e. positive) slope of the phase lines with height.

On the left-hand boundary of the unstable region, given by

$$N_0 d / U_0 = \left(\frac{2kd}{\tanh 2kd} - 1 \right)^{1/2} \le kd, \tag{9.20}$$

there is an additional steady solution with $c = 0$, and

$$n_1 = -n_3 = \left(\frac{N^2}{U^2} - k^2 \right)^{1/2}. \tag{9.21}$$

This motion consists of two steady plane waves continuously radiating energy to infinity above and below the shear layer, with phase lines for both having negative slope (from "top-left to bottom-right"). This motion above and below the shear layer is the same as that produced by flow over sinusoidal topography placed at the interface (see (10.11)). For these disturbances the phases of the interfaces are similar to those for the growing modes, with a forward tilt with height for displacement etc., so that the mutual forcing mechanism again applies. Here there is the difference that for this k and c each interface does not support a trapped mode, thereby avoiding resonance, and instead wave energy escapes to infinity, giving steady flow.

Steady radiating solutions also exist on a second curve shown in Figure 9.3. Here there are two solutions with speeds $\pm c$, which are essentially the same solutions by virtue of the symmetry of the system. The value of c varies from $\pm U / \sqrt{3}$ to zero along the curve, and $|n_1| \ne |n_3|$, so that the slopes of the phase lines of the radiating waves are still negative but unequal. The interfaces have the same phase for mutual forcing as before.

Hence this unstable system may also act as a continuous source of internal waves, extracting energy from the mean flow. An isolated vortex sheet in stratified flow shows similar behaviour (Lindzen, 1974; Craik, 1985), but this is not the same as the limit $d \to 0$ in (9.5), (9.13).

9.2.4 Over-reflection

If a plane internal wave is incident from below on the shear layer ((9.5), (9.13)), it produces (in general) a reflected wave and a transmitted wave with the same horizontal wavelength and frequency. Their amplitudes and phases may be readily obtained from the linear equations. An incident wave must satisfy $|U_0 - c| < N/k$, and a transmitted wave must satisfy $|U_0 + c| < N/k$. If the latter is not satisfied there is no transmitted wave, and the reflection coefficient (the ratio of reflected to incident wave energy) is unity. To have both transmitted and reflected waves we must therefore have $|c| < N/k - U_0$. If this condition is satisfied and $c \geq U_0$, there is some transmission and the reflection coefficient is less than unity. But if $|c| < U_0$, the reflection coefficient is greater than unity and the reflected wave has larger amplitude than the incident. This linear phenomenon is known as *over-reflection*. Further, the reflection coefficient increases to infinity as either of the two curves in Figure 9.3 for steady radiating waves is approached.

An examination of the interfaces in over-reflecting solutions for (9.5), (9.13) shows that they have the same phases as for unstable flows, with forward tilt with increasing height for displacement. Consequently, over-reflection is due to the same interaction mechanism that produces instability. The two phenomena are different, but for some systems over-reflection may contribute to instability.

In stratified flows, over-reflection only occurs in systems that are also unstable. As this example shows, this is because the same mechanisms are responsible for each. How instability and over-reflection coexist in practice has yet to be determined, and so far there are no satisfactory descriptions of observations of over-reflection of internal waves occurring in nature.

The above examples indicate that the processes of instability may produce two principal outcomes. In the case of Kelvin–Helmholtz instability, the disturbances grow to form stationary billows with local mixing and little or no propagation (Thorpe, 1987). In the case of Holmboe instability, on the other hand, experiments and numerical studies show that mixing is small and the main result is the generation of finite-amplitude, horizontally propagating waves (Lawrence et al., 1990; Smyth & Peltier, 1989; Smyth et al., 1988). If the system can also support vertically propagating waves, a potentially unstable region may generate such waves on a continuous basis, feeding energy into the wave field from the mean shear. Such processes may well cause a significant amount of the background internal waves observed in the atmosphere and ocean.

9.3 Instability in laminar boundary-layers: Tollmien–Schlichting waves

This digression is a famous example of fluid-dynamical instability that dates from the 1930s, and is significant because the introduction of viscosity into the system renders an otherwise stable flow to be unstable. It is described here because it is an important and different example of shear-flow instability, and it is a more difficult one to comprehend, being mostly described in the literature in mathematical terms, namely the solution of the relevant, somewhat complex, equations. The boundary-layer flow profile over a wing or flat plate is found to be unstable (for both theory and experiment) for Reynolds numbers greater than about 800. The length-scales involved are of order 1 mm, and no density stratification is assumed. It is shown here that this process can be understood as the interaction of two entities, in the same manner as above, except that one of these entities is a wave and the other is not.

Historically, the initial context was the onset of instability in boundary-layers on aircraft wings, causing turbulence and flow separation from the wing, with consequent loss of lift. Observations were initially made by Ludwig Prandtl, and mathematical theory was subsequently developed by his students Tollmien and Schlichting in the 1930s. Experiments confirming their theory were carried out in the USA by Schubauer & Skramstad (1947), first reported in classified literature in 1943. Appropriate references are the books by Schlichting & Gersten (2003) and Drazin & Reid (1981).

Here we consider a viscous fluid of uniform density that has velocity profile $U(z)$ in the undisturbed state. The linearised form of the equations (1.1), (1.4), with stream function ψ defined by (8.22) then give the vorticity equation

$$\left(\frac{\partial}{\partial t} + U\frac{\partial}{\partial x}\right)\omega - \psi_x\frac{d^2U}{dz^2} = \nu\nabla^2\omega, \qquad \omega = \nabla^2\psi, \qquad (9.22)$$

where ν denotes viscosity. If we look for disturbances in the form of normal modes in the system

$$\psi = \hat{\psi}(z)e^{ik(x-ct)}, \qquad (9.23)$$

substituting into (9.22) gives the Orr–Sommerfeld equation for $\hat{\psi}$

$$(U - c)\left(\frac{d^2\hat{\psi}}{dz^2} - k^2\hat{\psi}\right) - \frac{d^2U}{dz^2}\hat{\psi} = -\frac{i\nu}{k}\left(\frac{d^4\hat{\psi}}{dz^4} - 2k^2\frac{d^2\hat{\psi}}{dz^2} + k^4\hat{\psi}\right), \qquad (9.24)$$

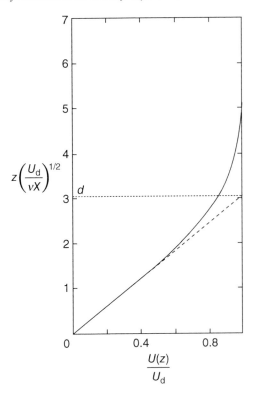

Figure 9.4 The velocity profile of a Blasius boundary layer over a flat plate, together with the idealised profile (9.25) with $\delta = 0$; the vertical scale varies with distance X along the plate. (From Baines et al., 1996, reproduced with permission.)

which is the governing equation for disturbances to the boundary-layer profile

$$U(z) = \begin{cases} U_d z/d, & 0 < z < d - \delta, \\ U_d \left[1 - \dfrac{(d + \delta - z)^2}{4d\delta} \right], & d - \delta < z < d + \delta, \\ U_d, & z > d + \delta, \end{cases} \qquad (9.25)$$

where d is a nominal boundary-layer thickness. This gives a good approximation to the Blasius profile when $\delta/d = 0.32$. The Blasius profile and curve for $\delta = 0$ is shown in Figure 9.4.

9.3.1 An inviscid mode

If (9.24) is simplified to the inviscid system ($v = 0$) with the free-slip boundary condition at $z = 0$, the idealised profile (9.25) with $\delta = 0$ has one free mode that propagates horizontally on the mean vorticity gradient, with a stream function of the form

$$\psi(z) = \begin{cases} U_a d\sinh(kz)/\sinh kd, & 0 < z < d, \\ U_a de^{-k(z-d)}, & z > d, \end{cases} \tag{9.26}$$

with eigenvalues (wave speeds)

$$c = U_a \left[1 - \frac{\tanh kd}{kd(1 + \tanh kd)} \right]. \tag{9.27}$$

In this simplified form, this is the same wave as that propagating on the vorticity gradient as described in §9.2, and this is the only inviscid wave in this system. In any realistic boundary-layer profile only one such wave mode is significant, and the presence of viscosity will imply that it is slowly damped.

9.3.2 Forced viscous flow

The presence of the above inviscid mode implies oscillatory motion at the boundary $z = 0$, of the form

$$u(x,0,t) = -\psi_z = u_0 e^{ik(x-ct)}, \qquad \text{where} \qquad u_0 = -\frac{kdU_a}{\tanh kd}. \tag{9.28}$$

This means that additional motion is required to negate this in order to satisfy the no-slip boundary condition. Close to the boundary, the mean flow profile (whether Blasius or approximations to it) is one of uniform shear, as in equation (9.25), and the appropriate additional motion is governed by equation (9.24). Hence if the additional stream function and vorticity are denoted by Ψ and ω respectively, so that

$$\omega = (\hat{\Psi}_z - k^2\hat{\Psi})e^{ik(x-ct)}, \tag{9.29}$$

equation (9.24) becomes

$$(U(z) - c)\omega = -\frac{kU_0}{\sinh kd}(\omega_{zz} - k^2\omega), \tag{9.30}$$

where Ψ satisfies the conditions $\Psi(0) = 0$, $\Psi_z(0) = -\psi_z(0)$, negating the inviscid term at the boundary. Equation 9.30 may be expressed in the form of Airy's equation

$$\frac{d^2\omega}{dZ^2} - Z\omega = 0, \qquad \text{where} \quad Z(z) = -(kdR_e)^{1/3}\left(i\left(\frac{z}{d} - \frac{c}{u_0}\right) + \frac{kd}{R_e} \right), \tag{9.31}$$

which has the Airy functions $Ai(Z)$ and $Ai(Ze^{i2\pi/3})$ as two linearly independent solutions. Of these two, only the latter approaches zero for large z, and is the appropriate form of the solution. Hence we may write

$$\omega = a_0 \, Ai\left(Z(z)e^{i2\pi/3} \right) e^{ik(x-ct)}, \tag{9.32}$$

and the solution of (9.29) for Ψ in terms of ω gives

$$\Psi(z) = -\frac{1}{2k}\left(\int_z^{\infty} \omega(Z(\xi)e^{\frac{2i\pi}{3}})e^{k(z-\xi)}\,d\xi + \int_0^z \omega(Z(\xi)e^{\frac{2i\pi}{3}})e^{-k(z-\xi)}\,d\xi\right) + Ae^{-kz},$$

(9.33)

where a_0 and A are constants. The latter may be determined by the boundary conditions at $z = 0$, namely

$$\Psi(0) = 0, \qquad \frac{d\Psi}{dz} = u_0, \qquad (9.34)$$

which give

$$A = \frac{u_0}{2k}, \qquad a_0 = \frac{u_0}{\int_0^{\infty} \text{Ai}(Z(\xi)e^{\frac{2i\pi}{3}})e^{-k\xi}\,d\xi}. \qquad (9.35)$$

The integral in the denominator for a_0 can be zero for certain profiles of Ψ or ω. In fact, there is an infinite set of such profiles for given k, u_0, which represent decaying free viscous modes in uniform shear. Mathematically, these modes are manifested as eigenfunctions of the Orr–Sommerfeld equation with the no-slip boundary condition, with the corresponding speeds (and decay rates) given as eigenvalues. They are independent dynamical entities. Any flow structure forced at the boundary can excite some of these modes, and for the present flow, this forcing is provided by the presence of the inviscid mode at the boundary.

The least damped modes are the most important in this situation. The speeds of these modes are approximately constant (i.e. independent of the values of parameters) if $\kappa = (\nu k^2/U_z)^{1/3}$ is small, and the speed of the first such mode is given by

$$c = c_{r1} + ic_{i1} = \left(\frac{\nu U_z^2}{k}\right)^{1/3}\frac{4.13 - 1.06i)}{(kdR_e)^{1/3}}. \qquad (9.36)$$

A typical picture of the velocity structure of this mode is shown in a plot of the streamfunction in Figure 9.5. This shows that it has a relatively simple structure, with zero velocity at $z = 0$ and positive vertical velocity up to levels above $z = d$.

9.3.3 Instability due to interaction between a wave and (an otherwise decaying) mode

If one equates the speed c_{r1} from (9.36) with the speed c of the inviscid mode in (9.27), one obtains a relation between kd and R_e:

$$R_e = \frac{69.93}{kd(1 - (1 - e^{-2kd})/(2kd))^3}. \qquad (9.37)$$

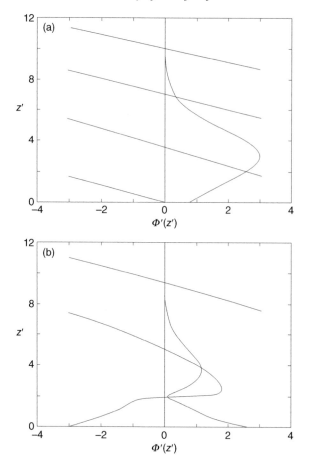

Figure 9.5 Amplitude and phase of the stream function of a decaying viscous mode in a boundary-layer over a flat plate, on the same scale as for Figure 9.4. (From Baines et al., 1996, reproduced with permission.)

This relation is plotted in Figure 9.6 as the curve AB, on top of the curves that describe the theoretical regions of instability and growth rates from complete linear calculations for a Blasius boundary-layer profile. That curve AB lies close to the line of maximum growth rate for a wide range of Reynolds numbers demonstrates the validity of the interaction mechanism described here.

A schematic diagram illustrating the instability mechanism is shown in Figure 9.7. Here the flow pattern is observed from a frame of reference moving with the growing wave. The mean velocity profile is shown on the left, and the dark line with arrows denotes a streamline near the level of maximum amplitude of the inviscid mode. The dashed lines then show the vertical motion associated with the viscous mode at the points of maximum amplitude. The phase of this motion increases

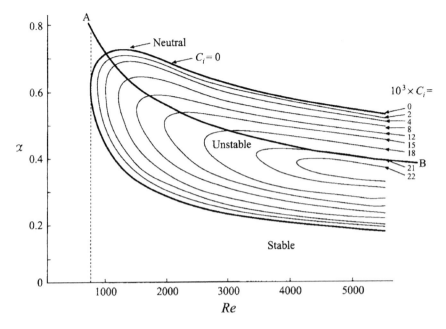

Figure 9.6 Regions of instability of a Blasius boundary layer on the (R_e, α)-plane. The line AB denotes (9.37), and the vertical dashed line denotes the critical value of Reynolds number. (Modified from Schlichting, 1968.)

the vertical displacement of the streamlines, and hence of the overall amplitude of the disturbance. The viscous "decaying mode" is no-longer decaying, but grows together with the inviscid wave mode on the vorticity gradient. For more details, see Baines et al. (1996).

9.4 The stability of internal waves

The preceding sections have been concerned with the stability of steady, horizontal flows. As described in Chapter 8, density-stratified fluids can also contain wave motions with frequencies less than the buoyancy frequency N, and these can also be unstable in ways that are independent of the processes described above. It is a common experience of experimenters that simple harmonic waves generated at one end of a stratified channel do not retain their initial structure as they propagate, and rapidly degenerate into chaotic patterns.

For the simplest form of this process, it is reasonably straightforward to generate a sinusoidal wave beam of given frequency ω in a uniformly stratified fluid, which propagates at an angle θ to the horizontal given by

$$\sin \theta = \frac{\omega}{N},$$

(9.38)

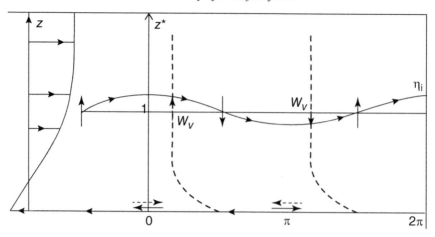

Figure 9.7 Schematic diagram of the interaction process that causes growth in a Tollmien–Schlichting wave. The flow is drawn in the frame of reference moving with the wave, so that the pattern is stationary but growing. The mean velocity profile in this frame is shown on the left. η_i denotes the vertical displacement of the vortical region of the inviscid partial mode propagating on the vorticity gradient. In this partial mode the fluid at the boundary $z = 0$ has the velocity denoted by the solid arrows, and the no-slip condition forces a viscous response represented by the dashed arrows. The dashed curves show the phase of this viscous response for the maxima in vertical velocity w_V. Since this is close to the maxima in η_i, it increases the amplitude of the vortical partial mode by advection. (From Baines et al., 1996, reproduced with permission.)

pointing upward or downward. Such structures are termed *internal wave beams*, and are themselves subject to instabilities at modest amplitude. The instability process derives from triad interactions, in which the primary wave interacts with two other waves (of different frequency and wavenumber) that have initially very small (near zero) amplitude, but grow to comparable amplitude due to the interaction, and tend to degrade the initial wave. This constitutes the process of *triadic resonant instability*, a process that was first discovered in the 1960s (Phillips, 1966; McEwan, 1971, 1973), and is still an active topic of research. The process is prominent in the ocean, in particular. Recent reviews have been given by Staquet & Sommeria (2002) and Dauxois et al. (2018).

10

Stratified flow over two-dimensional obstacles: linear and near-linear theory

> Countless others have written on this theme
> and it may be that I shall pass unnoticed amongst them;
> if so, I must comfort myself with the greatness and splendour of my rivals,
> whose work will rob my own of recognition.

TITUS LIVIUS (LIVY) *The Early History of Rome, Book I*

In this chapter we deal with two-dimensional flows, in which neither the flow nor the topography has any variation with the y-coordinate, and the flow is quasi-linear. We consider the flow resulting from the introduction of topography into a variety of fluids with horizontally uniform velocity and density profiles, representative of real conditions. An important special case is that where the initial flow has uniform buoyancy frequency N and uniform fluid velocity U, relative to the two-dimensional topography. This is the opposite extreme in density stratification to that of the single-layer flow discussed in Chapters 2 and 3, where all the density variation occurs at a single interface.

How relevant *are* the properties of flow with initially uniform U and N over topography to large-scale flows such as that of the atmosphere? Apart from the restrictions to time-scales less than a pendulum day and space-scales less than a few tens of kilometres (as discussed in Chapter 1), in the atmosphere there is the effect of the surface stress and the consequent turbulent boundary-layer and strongly sheared lower wind profile. The atmospheric boundary layer varies diurnally, and depends on factors such as surface roughness, neighbouring topography, the magnitude and duration of solar heating or of radiation from the ground, evaporation and so on.

For present purposes we exclude situations where solar heating and low-level convection are significant, leading to deep surface mixed layers. Instead, we concentrate on situations where the atmospheric boundary-layer may be loosely classified into three layers, as shown in Figure 10.1. First, there is a *lower* or *inner* or *surface* layer, where turbulent shear stresses are important, which has a typical thickness

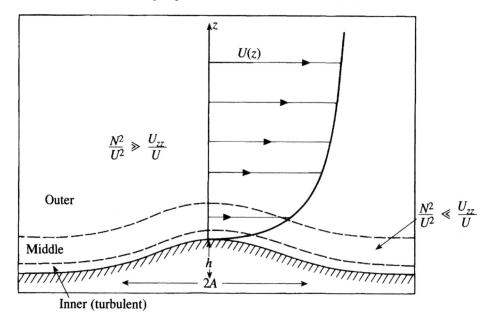

Figure 10.1 Schematic diagram of the three conceptual layers of the mean atmospheric wind profile near the ground. In the lowest layer, turbulent shear stresses
are important; in the middle layer, shear effects dominate stratification; and in the
uppermost layer, stratification effects dominate shear.

of several metres. Second, there is a middle layer where shear dominates over the
mean stratification, so that

$$\frac{\mathrm{d}^2 U}{\mathrm{d}z^2}/U \gg N^2/U^2$$

(cf. (8.25)). This layer has a typical thickness of 10–50 metres. Above these two
layers there is the *upper* or *outer* layer, where stratification dominates over shear,
and the above inequality is reversed. This region contains most of the atmosphere,
and here the mean wind normally varies relatively slowly with height.

The flows described in this and the next chapter may be regarded as applying
to this uppermost layer. The first two layers are relatively thin (generally <100 m
in total), and their height is much less than most topographic features of interest.
Furthermore, the motion in these lower two layers is essentially dependent on
the motion in the topmost layer. The reverse does not apply – the motion of the
uppermost layer is determined by large-scale forcing, and is not affected by the local
properties of the bottom surface and the lower two layers, except that the latter may
cause a slight change in the obstacle shape as seen by the flow in the outer layer.
Hence, it is appropriate to determine the motion of the upper layer, independent
of the surface conditions, and then to infer the consequent behaviour of the lower

layers if desired. This division of the flow into three dynamically different layers forms a convenient starting point for analytical studies of the "triple deck" type (see Hunt et al., 1988).

There is a significant complication to this picture of three-layer structure when the flow separates from the lower boundary. This may occur downstream of orography, causing a turbulent wake that may extend up to the maximum topographic height. The properties of such wake regions are also mostly dependent on the flow of the upper layer, although boundary-layer properties affect details such as the location of separation points. In complex three-dimensional terrain there may be many overlapping wakes, as discussed in Chapter 13, and in this event the simple three-layer structure of the boundary-layer becomes inappropriate.

We therefore proceed to discuss the effects of topography on stratified flow that is essentially inviscid, and where the effects of shear in the mean flow are not large. For the purposes of this chapter we divide the flow into two main types. The first is that of *infinite depth*, where the stratified fluid is effectively infinitely deep with an upper radiation condition, so that internal waves propagating upward away from the region of interest are not reflected back down again. A fluid in which $U(z)$ and $N(z)$ vary very little with height has this property. The second main type is flow of *finite depth*, where the fluid has an upper boundary or region that reflects all internal waves that are of interest. The simplest forms of such a boundary are a rigid boundary, or a surface of fluid particles of uniform density above which the fluid is homogeneous (see §1.2). There are other possibilities when U and N vary with height, such as critical layers, as described in Chapter 8, and hybrid systems, in which one part of the horizontal wavenumber spectrum "escapes" to high levels, and the other part is reflected. This hybrid behaviour is sensitive to the vertical structure of $U(z)$ and $N(z)$, and involves trapped and leaky modes as described in §8.5. It is quite common in the atmosphere, and examples are discussed in §10.5.1.

Infinite-depth systems are simpler than finite-depth systems because the latter has an additional parameter D, the height of the upper reflecting boundary. We discuss infinite-depth systems first. As shall be seen later, this leads us to examine finite-depth systems in order to complete the discussion of infinite-depth systems.

10.1 Observations of flows of infinite depth

We begin with a description of the observed properties of infinite-depth systems, before proceeding to a theoretical discussion in the following sections. These observations have necessarily been made in the laboratory, since field observations cannot be made in controlled conditions.

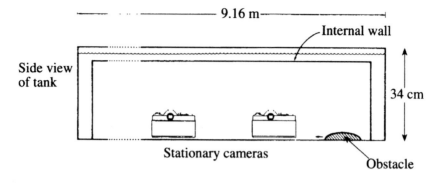

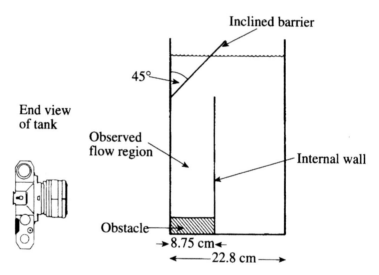

Figure 10.2 Schematic side- and end-views of the "infinitely deep" tank used for the observations described in §10.1. (From Smith & Grønås, 1993.)

Experimental setup

Laboratory observations in stratified fluid of simulated infinite depth have been reported by Baines & Hoinka (1985). It is appropriate to first describe how these experiments were carried out, since an infinitely deep tank does not exist in practice and must be simulated in some way. A schematic diagram of the tank used is shown in Figure 10.2. The basic idea was to arrange the geometry of the tank in such a manner that vertically propagating waves were reflected out of the working area and did not return. The tank (length 9.16 m, width 22.8 cm) was filled with uniformly stratified fluid to a depth of about 34 cm.

In a typical experiment, an obstacle was towed along the bottom of a channel within the tank from near one end to the other. This channel (the working section, width 8.75 cm) occupied slightly more than 1/3 of the width of the tank, and was separated from the remainder by a vertical internal wall of height 25 cm, with a gap of 5 cm at each end. This gave a "racetrack" geometry, so that fluid could flow around the internal wall. Above the working channel, a barrier angled at 45° to the horizontal and extending along the whole length of the tank was inserted, as shown in Figure 10.2. The motion of a two-dimensional obstacle along the bottom of the channel produces waves of the form $e^{i(kx+nz-\omega t)}$, that propagate upward. An analysis of the internal wave properties in this geometry shows that some of this wave energy will be diffracted over the internal wall into the wider section of the tank. Other waves encountering the angled barrier will be reflected over the internal wall if $0 < \omega/N < 1/\sqrt{2}$, but will be reflected back down into the working channel if $1/\sqrt{2} < \omega/N < 1$. Reflected waves in the latter range of frequencies are expected to be insignificant for several theoretical reasons; in particular, their group velocity is not large and vanishes at each end of this range. Hence they should only become apparent at some distance downstream of the obstacle (if at all), and they should also have a significant y-component. In practice, no effects that could be attributed to these waves were noticed in the experiments. Empirical tests with periodic topography showed that the reflection of vertically propagating waves was indeed negligible, so that this configuration was taken to be an effective simulation of an upper radiation condition.

For purely horizontally propagating columnar modes, on the other hand, this tank geometry does not give a perfect simulation of the "infinite-depth" situation. The vertical spectrum of these motions in this tank was found to be discrete rather than continuous (Baines, 1994), with a largest mode speed of about 10 cm/s (with $N \approx 1$ rad/s). This discrete spectrum could have been expected because there must be a limit to the maximum vertical wavelength and speed in a finite tank. It has only a minor effect on the simulations discussed below.

The flow was made visible by "filling" the fluid with neutrally buoyant polystyrene beads that have a range of densities covering that of the fluid, and then by photographing the motion using stationary, moving and video cameras.

Homogeneous fluids

We consider the flow of homogeneous fluids over a two-dimensional obstacle towed through fluid at rest, at constant speed at large Reynolds number. On a smooth surface a laminar boundary-layer with thickness of order $(vx/U)^{1/2}$ forms, where x is the distance along the obstacle from the leading edge (or rather, from the leading edge of the moving tray on which the obstacle is placed). If $R_e = Ux/v$ becomes greater than about 5×10^6, this boundary-layer becomes unstable because of the growth of disturbances termed Tollmien–Schlichting waves (see §9.3), caused by the interaction between a wave on the vorticity gradient and decaying sheared viscous modes, and by a complex process it then becomes turbulent. In many laboratory experiments R_e remains small enough for the boundary-layer to be laminar, but in the atmosphere it is almost always large enough for the flow to be turbulent. The question of whether or not flow separation occurs on the lee side of the obstacle is complicated by this transition to turbulence, as it is well known that transition delays separation. Nevertheless, experiments indicate that if the lee-side slope of the obstacle is large enough ($\approx 14°$), this boundary-layer separates from the obstacle to give a turbulent, unsteady, recirculating separation "bubble", and a downstream wake where the fluid has reduced momentum (Finnigan, 1988; Taylor, 1988). The details of these regions vary greatly with the shape of the obstacle and the surface roughness, but close to the obstacle they have a depth comparable with the obstacle height, h_m. If the upstream face of the obstacle is sufficiently steep, flow separation may also occur on the upstream side.

Above the boundary-layer and the associated separation bubble and wake regions, the flow is effectively inviscid potential flow (i.e. it has no vorticity and is described by a velocity potential satisfying Laplace's equation). Solving for the potential flow above the vortical region is not simple because it depends on the shape of the wake and the exchange of fluid with it. The wake region is generally turbulent, and its extent downstream depends on how "streamlined" the obstacle is. An obstacle with very gentle maximum slope dh/dx will have a negligible wake compared with that from a bluff body with large maximum slope on the lee side. The fluid within the separation "bubble" close to the obstacle has greatly reduced mean velocity relative to the body, but it only stays there for a limited period of time that depends on the scale of the turbulent eddies. After this time it leaves the bubble where it is replaced by new fluid from the upstream side, and it trails downstream as part of the extended wake region with reduced mean velocity. This process implies a transfer of momentum from the obstacle to the fluid. It may be represented by a drag force on the obstacle, expressed as

$$F_D = \frac{1}{2}\rho_0 C_D U^2 h_m, \tag{10.1}$$

per unit span, where C_D is the drag coefficient number[1] (see, for example, Batchelor, 1967). The drag C_D is small for streamlined bodies without flow separation, and is approximately unity for bluff bodies where separation occurs. Many topographic (or orographic) shapes on the surface of the Earth may be described as "bluff", so that they have wakes of this character in neutrally stratified conditions. The same applies to most obstacle shapes used in laboratory experiments, including the ones described here.

---

[1] It is common practice to define C_D with the factor 1/2 in (10.1) for obstacles, but without it for drag on surfaces (cf. (2.95))

Stratified fluids

In the experiments with near-inviscid stratified fluid, the relevant physical quantities are: the towing speed U, the uniform buoyancy frequency N, the maximum obstacle height h_m, the obstacle half-width scale A, and the kinematic viscosity, ν. If the obstacle is not symmetrical, A is replaced by A_u and A_d, for upstream and downstream respectively. The quantity U/N represents a natural length-scale associated with the flow. These quantities give the dimensionless numbers Nh_m/U, NA/U and h_m/A, only two of which are independent, and the Reynolds number, $R_e = Uh_m/\nu$. We are primarily concerned with situations where the Reynolds number is large, and for the most part $R_e > 1000$ in these experiments. At smaller R_e the Reynolds number affects the character of the stratified wake, and a variety of flow types have been observed for a circular cylinder (Boyer et al., 1989).

With R_e large, the most important of the above parameters is Nh_m/U, the "Nhu" of Chapter 1, which will usually be abbreviated to Nh/U. We use this parameter to classify the flow for obstacles of a given shape. If $NA/U \gg 1$, the flow is hydrostatic, but this is not the case in these experiments unless Nh/U is also large.

The number Nh_m/U also has a broader significance in that it can be regarded as (the square of) a Froude number for stratified flow over small obstacles, as described in §10.2 (Mayer & Fringer, 2017).

The above description of the flow for neutral stratification applies to the case $Nh/U = 0$. The character of the flow for various values of Nh/U at large times ($Ut \approx 600 \, \text{cm}$) is shown in Figure 10.3 for an obstacle of "Witch of Agnesi" shape[2] $z = h(x) = h_m/[1 + (x/a)^2]$, which we may take to be representative. For Nh/U small but finite, the effects of stratification become manifest at length-scales that are long compared with h_m.

The first frame in Figure 10.3 with $Nh/U = 0.47$ shows a lee-side separation "bubble" and the associated wake, with some stratification effects evident in the asymmetric upper flow. The lee-side flow is seen to descend into a lee-wave trough that is too long to be visible. In the second frame ($Nh/U = 1.08$), the scale of the lee waves (a general term denoting waves on the lee side) is smaller, and the trough with the forward tilt with height (implying upward energy propagation) is evident. As Nh/U increases further, the scale of the lee waves decreases until $Nh/U = 1.5$, where they steepen and break. This is observed to happen shortly after the flow has commenced, with the fluid toppling over in a forward sense. This results in a "stagnant patch" of fluid that is buried within the wave field, in a fixed position above and behind the obstacle, as seen in the third frame of Figure 10.3 ($Nh/U = 1.56$). Fluid flows around this patch, both above and below, and the particles within it

---

[2] The much-invoked curve "*Witch of Agnesi*" was studied by Italian mathematician Maria Agnesi in 1748. The name "witch" derives from a mistranslation of the term *averisera* ("versed sine curve") from the Latin *vertere*, "to turn".

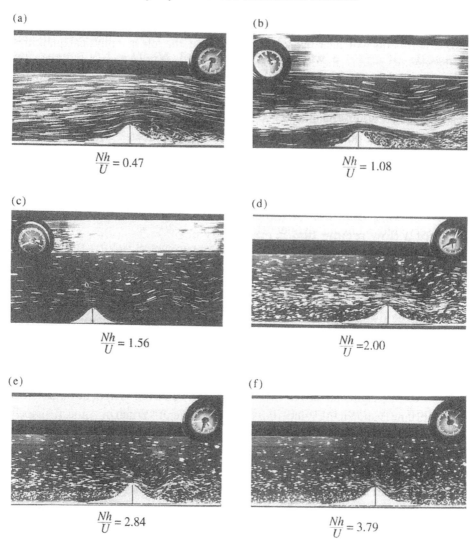

Figure 10.3 Near-steady-state flows over the "Witch of Agnesi" (with $h_m/A = 1.63$) relative to the topography for given values of Nh/U, in simulated infinite depth. (From Baines & Hoinka, 1985, reproduced by permission of the American Meteorological Society.)

remain there for a considerable time. As will be seen later, this phenomenon of wave-breaking and the formation of a "stagnant patch" marks the onset of a flow regime that is essentially non-linear and is governed by a "hydraulic transition".

The effects of different obstacle shapes are shown in Figures 10.5 and 10.6. The general flow patterns and the dependence on Nh/U are much the same as those shown in Figure 10.3, with the possible exception that upstream blocking appears

Figure 10.4 Upstream motion for the "Witch of Agnesi" as in Figure 10.3, for $Nh/U = 0.98$. The flow was recorded by a stationary camera at successive times at a single location, and joined to give a representation of an instantaneous flow by assuming no significant change to the pattern with time. The obstacle is beginning to appear at the right-hand end.

to commence at smaller values of $Nh/U (= 1.44)$ in Figure 10.6, where A_u/A_d is small.

Upstream motions

Stratified flow over an obstacle also produces motions on the upstream side. These are not as conspicuous as those on the downstream side in Figures 10.3–10.6, because they have smaller amplitude and the motion is mostly horizontal. They are best seen with stationary cameras, and take the form of horizontally propagating columnar motions as described in §§8.3, 8.4. Figure 10.4 shows an example of the extent of upstream propagation of columnar motion in deep fluid, for $Nh/U = 0.98$. As described above, the vertical spectrum of these modes in this tank is discrete. For this reason, these modes are not generated upstream if $Nh/U < 0.5$ (for the obstacles in Figures 10.3–10.6), which happen to be the values for which $U \geq c_1$, the speed of the fastest mode. Detailed observations of these upstream motions show that modes 1 and 2 are recognisable as distinct modes, but the higher modes (when they are generated) blur together to give an apparent continuum. This blurring is partly because the fronts of these columnar motions are not sharp but disperse as $(Nt)^{1/3}$ (in the manner described by McEwan & Baines, 1974), spreading over distances of up to 2 metres (for mode 1) before reaching the end of this tank. If an experiment is repeated with larger Nh/U, the amplitude and the dominant vertical wavenumber of the upstream disturbances both increase. This may be seen in Figure 10.7a, which shows velocity profiles observed immediately upstream of the obstacle near the end of the runs, and coinciding (roughly) with the conditions for Figure 10.3.

When Nh/U exceeds 2, the upstream motions have become large enough to completely block the flow at low levels, with the depth of blocked fluid $z_r \approx h_m/2$. For example, a region of stagnant upstream fluid may be seen in the last three frames of Figure 10.3. In the first of these (where $Nh/U = 2.0$), the upstream stagnant region

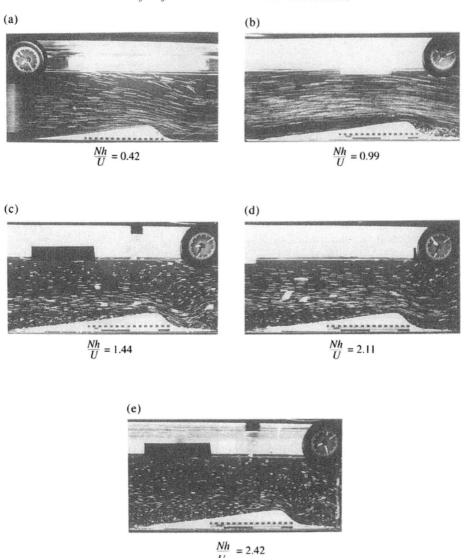

(a)

(b)

$\frac{Nh}{U} = 0.42$

$\frac{Nh}{U} = 0.99$

(c)

(d)

$\frac{Nh}{U} = 1.44$

$\frac{Nh}{U} = 2.11$

(e)

$\frac{Nh}{U} = 2.42$

Figure 10.5 As for Figure 10.3, but for an asymmetric obstacle with a plane ramp upstream (slope 0.12), "Witch of Agnesi" downstream with $A_d = 4.0$ cm. (From Baines & Hoinka, 1985, reproduced by permission of the American Meteorological Society.)

is barely visible and shows some horizontal variation, probably because the associated upstream disturbances are not yet fully developed. In Figure 10.3e,f, where the obstacle is moving more slowly, the upstream columnar modes have rendered the motion horizontally uniform, with a clear-cut blocked region of stagnant fluid.

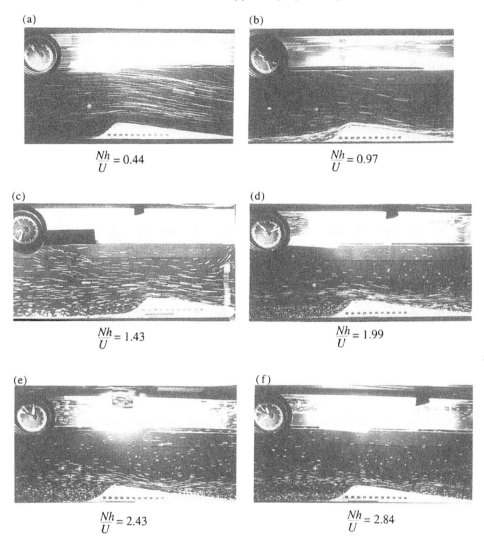

Figure 10.6 As for Figure 10.5, but with the obstacle reversed. (From Baines & Hoinka, 1985, reproduced by permission of the American Meteorological Society.)

Above this there is a strong "jet" of fluid that passes over the obstacle, beneath the "stagnant patch", and upon which there are significant waves on the lee side.

An analysis of the changes in density that must accompany these observed upstream velocity profiles (of the form described at the end of §8.3) shows that the density gradient must be reduced in the stagnant fluid, but increased in the jet. The motion at higher levels above the jet is much weaker. As Nh/U increases beyond 2 this pattern remains largely unchanged, and the observed depth z_r of the

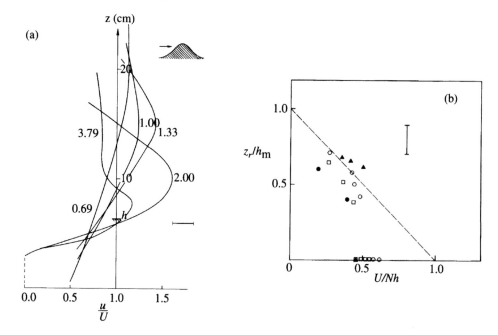

Figure 10.7 (a) Representative horizontal velocity profiles measured immediately upstream of the obstacle in near-steady state, for the obstacle and experiments as in Figure 10.3. The numbers refer to Nh/U values. (From Baines & Hoinka, 1985: reproduced by permission of the American Meteorological Society.) (b) The observed depth z_r of the blocked layer immediately upstream of various obstacles for various conditions and times in infinite-depth flows. The scatter reflects the accuracy of measurement and the long time to reach steady state. ■, •: "Witch of Agnesi" (•: data courtesy of W. Snyder); □: obstacle of Figure 10.5.

upstream stagnant layer shows a tendency to increase towards h_m as $Nh/U \rightarrow \infty$ (Figure 10.7b). The causes of these upstream motions are addressed in the following sections.

Lee-side separation

As Nh/U increases from zero, the stratification acts to inhibit (but not necessarily prevent) the separation of the flow from the lee side of the obstacle, and increasing suppression of separation can be seen in Figures 10.3a–e. A close inspection of these and other observations shows that the separation is to a large extent controlled by the dimensionless number NA_d/U (for the Witch of Agnesi of Figure 10.3 we take $A_d = 2a$, where a is specified by (10.14)). The internal waves have a typical length of $2\pi U/N$, and boundary-layer separation is suppressed when the lee-side half-length of the obstacle coincides with a half wavelength; that is, $NA_d/U = \pi$. In general, boundary-layer separation occurs when $NA_d/U < \pi$, and complete at-

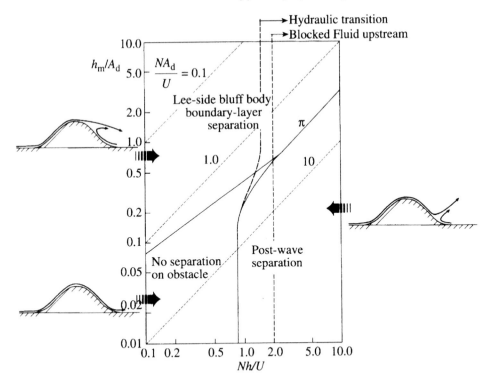

Figure 10.8 Lee-side separation properties as a function of the mean lee-side slope (h_m/A_d) and Nh/U on logarithmic scales, showing the three distinct types of behaviour. The approximate boundaries for the onset of hydraulic transition flow and upstream blocking are also shown.

tachment is found when $NA_d/U \geq \pi$. However, separation may still occur further downstream beneath the first lee-wave crest if Nh/U is large enough, and this is termed "post-wave" separation. There are three main flow types with regard to separation: (i) boundary-layer separation, as in Figure 10.3a; (ii) complete attachment, as in Figure 10.6a; and (iii) "post-wave" separation, as in Figure 10.3f. The conditions under which these occur may be expressed in terms of the lee-side slope h_m/A_d and Nh/U, as shown in Figure 10.8. The boundaries for the onset of lee-wave overturning and upstream blocking are also shown. Note that the boundaries between these various regions on this diagram are only approximate, because the detailed shape of the obstacle may cause some variability, and A_d is not precisely defined.

For hydrostatic flows with gentle lee slopes where $h_m/A_d \ll 1$, and no wave-breaking occurs, lee-side separation is not observed in experiments and is not shown in Figure 10.8. However, it is possible in principle, as the theoretical solutions (discussed below in §10.3) show that the fluid is decelerated by an adverse pressure gradient on the lower part of the slope. This is the condition that can lead to boundary-layer separation, so that separation is possible in practice if the boundary is sufficiently rough.

Post-wave separation is controlled by two factors that may act together. First, when partial upstream blocking occurs, the fluid flowing over the obstacle is deficient in heavy low-level fluid. Hence it is generally lighter than the heaviest fluid on the downstream side, and so it may separate from the obstacle surface and find a mean level there above the ground. This can be seen in Figures 10.3f, 10.5e, and 10.6e,f. The second factor is the flow against an adverse pressure gradient below the lee-wave crest. This may occur on the obstacle surface or downstream of it, and under certain circumstances periodic isolated separation regions may be observed under the crests of the waves. One unusual example that occurred for the Witch of Agnesi for $Nh/U \approx 0.8$ is shown in Figure 10.9. Boundary-layer separation occurred on the main obstacle to produce a slightly larger effective shape, which was then reproduced periodically downstream with relatively stagnant fluid under the wave crests. Similar phenomena (which were termed "isolated mixed regions") have been observed behind a horizontal cylinder in approximately the same parameter range by Boyer et al. (1989).

Some numerical simulations (e.g. Doyle & Durran, 2002, 2004) and field observations (Grubišić & Billings, 2008) of flow behind substantial mountain ranges have observed non-stagnant recirculating regions under lee-wave crests (i.e. in the shaded regions of Figure 10.9) termed "rotors", with reversed flow at ground level. Strong stratification above them enhances this circulation and surface roughness inhibits it. Velocities within the rotor are generally somewhat smaller than those above them, implying that they are driven by turbulent stresses from the overlying fluid (which are minimal in most laboratory experiments with modest Reynolds number). The observed flows in the Owens Valley (California) show recirculating velocities up to 20 m/s, though in most observed events the velocity is much less than this. Turbulence within the rotor is a maximum near the rising region above the surface separation point.

Figures 10.3–10.7 display a range of different dynamical phenomena, and most of the rest of this chapter is devoted to a theoretical explanation and dynamical description of these disturbances. Some of them may be interpreted in terms of linear wave dynamics as described in Chapter 8. In particular, waves are found on the downstream side, and columnar motions are found upstream and downstream. This suggests that these flows may be described by using perturbation theory with small-amplitude forcing, particularly when Nh/U is small, and this is the natural starting point for a theoretical discussion.

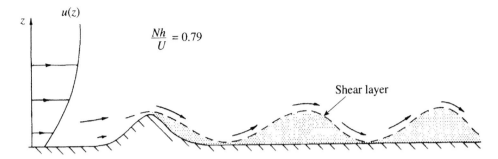

Figure 10.9 Schematic representation of the stationary lee-side humps observed in the lee of the "Witch of Agnesi" of Figure 10.3 when $Nh/U \approx 0.8$. (From Baines & Hoinka, 1985: reproduced by permission of the American Meteorological Society.)

10.2 Infinite-depth flows: theory for small Nh/U

Three different representations of topography for small Nh/U have been discussed in the literature: small-amplitude topography, finite topography with weak stratification, and a momentum source, and we describe these in turn.

10.2.1 Small-amplitude topography with the lower boundary as a streamline

We investigate the effect of flow over topography of the form $z = h(x)$, where the lower boundary is a streamline and h_m is assumed to be sufficiently small so that equations of motion may be linearised. Formally, this requires $|\mathbf{u}'/U| \ll 1$, where $\mathbf{u}'$ is the perturbation velocity due to the topography. A dimensionless criterion for this to be valid may be deduced from scale analysis. We have

$$w' \approx U\frac{dh}{dx} \approx Uh_m/A, \quad \text{and} \quad \frac{\partial w'}{\partial z} \approx Nh_m/A, \qquad (10.2)$$

since the vertical scale of the forced motion is N/U. From continuity we then have

$$u' \approx A\frac{\partial w'}{\partial z} \approx Nh_m, \quad \text{and therefore} \quad |u'/U| \approx Nh_m/U. \qquad (10.3)$$

Hence if $Nh/U \ll 1$ and $h_m/A \ll 1$, the linearised equations (8.18)–(8.21) are applicable. If the motion is commenced at $t = 0$, the exact lower boundary condition for $t > 0$ is

$$w = w' = (U + u')\frac{dh}{dx}, \quad \text{on} \quad z = h(x), \qquad (10.4)$$

where $h \to 0$ as $|x| \to \infty$. This equation is linear but non-separable in terms of x and z. This difficulty is avoided by taking a perturbation expansion in powers of h_m

about the horizontal level $z = 0$, and for the first term in this expansion we have

$$w = w' = U\frac{dh}{dx}, \quad \text{at} \quad z = 0. \tag{10.5}$$

This problem may be expressed in terms of a stream function ψ defined by (8.22), to give (8.23) with the boundary condition

$$\psi = Uh(x), \quad \text{at} \quad z = 0, \ t > 0. \tag{10.6}$$

For the initial conditions, the discussion of §§2.1 and 2.2 applies, but the details do not affect the final steady state and are not important here.

For steady flow ($\partial/\partial t = 0$), (8.23) gives

$$\frac{\partial^2}{\partial x^2}\left[\left(\frac{\partial^2}{\partial x^2} + \frac{\partial^2}{\partial z^2}\right)\psi + l^2(z)\psi\right] = 0, \tag{10.7}$$

where

$$l^2 = N^2/U^2 - \frac{d^2U}{dz^2}/U, \tag{10.8}$$

as in §8.5. The vertical displacement η of a streamline from its undisturbed position is then given by $\eta = \psi/U$. An analysis of the time-dependent form of this linearised problem shows that the commencement of motion generates a broad spectral range of transient waves, but no steady columnar modes (McIntyre, 1972). Hence the final steady state does not contain any permanent disturbances at upstream infinity, and (10.7) may be integrated twice with respect to x from $-\infty$ to give

$$\nabla^2\psi + l^2(z)\psi = 0. \tag{10.9}$$

In order to obtain the solution to (10.9) that corresponds to the time-dependent initial value problem where the motion is commenced from rest, we must invoke an upper boundary condition called the *radiation condition*, which specifies that there is no incoming wave energy from infinity, and that the solution is bounded there. This may be achieved by ensuring that each Fourier component of the wave motion in the solution has upward group velocity (see §8.4). If $l^2(z)$ is uniform with height, no variations in the background medium cause internal reflections of wave energy. The simplest example of this is when U and N are both uniform, so that $l = N/U$, and we will assume this in this section.

For these flows the parameter Nh_m/U may be expressed as (the square of) an inner-scale Froude number in the form $F_\delta = u_0/(g'\delta)^{1/2}$, where u_0 is the characteristic scale of the velocity of the perturbation, δ is its vertical length-scale, and g' is reduced gravity (Mayer & Fringer, 2017). With this scaling we have $g' = UN$, so that $Nh_m/U = F_\delta^2 = u_0^2/(g'\delta)$. Here there is no connotation of critical flow when F_δ equals unity, but this comparison indicates the broad significance of Nh_m/U.

Periodic topography

If we consider periodic topography where

$$h(x) = h_{\mathrm{m}} \cos kx, \tag{10.10}$$

(10.9) has the steady-state solution

$$\psi = \begin{cases} Uh_{\mathrm{m}} \cos[kx + (N^2/U^2 - k^2)^{1/2}z], & 0 < k < N/U; \\ Uh_{\mathrm{m}} \cos kx.e^{-(k^2-N^2/U^2)^{1/2}z}, & k > N/U. \end{cases} \tag{10.11}$$

For $k > N/U$, this solution is very similar to the potential flow solution that would be obtained if $N = 0$; the wavelength is so short that the fluid traverses the bumps before the buoyancy forces have time to act. For $0 < k < N/U$, the time-dependent solution is analytically complex, but the steady solution (10.11) is essentially attained at a given height after the vertically propagating wave front moving at the group velocity knU^3/N^2 (from (8.53) has passed. The phase lines slope upward and to the left but, as discussed in §8.4 and Figure 8.4, the direction of energy propagation is upward and to the right in the direction (k, n), where $n = (N^2/U^2 - k^2)^{1/2}$, because the horizontal component of the group velocity is the sum of the joint effects of wave propagation and advection by the mean flow.

The horizontal drag force on the topography due to the pressure acting on the surface, F_D, is given by

$$F_D = -\int \frac{\partial p}{\partial x} h(x)\mathrm{d}x = \int p \frac{\mathrm{d}h}{\mathrm{d}x}\mathrm{d}x, \tag{10.12}$$

where p or $\partial p/\partial x$ is evaluated on the surface $z = h(x)$. In the linearised form this is taken at $z = 0$, and substituting from (10.11) we obtain

$$F_D = \begin{cases} \frac{1}{2}\rho_0 U^2 h_{\mathrm{m}}^2 k(N^2/U^2 - k^2)^{1/2}, & 0 < k < N/U, \\ 0, & k > N/U, \end{cases} \tag{10.13}$$

per unit horizontal wavelength (see also Gill, 1982, p. 145). This wave drag is due to the fact that, when $0 < k < N/U$, the total pressure $p_0 + p'$ is increased on the upslope side of the topography, and decreased on the downslope side. It is a maximum over k for $k = N/U\sqrt{2}$, and is equal to the upward flux of momentum carried by the waves, $-\rho_0 \overline{u'w'}$. Since the perturbation pressure p' and horizontal velocity perturbation u' are related by $p' = -\rho_0 U u' = \rho_0 U \partial\psi/\partial z$, the total velocity is decreased on the upslope side and increased on the downslope side, with the extreme values occurring at the points of maximum slope.

Isolated topography

The solution for steady flow over isolated obstacles may be obtained by Fourier superposition of (10.11), weighted by the Fourier transform of $h(x)$ over the range of wavenumbers k. For example, the "Witch of Agnesi" profile,

$$z = h(x) = h_m/[1 + (x/a)^2], \tag{10.14}$$

for which the half-width scale $A \approx 2a$, has the Fourier transform

$$\hat{h}(k) = \int_{-\infty}^{\infty} h(x)e^{-ikx}dx = \pi h_m a \cdot e^{-|ka|}. \tag{10.15}$$

The general Fourier integral solution is

$$\psi = \frac{U}{2\pi} \text{Re} \int_{-\infty}^{\infty} \hat{h}(k) \cdot e^{i[kx+n(k)z]}dk, \tag{10.16}$$

where

$$n(k) = \begin{cases} k\left(\dfrac{N^2}{k^2U^2} - 1\right)^{1/2}, & |k| < N/U, \\[2mm] i\left(k^2 - \dfrac{N^2}{U^2}\right)^{1/2}, & |k| > N/U. \end{cases} \tag{10.17}$$

For the "Witch of Agnesi" this gives

$$\psi = Uh_m a \int_0^{N/U} e^{-ka} \cdot \cos[kx + (N^2/U^2 - k^2)^{1/2}z]dk$$

$$+ Uh_m a \int_{N/U}^{\infty} e^{-ka} \cos kx \cdot e^{-(k^2-N^2/U^2)^{1/2}z}dk. \tag{10.18}$$

If $Na/U \ll 1$, waves are generated in approximately equal amplitudes (for ψ and η) over the whole range of wavenumbers $0 < k < N/U$, and Figure 8.7 shows that these waves are spread over the range of elevations from horizontal to vertical in physical space on the downstream side of the obstacle.

The archetypical mountain when NA/U is small is the delta-function mountain

$$h(x) = h_m\delta(x/A), \tag{10.19}$$

for which $\hat{h}(k) = h_m A$. At large distances from this obstacle where $Nr/U \gg 1$, where $r^2 = x^2 + z^2$ and $x, z > 0$, the integral in (10.16) may be approximated by the method of stationary phase (Jeffreys & Jeffreys, 1962; Lighthill, 1978) to give

$$\psi \approx Uh_m \frac{NA}{U} \frac{1}{\sqrt{\pi}} \frac{z}{r} \frac{1}{(Nr/U)^{1/2}} \cos(Nr/U - \pi/4), \quad Nr/U \gg 1. \tag{10.20}$$

This gives the amplitudes of the waves whose wavenumbers are specified by (8.51) and Figure 8.7. The lines of constant phase are seen to be circles centred on the obstacle. Equation (10.20) may be used to obtain approximate expressions for the wave field for obstacles with $NA/U \ll 1$ by multiplying it by $\hat{h}(k^*)/h_m$, where $k^* = Nx/rU$.

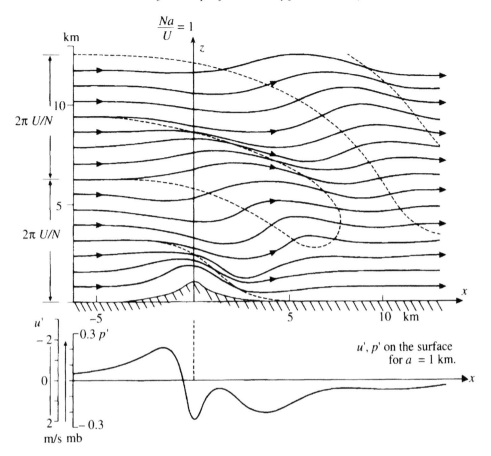

Figure 10.10 Streamlines for the linear boundary perturbation solution for flow
over the "Witch of Agnesi" with $Na/U = 1$. (Adapted from Queney, 1948.)

As NA/U increases, the term represented by (10.20) decreases in magnitude
because $\hat{h}(k^*)$ decreases, and the contribution to the integral from the range where
k is small (which is not given by stationary phase) becomes relatively larger. Hence,
for larger NA/U, lee-wave amplitudes at small elevations are reduced and those at
higher elevations are increased, and this is reflected in Figure 10.10 for $Na/U = 1$.

If $NA/U \gg 1$, the flow is in hydrostatic balance, and only the part of the
spectrum where $k \ll N/U$ has significant amplitude. The wavenumber surface is
now effectively that of Figure 8.8, which implies that the "lee" waves are only found
directly over the obstacle. For flow that is assumed to be hydrostatic, the solutions
may be expressed in the more general form (Drazin & Su, 1975)

$$\psi = \mathrm{Re}\, U[h(x) + \mathrm{i}f(x)].\mathrm{e}^{-\mathrm{i}Nz/U}, \qquad (10.21)$$

where Re denotes the real part and $f(x)$ is determined by the upper radiation condition and is the Hilbert transform of $h(x)$, defined by

$$f(x) = \frac{1}{\pi} \fint_{-\infty}^{\infty} \frac{h(x')}{x' - x} dx', \tag{10.22}$$

where $\fint$ denotes a Cauchy principal-value integral. For $h(x)$ given by (10.14), (10.22) gives

$$f(x) = -h_m(x/a)/[1 + (x/a)^2]. \tag{10.23}$$

This solution is periodic in the vertical, and is shown in Figure 10.11. On the surface of the topography the perturbation velocity $u' = Nf(x)$, so that $u = U + u'$ has a minimum on the upstream side and a maximum on the downstream side, as noted for periodic topography with $k < N/U$. Hydrostatic solutions of this form may be generalised to finite amplitude, as shown in §10.3.

The drag force on a single obstacle is again given by (10.12), and with some manipulation it may be expressed as (Gill, 1982; Blumen, 1965)

$$F_D = \frac{\rho_0 U^2}{2\pi} \int_0^{N/U} |\hat{h}(k)|^2 k(N^2/U^2 - k^2)^{1/2} dk, \tag{10.24}$$

which is effectively a Fourier integral of (10.13). Each Fourier component therefore contributes to this wave drag independently. Obstacle shapes that have a significant part of their spectrum near $k \approx N/U\sqrt{2}$ will tend to have a relatively large drag. Long obstacles with $NA/U \gg 1$, with $|\hat{h}(k)|$ small unless $k \ll N/U$, have relatively small wave drag that may be expressed (approximately) as

$$F_D \approx \frac{\rho_0 NU}{2\pi} \int_0^{\infty} k|\hat{h}(k)|^2 dk. \tag{10.25}$$

In this limit $F_D \approx (\pi/8)\rho_0 NU h_m^2$ for the "Witch of Agnesi" obstacle (10.14). If we express this in terms of a drag coefficient with $F_D = \frac{1}{2}\rho_0 C_{DW} U^2 h_m$, where C_{DW} is the boundary-perturbation-induced wave-drag coefficient, then in this example $C_{DW} \approx (\pi/4)Nh_m/U$.

It should be noted that this steady linear solution is only applicable if $h \to 0$ as $|x| \to \infty$ both upstream and downstream. If h has different values at $\pm\infty$, $\hat{h}(k)$ is singular at $k = 0$ (i.e. $\hat{h}(k) \approx 1/k$), giving a divergent Fourier integral. For an obstacle that covers the range $0 < x < L$, as L becomes large the potential flow becomes singular as $\log L$. This singularity is also reflected in the stratified solution, where $\psi \approx (Uh_m \sin Nz/U) \cdot \log L$ as $L \to \infty$.

This linear theory for boundary perturbations provides a satisfactory qualitative description of the lee-wave field (a quantitative test is described in §10.6 for finite depth), but it does not produce the horizontally propagating disturbances observed on the upstream side. If this linear solution is regarded as the first term in a power series expansion in Nh/U, an extension of the calculation to higher orders beyond

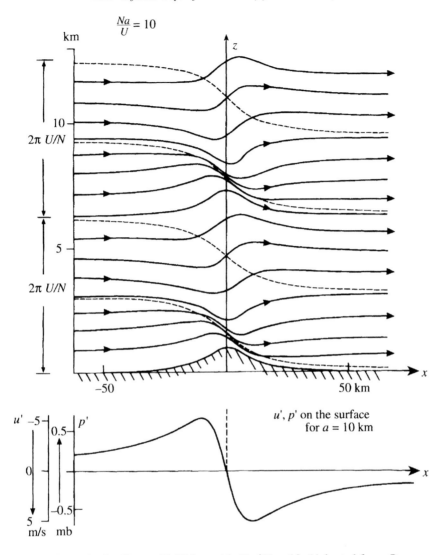

Figure 10.11 As for Figure 10.10 but with $Na/U = 10$. (Adapted from Queney, 1948.)

the first term does not produce these columnar motions either, in the steady-state limit (McIntyre, 1972). This implies that we must consider other theoretical models or processes in order to describe and account for these motions.

10.2.2 *Finite topography with weak stratification, with the lower boundary as a streamline*

An alternative approach that keeps the lower boundary condition on the surface of the finite obstacle and is analytically tractable if Nh/U is small is the limit of weak stratification (i.e. sufficiently small N). A perturbation theory with weak stratification about the state of potential flow over a finite obstacle (a semi-circle, with radius a), with small parameter Na/U and involving matched inner and outer asymptotic expansions, is quite consistent with the small-amplitude perturbation theory (Baines & Grimshaw, 1979). This theory produces lee waves in the far-field (in the outer expansion) that are essentially the same as those described above, and does not give columnar motions, or any other motions, far upstream. The solution also provides some details of how the flow develops with time in this limit, but it does not contribute further useful information about the nature of the steady state or the contrast between boundary-perturbation theory and experiment.

10.2.3 *The effects of frictional drag and lee-side separation: the obstacle as a momentum source*

The analysis of the previous two subsections makes the assumption that the surface of the topography is a streamline. As noted in §10.1, this is not the case if the lee-side slope is sufficiently steep, in which case, for sufficiently large Reynolds number, the flow separates on the lee side of the obstacle to give a turbulent wake. This has two specific effects on the fluid. Firstly, the fluid in this wake has reduced velocity, because it has passed through (or had contact with) the separation region or "bubble" on the lee side. Consequently, the obstacle transfers momentum to the fluid in the wake by a process that is distinctly different from lee-wave forcing by boundary perturbations, as described in §10.2.1 or §10.2.2. The magnitude of this momentum transfer may be expressed as a drag force on the obstacle that has the form of (10.1), where the drag coefficient for this effect alone may be written C_{DS} (S for separation), and it has a value somewhat less than unity. This value is observed to decrease with increasing stratification (Castro et al., 1990). If separation does not occur, a smaller but significant amount of momentum may be imparted to the fluid by the viscous boundary-layer alone, if the obstacle is sufficiently long. The second effect of separation is that the turbulence in the wake will partially mix the fluid in this region. These two effects (drag and mixing) may be expected to have an effect on the motions in the stratified fluid outside the wake, and we next analyse what these effects are.

As in §10.2.1 we assume that Nh/U is small, so that the disturbance is small and the equations may be linearised. Drag associated with a turbulent wake may be represented by a body-forcing term $\mathcal{B}$ in (8.18) (see, for example, Batchelor, 1967), and turbulent mixing by a corresponding term $\mathcal{G}$ in (8.20), so that they become

$$\rho_0 \left[\left(\frac{\partial}{\partial t} + U\frac{\partial}{\partial x} \right) u' + \frac{dU}{dz}w' \right] + \frac{\partial p'}{\partial x} = \mathcal{B}, \tag{10.26}$$

$$\left(\frac{\partial}{\partial t} + U\frac{\partial}{\partial x} \right) \rho' + \frac{d\rho_0}{dz}w' = \mathcal{G}. \tag{10.27}$$

These then give forcing terms in (8.23), which with U uniform becomes

$$\left(\frac{\partial}{\partial t} + U\frac{\partial}{\partial x} \right)^2 \left(\frac{\partial^2}{\partial x^2} + \frac{\partial^2}{\partial z^2} \right) \psi + N^2 \frac{\partial^2 \psi}{\partial x^2}$$

$$= - \left(\frac{\partial}{\partial t} + U\frac{\partial}{\partial x} \right) \frac{\partial}{\partial z} \frac{\mathcal{B}}{\rho_0} - \frac{g}{\rho_0} \frac{\partial \mathcal{G}}{\partial x}. \tag{10.28}$$

The body force $\mathcal{B}$ may be taken to have a spatial distribution extending over a region on the downstream side of the obstacle, within which the fluid is decelerated. For present purposes, however, we may simplify this distribution by taking it to be concentrated at one x-location only ($x = 0$), and distributed uniformly with z up to the level $z = h_m$. Different vertical distributions of this drag force over $0 < z < h_m$ (for example, with $\mathcal{B}$ decreasing upwards) but with the same vertical integral may be more realistic, but these differences have a small effect on the results. Hence we have

$$\mathcal{B} = \begin{cases} -\frac{1}{2}C_{DS}\rho_0 U^2 \delta(x), & 0 < z < h_m, \\ 0, & z > h_m, \end{cases} \tag{10.29}$$

where $\delta(x)$ is the Dirac delta function. In general C_{DS} is a function of the obstacle shape and of Nh/U, but here it may be taken to have a value of order unity (see Castro et al., 1990). If there is no separation and the drag is due to the viscous boundary layer alone, h_m in (10.29) may be replaced by δ_1, the boundary-layer displacement thickness. For a flat plate of length L, the drag is given by $F_D = 0.005L(\frac{1}{2}\rho_0 U^2)$, so that $C_{DS}\delta_1 = 0.005L$. If the mixing in the turbulent wake has efficiency α (where $0 < \alpha \leq 1$, and $\alpha = 1$ implies complete mixing), the term $\mathcal{G}$ may be similarly expressed as

$$\mathcal{G} = \begin{cases} \alpha U \frac{\rho_0 N^2}{g} \left(z - \frac{h_m}{2} \right) \cdot \delta(x), & 0 < z < h_m, \\ 0, & z > h_m. \end{cases} \tag{10.30}$$

Here, α is expected to be related monotonically to C_{DS}. The relative effects of $\mathcal{B}$ and $\mathcal{G}$ may be measured by comparing the magnitudes of the respective forcing terms in (10.28), giving

$$\frac{\text{``}\mathcal{G}\text{ term''}}{\text{``}\mathcal{B}\text{ term''}} \approx \frac{\alpha}{2C_{DS}}\left(\frac{Nh_{\mathrm{m}}}{U}\right)^2, \tag{10.31}$$

which is generally small when the linear equations are applicable. Further, the vertical structure of the forcing term due to mixing shows that it is concentrated at wavelengths $\approx h_{\mathrm{m}}$, which are much shorter than $\pi U/N$, so that it has little impact on the columnar motions and wave field. Hence we will ignore the $\mathcal{G}$ term from here on.

The solution to (10.28) in the limit of large time may be found by assuming that the forcing has $e^{\epsilon t}$ time-dependence, where $0 < \epsilon \ll N$, and then putting $\epsilon = 0$ (Lighthill, 1965; Janowitz, 1981). With $\mathcal{G} = 0$, the "steady-state" solution for ψ for fluid of infinite depth with a rigid boundary at $z = 0$ is

$$\psi = \begin{cases} \dfrac{C_{DS}U^2}{2\pi N}\left[\displaystyle\int_0^1 \dfrac{\sin m\frac{Nh_{\mathrm{m}}}{U}\cdot\sin m\frac{Nz}{U}}{m(1-m)}dm \right. \\ \qquad \left. + \displaystyle\int_1^\infty \dfrac{\sin m\frac{Nh_{\mathrm{m}}}{U}\cdot\sin m\frac{Nz}{U}}{m^2-1}e^{(m^2-1)^{1/2}Nx/U}dm\right], \qquad x < 0, \\[4ex] \dfrac{C_{DS}U^2}{2\pi N}\displaystyle\int_0^1 \sin m\frac{Nh_{\mathrm{m}}}{U}\sin m\frac{Nz}{U}\cdot\left(\dfrac{1}{m(1+m)}+\dfrac{2\cos[(1-m^2)^{1/2}Nx/U]}{1-m^2}\right)dm \\ \qquad + \dfrac{C_{DS}U^2}{2\pi N}\displaystyle\int_1^\infty \dfrac{\sin m\frac{Nh_{\mathrm{m}}}{U}\sin m\frac{Nz}{U}}{m^2-1}[2-e^{-(m^2-1)^{1/2}Nx/U}]dm, \qquad x > 0. \end{cases}$$
$$\tag{10.32}$$

This describes the flow due to a separated wake modelled as a momentum source. Note that this solution is consistent with Figure 8.5, and contains the purely columnar motions which are absent in the preceding models of §§10.2.1 and 10.2.2. Also, conditions at the lower boundary are not altered, as $z = 0$ is a streamline. Hence this motion is quite separate from the motion induced by boundary perturbations, as described in §10.2.1.

On the upstream ($x < 0$) side, the solution (10.32) consists of columnar motions with $0 < n < N/U$, and evanescent terms. On the downstream ($x > 0$) side, there are the expected columnar motions at all wavenumbers, and lee-wave-like terms. On both sides these integrals are divergent, with logarithmic singularities at $m = 1$, which corresponds to the vertical wavenumber $n = N/U$. These singularities represent a real physical phenomenon, which is as follows. The waves propagating horizontally against the stream with vertical wavenumber n have group velocity $N/n - U$ (directed upstream), so that waves with $n = N/U$ have zero group velocity relative to the source. Similar waves with n very close to N/U therefore advance very slowly upstream or downstream with time. These waves are generated continuously, and this allows the energy density to accumulate to high levels, given sufficient time.

We may use this expression for the group velocity to eliminate the waves that have not yet had time to reach a specific x-location from the integrals of (10.29). Hence, at an upstream position x (< 0) at time t the flow field will be given approximately by

$$\psi = \frac{C_{DS}U^2}{2\pi N} \int_0^{1/(1+|x|/Ut)} \frac{\sin m \frac{Nh_m}{U} \cdot \sin m \frac{Nz}{U}}{m(1-m)} dm, \qquad (10.33)$$

omitting the evanescent terms and the effects of dispersion on broadening the fronts of the columnar motions. The logarithmic "singularity" is very weak and is seen to give terms of order unity on any practical time-scale, such as those of the experiments of §10.1. For $Nh/U \ll 1$, the upstream velocity from (10.33) is then given by

$$\frac{u'}{U} \approx -\frac{C_{DS}}{2\pi} \frac{Nh_m}{U} \int_0^{1/(1+|x|/Ut)} \frac{\cos m \frac{Nz}{U}}{1-m} dm. \qquad (10.34)$$

In the lee waves of §10.2.1 we have $u' \approx Nh_m$, and these (10.34) upstream motions are of the same magnitude in dimensional terms. However, they are numerically smaller by the factor $C_{DS}/2\pi$, which is typically about 0.05 on the numbers given above. On the downstream side, it can be shown from (10.32) that $u' \approx -C_{DS}U$ at low levels where $0 < z < h_m$, reflecting the presence of the wake, but at higher levels where $z > h_m$, $u' = O(C_{DS}Nh_m)$.

Another method that has sometimes been used to represent an obstacle in a stratified fluid is to place sources and sinks of fluid in the flow (e.g. Wong & Kao, 1970). An isolated obstacle is represented by a doublet (i.e. a source-sink pair), and a source would represent an obstacle whose length increases continuously with time. The flow fields obtained at large distances are expected to correspond to those produced by the represented obstacle. A doublet does not produce columnar motions in the steady state, but a single source or sink does. A study of the linearised equations shows that a source of mass in a fluid flow has much the same effect as a source of momentum, except for the important wake region. For this reason such models are best used as representations of inflows and outflows, rather than of obstacles. Sinks of fluid in stratified flow have been used to represent withdrawal flows from reservoirs for example (Imberger & Patterson, 1990), and the resulting flow patterns are dominated by columnar motions.

The general character of the upstream flows described by the momentum-source model is quite consistent with the upstream motions described in §10.1. The only significant difference is the non-observance of columnar modes in the experiments when $Nh/U < 0.5$, which is due to the limitation on the maximum vertical wave length and the discrete columnar wave spectrum imposed by the tank geometry. Hence, although detailed quantitative comparisons have not been made because of the difficulty of quantifying the momentum source, the general similarities between theory and observation indicate that the momentum loss in the fluid due to lee-side separation and the turbulent wake can produce the observed upstream motions for $Nh/U < 1$.

We therefore conclude that for small Nh/U an isolated obstacle may be appropriately represented by the sum of two different types of forcing: a boundary perturbation and a momentum source. However, these factors are not independent. The extent of lee-side separation is affected by the flow field that is set up by the boundary perturbations, as discussed above, and lee-side separation in turn affects the apparent obstacle shape. For Nh/U small, we may ignore this interaction and heuristically add these two effects, in order to get a total effect of the flow over a real obstacle. In particular, the total drag coefficient may be written

$$C_D = C_{DW} + C_{DS}, \tag{10.35}$$

where C_{DW} is the drag coefficient due to the linear boundary perturbation alone, as described in §10.2.1. In stratified fluids, the value of the total drag coefficient is often observed to be somewhat greater than unity, implying that the drag exceeds that of the neutrally stratified case, because of the additional drag due to the lee waves and columnar motions.

The momentum-source effect due to separation is not restricted to small Nh/U. Equation (10.32) is not formally applicable for $Nh_m/U \approx 1$, but it should still be approximately valid. This is because the motion is mostly horizontal, with weak horizontal gradients, implying weak non-linear self-advective effects. However, the disturbances caused by boundary perturbations increase in magnitude and change character as Nh/U increases, and these come to dominate the motion as the following sections show.

10.3 Infinite-depth flows: finite-amplitude topography and "Long's model"

The boundary perturbation model of §10.2.1 may be extended to give exact solutions when Nh/U is not small. To do this we must assume that the lower boundary is a streamline so that there is no flow separation, that the flow is steady, and that the conditions far upstream are known. We start with (1.1)–(1.4) with $\partial/\partial t \equiv 0$, and assume that far upstream the velocity and density profiles are given by $U(z)$ and $\rho_0(z)$ respectively, with $U > 0$ and $d\rho_0/dz \le 0$ everywhere, and the lower boundary there is located at $z = 0$. We also assume that all streamlines extend far upstream, so that a particular streamline may be identified by its upstream elevation, z_0. We next introduce the variable $\delta(x, z)$, defined by

$$\delta(x, z) = z - z_0, \tag{10.36}$$

so that δ represents the vertical displacement of the streamline at the point (x, z) from its upstream elevation (note that $\delta = \eta$ in the steady-state limit of time-dependent flows where the flow properties far upstream are not altered by the disturbance). We may also define a total stream function $\Psi(z_0)$ by

$$\Psi(z_0) = -\int_0^{z_0} U(z')dz', \tag{10.37}$$

so that the velocity field is given by

$$\mathbf{u} = \left(-\frac{\partial\Psi}{\partial z}, \frac{\partial\Psi}{\partial x}\right) = U(z_0)\left(1 - \frac{\partial\delta}{\partial z}, \frac{\partial\delta}{\partial x}\right). \tag{10.38}$$

The equations (1.1)–(1.4) then give (Long, 1953)

$$\mathbf{u} \cdot \nabla(p + \frac{1}{2}\rho u^2 + \rho g z) = 0, \tag{10.39}$$

and

$$\mathbf{u} \cdot \nabla\left[\nabla^2\Psi + \frac{1}{\rho}\frac{d\rho}{d\Psi}\left(\frac{1}{2}(\nabla\Psi)^2 + gz\right)\right] = 0. \tag{10.40}$$

Expressing (10.40) in terms of δ then gives

$$\frac{\partial}{\partial s}\left(\nabla^2\delta + \frac{1}{q}\frac{dq}{dz_0}\left(\frac{\partial\delta}{\partial z} - \frac{1}{2}(\nabla\delta)^2\right) + \frac{N(z_0)^2}{U(z_0)^2}\delta\right) = 0, \tag{10.41}$$

where $q = \rho_0(z_0)U(z_0)^2/2$, and s is a curvilinear coordinate measured along a streamline. Since $\delta \to 0$ as $x \to -\infty$, this equation may be integrated to give

$$\nabla^2\delta + \frac{1}{q}\frac{dq}{dz_0}\left(\frac{\partial\delta}{\partial z} - \frac{1}{2}(\nabla\delta)^2\right) + \frac{N(z_0)^2}{U(z_0)^2}\delta = 0. \tag{10.42}$$

This equation describes the steady flow field for given upstream conditions, and it may be solved for a given obstacle shape, which implies a lower boundary condition

$$\delta = h(x), \quad \text{on the boundary} \quad z = h(x). \tag{10.43}$$

We discuss the infinite-depth situation here, and treat the important variations of the corresponding finite-depth case in §10.5. In order that the solution to this steady-state problem represents a flow field in which the horizontal variations are created by flow over the obstacle, conditions must be imposed on the solution at large distances, which are tantamount to radiation conditions on the wave field. Mathematically, it is sufficient to require that the disturbances are on the downstream side, so that (Miles, 1968)

$$\delta = \begin{cases} O(1/r^{1/2}), & r \to \infty, & 0 \le \theta \le \pi/2, \\ o(1/r^{1/2}), & r \to \infty, & \pi/2 \le \theta \le \pi, \end{cases} \tag{10.44}$$

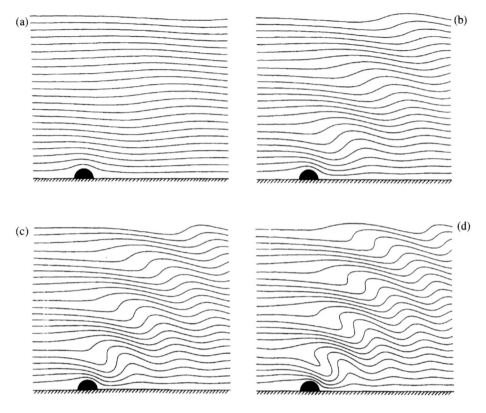

Figure 10.12 Streamlines showing Long-model solutions for uniform upstream flow over a semi-circle. (a) $Nh/U = 0.5$; (b) $Nh/U = 1.0$; (c) $Nh/U = 1.27$; (d) $Nh/U = 1.5$. (From Miles, 1968, reproduced with permission.)

where $O(x)$ means the same order of magnitude as x, $o(x)$ means of order much smaller than x, and r and θ are polar coordinates. However, this does not guarantee that the resulting solution is physically realisable. As is seen below, a given obstacle height and shape may be incompatible with given upstream velocity and density profiles.

Equation (10.42) is non-linear, and to proceed further we must specify q. The only choice for which useful solutions have been found is for $dq/dz = 0$, so that the vertical profile of the kinetic energy of the upstream flow is uniform. The equation then becomes linear and takes the form of a Helmholtz equation:

$$\nabla^2 \delta + \kappa^2 \delta = 0, \tag{10.45}$$

where

$$\kappa^2 = N(z_0)^2/U(z_0)^2 = -g\frac{d\rho_0}{dz_0}/2q. \tag{10.46}$$

Here κ is constant if $d\rho_0/dz$ is constant. If ρ is approximately constant so that the Boussinesq approximation is applicable, $N(z)$ and $U(z)$ are effectively constant. The linearity of (10.45) sets it apart from the general form (10.42) and endows the flows with uniform q the special character noted at the start of the chapter. This linear equation for the non-linear flow field, with boundary conditions (10.43)–(10.44), is often termed "Long's model", and it may be solved using standard techniques. Note that non-linear advection is still important, and this flow may be regarded as an extension of potential flow to stratified fluids.

Figure 10.12 shows solutions to (10.43)–(10.45) for flow over a semi-circle of radius h_m for a range of values of $\kappa h_m = Nh/U$, obtained by analytic methods (Miles, 1968). As Nh/U increases, the lee waves become progressively steeper, and they reach a point where the streamlines become vertical when $Nh/U = (Nh/U)_c = 1.27$, as shown in Figure 10.12c. For larger values of Nh/U the flow becomes statically unstable, as shown in Figure 10.12d. As one might expect, these statically unstable solutions are dynamically unstable (Laprise & Peltier, 1989a), and the vertical streamline criterion marks the approximate limit of applicability of these solutions.

These solutions have been extended to obstacles of general semi-elliptical shape by Huppert & Miles (1969). With $\gamma = h_m/a$, where a is the half-width of the obstacles, these shapes covered the whole range of "aspect ratios" from the flat semi-ellipse ($\gamma = 0$) to the semi-circle ($\gamma = 1$) to the vertical barrier ($\gamma = \infty$). The flow fields are visually similar to those of Figure 10.12, and the limiting values of Nh/U, above which the flow field was statically unstable, ranged from 0.67 (when $\gamma = 0$) to 1.73 (when $\gamma = \infty$), as shown in Figure 10.13. As a general rule, longer obstacles produce steeper lee waves. The drag force F_D and drag coefficient C_{DW} are shown in Figures 10.14a,b respectively, as functions of Nh/U, for the "squat semi-ellipses", where $0 \le \gamma \le 1$. We see that C_{DW} varies approximately as $(Nh/U)^3$ for given γ, unless the flow is hydrostatic with $\gamma \ll 1$ and $Na/U \gg 1$, in which case $C_{DW} \approx Nh/U$. The maximum value of C_{DW} for statically stable flow for all squat semi-ellipses was found to be approximately 3.0.

When $h_m/a \ll 1$, so that the flow is hydrostatic, (10.45) simplifies to

$$\frac{\partial^2 \delta}{\partial z^2} + \kappa^2 \delta = 0, \tag{10.47}$$

where again, $\kappa = N/U$ for a Boussinesq fluid. This equation has the general solution satisfying the lower boundary condition (10.43)

$$\delta(x, z) = h(x) \cdot \cos \kappa[z - h(x)] + f(x) \cdot \sin \kappa[z - h(x)] \tag{10.48}$$

$$= \mathrm{Re}\,[H(x) \cdot e^{-i\kappa(z-h)}], \tag{10.49}$$

where $H(x) = h + if$. This is a generalisation of (10.21), and again the real function

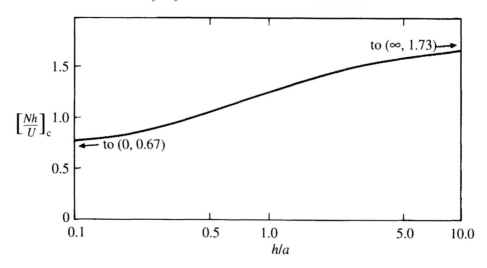

Figure 10.13 Limiting values of Nh/U for semi-elliptical obstacles for the range of values of $\gamma = h_m/a$, from Long's model. (From Huppert & Miles, 1969, reproduced with permission.)

$f(x)$ must be determined from the upper radiation condition of no incoming energy. The requirement that all Fourier components of (10.48) have upward group velocity yields the equation for $f(x)$ (Lilly & Klemp, 1979)

$$H(x) \cdot e^{i\kappa h(x)} = \frac{i}{\pi} \int_{-\infty}^{\infty} \frac{H(x') \cdot e^{i\kappa \cdot h(x')}}{x' - x} dx',$$ (10.50)

which is a finite-amplitude generalisation of (10.22). Bernoulli's equation for this hydrostatic flow has the form

$$\frac{1}{2}\rho_0 u^2 + \frac{1}{2}\rho_0 N^2 \delta^2 + p = \frac{1}{2}\rho_0 U^2,$$ (10.51)

and by substituting for p in (10.12) we obtain the drag force F_D as

$$F_D = \rho_0 U N \int_{-\infty}^{\infty} \left(1 - \frac{Nf}{2U}\right) f \frac{dh}{dx} dx.$$ (10.52)

The horizontal velocity on the surface of the topography is given by $u = U + Nf(x)$. Numerical solutions of (10.50) for $f(x)$ for the "Witch of Agnesi" (Lilly & Klemp, 1979) show that the minimum in u on the upstream face decreases more slowly with Nh/U than the trend given by linear theory (10.23), reaching $0.69U$ at the limiting height, and moves upstream slightly. The maximum in u on the downstream side also moves upstream (closer to the crest) as Nh/U increases, and increases more rapidly than linear theory to a value of $1.73U$.

The solution (10.48) for the "Witch of Agnesi" at the limiting value $Nh/U =$

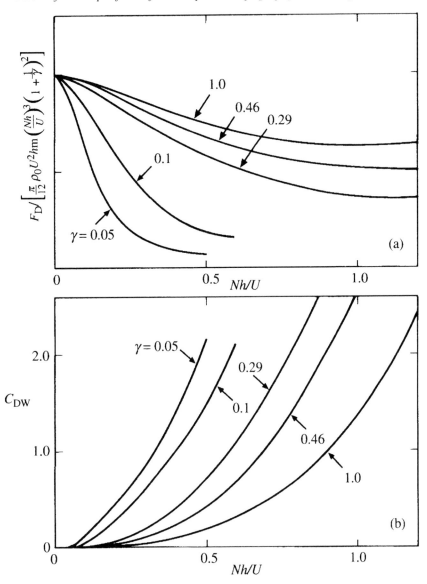

Figure 10.14 (a) Drag and (b) drag coefficient for semi-elliptical obstacles for various ϵ as a function of Nh/U, obtained from Long's model. (Adapted from Huppert & Miles, 1969, reproduced with permission.)

0.85 is shown in Figure 10.15. As for the linear solution (10.18), this solution is periodic in the vertical coordinate, and the disturbance is situated directly over the topography. Corresponding hydrostatic solutions for flow over asymmetric "Witches

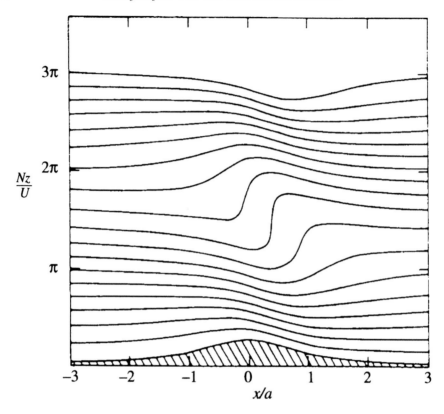

Figure 10.15 The hydrostatic Long-model solution for flow over the "Witch of Agnesi" for the value $Nh/U = 0.85$. Note that in this and other illustrations of hydrostatic solutions the vertical scale has been stretched for clarity; the flow structure is much more elongated horizontally in practice. (From Lilly & Klemp, 1979, reproduced with permission.)

of Agnesi" of the form

$$z = h(x) = \begin{cases} h_{\mathrm{m}}/[1 + (x/a_{\mathrm{u}})^2], & x < 0, \\ h_{\mathrm{m}}/[1 + (x/a_{\mathrm{d}})^2], & x > 0, \end{cases} \qquad (10.53)$$

where $2a_{\mathrm{u}}$ and $2a_{\mathrm{d}}$, the "half-widths" on the upstream and downstream sides respectively, have limiting values of Nh/U as shown in Figure 10.16 (Baines & Granek, 1990). These lie in the range

$$0.5 < \left(\frac{Nh}{U}\right)_c < 1, \qquad (10.54)$$

with the smaller values (i.e. steeper waves) occurring when the downstream face is steeper. Critical values for other realistic obstacle shapes are also expected to lie in the range given by (10.54) for hydrostatic flow.

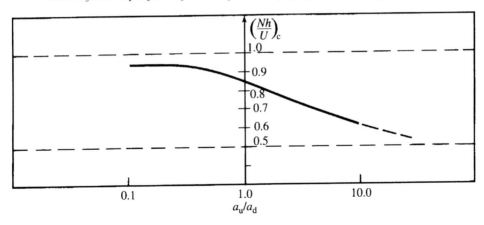

Figure 10.16 $(Nh/U)_c$ for hydrostatic asymmetric obstacles with two-sided "Witch of Agnesi" profiles, as a function of the ratio a_u/a_d of the upstream and downstream half-widths. (From Baines & Granek, 1990.)

Hence, for situations where q and $d\rho/dz$ are independent of height far upstream, there are steady solutions to the complete non-linear inviscid equations for flow over topography where the bottom streamline is specified. These solutions are non-linear extensions of the small-height perturbation solutions of §10.2.1, and apply for Nh/U less than a maximum value $(Nh/U)_c$, above which the solution is not dynamically stable. Are these steady Long-model solutions realised when the flow is commenced from rest? Is it possible, for instance, that in the time-development of the flow, upstream columnar motions (consistent with Figures 8.5 or 8.6) will be generated that will alter the local upstream profiles of U and N so that (10.45) will not be applicable? Solutions of the resulting form of (10.42), with the additional non-linear terms, may then have quite different character.

Numerical experiments that addressed this question have been described by Pierrehumbert & Wyman (1985), using the Boussinesq, hydrostatic "primitive equation" model of Orlanski & Ross (1977). This model has open boundary conditions upstream and downstream for waves with a specified speed, and a wave-absorbing "sponge" layer at the top. The lower boundary was (inviscid) free-slip, precluding flow separation, with Gaussian topography that has $(Nh/U)_c = 0.75$ for hydrostatic flow. The fluid was impulsively set into motion at a given time ($t = 0$), and for all values of Nh/U less than this limiting value, the flow tended to the Long-model solution at large times, after the expected wave transients had dispersed. Further experiments of a similar nature by Bacmeister (1987) and subsequently, with shorter obstacles using the non-hydrostatic anelastic model of Clark (1977), again with a free-slip lower boundary precluding separation, gave the same conclusion. However, for larger values of Nh/U approaching $(Nh/U)_c$, the approach to the steady

"Long's-model" state was slow, and the flow contained upstream transients with small k, with structure similar to the permanent motions observed in §10.1.

We may infer from these studies that when $Nh/U < (Nh/U)_c$, the observed steady upstream columnar disturbances described in §10.1 are due to the lee-side separation and frictional drag effect, modelled by a momentum source as in §10.2.3. The solutions due to these two forcing processes may be heuristically added when Nh/U is small, but they interact for larger values. In particular, the presence of the columnar motions is observed to oppose the steepening of the lee waves and to increase $(Nh/U)_c$; overturning in the experiments is not observed to occur until Nh_m/U attains larger values than those predicted by Long's model (Baines & Hoinka, 1985). For instance, $(Nh/U)_c$ values from Long's model for the semi-circle and the semi-ellipse with $\gamma = 0.35$ are 1.27 and 0.94 respectively, whereas the observed value of Nh/U for overturning for both cases is approximately 1.5.

10.4 Infinite-depth flows with $Nh/U > (Nh/U)_c$: numerical studies

What happens when $Nh/U > (Nh/U)_c$? We must distinguish between two situations: real flows such as in laboratory experiments, and theoretical or numerical ones where the lower boundary is assumed to be a streamline. We restrict consideration initially to this second class of flows, so that separation and wake effects are excluded (most numerical model studies have made this assumption). It is instructive to examine the evolution of a flow from an unstable Long's-model state where Nh/U is only slightly above the limiting value. Figure 10.17 shows such an evolution, computed by Laprise & Peltier (1989b) using the non-hydrostatic anelastic model of Clark (1977). The obstacle is the Witch of Agnesi with $h_m/a = 0.095$ so that the initial flow is hydrostatic, and $Nh/U = 0.95$, which exceeds the limiting value of 0.85, so that the flow is unstable (the initial Long-model solution was modified to have reduced amplitude at higher levels, for computational simplicity and realism, so that only the lowest wave was statically unstable).

The most striking features of this development are as follows. The region where the streamlines are statically unstable evolves into a deep, persistent region where the density is nearly homogeneous with very weak mean flow. Below this region, the character of the flow changes to resemble a hydraulic type of transition between the upstream and downstream flow states over the obstacle, from a deep, slowly moving stream to a shallower, rapidly flowing one. This change in the flow is initiated by an identifiable growing mode arising from the unstable initial state, trapped below the homogeneous region and gaining kinetic energy from the mean shear (Laprise & Peltier, 1989a). It is evident in frames (d), (e) and (f) that columnar disturbances are propagating upstream and downstream away from the obstacle on this flow, and that these will permanently change the upstream (and possibly downstream) velocity

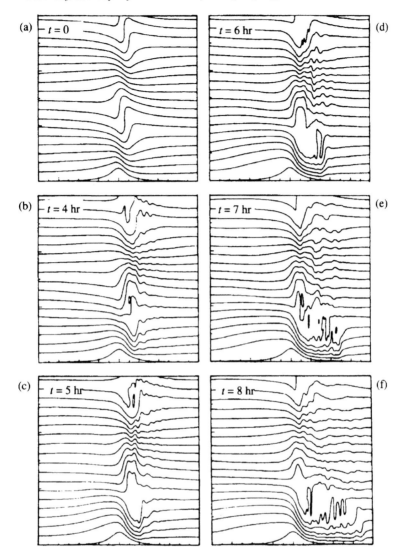

Figure 10.17 The evolution with time commencing with the unstable hydro-static Long-model solution over the "Witch of Agnesi" with $Nh/U = 0.95$ and $h_m/a = 0.095$. Note the evolution to a low-level hydraulic flow, and the upstream-propagating disturbance. (From Laprise & Peltier, 1989b: reproduced by permission of the American Meteorological Society.)

and density profiles. There are also changes in the flow at higher levels, but these have little interaction with the flow at low levels and are largely inconsequential.

Other numerical simulations with $Nh/U > (Nh/U)_c$, where the flow is initialised from a state of rest, also evolve to give the same type of flow state (Clark &

Peltier, 1977; Peltier & Clark, 1979; Pierrehumbert & Wyman, 1985). Initially lee waves develop, become unstable and break, to give a semi-stagnant homogeneous region over the obstacle with a hydraulic type of flow below. The time taken to reach an approximately steady form of this state is usually much larger than those suggested by simple scaling, such as a/U. This phenomenon (or variations of it) is primarily responsible for downslope windstorms, which are anomalously strong winds, observed on occasions on the lee sides of mountain ranges such as the Rocky mountains (see §11.8).

The homogeneous stagnant region over the obstacle may be regarded as a critical layer of finite thickness, and from Chapter 8 we may expect this region to be a substantial reflector of wave energy incident from below. Since the flow beneath it is insulated from the flow at higher levels, the low-level flow is now essentially a "finite-depth" system with a pliant upper surface. This upper surface only exists over the lee side of the obstacle (i.e. it is not present upstream or further downstream), but this is in fact where it really matters. The low-level stratified flow now has a hydraulic character, similar in many respects to that of a single layer as described in Chapter 2. It has a sub- to supercritical transition over the obstacle, but is more complex, with variable velocity and density profiles in addition to its varying total depth. It is appropriate to defer the discussion of the detailed mechanics of this finite-depth system to §11.4.2.

The numerical flow depicted in Figure 10.17 is qualitatively very similar to the observations of the experiments described in §10.1 when $Nh/U > 1.5$ (where overturning occurs for the obstacles used), so that, in this parameter range, the same physical processes appear to be happening in each system. In particular, both show the generation of upstream disturbances. The momentum-source model of §10.2.3 should be irrelevant to these upstream motions, for two reasons. First, the increased amplitude of the boundary-forced perturbations tends to suppress lee-side separation. Second, if lee-side separation does occur, the descending supercritical stream will inhibit upstream propagation of disturbances forced by the lee-side momentum source. Hence, in these experiments, the onset of overturning marks a boundary between the nature of the sources of the upstream columnar motion, from a separated wake as in §10.2.3 for $Nh/U < 1.5$, to a hydraulic one as in §2.3.2 for $Nh/U > 1.5$. As Nh/U increases above the limiting value, the upstream motion increases in amplitude until upstream blocking ensues for $Nh/U > 2$, as described in §10.1. As it increases further, the depth of this blocked region approaches h_m, and the vertical scale of the hydraulically controlled stream above the blocked region becomes progressively thinner. But no change to this general character of the flow was apparent for Nh/U values up to 4, and it appears to be maintained up to quite large values.

When transition to a hydraulic flow state occurs, it is accompanied by a marked increase in the drag (and drag coefficient), typically by a factor of 2 to 3. This is due to the increased difference between the pressure on the upstream and downstream faces of the obstacle, in turn related to the difference between the sub- and supercritical flow states. We return to this question in §11.5.

10.5 Linear theory for small $N h_m/U$: finite depth

A finite-depth system has the additional length-scale D, the depth of the fluid or height of the upper boundary. We will take the upper boundary to be rigid in most cases. Since the total nett horizontal flow is fixed, the upper boundary condition (1.24) is then

$$\psi = 0, \quad \text{on } z = D. \tag{10.55}$$

The general properties of waves in a finite-depth system with rigid upper and lower boundaries are described in §8.3. We assume that $U > 0$ in the range $0 \le z \le D$, so that critical layers are avoided, and initially also assume that U and N are constant so that $l = N/U$. This system has the modes and wave speeds given by (8.31)–(8.33), and in the long-wave limit ($k \rightarrow 0$), the fastest mode has the speed $c_1 = ND/\pi$ relative to the fluid. Compared with the infinite-depth case we now have the additional parameter K, defined by

$$K = \frac{ND}{\pi U}, \tag{10.56}$$

which is the reciprocal of the Froude number with respect to the fastest internal wave mode. The equations for an upper boundary at a material surface with uniform fluid above are given in §§1.2, 8.3, and the overall results are similar to those for a rigid one, apart from the presence of the "external" or surface mode. We describe the finite-depth forms of the two models of §10.2.1 and 10.2.3: the boundary perturbation model for small-amplitude topography, and the momentum-source model. Time-dependent effects are more important here, and are given correspondingly greater attention.

10.5.1 Small-amplitude topography with the lower boundary as a streamline

Following §10.2.1, we again consider the solution arising from a time-dependent perturbation problem where $Nh/U \ll 1$ and the flow is governed by (8.23) with U constant; that is,

$$\left(\frac{\partial}{\partial t} + U \frac{\partial}{\partial x} \right)^2 \nabla^2 \psi + N^2 \frac{\partial^2 \psi}{\partial x^2} = 0, \tag{10.57}$$

with the lower boundary condition (10.6). For the initial conditions we assume that the fluid is impulsively set into motion at $t = 0$, corresponding to the experimental state of flow over a towed obstacle with a sudden start. This gives potential flow with horizontal density surfaces at $t = 0$, so that

$$\nabla^2 \psi = 0, \quad \text{at } t = 0.$$
$$\frac{\partial}{\partial t} \nabla^2 \psi = 0. \tag{10.58}$$

Provided that K is not an integer, the solution (modified from McIntyre, 1972) is

$$\psi(x, z, t) = \psi_s(x, z) + \sum_{j=1}^{\infty} \psi_j(x, t) \sin \frac{j\pi z}{D}, \tag{10.59}$$

where

$$\psi_s(x, z) = \frac{U}{2\pi} \int_{-\infty}^{\infty} \hat{h}(k) e^{ikx} \left[\frac{\sin[(N^2/U^2 - k^2)^{1/2}(D - z)]}{\sin[(N^2/U^2 - k^2)^{1/2}D]} \right] dk,$$

$$\psi_j(x, t) = \frac{Uj}{2N^2 D^2} \int_{-\infty}^{\infty} \hat{h}(k) e^{ikx} v_j^3 \left(\frac{e^{-ik(U+v_j)t}}{U + v_j} - \frac{e^{-ik(U-v_j)t}}{U - v_j} \right) dk. \tag{10.60}$$

Here v_j is the speed of the jth internal wave mode in fluid at rest, given by (8.33) with $U = 0$, and the integrals are Cauchy principal values. If $K = j$ for some positive integer j, the mean flow is critical with respect to the jth mode in the long wavelength limit; here there is no steady-state solution, and in the solution to the above initial value problem ψ_j grows with time as $(Nt)^{1/3}$, in the same manner as for single-layer flows in §2.2.2. When K is not an integer, the large-time limit of (10.60) contains terms localised to the region of the obstacle, plus steady lee waves and upstream and downstream transients for modes with $j < K$, and only transients for $j > K$. Both ψ_s and the ψ_j contribute to the steady disturbance. Some comparisons with experiment are described in §10.6.

In the steady state, the drag force on the obstacle resulting from (10.60) is

$$F_D = \frac{1}{2} C_{DW} \rho_0 U^2 h_m = \frac{\pi \rho_0 U^2}{D^3} \sum_{j=1}^{j_K} j^2 |\hat{h}(k_j)|^2, \tag{10.61}$$

where C_{DW} and $\hat{h}(k)$ are defined as in §10.2.1,

$$k_j = \frac{\pi}{D}(K^2 - j^2)^{1/2}, \tag{10.62}$$

and j_K is the largest integer less than K. This drag is proportional to the sum of the squares of the amplitudes of the lee-wave modes, and is analogous to (2.33) for a single layer.

When $N(z)$ is not uniform, expressions analogous to (10.60) may be derived. But if $U(z)$ is not uniform the problem is more complex, because of the possibility of singular modes with critical layers. This difficulty may be circumvented by assuming a shear flow where U is uniform in layers, as described in §8.1.

Hydrostatic flows

It is not easy to discern the general character of (10.60) directly, since it involves some cancellation between the integral terms. The hydrostatic limit, on the other hand, is more instructive. The governing equation here is (8.54) with U constant, namely

$$\left(\frac{\partial}{\partial t} + U\frac{\partial}{\partial x}\right)^2 \frac{\partial^2\psi}{\partial z^2} + N^2\frac{\partial^2\psi}{\partial x^2} = 0, \tag{10.63}$$

and with the same boundary conditions and initial conditions (10.58) as above, the solution (again provided that K is not an integer) is (Baines & Guest, 1988)

$$\psi(x,z,t) = Uh(x)\frac{\sin K\pi(1 - z/D)}{\sin K\pi}$$

$$- U\frac{K}{\pi}\sum_{j=1}^{\infty}\left(\frac{h[x - (1 - K/j)Ut]}{j - K} - \frac{h[x - (1 + K/j)Ut]}{j + K}\right)\frac{\sin j\pi z/D}{j}.$$

$$\tag{10.64}$$

Here the steady and transient parts of the solution are quite distinct. The steady part is localised over the obstacle, as for the infinite-depth case, and the transient part for each mode consists of two components, the one with larger amplitude propagating against the stream and the smaller propagating with it. Each term has the form of the obstacle itself. The situation is a many-mode analogue of the single-layer case (2.13).

For long obstacles, (10.64) implies that the flow will take a long time to reach steady state, particularly if a mode is close to being critical. For obstacles of semi-infinite length, the initial potential-flow solution exists (unlike the infinite-depth case), and the solution (10.64) is still valid. Hence, if $h \to h_0 \neq 0$ as $x \to \infty$, the upstream-propagating transients in (10.64) will never leave the vicinity of the obstacle, and they will constitute steady upstream columnar motions.

As K approaches an integer $j = n$, the steady part and the upstream-propagating transient with $j = n$ both become singular but with opposite signs. The solution to the above initial-value problem when $K = n$ is then

$$\psi = \frac{U^2 t}{n\pi}\frac{dh}{dx}\sin\frac{n\pi z}{D} + Uh(x)\left[\left(1 - \frac{z}{D}\right)\cos\frac{n\pi z}{D} + \frac{1}{n\pi}\sin\frac{n\pi z}{D}\right] \tag{10.65}$$

$$+ \text{ propagating terms,}$$

where the latter are the terms of (10.64) with $j \neq n$ and the downstream-propagating transient with $j = n$. Equation (10.65) is the stratified analogue of (2.14), and shows that the solution grows linearly with time with this resonant forcing. As for a single layer, the singularities and the growth rate with time when $K = j$ are more severe in the hydrostatic case because wave dispersion is absent.

For steady flow with $K \neq n$ the drag vanishes, because the pressure distribution over the obstacle is symmetric about the upstream and downstream sides, or equivalently, in (10.61) the forcing of each lee-wave mode vanishes because $\hat{h}(k_j)$ vanishes (for $j \leq j_K$) unless $k_j = 0$. On the other hand, when resonance occurs the drag is non-zero, and from (10.12), (10.65) it increases linearly with time.

If $N(z)$ is not uniform, the above solution may be generalised as follows. If a uniform flow with velocity U is suddenly commenced at $t = 0$ over the obstacle $z = h(x)$, the general solution to (10.63) may be written

$$\psi(x, z, t) = \sum_{j=0}^{\infty} A_j(x, t)\psi_j(z),$$
(10.66)

where the ψ_j are a complete set of orthogonal eigenfunctions with eigenvalues $v_j > 0$, satisfying

$$\frac{d^2\psi_j}{dz^2} + \frac{N^2(z)}{v_j^2}\psi_j = 0,$$
(10.67)

where $\psi_j = 0$ on $z = 0, D$, for a rigid upper boundary. For the case of a pliant upper boundary at $z = D$, where the density is uniform for $z > D$, the boundary conditions are

$$\psi_j = 0, \quad \text{on } z = D,$$
$$v_j^2\frac{d\psi_j}{dz} = g\frac{\Delta\rho}{\rho}\psi_j, \quad \text{on } z = D,$$
(10.68)

where $\Delta\rho$ is the density jump (if any) at $z = D$. The orthogonality conditions for the ψ_j are

$$I_j\delta_{ij} = \int_0^D N^2\psi_i\psi_j dz = -v_j^2 \int_0^D \frac{d\psi_i}{dz}\frac{d\psi_j}{dz} dz,$$
(10.69)

defining I_i, where δ_{ij} is the Kronecker delta function. The internal wave speeds in this flow are given by $c_j^{\pm} = U \pm v_j$ (for the upstream(−) and downstream(+) propagating components of the jth mode), and the amplitude functions A_j are given by

$$\frac{1}{v_j^2}\left(\frac{\partial}{\partial t} + U\frac{\partial}{\partial x}\right)^2 A_j = \frac{\partial^2 A_j}{\partial x^2} + \left(\frac{\partial}{\partial t} + U\frac{\partial}{\partial x}\right)^2 B_j,$$
(10.70)

where B_j is the projection of the topographic forcing on the jth mode, and is given by

$$I_j B_j = \frac{d\psi_j(0)}{dz}.Uh(x), \qquad t > 0. \tag{10.71}$$

In particular, solution (10.64) with the same initial and boundary conditions may be generalised to the form

$$\psi = Uh(x)g(z) + \sum_{j=1}^{\infty} a_j \psi_j(z) \left(\frac{h[x - (U - v_j)t]}{U - v_j} - \frac{h[x - (U + v_j)t]}{U + v_j} \right), \tag{10.72}$$

where $g(z)$ is given by

$$\frac{d^2 g}{dz^2} + \frac{N^2(z)}{U^2} g = 0, \qquad \text{with} \quad g = 1 \text{ on } z = 0, \quad g = 0 \text{ on } z = D, \tag{10.73}$$

and the a_j by

$$a_j = \frac{U(U^2 - v_j^2) \int_0^D N^2 \left[1 - \frac{z}{D} - g(z) \right] \psi_j dz}{2 v_j \int_0^D N^2 \psi_j^2 dz}. \tag{10.74}$$

(The a_j do not vanish when $U = v_j$ because this cancels with a singularity there in $g(z)$.) This procedure cannot be readily extended to the case where $U(z)$ varies with height, for two main reasons. First, if U varies continuously, the forcing may excite waves (both steady or transient) with speeds within the range of values of U, that are absorbed in critical layers. Second, although complete sets of wave modes exist as described in §8.3, there is no simple orthogonality relationship between them like (10.69), and one must be constructed using a Schmidt procedure (e.g. Morse & Feshbach, 1953, part I). This is because the z-dependence in (8.23) cannot be separated from x and t.

Steady-state solutions for general U, N profiles

Steady-state solutions for variable $U(z)$ and $N(z)$ are more readily obtained than time-dependent ones. The governing equation for linearised topography is (10.9), namely

$$\nabla^2 \psi + l^2(z)\psi = 0, \qquad \text{where} \quad l^2 = N^2/U^2 - \frac{d^2 U}{dz^2}/U, \tag{10.75}$$

which is free of singularities if U is positive everywhere. For flows of finite depth, the boundary conditions are

$$\psi = \begin{cases} Uh(x), & z = 0, \\ 0, & z = D. \end{cases} \tag{10.76}$$

As discussed above, the radiation condition implies that there is no finite steady part to the solution at upstream infinity, so that $\psi \to 0$ as $x \to -\infty$. Such flows tend to be dominated by lee waves, and we consider two representative special cases: uniform flow with a density interface, as in Chapter 2, and uniform stratification.

(i) Density interface: $l^2 = g'\delta(z - d_1)$.

Here $g' = g\Delta\rho/\rho$, and $\Delta\rho$ is the discontinuity in density at the interface at the (undisturbed) height d_1 and $\Delta\rho \ll 1$. Defining the Fourier transforms $\hat{h}(k), \hat{\psi}(k, z)$ of $h(x)$, ψ with respect to x as in (10.15), we obtain

$$\hat{\psi}(k, z) = U\hat{h}(k) \left[H_{\text{eav}}(d_1 - z)\frac{\sinh k(d_1 - z)}{\sinh kd_1} + B_A(k)F(k, z) \right], \qquad (10.77)$$

[where H_{eav} denotes the Heaviside step function],

$$F(k, z) = \begin{cases} \dfrac{\sinh kz}{\sinh kd_1}, & 0 < z < d_1, \\ \dfrac{\sinh k(D - z)}{\sinh k(D - d_1)}, & d_1 < z < D, \end{cases} \qquad (10.78)$$

and

$$B_A = \frac{\sinh k(D - d_1)}{\sinh kD - \sinh kd_1 \sinh k(D - d_1) \cdot g'/kU^2}. \qquad (10.79)$$

This gives the solution for flow over periodic topography of the form $\hat{h}(k)e^{ikx}$. The solution for ψ for the obstacle $h(x)$ is then given by

$$\psi(x, z) = \frac{1}{2\pi} \int_{-\infty}^{\infty} \hat{\psi}(k, z)e^{ikx}\,dk. \qquad (10.80)$$

If the flow is subcritical, the denominator of B_A has zeros on the real k-axis at $k = \pm k_1$ (defining k_1), and investigation of the various possibilities shows that the contour of integration must be taken below these poles in order to satisfy the condition of zero disturbance at upstream infinity. For x large and positive, the contributions to the integral are small except near these singularities, and evaluating these terms by standard complex variable methods gives

$$\psi \approx iU\frac{\sinh k_1(D - d_1)}{C_A(k_1)}[\hat{h}(k_1)e^{ik_1x} - \hat{h}(-k_1)e^{-ik_1x}] \cdot F(k_1, z), \qquad (10.81)$$

as $x \to \infty$, where

$$C_A = D\left\{\cosh k_1 D - \frac{\sinh k_1 D}{k_1 D}\left[\frac{k_1 d_1}{\tanh k_1 d_1} + \frac{k_1(D - d_1)}{\tanh k_1(D - d_1)} - 1\right]\right\}. \qquad (10.82)$$

For symmetric obstacles $\hat{h}(k)$ is real and symmetric, and (10.81) becomes

$$\psi \approx -2U\hat{h}(k_1)\frac{\sinh k_1(D - d_1)}{C_A(k_1)}\sin k_1 x \cdot F(k_1, z). \qquad (10.83)$$

Hence if $\hat{h}(k_1)$ is positive, the phase of this wave is such that maximum downward flow coincides with the centre of the obstacle ($x = 0$). Although this expression is for x large, the wave usually dominates the solution for $x > 0$, so that it gives a good approximation to it close to the obstacle.

(ii) Uniformly stratified flow: $l^2 = N^2/U^2$.

The development here parallels the two-layer case very closely. The solution for $\hat{\psi}(k, z)$ is

$$\hat{\psi}(k, z) = \begin{cases} U\hat{h}(k)\dfrac{\sin(l^2 - k^2)^{1/2}(D - z)}{\sin(l^2 - k^2)^{1/2}D}, & 0 < |k| < l, \\[4mm] U\hat{h}(k)\dfrac{\sinh(k^2 - l^2)^{1/2}(D - z)}{\sinh(k^2 - l^2)^{1/2}D}, & l < |k| < \infty, \end{cases} \tag{10.84}$$

which gives the solution for periodic topography, and the solution for ψ for the isolated obstacle is then given by (10.80). For subcritical flow the integrand (10.84) has zeros at $k = \pm k_j$, (see (10.62) and associated text) on the real axis, and the radiation condition again implies that the contour be taken below them. Consequently there may be several subcritical lee-wave modes, which have the form

$$\psi \approx -2\pi U \sum_{j=1}^{j_K} \frac{j}{k_j D^2} \sin \frac{j\pi z}{D} \cdot \frac{1}{2\mathrm{i}}[\hat{h}(k_j)\mathrm{e}^{\mathrm{i}k_j x} - \hat{h}(-k_j)\mathrm{e}^{-\mathrm{i}k_j x}], \tag{10.85}$$

as $x \to \infty$, which for symmetric obstacles becomes

$$\psi \approx -2\pi U \sum_{j=1}^{j_K} \frac{j}{k_j D^2} \hat{h}(k_j) \sin k_j x \sin \frac{j\pi z}{D}. \tag{10.86}$$

If $\hat{h}(k_j) > 0$, the phase is such that the maximum downward motion at the lowest level coincides with the centre of the obstacle.

It sometimes happens in the atmosphere that N may be large at low levels, particularly at night due to radiative cooling at and near the ground, and/or that U increases with height, often with $U'' < 0$. For these reasons it is possible that l^2 may be relatively large at low levels, and then decrease with height to some minimum value, l_m^2 say. This system is represented in §8.5 by the simple prototype (8.66). It can support a number of discrete trapped modes whose wavelength is short enough to satisfy $k^2 > l_m^2$, as typified by (8.68), and these waves will be excited by the topography if there is sufficient forcing at these wavenumbers. For the other Fourier components where $k^2 < l_m^2$, the topography will force a continuous spectrum of *leaky modes*, where all of the energy eventually leaks into the upper region. Because of this partial trapping, the energy of these modes is spread further downstream at low levels than it would be if $l = l_m$ everywhere.

10.5.2 The momentum-source model

The presence of an upper boundary changes the form of the solution for a momentum source, as described in §10.2.3, and may change the value of C_{DS}, but it does not change its fundamental character. With equations (10.26)–(10.29), and assuming that K lies in the range $j_K < K < j_K + 1$ where j_K is an integer, the steady solution corresponding to (10.32) for this finite-depth configuration is (adapted from Janowitz, 1981)

$$
\psi = \begin{cases}
\dfrac{C_{DS}U^2K}{2\pi N}\left[\displaystyle\sum_{j=1}^{j_K}\dfrac{\sin\frac{j\pi h_m}{D}}{j(K-j)}\cdot\sin\dfrac{j\pi z}{D} + \sum_{j=j_K+1}^{\infty}\dfrac{\sin\frac{j\pi h_m}{D}}{j^2-K^2}\cdot e^{|k_j|x}\cdot\sin\dfrac{j\pi z}{D}\right], & x \le 0, \\[2em]
\dfrac{C_{DS}U^2K}{2\pi N}\left[\displaystyle\sum_{j=1}^{j_K}\dfrac{\sin\frac{j\pi h_m}{D}}{j(K+j)}\left(1 + \dfrac{2j}{K-j}\cdot\cos k_j x\right)\cdot\sin\dfrac{j\pi z}{D}\right. \\[1.5em]
\left. \quad\quad + \displaystyle\sum_{j=j_K+1}^{\infty}\dfrac{\sin\frac{j\pi h_m}{D}}{j^2-K^2}\cdot(2 - e^{-|k_j|x})\cdot\sin\dfrac{j\pi z}{D}\right], & x \ge 0.
\end{cases}
$$

$$(10.87)$$

As with the boundary perturbation model, the discrete spectrum of this system implies that there is no logarithmic singularity in t and instead resonance occurs when K is an integer, where the corresponding solution grows linearly with time.

10.6 Comparison between linear theory, and observations and numerical results for finite depth and small Nh/U

At this point we pause in the theoretical development and examine how well the above linear models describe laboratory observations and results from numerical experiments. We expect these models to be valid when Nh/U is small, and the value of K is not close to an integer. Our conceptual model for flow over an obstacle then consists of a superposition of the effects of boundary displacement as described in §10.5.1, and a momentum source due to lee-side separation as described in §10.5.2. This combination of models gives a reasonable qualitative description of the flow, but when comparing the detailed descriptions with observations there are two problems. Firstly, the value of C_{DS} is generally not known precisely. It is expected to be less than unity, and may well depend on K and the other flow conditions; a representative value may be taken to be 0.3. Secondly, the momentum-source model with the $\delta(x)$-dependence gives a crude description of the flow close to this source, which in practice is located at or close to the downstream end of the obstacle. The vertical velocities w' in the lee-wave modes in the momentum-source model are $w' = O(Uh_m/D)$, which is small. However, their form is dependent on the assumed δ-function form of the forcing, and hence they are not a robust part of the solution. The part of the momentum-source solution that *is* robust is the horizontal velocity component upstream and downstream, and it is the upstream part that we may qualitatively compare with experiment.

Laboratory observations of flow over obstacles with small Nh/U have been described by Baines (1979a), and compared with flow fields described by (10.60). These experiments covered Nh/U values up to about 0.8 (although most were much less than this), h_m/D values up to about 0.1, and K values from zero up to 3. Obstacle shapes used were semi-circular, with h_m/a values of 1, and long, semi-elliptical (or nearly so) obstacles with $h_m/a = 0.04$ or 0.095. In the figures relating to these experiments the obstacle height h_m is replaced by h_e, the *effective* height, which is the actual obstacle height *plus* the displacement thickness of the viscous boundary-layer on the obstacle. This is significant for long obstacles.

For $K < 1$, the observations were in very good agreement with the linear pertur- bation theory as described by (10.60), even for K very close to unity. No disturbances were observed on the upstream side, even for K values up to 0.96. Over the obstacle the flow reached or approached a steady state, and the transient motions that were observed downstream became more detached from the obstacle as time progressed. When K was very close to unity, steady state over the obstacle was not attained in the time available, but the time-dependent form of (10.60) gave excellent agreement with the observed flow. (An example is shown in the first frame of Figure 10.21, below.)

For $K > 1$, lee-wave modes with mode number $n < K$ were observed on the downstream side and depending on K and Nh/U, corresponding columnar modes were sometimes observed on the upstream side. Figure 10.18 shows examples of the downstream flow for both long and short obstacles (taken with a camera that was stationary relative to the obstacle), where the mode-1 lee wave is evident. Linear boundary perturbation theory was found to give a good description of the lee waves forced by flow over the long obstacles, with regard to their wavelength, amplitude and phase. An example of these results is shown in Figures 10.19 and 10.20. For the semi-circular obstacles the agreement in amplitude and phase was less satisfactory, and this can be attributed to the conspicuous separation of the flow from the lee side, as seen in Figure 10.18a. In fact, the observed waves correspond better to those forced by a longer obstacle with a shape including the separation region.

The flow patterns over the obstacle are best seen by a camera that is stationary relative to the laboratory (i.e. the undisturbed fluid), and examples are shown in Figure 10.21. These flow patterns were approximately stationary, although a downstream transient is still visible for $K = 0.865$, as are upstream transients for $K = 1.01$, 1.12 and 2.28. The upstream columnar motions (where they occur) are seen to be closely connected to the circulation over the obstacles. The extent to which the time-dependent boundary perturbation model is able to describe the

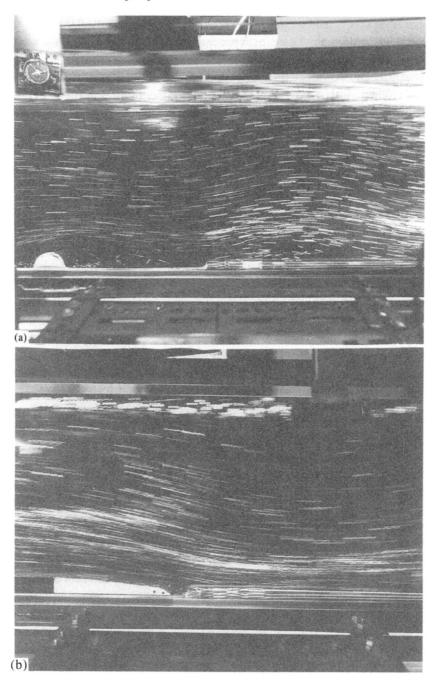

Figure 10.18 Examples of steady flow fields relative to the moving obstacle. (a) Semi-circle with $Nh/U = 0.41$, $K = 1.47$; (b) semi-ellipse with $h_m/a = 0.095$, $Nh/U = 0.27$, $K = 1.12$. (From Baines, 1979a: reproduced by permission of *Tellus*.)

observed flow over a long obstacle is illustrated in Figure 10.22 for $1 < K < 2$. The agreement is seen to be poor for K close to 1, but it improves as K increases and becomes reasonable for $K \geq 1.36$ (although this value is dependent on h_m/D).

For the upstream motions, examples of the behaviour for long obstacles and the comparison with boundary perturbation theory are shown in Figure 10.23. As time increased from zero, at a given upstream location the amplitude of mode 1 for long obstacles was observed to attain an initial maximum and then decrease to steady values, but for short obstacles this behaviour was more nearly monotonic. The amplitudes of the steady upstream motions are shown in Figure 10.24 as a function of K and Nh/U for obstacles with $h_m/A \ll 1$. For given K, if h_m/D is sufficiently small there is no permanent upstream motion. The linear boundary perturbation model does not describe these upstream motions at all.

Numerical computations by Lamb (1994) with a non-linear inviscid model with $K = 1.5$ and the lower boundary as a streamline show agreement with linear boundary perturbation theory for upstream motions if $Nh/U \leq 0.05$, but departure from it becomes noticeable for $Nh/U > 0.25$. For $Nh/U \approx 0.6$, the initial peak in mode 1 amplitude was larger than the linear one and lagged behind it. However, the upstream motion showed little resemblance to that seen in Figure 10.23, although the agreement was closer if the boundary-layer displacement thickness was added to modify the obstacle shape.

The above suggests that the discrepancy between perturbation theory and the observations in Figures 10.22 and 10.23, and the upstream motions of Figure 10.24, are probably due to the momentum-source effect of §10.5.2. In this case, the observed upstream amplitudes of the disturbances may be used to infer a value for C_{DS}. However, the agreement is at best qualitative. The lag of the observed upstream disturbances (behind the predictions of linear theory) seen in Figure 10.23 may be due to a momentum source connected with separation beneath the first lee-wave crest downstream, rather than over the lee side of the obstacle, but this is speculation and further investigation is needed.

The momentum-source model has been compared with some observations where Nh/U is *not* small. Janowitz (1981) and Castro & Snyder (1988) have compared its predictions with observed flows with steep topography such as triangular profiles and vertical barriers, where lee-side separation is substantial. Provided C_{DS} is appropriately chosen (a single adjustable parameter), the amplitudes of the observed upstream modes can be fitted reasonably well. However, the model becomes heuristic when Nh/U is not small, because here upstream motions may be generated by non-linear effects that can only be crudely parametrised by a momentum source, as described below. In spite of this, for steep obstacles with substantial turbulent wakes, this type of model does seem to give a better *qualitative* description of the observations than the linear perturbation model or Long's model (see §10.7).

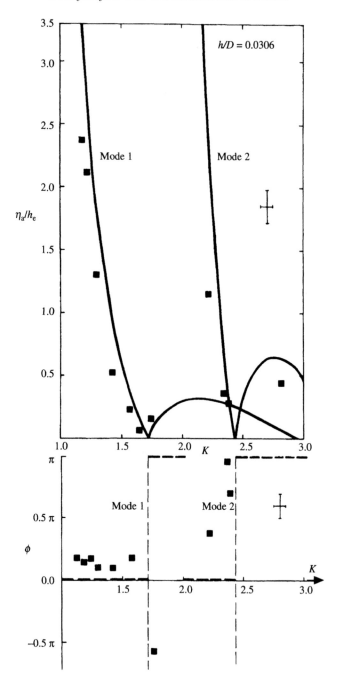

Figure 10.19 Observed properties of lee waves in finite depth, compared with linear perturbation theory. Amplitudes η_a scaled with effective topographic height h_e and phase ϕ for $h_e/D = 0.031$ for modes 1 and 2, where displacement $\eta = -\eta_a \sin(kx - \phi) \sin n\pi z/D$, for a long flat obstacle with $h_e/A = 0.01$.

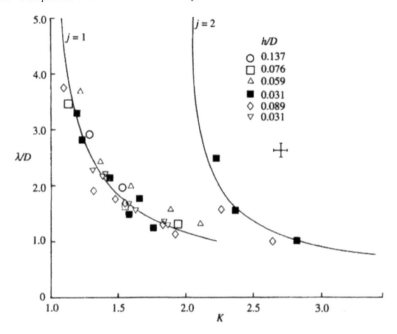

Figure 10.20 Wavelengths of lee waves for a variety of obstacles, compared with linear theory. (From Baines, 1979a: reproduced by permission of *Tellus*.)

Trustrum (1971) has shown that an Oseen model (which is generically similar to a momentum-source model) gives a satisfactory qualitative description of the observed flows of Davis (1969) over a vertical barrier. This does not require any adjustable parameter, but it must be based on heuristic arguments.

A numerical study by Hanazaki (1989) of flow over a vertical barrier with $h_m/D = 1/4$ and $R_e = 20$ describes similar agreement with the momentum-source model, in spite of the smallness of R_e. At this value of the Reynolds number there is a separated wake but it is not conspicuously unsteady. Nonetheless, upstream modes and lee waves are produced, and the momentum-source model gives a reasonable qualitative description of the upstream flow with a suitable choice of C_{DS}, but no value gives correct amplitudes of both the upstream modes and the lee waves.

We may conclude that for $Nh/U \ll 1$, for smooth topography where lee-side separation is negligible and K is not very close to an integer value, linear perturbation theory gives a reliable description of the flow over topography and the resulting lee waves. If flow separation is present this model must be supplemented by an additional effect of a momentum source due to lee-side separation or friction, that alters the local flow and produces relatively weak upstream motions, but this effect needs closer study for quantitative understanding. The momentum-source model is apparently successful in describing *qualitatively* the flow over larger obstacles

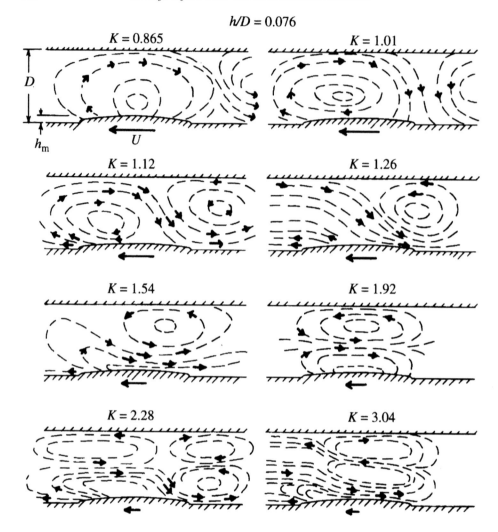

Figure 10.21 Instantaneous observed streamline patterns produced by a moving obstacle (the semi-ellipse of Figure 10.18b) as seen by a stationary observer, near the end of each run. Arrow lengths indicate velocity magnitudes relative to the obstacle speed U, shown below each figure. (From Baines, 1979a: reproduced by permission of *Tellus*.)

with $Nh/U = O(1)$ where significant lee-side separation occurs, but in general this is a crude and superficial model of the physics of these flows. A more complete description of observations for finite Nh/U is given below in §11.5.

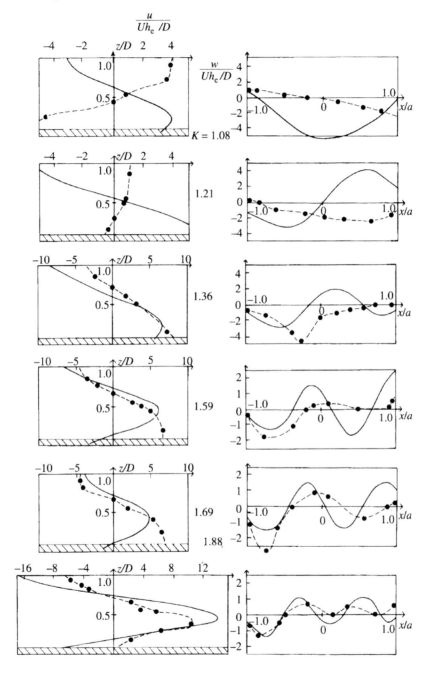

Figure 10.22 Observed horizontal velocity profiles over the centre-point ($x = 0$) and vertical velocity profiles at mid-depth ($z = D/2$) for the long flat obstacle ($h_e/A = 0.01$), after time $\pi Ut/D = 90.0$. The points denote observations and the continuous lines the predictions from linear boundary perturbation theory for the same time. The value of x is in units of D/π. (From Baines, 1979a.)

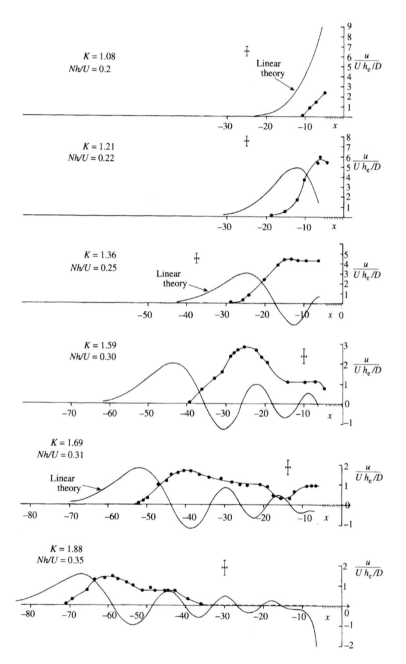

Figure 10.23 Upstream horizontal velocity u observed near the top boundary for the same runs as in Figure 10.22, at time $\pi U t D = 83.0$. This represents the amplitude of mode 1 upstream, excepting close to the obstacle (on the right). The continuous line denotes linear boundary perturbation theory at the same time. (From Baines, 1979a: reproduced by permission of *Tellus*.)

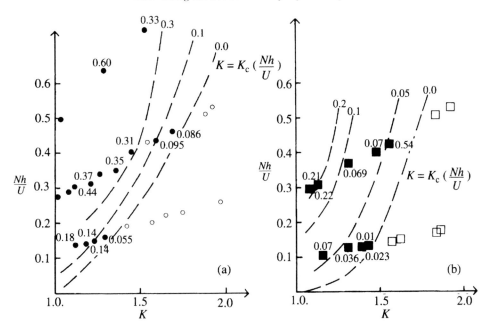

Figure 10.24 Values of the upstream velocity amplitude u_1/U for mode 1, observed near the end of the runs as in Figure 10.23, for (a) long obstacles ($h/A = 0.01, 0.024$) and (b) semi-circular obstacles ($h/A = 1$). (From Baines, 1979a: reproduced by permission of *Tellus*.)

10.7 Long-model solutions for finite depth

Following on from §10.3, Long-model solutions for finite-amplitude topography may be obtained for finite-depth situations where the lower boundary is a streamline. We consider both rigid and pliant upper boundaries.

10.7.1 Rigid upper boundary

This involves solving (10.45) with $\kappa = N/U$ constant with the boundary conditions

$$
\begin{aligned}
\delta &\to 0 && \text{as } x \to -\infty, \\
\delta &= 0 && \text{on } z = D, \\
\delta &= h(x) && \text{on } z = h(x),
\end{aligned}
\tag{10.88}
$$

where $h(x) \to 0$ as $x \to -\infty$. If the Boussinesq approximation is made, the condition that ρU^2 is constant upstream is equivalent to U being constant, and $U\delta$ may be identified as the stream function ψ for the perturbations. The general solution for $j_K < K < j_K + 1$ may be expressed in the form (Miles, 1968)

$$\delta(x,z) = \int_{S'} \left[\frac{\partial G}{\partial n}\delta(\xi,\eta) - G(\xi,\eta)\frac{\partial \delta}{\partial n} \right] dl, \qquad (10.89)$$

where S' is the surface of the obstacle expressed as $\xi = h(\eta)$, n denotes the outwardly-directed normal to this surface, and $G(x,z/\xi,\eta)$ is the Green's function, given by

$$G = -\frac{2}{\pi}H_{\text{eav}}(x - \xi)\sum_{j=1}^{j_K} \frac{\sin k_j(x - \xi)}{k_j} \cdot \sin\frac{j\pi\eta}{D} \cdot \sin\frac{j\pi z}{D}$$

$$+ \frac{1}{\pi}\sum_{j=j_K+1}^{\infty} \frac{e^{-|k_j(x-\xi)|}}{k_j} \cdot \sin\frac{j\pi\eta}{D} \cdot \sin\frac{j\pi z}{D}, \qquad (10.90)$$

where H_{eav} is the Heaviside step function, and again k_j is given by (10.62). From (10.89) and the structure of this Green's function, it is apparent that this steady solution contains sinusoidal waves on the downstream side if $j < K$, and evanescent terms on both sides for $j > K$. It is singular when K is an integer for the same reasons that the linear solutions are. The solutions resemble the steady boundary perturbation solutions when $H = h_m/D$ (or Nh/U) is small, and as it increases the lee waves become steeper. For each value of K, as $|H|$ increases, a point may be reached where $u = 0$ somewhere in the flow (this occurs on $z = D$ for $H > 0$, $1 < K < 2$); for larger values of $|H|$, the solution contains regions of closed streamlines containing statically unstable flow, corresponding to the statically unstable waves in the infinite-depth case. Flows with these regions of closed streamlines violate the assumptions of the model, which require all streamlines to begin at upstream infinity. Hence the boundary in (K,H)-space denoting the onset of this phenomenon marks an upper limit for the applicability of this model. The character and shape of such boundaries for non-hydrostatic flows are similar to those for hydrostatic flows (e.g. Baines, 1977), and we take the latter as a prototype and discuss it in more detail.

The hydrostatic solution is obtained by solving (10.47) with boundary conditions (10.88), and is

$$\psi = -Uz + U\delta = -Uz + Uh(x)\frac{\sin K\pi(1 - z/D)}{\sin K\pi(1 - h(x)/D)}, \qquad (10.91)$$

which is an extension to finite amplitude of the steady linear perturbation solution (10.64), in the same way that the infinite-depth solution (10.49) generalises (10.21). The corresponding density field is

$$\frac{\rho}{\rho_0} = 1 + \frac{\Delta\rho}{\rho_0}\left(\frac{1}{2} - \frac{z}{D} + \frac{h(x)}{D}\frac{\sin K\pi(1 - z/D)}{\sin K\pi(1 - h(x)/D)} \right). \qquad (10.92)$$

Equation (10.91) describes a forced response that is only present over the obstacle, and has a standing wave character if $K > 1$. The limits to the applicability of (10.91) are shown in Figure 10.25 on a (K, H)-plane, for both positive h (obstacles) and negative h (depression). Note that these properties are all independent of the shape of the obstacle (in contrast with the infinite-depth case as shown in Figure 10.16, and only depend on K and H. There are three types of region in this diagram, but two (shaded and dark) are only of passing interest. Stable solutions with all streamlines commencing upstream are only possible in the clear regions of the diagram, which are bounded by the two curves (denoted H_s) where static instability occurs. Solutions in this statically stable region do not become critical anywhere with respect to any of the internal wave modes (see §11.4.1). In other words, modes that are sub- or supercritical upstream (as determined by the value of K) remain so throughout the flow field. For $K < 1$, solutions are possible for all positive h_m up to the maximum (D). Note the analogous situation with depressions or "holes", which can be almost bottomless when K is small.

In the shaded regions, (10.91) describes statically unstable flows with closed streamlines, and as the dark region (at H_∞) is approached, velocities in the flow field approach infinity. Hence, no physical significance can be attached to (10.91) in the dark regions. Flow in the shaded regions of Figure 10.25 violates the assumptions for its derivation, but one may still ask whether the solution (10.91) is physically meaningful if the regions with closed streamlines are regarded as containing stationary, homogeneous fluid that gives the same pressure field. This seems not to be the case for two reasons. First, in the recirculating regions that form adjoining the upper boundary (for example), the fluid density takes values outside the range of densities upstream, and this will still apply if the fluid there is mixed and homogenised. Hence this flow cannot be generated from the same initial conditions as those upstream, because fluid with this lighter density is not present in the flow. Secondly, these flows are subject to shear instability, since $R_i < 1/4$ in parts of this flow (Long, 1955).

10.7.2 *Pliant upper boundary*

We next consider the same system as in §10.7.1 but with a pliant upper boundary (see §1.2), where the density is continuous and the flowing stratified fluid is surmounted by homogeneous fluid (which may be stationary or in motion, but the motion has no effect on the fluid below). The topmost streamline has elevation $z = D$ far upstream. For steady flows the Bernoulli equation (10.39) implies that

$$\frac{1}{2}\rho \mathbf{u}^2 + p + \rho g z = \text{constant} \qquad (10.93)$$

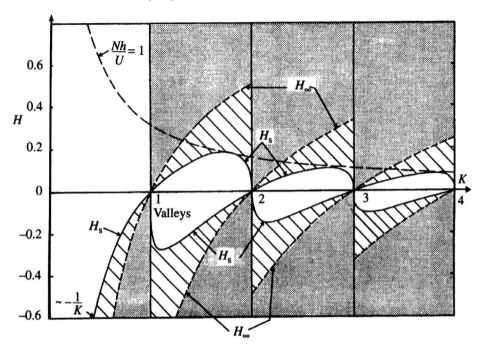

Figure 10.25 Regions of validity of the hydrostatic Long-model solution (10.91) for finite depth, plotted in terms of $K = 1/F_0$. The solution is valid in the clear regions where it is statically stable, but not outside them. In the shaded regions it gives flows with regions of closed streamlines which are not realised in practice, and in the dark regions the solution is singular. Note the extension to "valleys" or depressions, which is not symmetric in h_m. The curve $Nh/U = 1$ is included for comparison. (From Baines & Guest, 1988, reproduced with permission.)

on all streamlines, and this topmost one in particular. Since the pressure is continuous and is hydrostatic in the overlying layer, $p + \rho g z$ is constant on the topmost streamline, and hence u is constant. The upper boundary condition is therefore

$$u = U, \quad \frac{\partial \delta}{\partial z} = 0 \quad \text{at} \quad z = D + \delta_T, \tag{10.94}$$

where $\delta_T(x)$ is the vertical displacement of the topmost streamline from its upstream level, and this function must be determined. One may derive a solution corresponding to (10.89), (10.90) for this system, but we concentrate here on the hydrostatic limit for long obstacles, where the solution again takes a simple form.

In the long-wave (hydrostatic) limit, wave modes in this uniformly stratified layer with a pliant upper boundary have the form

$$\delta = e^{ik(x - c_n t)} \cdot \sin\left(n + \frac{1}{2}\right)\frac{\pi z}{D}, \tag{10.95}$$

where the wave speeds c_n are given by

$$c_n = \frac{\pm ND}{(n + \frac{1}{2})\pi}, \qquad n = 0, 1, 2, 3, \ldots \tag{10.96}$$

Compared with the rigid-lid case, this system has an additional mode ($n = 0$) that corresponds to the mode of a single homogeneous layer. A Froude number F_0 for the undisturbed flow based on this fastest mode is defined by

$$F_0 = U/c_0 = \pi U/2ND. \tag{10.97}$$

The general solution to (10.47) with upper conditions (10.94) is (Smith, 1985)

$$\delta(x, z) = \delta_T(x) \cos \frac{N}{U}[z - D - \delta_T(x)], \tag{10.98}$$

where $\delta_T(x)$ is given by

$$h(x) = \delta_T(x) \cos \frac{N}{U}[h(x) - D - \delta_T(x)]. \tag{10.99}$$

This solution represents realistic flows provided that h_{m}/D is less than a maximum that depends on the Froude number F_0. This maximum is shown in Figure 10.26 for both positive and negative topography (i.e. depressions). For values of $|h_{\mathrm{m}}/D|$ below this maximum, the solution is single-valued in terms of the height $h(x)$, so that the upstream and downstream flows are the same if $h(-\infty) = h(+\infty)$.

As shown in Figure 10.26, the boundary curve for solutions of the form (10.98) is made up of alternating sections of different types. For the first type, denoted by a *solid* line, the limit is hydraulic in character, in the same manner as for a single homogeneous layer as described in Chapter 2 and shown in Figure 2.12. Differentiating (10.99) gives (cf. (2.59))

$$\left(\frac{1 - \cos \phi}{1 - \frac{N\delta_T}{U} \sin \phi} - 1 \right) \frac{d\delta_T}{dx} = -\frac{dh}{dx}, \tag{10.100}$$

where $\phi = (D + \delta_T - h)N/U$. The limit of the Long-model solution is reached when the bracketed term vanishes at $h = h_{\mathrm{m}}$, where $dh/dx = 0$. This implies that the flow is critical there, in that a wave mode has zero propagation speed. When this occurs, a second solution is possible in which the flow undergoes a hydraulic transition to a different, supercritical state downstream. As Figure 10.26 shows, there is a section of this hydraulic-type boundary curve for each of the modes $n = 0, 2, 4, \ldots$ for $h_{\mathrm{m}} > 0$, and for the modes $n = 1, 3, 5, \ldots$ for $h_{\mathrm{m}} < 0$, as identified by their contact points on the axis $h_{\mathrm{m}} = 0$. An example of one of these flows, taken at the lower extremity of the boundary curve for mode $n = 0$ at $F_0 = 1/3$, is shown in

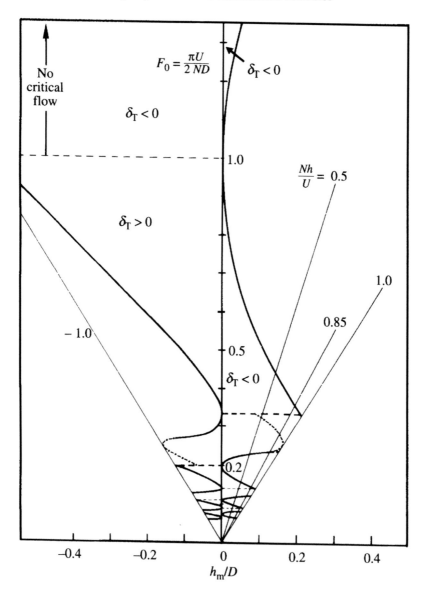

Figure 10.26 The limits of applicability of the Long-model solution for uniformly stratified hydrostatic flow with a pliant upper surface, plotted in terms of F_0 and h_m/D. The heavy solid lines denote the boundaries where the solution is limited by a hydraulic transition, in the same manner as for a homogeneous layer as shown in Figure 2.12. The dot–dashed lines denote the boundaries where the solution is limited by static instability, in the same manner as for the boundaries for finite-depth flow in Figure 10.25. Sections of these different boundary types alternate with decreasing F_0, for both positive and negative topography. Curves of constant Nh/U are shown for comparison and later reference. (Adapted from Baines & Granek (1990): reproduced by permission of Kluwer Academic Publishers.)

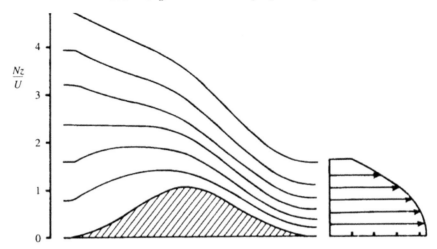

$\dfrac{Nz}{U}$

Figure 10.27 An example of a hydraulic solution from Long's model for $F_0 = 1/3$, $h_m/D = 0.21$, at the lower limit of the uppermost solid curve of Figure 10.26. The moving stratified fluid is surmounted by homogeneous fluid which may or may not be in motion. (From Smith, 1985: reproduced by permission of the American Meteorological Society.)

Figure 10.27. On these hydraulic boundary curves the drag force on the obstacle may be expressed as (Smith, 1985)

$$F_D = \frac{\rho_0 N^2}{6}(d_u - d_d)^3, \tag{10.101}$$

where d_u and d_d are the depths of the stratified layer upstream and downstream respectively. This expression is similar to (2.79) for a single layer.

The hydraulic boundaries in Figure 10.26 are joined by dot–dashed lines that mark the second type of limit to the validity of (10.98). Here the solution is limited by static instability of the flow, in the same manner as is the rigid-lid solution described in §10.7.1.

In order to obtain solutions for larger H outside the boundaries marked in Figure 10.25 or 10.26, we must develop a more general hydraulic model, and this is described in §§11.2 and 11.3. The hydraulic behaviour in this pliant-surface model is relevant to the infinite-depth case (particularly when $F_0 > 1/3$), as described in §11.5.

10.7.3 *Upper surface with an infinitely deep stratified upper layer*

Durran (1992) has described Long-model solutions for the system of §10.7.2 where the upper layer is stratified, rather than homogeneous. This variation introduces some different and significant features. Specifically, U is uniform and $N = N_L$

below height D and $N = N_U$ above in undisturbed flow, so that (10.47) applies in each layer. If the density is continuous at the junction, continuity of displacement and pressure imply that δ and $\partial \delta / \partial z$ are continuous there. If $\eta_T(x)$ denotes the vertical displacement of the interface from its mean position $z = D$, the solution has the form

$$
\begin{aligned}
\delta_L(x, z) &= h(x) \cos \kappa_L(z - h) + e(x) \sin \kappa_L(z - h), \\
\delta_U(x, z) &= \eta_T(x) \cos \kappa_U(z - \eta - D) + f(x) \sin \kappa_U(z - \eta - D),
\end{aligned}
\tag{10.102}
$$

for the lower and upper layers respectively, where $\kappa_L = N_L/U$ etc. The boundary conditions then give

$$
h(x) = \eta_T \cos \phi - \frac{N_U}{N_L} f \sin \phi, \qquad e(x) = \eta_T \sin \phi + \frac{N_U}{N_L} f \cos \phi, \tag{10.103}
$$

where $\phi = \kappa_L(\eta_T + D - h)$, and the solution is then completed by the upper radiation condition. This is given by (10.50), with κ replaced by κ_U and h by η_T. Equation (10.102) reduces to (10.48) if $N_U = N_L$, and to (10.98) if $N_U = 0$.

These solutions are sensitive to the values of N_U/N_L and D. They agree with those obtained from the linear perturbation procedure when $N_L h_m/U \ll 1$, but show substantial quantitative departures from it as $N_L h_m/U$ increases to about 0.5. Hence linear theory is not a reliable guide to flows with non-uniform density structure for large-amplitude forcing. In many cases the upper limit of the value of this parameter for which Long-model solutions are possible is determined by hydraulic considerations as in §10.7.2, but the situation is complicated by the fact that this system contains vertically propagating waves above, rather than just purely horizontally propagating modes. Differentiating the first of (10.102) gives

$$
(\mathfrak{A} - 1) \frac{d \eta_T}{dx} + \mathfrak{B} \frac{df}{dx} = -\frac{dh}{dx} \tag{10.104}
$$

where

$$
\mathfrak{A} = \frac{1 - \cos \phi}{1 - \kappa_L \eta_T \sin \phi - \kappa_U f \cos \phi}, \qquad \mathfrak{B} = \frac{N_U}{N_L} \cot\left(\frac{\phi}{2}\right) \mathfrak{A}. \tag{10.105}
$$

The quantity $f(x)$ depends on the total function η_T through the radiation condition, so that (10.104) does not have the same simple interpretation as (10.100). Nonetheless, (10.104) appears to determine the limit of Long's model through the possibility of a second solution with a hydraulic type of transition. This transition need not be confined to where $dh/dx = 0$, and does not (necessarily) involve overturning regions in the fluid.

11

Stratified flow over two-dimensional obstacles: non-linear hydraulic models with applications

And though use had not yet taught his wing a permanent and equable flight,
there are parts of it which exhibit his best manner in full vigour.
I had once the pleasure of examining it with Mr. Edmund Burke,
who confirmed me in this opinion.

JAMES BOSWELL, *The Life of Samuel Johnson*,
on the latter's first literary work, a translation.

Following on from Chapter 10, this chapter is also concerned with two-dimensional flows, but focuses on situations where non-linear effects are important. In some applications described below the flow is actually three-dimensional, but two-dimensional hydraulics captures the essential dynamical features.

11.1 Models with non-linearity and dispersion

So far in our survey of models of flow over two-dimensional topography we have described linear perturbation models and their extension, Long's model, and linear models with a momentum source. Before examining general non-linear hydrostatic models, it is instructive to examine weakly non-linear models that incorporate linear wave dispersion, as we have done for one- and two-layer systems.

An extension of the initial-value problem of §10.5.1 to second (and higher) order in the perturbation series for small Nh/U, for finite depth for the case of constant U and N, shows that the solution has essentially the same character as the first-order (linear) solution, with one significant exception: the development of second-order columnar modes that are forced by the transient "tails" of the linear lee-wave modes (McIntyre, 1972). These columnar modes have vertical wavenumber $2j\pi/D$ for the jth lee-wave mode, and propagate upstream and downstream from the tail as the wavetrain grows and the tail moves downstream. The upstream-propagating component of the jth mode may extend upstream of the obstacle if $2j < K$. The amplitudes of these modes are numerically very small in laboratory experiments (for example), and they would disappear in the large-time limit if only a small amount of dissipation were present in the fluid. However, they may complicate inviscid numerical computations where the lee-wave tails are generated.

In Chapter 2, for fairly long obstacles, the effects of non-linearity and dispersion for a single layer with F_0 near unity and h_m small are described by a forced Korteweg–de Vries (KdV) equation. For two-layer flows in Chapter 3, this was generalised to an extended KdV equation containing a cubic non-linear term, that was significant only when the quadratic term was small or zero. A corresponding analysis may be done for finite-depth situations with continuous stratification with $N = N(z)$ and U uniform, as follows.

In §10.5.1, the linear perturbation solution (10.72) for hydrostatic flow over topography becomes singular when $v_j \approx U$, so that $c_j^- \approx 0$, and the jth mode is resonantly forced. The amplitude of the jth mode, $A_j(x,t)$, may be expressed as the sum of the upstream- and downstream-propagating parts, i.e., $A_j = A_j^- + A_j^+$, where A_j^- becomes singular when $v_j = U$. For moderately long obstacles where h_m is small, the growth of this term is controlled jointly by non-linear advection and linear dispersion, whereas the behaviour of A_j^+ and the other modes is still linear. As for the one- and two-layer systems, the behaviour of A_j^- (henceforth denoted by **A**) is again described by the forced KdV equation (Grimshaw & Smyth, 1986)

$$\mathbf{A}_t + (U - v_j)\mathbf{A}_x - \mu \mathbf{A}\mathbf{A}_x - \lambda \mathbf{A}_{xxx} = \frac{1}{2}v_j \mathbf{B}_{jx}, \qquad (11.1)$$

where $\mathbf{B}_j$ is given by (10.71), and μ and λ by (10.69) and

$$
\begin{aligned}
I_j \mu &= \frac{3}{2}v_j^2 \int_0^D \left(\frac{d\psi_j}{dz}\right)^3 dz, \\
I_j \lambda &= \frac{1}{2}v_j^3 \int_0^D \psi_j^2 dz.
\end{aligned}
\qquad (11.2)
$$

Note that λ is always positive, but μ may be positive, zero or negative depending on the form of ψ_j, and hence on $N(z)$. Provided $\mu \neq 0$, (11.1) is equivalent to (3.38) after a suitable transformation of the variables, so that its solutions have the same character as those of the canonical KdV equation.

If $\mu = 0$ the non-linear term in (11.1) vanishes, and the balance assumed in its derivation is then not valid. This occurs when N is uniform, and a different treatment is needed. In the steady state the governing equation is linear (Long's model), but the development of the flow with time is generally non-linear. For simplicity and definiteness we assume that the upper boundary at $z = D$ is rigid. ψ_j, v_j in (10.66),

(10.67) then take the form

$$\psi_j(z) = \sin\frac{j\pi z}{D}, \qquad v_j = \frac{ND}{j\pi}, j = 1, 2, 3, \ldots, \qquad (11.3)$$

On the assumptions that h_m is small, the obstacle is fairly long, the jth mode is close to resonance (so that $U \approx v_j$), and the stratification departs only slightly (if at all) from uniform N, Grimshaw & Yi (1991) have derived an equation for the behaviour of the growth with time of $\mathbf{A}(x,t) = \mathbf{A}_j^-$, which has the form

$$\int_{-\infty}^{x} \kappa(\mathbf{A}, \mathbf{A}')\frac{\partial\mathbf{A}}{\partial t}dx' + \frac{D(U - v_j)}{v_j^2}\mathbf{A} - \mathbf{C}(\mathbf{A}) - \frac{v_j^3 D}{2N^2}\mathbf{A}_{xx} - \frac{v_j^2}{N}\left(1 + \frac{N\mathbf{A}}{v_j^2}\right)h(x) = 0,$$

$$(11.4)$$

which we may term the *Grimshaw–Yi* equation. Here $\mathbf{A}' = \mathbf{A}(x',t)$, and $\mathbf{C}(\mathbf{A})$ is a complex non-linear term that is essentially quadratic and depends on the variable inertia of the fluid and the non-uniformity of N; for present purposes we may take it to be zero. The kernel $\kappa(\mathbf{A}, \mathbf{A}')$ is defined by introducing the variable ξ by

$$\xi = z - \eta(x, z, t) = z - \frac{\mathbf{A}(x,t)}{v_j}\sin\frac{j\pi z}{D}, \qquad (11.5)$$

so that η is the vertical displacement of a streamline at (x, z, t) from its elevation far upstream, ξ, due to the $\mathbf{A}_j^-$ term alone. Equation (11.5) defines the inverse relationships $z = z(\xi, \mathbf{A})$, and $z' = z(\xi, \mathbf{A}')$. We then have

$$\kappa(\mathbf{A}, \mathbf{A}') = \int_0^D \left(\frac{\partial z}{\partial\mathbf{A}}\frac{\partial z'}{\partial\mathbf{A}'} - z\frac{\partial z}{\partial\mathbf{A}}\frac{\partial^2 z'}{\partial\xi\partial\mathbf{A}'} - z'\frac{\partial z'}{\partial\mathbf{A}'}\frac{\partial^2 z}{\partial\xi\partial\mathbf{A}}\right)d\xi. \qquad (11.6)$$

Although the forcing in (11.4) is small (of order h_m), the magnitude of $\mathbf{A}$ is not necessarily small and may become of order unity. As a consequence of this, within a finite time the disturbance amplitude may increase to the point where $|\mathbf{A}| > v_j^2/N$, so that $u = 0$ at some point, and the flow verges on static instability. When this condition is reached, the transformations based on (11.5) break down in the vicinity of this point, and (11.4) is no longer applicable.

Some numerical solutions of (11.4) have been obtained by Grimshaw & Yi for Gaussian-shaped obstacles where the uniform flow is suddenly commenced from rest, as in §10.5.1. The solutions may be expressed in terms of the parameter

$$P = (U - v_j)D/(v_j h_m) = (jF_0 - 1)/H, \qquad (11.7)$$

where $F_0 = 1/K$. The values of P relate to directions on the (K, H)-plane of Figure 10.25 at the critical points $(K = 1, 2, \ldots, H = 0)$ for the various modes. Figure 11.1 shows the scaled vertical displacement $\eta(x,t)/(D/j) = -\mathbf{A}/(ND^2/\pi)$ for two cases for subcritical flows ($P < 0$) over an obstacle. In the directions in which no stable solution to Long's model (as described in §10.7.1) exists, corresponding

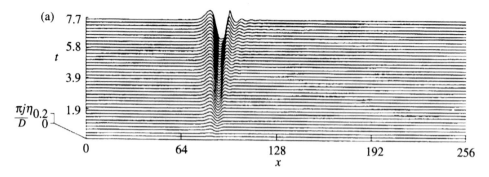

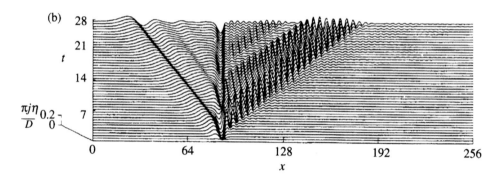

Figure 11.1 Examples of solutions to the Grimshaw–Yi equation (11.4) for the behaviour of a nearly critical mode (mode number $j = 1,2,...$) in uniformly stratified flow over a Gaussian obstacle at $x = 85$, in terms of dimensionless x and time t, for two representative values of P (11.7). (a) $P = -4/\pi$, where the solution grows with time to become statically unstable, and (b) $P = -12.5/\pi$, where the solution apparently remains stable. (From Grimshaw & Yi, 1991, reproduced with permission.)

to $-2 < P \le 0$, the numerical solutions reach the point of static instability after a finite time, and an example is shown in Figure 11.1a. In the directions where a stable Long's model solution does exist ($P > 0$ and $P < -2$), the solutions of (11.4) generally tend to a stable super- or subcritical solution (shown in Figure 11.1b) as appropriate, except for a range of directions with P close to -2 where the solutions become statically unstable after a long time. In the latter regions, numerical studies by Lamb (1994) with a non-hydrostatic numerical model have shown similar behaviour, but the overturning is a transient that eventually disappears, leaving the Long-model solution. Hence the general conclusion from these analytical and numerical studies is that the flow tends to the solution of Long's model, where it exists and is stable, but this may take a very long time. (However, see §11.5.)

11.2 Non-linear hydraulic flow theory for finite depth

This section presents a classical layered model for stratified hydraulics. It should be noted that given the development of mathematical techniques in recent years, there are other approaches to this type of problem. A relevant example is *flow-solve* by Winters & De La Fuente (2012). A different approach by Winters & Armi (2014) is to impose slowly varying properties on the uppermost streamline and integrate downward; one then finds out what topography one has solved for.

11.2.1 Equations for stratified flow hydraulics

For sufficiently high obstacles the flow is strongly influenced by non-linear advective effects, and the preceding models are inappropriate. The theoretical analysis of these effects is now well developed for "hydraulic" flows, where the horizontal length-scale of the obstacle is large compared with the other length-scales D and U/N. The hydraulic model is essentially inviscid, with the bottom boundary as a streamline. Because of the proportionally greater length of the obstacle and the gentle topographic slopes, lee-side separation effects are expected to be small and are omitted. With less justification, frictional effects are also omitted. This model predicts that these non-linear flows over topography produce disturbances that permanently alter the upstream flow state if the obstacle height is large enough, and that these upstream disturbances can have much larger amplitude than those described by the linear momentum-source model of §10.5.2. Some general flow properties have been described by Killworth (1992) and Pratt & Helfrich (2005).

For continuous stratification, the hydraulic effects of large-height topography are conveniently described by a layered model. Here the stratification is approximated by a suitable number of layers, in each of which the density and horizontal velocity are assumed to be uniform. These models are therefore an extension of the one- and two-layer hydraulic models of Chapters 3 and 4 to many layers. Linear disturbances in such a layered model were described in §8.1, and a descriptive sketch with notation is shown in Figure 8.1. The equations governing the hydrostatic motion of n incompressible layers are

$$\frac{\partial u_i}{\partial t} + u_i \frac{\partial u_i}{\partial x} = -\frac{1}{\rho_i}\frac{\overline{\partial p}}{\partial x}, \qquad i = 1, 2, 3, \ldots n, \qquad (11.8)$$

$$\frac{\partial d_i}{\partial t} + \frac{\partial}{\partial x}(u_i d_i) = 0, \qquad i = 1, 2, 3, \ldots n, \qquad (11.9)$$

where u_i, d_i and ρ_i denote the horizontal velocity, thickness and density of the ith layer respectively, numbered upward from the bottom, and the overbar denotes the vertical average in the ith layer. If the pressure at the top of the nth layer is denoted by $p_s(x,t)$, then for hydrostatic flow the pressure at a point within the ith layer is given by

$$p(x, z, t) = p_s + g \sum_{j=i+1}^{n} \rho_j d_j + g \rho_i \left(\sum_{j=0}^{i} d_j - z \right), \tag{11.10}$$

where $d_0 = h(x)$ denotes the height of the topography above the level surface $z = 0$. Equations (11.8) and (11.10) together give

$$\frac{\partial u_i}{\partial t} + \frac{\partial}{\partial x} \left(\frac{1}{2} u_i^2 + g \sum_{j=0}^{i} d_j + g \sum_{j=i+1}^{n} \frac{\rho_j}{\rho_i} d_j + \frac{p_s}{\rho_i} \right) = 0, \quad i = 1, 2, \dots, n. \tag{11.11}$$

As before, we consider two types of upper boundary: rigid and pliant. For a rigid upper boundary, we have

$$\sum_{j=0}^{n} d_j = D = \text{constant}, \tag{11.12}$$

and for a pliant upper boundary where the nth layer is surmounted by a deep layer of density ρ_{n+1}, p_s is given by

$$p_s = P_s - g \rho_{n+1} \left(\sum_{j=0}^{n} d_j - D \right), \tag{11.13}$$

where P_s is the value of p_s far upstream, and

$$D = \sum_{j=0}^{n} D_j, \tag{11.14}$$

where D_j is the value of d_j far upstream.

If we assume that the flow is steady and that $u_i = U_i$ far upstream, then (11.9), (11.11) integrate to give

$$d_i u_i = D_i U_i, \tag{11.15}$$

$$\frac{1}{2}(u_i^2 - U_i^2) + g \sum_{j=0}^{n} \rho_{ij}(d_j - D_j) + \frac{1}{\rho_i}(p_s - P_s) = 0, \tag{11.16}$$

where

$$\rho_{ij} = \begin{cases} 1, & j \leq i, \\ \rho_j/\rho_i, & j > i. \end{cases} \tag{11.17}$$

Differentiating (11.15) and (11.16) and eliminating du_i/dx gives

$$-t_i \frac{dd_i}{dx} + \sum_{j=1}^{i} \rho_j \frac{dd_j}{dx} + \rho_i \sum_{j=i+1}^{n} \frac{dd_j}{dx} + \frac{1}{g} \frac{dp_s}{dx} = -\rho_i \frac{dh}{dx}, \quad i = 1, \dots, n, \tag{11.18}$$

and for a pliant upper boundary, (11.13) gives

$$\frac{dp_s}{dx} = -\frac{\rho_{n+1} t_n}{\Delta_n \rho} \frac{dd_n}{dx},$$ (11.19)

where

$$t_i = \frac{\rho_i u_i^2}{g d_i}, \qquad \Delta_i \rho = \rho_i - \rho_{i+1}.$$ (11.20)

After some manipulation (subtracting successive equations), (11.18) and (11.19) may be expressed in the matrix form

$$
\begin{bmatrix}
\Delta_1\rho - t_1 & t_2 & 0 & 0 & 0 \\
\Delta_2\rho & \Delta_2\rho - t_2 & t_3 & 0 & 0 \\
\Delta_3\rho & \Delta_3\rho & \Delta_3\rho - t_3 & 0 & 0 \\
& & \cdot & t_i & \\
\Delta_i\rho & \Delta_i\rho & \Delta_i\rho & \Delta_i\rho - t_i & 0 \\
& & & \Delta_{i+1}\rho \quad \cdot & 0 \\
& & & \cdot & t_n \\
\Delta_n\rho & \Delta_n\rho & \Delta_n\rho & \Delta_n\rho & \Delta_n\rho - t_n
\end{bmatrix}
\times \frac{d}{dx}
\begin{bmatrix}
d_1 \\ d_2 \\ d_3 \\ \\ d_i \\ \\ \\ d_n
\end{bmatrix}
$$

$$
= -\frac{dh}{dx}
\begin{bmatrix}
\Delta_1\rho \\ \Delta_2\rho \\ \Delta_3\rho \\ \\ \Delta_i\rho \\ \\ \\ \Delta_n\rho
\end{bmatrix}
$$ (11.21)

This shows that at a point where $dh/dx = 0$, we must have either

$$\frac{dd_i}{dx} = 0, \quad \text{for all } i,$$ (11.22)

or the determinant of the matrix of coefficients must vanish. By subtracting successive columns (the second from the first, the third from the second etc.), the determinant may be expressed as

$$
\det
\begin{bmatrix}
\Delta_1\rho - t_1 - t_2 & t_2 & 0 & & & 0 \\
t_2 & \Delta_2\rho - t_2 - t_3 & 0 & & & 0 \\
0 & t_3 & & & & \\
 & & & t_i & & \\
0 & 0 & \Delta_i\rho - t_i - t_{i+1} & & & 0 \\
 & & t_{i+1} & & & \\
 & & & & \cdot & \\
 & & & & \cdot & t_n \\
0 & 0 & 0 & & t_n & \Delta_n\rho - t_n
\end{bmatrix}
= 0.
$$

(11.23)

This tri-diagonal matrix is the same as the long-wave limit of that in (8.8), the equation for internal wave modes and speeds, since T_i is identified with t_i, and $c = 0$. Consequently, (11.23) is the condition for an internal wave speed to vanish, and if it is satisfied it implies that $c_j = 0$, where c_j is the speed of some internal wave mode (mode j), propagating on the given flow state. In the terminology of hydraulics, the flow is said to be "critical" with respect to this mode, or that the mode is critical, at this location. There are n layers with a pliant upper surface having $2n$ internal wave modes, with $2n$ possible critical conditions. These modes are bracketed in n pairs, each of which may be regarded as the same mode propagating in opposite directions, but with different structure. For the case of a rigid upper boundary, equations (11.21), (11.23) still apply but with the bottom row (or equation) omitted, so that there are $2(n-1)$ critical conditions.

We therefore have a generalisation to n-layered systems (and by implication, to an arbitrarily close approximation to continuous stratification) of the "hydraulic alternative" of one- and two-layer systems, discussed in Chapters 2, 4 and 5: at a point where $dh/dx = 0$, either the horizontal gradient of the flow properties vanishes, or some internal wave mode is critical there. Equation (11.21) also implies that at a point where $dd_j/dx = 0$ for all i, or the flow is critical, we must also have $dh/dx = 0$. Hence, steady critical flow cannot occur over a sloping region of topography.

We may use these equations to construct a theory of hydraulic flow over topography by generalising the results for two-layer systems. Before doing this, however, we must first discuss hydraulic jumps for stratified systems.

11.2.2 Modelling hydraulic jumps

Hydraulic jumps in stratified flow are discussed in §§7.3 and 7.4. In many situations the essence of these jumps may be roughly approximated by one- and two-layer models (where the velocity in each layer is uniform or linearly sheared), which we generalise here. Stratified hydraulic jumps are turbulent if their amplitude is sufficiently large, and the associated mixing presents major problems in modelling them. For present purposes we model a hydraulic jump as a discontinuity between two horizontal flow states that remains steady in a frame of reference moving with the jump. For a layered model of jumps, we then make three assumptions, as follows. We assume:

(i) Each layer maintains its identity, density, and mass and momentum flux through the jump. This implies that mixing between these layers in the jump is negligible, which will be the case if the layers are immiscible or the jump is sufficiently weak. With this assumption ρ_i is constant through the jump, and if U_i and D_i denote velocity and layer thickness upstream and u_i and d_i the same downstream, for the ith layer in a frame of reference in which the jump is stationary, we have

$$U_i D_i = u_i d_i. \tag{11.24}$$

(ii) The flow in the jump is hydrostatic. Integrating the equation for conservation of momentum (§1.3) obtained from (11.8), (11.9) across the jump gives

$$d_i u_i^2 - D_i U_i^2 + \frac{1}{2} g(d_i^2 - D_i^2) + g \sum_{\substack{j=1 \\ j \neq i}}^{n} \rho_{ij} \overline{d}_i (d_j - D_j) + \frac{\overline{d}_i}{\rho_i} (p_s - P_s) = 0, \tag{11.25}$$

where P_s and p_s are the upstream and downstream pressures at the top of the nth layer, and $\overline{d}_i$ denotes the mean value of d_i in the jump.

(iii) The third assumption is that $\overline{d}_i$ is given by

$$\overline{d}_i = \frac{1}{2}(D_i - d_i), \tag{11.26}$$

the mean of the upstream and downstream thicknesses.

Combining (11.24)–(11.26) and eliminating the u_i, we obtain equations relating conditions on the downstream side of the jump to those upstream (Su, 1976)

$$\frac{2\xi_i U_i^2}{g D_i(1 + \xi_i)(2 + \xi_i)} = \sum_{j=1}^{n} \rho_{ij} \frac{D_j}{D_i} \xi_j + \frac{1}{g\rho_i D_i}(p_s - P_s), \quad i = 1, 2, \ldots, n, \tag{11.27}$$

where

$$\xi_i = \frac{d_i}{D_i} - 1. \tag{11.28}$$

For a travelling jump propagating into a given flow profile, (11.27) defines the jump speed as a function of its amplitude, and the downstream structure.

The above assumptions are the same as those employed by Yih & Guha (1955) and others. Ideally, it would be preferable to use the vorticity balance (FVS) model (described in §4.4), but this is more difficult to formulate for a general case. Presumably this difficulty could be circumvented by an iterative process (beginning with YG), but in general, the differences between FVS and YG speeds are small unless the jump amplitudes are large (Baines, 2016). For the (mostly) equally spaced layered models described in the following sections, the upstream disturbances are mostly of the rarefaction form (or a mixture of jump and rarefaction, with mostly rarefaction), so that any jump component has small amplitude.

11.2.3 A procedure for obtaining steady hydraulic flow states over topography of finite height

We next describe a general procedure for calculating the properties of hydrostatic flows over obstacles of finite height, when these are introduced into a known, stable, unidirectional stratified shear flow. It is based on generalising the results for two-layer systems described in Chapter 4, and as was done there, it concentrates on the flow conditions upstream and over the obstacle. From this, one may infer details of the downstream flow. Various steps in this procedure may be carried out numerically, and algorithms for these are summarised in Baines (1988). The procedure has been applied to a range of examples, as described in the following section.

We begin by defining an overall initial Froude number F_0 for the flow in the absence of topography, by

$$F_0 = \frac{\overline{U}}{\overline{U} - c_1}, \qquad \text{where} \quad \overline{U} = \frac{1}{D} \int_0^D U(z)dz = \frac{1}{D} \sum_{i=1}^n U_i D_i. \qquad (11.29)$$

The U_i are all assumed to be positive, and c_1 is the velocity of the fastest linear wave mode propagating against the stream, in the frame of the obstacle (if $c_1 < 0$ so that the wave propagates upstream, $F_0 < 1$, etc.). We then obtain (in conceptual terms) the steady-state flow that results from a succession of infinitesimal increases in the height of the obstacle, from zero to some finite value. This involves calculating the flow upstream and over the obstacle at each incremental height level until the desired obstacle height is reached. This gives a solution for the flow for this obstacle height, but it may not be unique, as examples in Chapters 3 and 4 have shown. The procedure is described in three stages.

Stage 1: We assume initially that $c_1 < 0$, so that $0 < F_0 < 1$, and at least one internal wave mode can propagate in the upstream direction (the case of $F_0 > 1$, $c_1 > 0$ is discussed at the end of this section). The introduction of an obstacle of very small height will give a disturbance that may be described by linear hydrostatic theory, as in §10.5. After the transients for each mode have had sufficient time to propagate away upstream and downstream, the remaining disturbance will be confined to the region of fluid over the topography, and may be regarded as a superposition of subcritical or supercritical patterns for each mode, in a manner similar to (10.64). At each x-value the local flow state depends only on the local obstacle height; it satisfies the condition $\mathrm{d}d_i/\mathrm{d}x = 0$ (for all i) at the crest, and it returns to the original undisturbed flow on the downstream side by passing back through the same flow states as those on the upstream side, as $h \to 0$. This flow type is shown schematically in Figure 11.2a, and it may be calculated from (11.15), (11.16).

If the height of the obstacle is then increased further by a succession of small amounts, we expect subcritical modes to move closer to criticality (i.e. the local value of c_j increases) at the obstacle crest. The character of the flow remains the same until a critical height h_c is reached (also obtainable from (11.15), (11.16)), where such a mode becomes critical there. From (11.21)–(11.23), we saw that at points where $\mathrm{d}h/\mathrm{d}x = 0$, with steady flow we must have

$$\frac{\mathrm{d}h}{\mathrm{d}x} = 0 \Rightarrow \frac{\mathrm{d}d_i}{\mathrm{d}x} = 0 \quad \text{for all } i, \textbf{ or} \qquad c_j = 0, \quad \text{for some } j, \tag{11.30}$$

and that conversely either of these conditions must imply that $\mathrm{d}h/\mathrm{d}x = 0$. When $h_\mathrm{m} < h_\mathrm{c}$, the first of these alternatives is the one that occurs, but when $h_\mathrm{m} = h_\mathrm{c}$, either is possible. Accordingly, the flow pattern may then take one of two forms: it may return to the upstream flow as before, or the flow may undergo a hydraulic transition at $h = h_\mathrm{m}$, and be supercritical on the downstream side for mode j.

Stage 2: For further increases in h_m beyond h_c, the nature of the flow must be different from that for $h_\mathrm{m} < h_\mathrm{c}$. This may be seen by noting that, if the flow were to remain the same as in Stage 1 for $h_\mathrm{m} > h_\mathrm{c}$, it would imply that $c_j = 0$ over the sloping topography on the upstream and downstream sides, which is not possible (leaving aside the possibility of an inflexion point in $h(x)$). From the example of one- and two-layer systems, if h_m is increased above h_c by a small amount, the flow will adjust locally so that it is again critical at the obstacle crest for the same jth mode. This will cause a small linear disturbance to be sent upstream, in the form of a columnar disturbance mode of mode j and travelling at speed c_j, that will alter slightly the oncoming velocity and density profiles in the new steady state. This is shown schematically in Figure 11.2b. Over the obstacle, mode j now changes from being subcritical on the upstream side to supercritical on the downstream side. This flow may be calculated (in principle) by adding a small increment of columnar mode j to the upstream flow, and then using (11.15), (11.16) with the new upstream

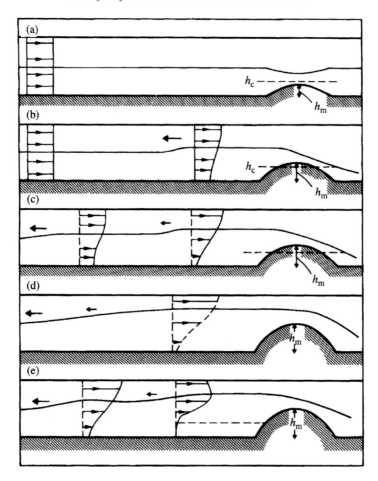

Figure 11.2 Schematic diagram illustrating the development of the various flow states that are realised as the height of a long obstacle is slowly increased, in small increments. The obstacle is introduced into a flow with an initially uniform velocity profile (but an unspecified density profile) with $F_0 < 1$, and the sketches show upstream velocity profiles and a typical streamline. (a) $h_m < h_c$, the critical height for this value of F_0; steady states have the same flow upstream and downstream. (b) h_m has increased to slightly above h_c; there is now critical flow and a hydraulic transition at the obstacle crest that causes a columnar disturbance mode of small amplitude to be sent upstream, altering the upstream flow state. (c) A further increase in h_m causes the process to be repeated; the second disturbance may travel faster (resulting in a jump) or slower (resulting in a rarefaction) than the previous one. (d) If h_m is sufficiently large the flow may become critical just upstream and supercritical over the obstacle, as shown here, or (e) a fluid layer may become blocked; this usually involves more than one upstream mode. Typically, the flow may pass through states (a–d) to reach state (e), where a hydraulic transition for a second mode is shown at the obstacle crest. The evolution depicted here may be used to calculate the steady-state flow for given parameters. (From Baines, 1988, reproduced with permission.)

flow to find the new critical height which is the appropriate value of h_m, and hence to obtain the flow field that must now have a hydraulic transition at $h = h_m$.

If the obstacle height is then increased further by an infinitesimal amount, this process will be repeated: the flow will adjust to a slightly different steady state at the obstacle crest, again with $c_j = 0$ there, and this will cause a second linear columnar disturbance to propagate upstream. This second disturbance will propagate on a slightly different flow from the first, because the latter has altered the upstream velocity and density profiles. Consequently, its structure and wave speed will also be slightly different from those of the first disturbance. This process is illustrated in Figure 11.2c.

At this point we must distinguish between two different possibilities. The propagation speed of the new disturbance may be written $c_j + \Delta c_j$, where c_j is negative (i.e. directed upstream), and Δc_j denotes the difference in speed between the second disturbance and the initial one. This difference Δc_j may be positive or negative. If it is positive or zero, the second disturbance propagates more slowly and will never catch up to the first one, and we may again use (11.15), (11.16), with modified upstream flow, to obtain the flow field as before. However, if $\Delta c_j < 0$, the second disturbance will travel faster than the first, and will catch up with it and increase its amplitude. In effect, this will form an infinitesimal hydraulic jump. As discussed above, a hydraulic jump travels at a speed that depends on its amplitude, and its formation imposes the "jump conditions" (11.27), which determine the flow on the downstream side in terms of the upstream flow and the jump amplitude.

Once the jump has formed, this flow downstream of it will in general be slightly different from that which was initially present behind the second upstream disturbance. This difference will then be communicated back to the obstacle and cause further adjustments there. These changes will in turn affect the jump, and the flow will reach a steady state when the jump amplitude is adjusted to be consistent with a critical flow state at the obstacle crest.

If the obstacle height is increased still further and successive values of Δc_j all have the same sign, these processes will be repeated. The result in the first case ($\Delta c_j > 0$) will be a succession of upstream disturbances that become increasingly spread out, forming a *rarefaction* (generalising the term used for two-layer disturbances in Chapter 4) which is effectively a process of *amplitude dispersion*, and the result in the second case ($\Delta c_j < 0$) will be a progressively larger hydraulic jump. These conditions may be summarised by saying that

$$\frac{dc_j}{d\alpha} \begin{cases} < 0 & \Rightarrow \text{hydraulic jump,} \\ > 0 & \Rightarrow \text{rarefaction,} \end{cases} \tag{11.31}$$

where c_j is negative and α denotes the amplitude of the jth columnar mode. Note that these are cumulative variables, in the sense that following disturbances propagate on and add to previous ones. The structure of the corresponding eigenfunction also changes continuously. Expressions for $dc_j/d\alpha = 0$ may be obtained in terms of the local mean flow properties and the structure of the associated eigenfunction, and the details are given in Baines (1988).

Different numerical procedures are required in order to calculate the flows in these two situations. In the rarefaction case, the upstream columnar disturbance may be increased by a succession of small increments, for each of which the structure of the changing jth eigenfunction must be obtained. As the upstream disturbance builds up, (11.15), (11.16) may be used to obtain the increasing critical obstacle heights, and the associated flow. In the upstream jump case, the upstream disturbance is assumed to be concentrated into a single hydraulic jump. For a given jump amplitude, the subsequent flow over the obstacle with the (necessary) critical condition at $h = h_m$ is obtained from (11.15), (11.16), with the conditions downstream of the jump used as the upstream conditions for flow over the obstacle. It often occurs that $dc_j/d\alpha < 0$ when α is small, but changes to become positive as α increases (an example is found in two-layer flows with $r \ll 1$, in Chapter 4). When this happens, it is necessary to use a combination of these procedures: the upstream jump alone for α small, and a jump of maximum height followed by a rarefaction, for larger α. Practical details are given in Baines (1988). It appears that the difference between the results of these two procedures (for example, the resulting upstream profiles for a given obstacle height) is small unless the amplitude of the upstream disturbance is large, and in many-layered situations the rarefaction procedure (which has been found easier to use) may be substituted for the jump model.

Stage 3: The above procedures may be followed to obtain the flow over progressively higher obstacles until one of three things happens. These are:

(i) the flow immediately upstream of the obstacle may become critical (with respect to the jth mode, so that $c_j = 0$ just upstream);

(ii) the velocity U_i of some fluid layer (usually, but not always, that of the lowest layer, U_1) may become zero just upstream; or

(iii) the $(j + 1)$th mode, which was initially supercritical, may become just subcritical in the upstream flow. We discuss each of these situations in turn.

(i) *Critical flow upstream.* When rarefactions occur upstream, the value of c_j increases (i.e. its magnitude decreases) with increasing disturbance amplitude (and obstacle height), and critical flow occurs when c_j increases to zero. This is shown schematically in Figure 11.2d. If h_m is increased further, the jth mode is no longer critical at $h = h_m$, and the flow reverts to the other alternative in (11.30), where it is symmetric about the obstacle crest. As shown in §4.8, this may imply dissipative flow structures on the downstream side. For further increase in h_m the flow may be calculated by using (11.15), (11.16), until another mode (the $(j-l)$th) becomes critical at $h = h_\mathrm{m}$, where Stage 2 may be repeated for this mode.

(ii) *Blocking.* Blocking occurs when the velocity of an upstream layer of fluid in steady-state flow is reduced to zero by the upstream disturbances (shown schematically in Figure 11.2e), so that the fluid in this layer cannot pass over the obstacle and the layer terminates there. Two observational facts are significant here. First, the introduction of an obstacle into a unidirectional flow does not cause the direction of the flow to reverse at any level, so that the nett velocity of any layered flow can never become negative. Second, if the lowest layer (or layers, or region of stratified fluid) becomes blocked upstream for a given value of h_m, increasing h_m further will not cause it to become unblocked, although its thickness may be altered. This implies that the upstream disturbances are now more complicated, and generally involve more than one mode. As h_m increases so that the lowest layer becomes blocked, the thickness d_1 of the latter decreases to zero over the obstacle crest. Equation (8.10) for the mode speeds then shows that this implies that $c \rightarrow u_1$ for *two* modes (or rather, the upstream- and downstream-propagating forms of the same mode), and these modes disappear from the flow over the obstacle when the layer disappears. If one of these two modes is the one that is critical at the obstacle crest (as happens when F_0 is small for $N_2 = 3$ or 4, for example), then as d_1 vanishes so does u_1, but if some other mode is critical, u_1 at the crest remains finite.

It is possible that the upstream disturbances may cause a layer other than the bottom layer to be brought to rest when layers at lower levels are still in motion, and examples are given in the next section. This is not "blocking" by the above definition because the stationary layer still extends downstream, but it has the same cause. A further increase in h_m then implies that *two* modes are generated upstream. The first of these is controlled by a critical condition at the obstacle crest as before, and the amplitude of the second (and probably faster) mode is determined by the condition that the stationary layer remain stationary because it cannot have a negative velocity (which it would otherwise have if the first mode alone were present). This second mode may be termed a "*passenger mode*". This process has been assumed in calculating some regions of the diagrams shown below, but the phenomenon has yet to be observed and described experimentally. Passenger modes must also occur when upstream disturbances are forced with lower blocked layers present, and the thickness of these layers generally decreases as h_m increases (with F_0 constant).

In practice, implementing these steps with blocked layers was found to be difficult and tedious, so that blocked flows have not been explored in great detail. Some simplifying approximations may be made, and the details are given in Baines (1988).

(iii) *Higher-order mode becomes critical.* As the amplitude of the jth mode increases upstream due to increasing obstacle height, it is possible that the initially supercritical $(j+1)$th mode may become subcritical on this altered flow profile. This implies that this $(j+1)$th mode must then be generated upstream instead of the jth. An example is discussed below. This type of interaction is a possible cause of oscillatory behaviour in non-hydrostatic systems (see §11.5), and deserves further study.

In the above discussion it has been assumed that F_0 lies in the range $0 < F_0 < 1$. The other possibility is that $F_0 > 1$, so that $c_1 > 0$ and all modes are supercritical. If h_m is small, linear theory is again applicable, and the above discussion for Stage 1 of the "procedure" applies. Whether or not a critical value $h_m = h_c$ is reached as h_m increases depends on the flow profile, and whether the local value of c_1 at the obstacle crest increases or decreases as h_m increases. In uniform flow with two-layer systems, for example, if the layers are of equal thickness ($r = 0.5$), no upstream disturbances occur at all. If $r \ll 1$, on the other hand, a critical height is reached, and higher obstacles produce upstream jumps.

To summarise briefly, this hydraulic model of flow over topography is based on determining the form of the upstream disturbance (if any) that would result from a small increase in obstacle height above that for which the flow is known. One then finds the obstacle height and flow that result from adding a small increment of this disturbance to the upstream flow profile.

The model has been extended by Nielsen et al. (2004) to include the effects of drag on the bottom surface, and entrainment of mass and momentum from passive overlying fluid. This makes the problem much more complicated. The solutions obtained have similar form to those described here, but are quantitatively different. In particular, with both entrainment and/or bottom drag included, the critical control sections tend to lie on the downstream side of the topography rather than at the peak.

With continuous stratification, the model could also be improved by using a more "modern" model for upstream internal hydraulic jumps where they occur, as described by White & Helfrich (2014).

11.3 Applications of the hydraulic theory

We next consider some examples of flows over topography obtained with the model of §11.2. All of these flows have uniform velocity U initially (i.e. in the absence of the obstacle), and we concentrate in particular on the flow of several homogeneous layers with equal density increments and initial thickness, for a varying number of layers, n_1. Fluids composed of many thin layers may be used to approximate a continuously stratified fluid, but for flow over obstacles the approximation is sometimes imperfect, as we shall see. Except for some aspects of three-layered flows, the details have not been confirmed experimentally, and should be regarded as resulting from a logical application of the procedure of §11.2, rather than as definitive solutions. This applies in particular to situations where layers are blocked, or more than one mode is critical.

As a starting point we note the behaviour of two-layer systems with a rigid lid as shown in Figure 4.11, particularly the case of two equal layers in Figure 4.11c. This diagram has a characteristic shape that is repeated many times in the figures below. If $F_0 > 1$, the flow is always supercritical everywhere. When $F_0 < 1$, the upstream disturbances (where they occur), are all rarefactions, and if $H = h_m/D$ is sufficiently large, the flow is either blocked (when $F_0 < 0.25$) or critical upstream (when $0.25 < F_0 < 1$).

11.3.1 Three equal layers: rigid upper boundary

Figure 11.3 shows the Froude number/obstacle height diagram for steady-state flow of three layers with equal density increments and initial thicknesses, with a rigid lid. If $F_0 > 1$ the flow is always supercritical, so that only the range $F_0 < 1$ is interesting. The undisturbed flow has two internal wave modes that propagate against the stream: a fast mode (mode 1) where the vertical displacements of both interfaces are in phase; and a slow (sausage) mode (mode 2), where their displacements are opposite. In this flow the fast mode is critical when $F_0 = 1$ (by definition), and the slow mode is critical when $F_0 = 0.575$. For $F_0 < 0.575$, as H increases the flow is initially subcritical for both modes, until the curve AB′ is reached, where the slow mode becomes critical at the obstacle crest. As H is increased further, an upstream rarefaction of mode 2 structure is generated, and this grows until either the curve AB is reached, where the lowest layer becomes blocked, or BB′ is reached, where mode 2 becomes critical upstream. The similarity between the triangular region ABB′ of the diagram, and Figure 4.11c for two layers, should be noted. As H is increased further, there is no change in the upstream profile or the nature of the flow until curve DBG′ is reached. Here the flow has again become critical over the obstacle crest, for the fast mode on BG′, and for the two-layer flow on DB. On BG′, the upstream disturbance is again a rarefaction, with the lowest layer becoming blocked on curve BFGC, and mode 1 becoming critical upstream at CG′. To the right of DBFG, where only the upper two layers flow over the obstacle, there are two-layer upstream jumps in the region DBFE (leaving aside the complication of the blocked lowest layer), with rarefactions in the region FEG, as in Figure 4.11b for example. In each of the three shaded regions the upstream disturbance increases as H increases, but in the unshaded regions of the diagram it does not.

This rather complete description of flow regions on the (F_0, H)-plane illustrates

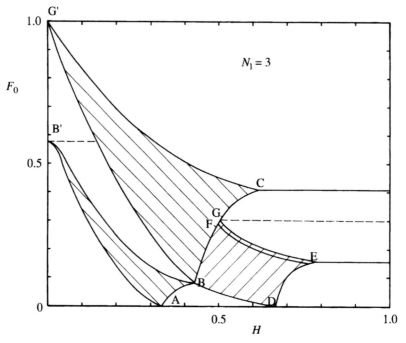

Figure 11.3 (F_0, H)-diagram showing the regions for various flow types for three equal layers (three equal initial thicknesses with two equal density increments) with a rigid upper boundary. The flow is critically controlled with hydraulic transition at the obstacle crest in the shaded regions, for the slow mode to the right of B'A, and the fast mode to the right of G'B. These modes become critical upstream on G'C and B'B. Layer 1 is blocked to the right of ABC, and layer 2 to the right of DE. In the shaded regions the upstream disturbance increases with increasing H, but in the unshaded regions it does not. (From Baines & Guest, 1988, reproduced with permission.)

the overall patchwork nature of these diagrams for layered flows in general, and how the pattern of two-layer flows (in Figure 4.11c) tends to be repeated, three times in this diagram. Figure 11.4 is the same diagram but showing contours of the upstream thickness of the lowest layer scaled with its original value, $d_1/(D/3)$. The points marked a, b, c and d refer to the same parts of Figure 11.5, which shows some representative laboratory experiments with three-layer flows with the same parameter values. There is generally good agreement between the observations and the hydraulic model predictions: upstream rarefactions are evident in (a) and (c), and the blocked lowest layer may be seen in (b) and (d). Note that the lowest layer thickness increases (and its velocity decreases) with increasing H up to the line where it becomes blocked, whereafter its thickness either remains constant, or decreases due to the changing flow in the layers above.

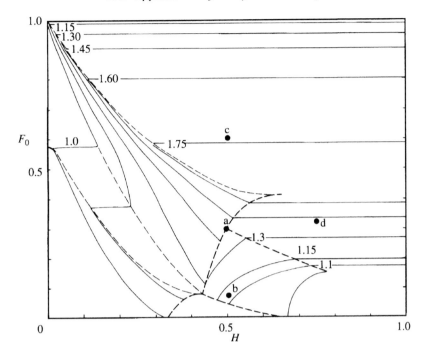

Figure 11.4 As for Figure 11.3, but showing upstream lower layer thickness $d_1/(D/3)$. The points a, b, c and d denote the flow states for the corresponding experiments in Figure 11.5. (From Baines & Guest, 1988, reproduced with permission.)

11.3.2 Three layers with a thinner upper layer: rigid upper boundary

In contrast to the above example, Figure 11.6 shows the (F_0, H)-diagram for the same three-layer system except that the initial depth of the topmost layer $(D/5)$ is now one-half that of each of the other two $(2D/5)$. This may be taken to be a crude approximation to a continuously stratified oceanic system with a shallow thermocline and a deep stratified region. The diagram and flow regimes are essentially very similar to Figure 11.3, except that there are two regions with small F_0 (centre bottom of diagram) where the topmost layer has been brought to rest, and most or all of the flow is carried by the central layer. The requirement that layer velocities cannot become negative implies that both (fast and slow) modes must be generated upstream in the first (vertically shaded) of these two regions. The figure has been computed on the assumption that mode 2 is forced by the critical condition at the obstacle crest, and that mode 1 is a "passenger mode", forced at an amplitude determined by the zero velocity condition for the stationary layer. In the second of these regions where only the middle layer is moving, the two-layer flow over the obstacle is subcritical.

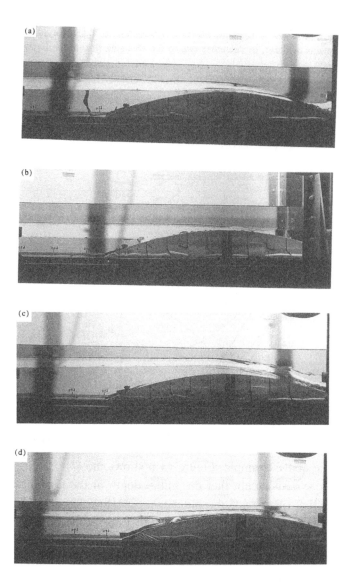

Figure 11.5 Examples of the flow of three initially equal layers over a towed obstacle. The middle layer is water, and the top and bottom layers are a mixture of kerosene and freon, with densities such that the system may be assumed to approximate one with a rigid lid. The flow is from left to right relative to the obstacle, and has reached a steady state in the field of view. (a) $F_0 = 0.3$, $H = 0.5$; (b) $F_0 = 0.08$, $H = 0.5$; (c) $F_0 = 0.6$, $H = 0.5$; (d) $F_0 = 0.32$, $H = 0.75$. The lowest layer is almost blocked in (a), blocked in (b) and (d) and not blocked in (c), consistent with Figure 11.4. (From Baines & Guest, 1988, reproduced with permission.)

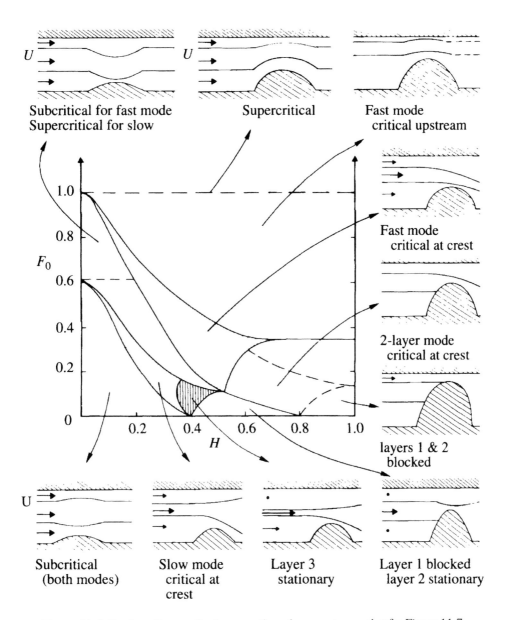

Figure 11.6 Regime diagram for the same three-layer system as that for Figure 11.7 except that the initial layer depths are $0.4D$, $0.4D$ and $0.2D$ (numbered upwards), rather than $D/3$ each. In the shaded regions the flow is critical at the obstacle crest with respect to one of the modes, and the resulting upstream disturbances are of rarefaction type. (From Baines, 1990: reproduced by permission of ASCE.)

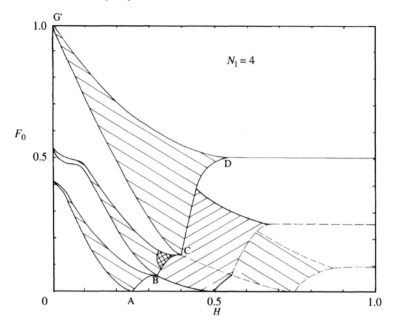

Figure 11.7 Flow regimes on the (F_0, H)-diagram for a system with four initially equal layers (i.e. three equal density increments, and four equal layer thicknesses in the absence of topography) with a rigid upper boundary, in the same manner as Figure 11.6. Shaded regions denote those with a hydraulic transition at the obstacle crest, and the lowest layer is blocked to the right of ABCD, etc. In the double-shaded region, layer 4 is stationary but layer 1 is not. (From Baines & Guest, 1988, reproduced with permission.)

11.3.3 Four equal layers: rigid upper boundary

Figure 11.7 shows the regions of the (F_0, H)-diagram for four equal layers with original thickness $D/4$, and equal density increments. The diagram reflects the fact that this system has three modes (propagating leftwards), and in the shaded regions the flow is critical at the obstacle crest with respect to one of them. The increasingly patchwork nature of the diagram is obvious, with the "two-layer pattern" being repeated six times. In the double-hatched region the topmost layer (layer 4) is stationary but the others are not (as in Figure 11.6), so that two modes (the middle and the fastest) must be generated upstream here, the fastest being a "passenger mode".

11.3.4 Three equal layers, with a pliant upper boundary

With a pliant upper boundary, as described in §10.7.2, the uppermost layer is bounded above by an infinitely deep homogeneous fluid of slightly lower density.

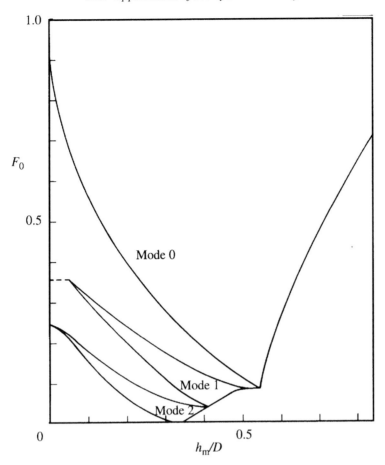

Figure 11.8 Flow regimes in the (F_0, H)-diagram for three initially uniform layers with a pliant upper surface, calculated to the point where the lowest layer becomes blocked using the rarefaction procedure. (From Baines & Granek (1990): reproduced by permission of Kluwer Academic Publishers.)

Any motion in this overlying fluid has negligible effect on the flow below. The system now has an extra mode (mode 0), which involves all three interfaces moving together. As an example, the situation for three initially equal layers in uniform motion is shown in Figure 11.8. Note that F_0 is now defined in terms of mode 0, rather than mode 1. Mode 0 gives the system a mode of behaviour rather like that of a single layer, with the critical curve continuing for $F_0 > 1$ (but not shown) in the same manner. Another difference from the rigid lid case is that mode 1 does not become critical (and hence does not force upstream motions) until H increases to a finite value in excess of about 0.05 (cf. Figure 10.26).

11.4 The approach to continuous stratification

There are two main methods of mathematically modelling the flow of continuously stratified fluid – the first is a layered model as above, with many layers, and the second involves the representation of the flow in terms of a Fourier series of relevant modes. Here results from a layered model are described first, followed by the modal model, for uniformly stratified fluid. A major point of interest centres on the conditions that can create upstream-propagating disturbances, to the extent that they can result in totally blocked flow at low levels.

11.4.1 Many equal layers, with a rigid upper boundary

We consider a system of $n_1 = 64$ equal layers with equal density increments, which may be regarded as a density profile that closely approximates continuous uniform stratification. As in the preceding examples the above procedure is applied to obtain steady-state flows. This system has a rigid upper boundary; we will contrast this with the case of a pliant upper boundary below. If $N^2 = g\Delta\rho/(\bar{\rho}D)$, where $\Delta\rho$ is the total density variation, then the speed of the lowest internal wave mode is (very nearly) ND/π, so that $F_0 = U\pi/ND = 1/K$. The critical heights for this 64-layer system may be determined from Stage 1 of §11.2, and these are shown in Figure 11.9. For $F_0 > 1$ the flow is again everywhere supercritical for all obstacle heights, and for $F_0 < 1$ the boundary of subcritical flow is found to have the sawtooth form

$$H = \frac{1}{2}(1 - nF_0), \quad \text{for } \frac{1}{n+1} < F_0 < \frac{1}{n}, \quad \text{i.e. } n < K < n+1, \quad n = 1, 2, 3, \ldots.$$

$$(11.32)$$

This curve was obtained numerically for $F_0 > 0.2$, but must deviate from this form as F_0 becomes small, because of the discrete layer thickness. However, equation (11.32) may be presumed to hold down to zero F_0 in the limit $n_1 \to \infty$, and the line $Nh/U = \pi/2$ is seen to be an envelope for this sawtooth boundary. We may compare (11.32) with the curve marking the limit of static stability of the Long-model solutions for hydrostatic stratified flow, given in Figure 10.25, and marked in Figure 11.9 as a dashed curve. The curve for Long's model is seen to lie inside the hydraulic one, following it closely except near the peaks.

A comparison between velocity profiles for the layered model and Long's continuous model is shown in Figure 11.10, for the two points marked "×" in Figure 11.9. In Figure 11.10a, inside the statically stable region for Long's model, the two velocity profiles are seen to agree quite well (although the discreteness of the layers is quite evident near the top). However, in Figure 11.10b the profiles are quite different, and this is attributed to the inadequacy of Long's model beyond its limit of viability.

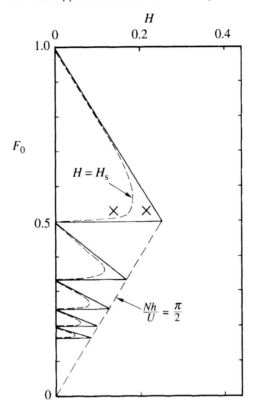

Figure 11.9 The (F_0, H)-diagram for 64 equal layers with a rigid upper boundary. The solid sawtooth line denotes the criterion for critical flow, from the hydraulic model. The dashed curve inside it denotes the limit of the solution for Long's model for continuous stratification, replotted from Figure 10.25. Inside this curve the hydraulic and Long-model solutions agree, but outside it they differ, as shown in Figure 11.10 for flow conditions marked by the crosses. (From Baines & Guest, 1988, reproduced with permission.)

Independently of these computations, it may be readily shown that where the solution of Long's model for continuous stratification is statically stable, the flow cannot be critical. We use the velocity and density profiles given by (10.91), (10.92) to provide the mean velocity $U(z)$ and buoyancy frequency $N(z)$ profiles at locations where $h = 0$. We then substitute these into the equation for small disturbances ((8.25) with $k = 0$), and look for conditions where there are solutions with $c = 0$, implying criticality. We may do this by putting $c = 0$ in (8.25), and looking at the conditions that the corresponding solutions require. Equation (8.25) then reduces

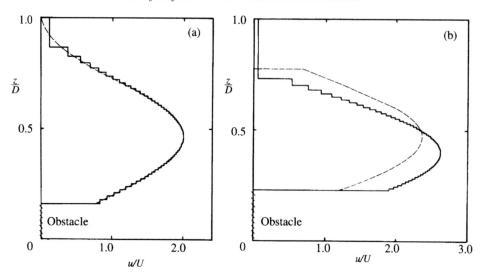

Figure 11.10 Comparison between the velocity profiles at $h = h_m$ for the 64-layered solution and the Long-model solution (dashed) for (a) $F_0 = 1/K = 0.53$, $H = 0.161$, and (b) $F_0 = 0.53$, $H = 0.229$, marked by crosses in Figure 11.9. (From Baines & Guest, 1988, reproduced with permission.)

to

$$\frac{d^2\hat{\psi}}{dz^2} + \left(\frac{N_0^2}{U_0^2}\right)\hat{\psi} = 0, \tag{11.33}$$

where N_0 and U_0 denote the *initial* uniform upstream values of $N(z)$ and $U(z)$, so that $K = N_0 D/\pi U_0$. With the lower boundary condition $\hat{\psi} = 0$ on $z = h(x)$, (11.33) has the solution

$$\hat{\psi} = \sin\frac{N_0}{U_0}(z - h). \tag{11.34}$$

The rigid boundary condition $\hat{\psi} = 0$ on $z = D$ then implies that for critical flow at $h = h_m$ we must have

$$\frac{N_0}{U_0}(D - h_m) = n\pi. \tag{11.35}$$

This gives a sawtooth curve that has twice the amplitude for H as does (11.32), so that the flow inside the latter cannot be critical. The difference between (11.35) and (11.32) emphasises the point that (10.91), (10.92) do not describe a real flow beyond this boundary.

As was the case with $n_1 = 2$, 3 or 4 evenly spaced layers with the same density increments (approximating uniform stratification), the upstream disturbances generated when H is increased above the critical height travel more slowly as their

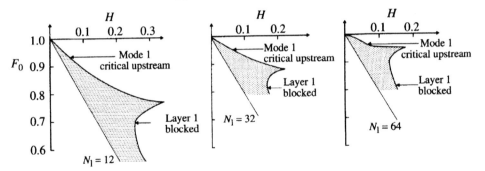

Figure 11.11 The dependence of the subcritical flow regimes on the number of layers, N_1, for F_0 near 1. (From Baines & Guest, 1988, reproduced with permission.)

amplitude increases, so that the disturbances are of the rarefaction type. As shown in the above examples, if $F_0 > 0.25$ (i.e. $n_1 = 2$), or 0.42 (i.e. $n_1 = 3$) or 0.51 (i.e. $n_1 = 4$), the lowest layer is never blocked for any value of H, and the maximum amplitude of the upstream disturbance occurs when the speed of the fastest upstream mode has decreased to zero. As n_1 increases further, this range of F_0 contracts upwards as shown in Figure 11.11, but even for $n_1 = 64$ it is still significant and covers the range $0.95 \leq F_0 < 1$. This indicates that there is a fundamental difference between the behaviour of layered and continuously stratified flows in these circumstances. Figure 11.12a (below) shows the maximum upstream disturbance to the 64-layered velocity profile when $F_0 = 0.98$, which shows that this phenomenon occurs because slowly moving layers become very thick, with relatively large velocity steps. This magnifies the effects of their discreteness. It implies that in general, small variations or differences in the stratification between two situations may cause substantial differences to the eventual steady-state flow.

Figure 11.12 is the same diagram as Figure 11.9, and shows the regions of (F_0, H)-space where various flow types are obtained by using the hydraulic model, up to the point where the lowest layer becomes blocked. If the jth columnar mode is present upstream with small amplitude ϵ, it may be shown (by substituting (8.47), (8.48) into (8.25)) that the structure and speed of the same jth mode propagating on this flow are unchanged (i.e. the modes are effectively sinusoidal) with error of $O(\epsilon^2)$. Hence, as H increases into the hydraulically controlled region we have $dc/d\alpha = 0$ initially, but computations show that eventually it becomes positive. This decreases the upstream wave speeds and enables the upstream modes to become critical when $F_0 \leq 1/j$, as shown in Figure 11.11 for mode $j = 1$. Figure 11.12 shows that it also occurs for the higher modes $j = 2$ and 3.

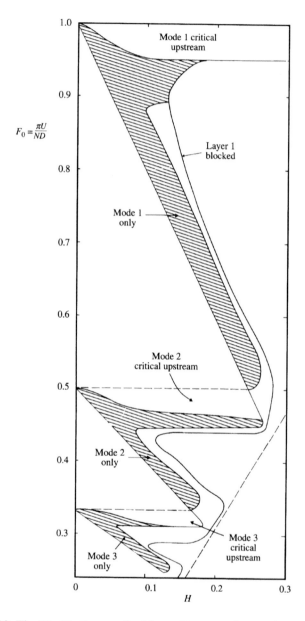

Figure 11.12 The (F_0, H)-diagram for 64 equal layers with a rigid upper boundary. In the shaded region the flow has a hydraulic transition at $h = h_m$, but only one mode is present. To the right of this shaded region, for $F_0 < 0.95$, at least one other mode is present. The curve denoting that the lowest layer is blocked is only approximate for $F_0 < 0.4$. (From Baines & Guest, 1988, reproduced with permission.)

Figure 11.12 also shows that the amplitude of the upstream disturbance increases very rapidly as H increases beyond the minimum value for critical flow. In the range $0.5 < F_0 < 0.95$, for example, only a small increase in H causes the upstream disturbance to increase to the curve where the bottom layer becomes blocked. However, in the range $0.5 < F_0 < 0.89$, a new phenomenon occurs before this point is reached. Here the increased amplitude of mode 1 upstream causes mode 2 to become (just) subcritical there, as described in Stage 3 of §11.3. This causes mode 2 to be forced upstream rather than mode 1, but a small increase in mode 2 upstream renders it supercritical there again. In the procedure of §11.2, which proceeds in discrete steps and assumes that steady flows are achieved, modes 1 and 2 increase alternately. In actuality this means that they increase together, with both being critical at the obstacle crest and mode 2 just subcritical upstream. This occurs in the unshaded region up to the point where the lowest layer becomes blocked.

Figure 11.13 shows the upstream velocity profiles attained at the point of blocking [in (b), (c), (d) and (e)], and the closest it gets for $F_0 = 0.98$ [in (a)]. For $F_0 = 0.89$ [in (b)], from the shape of the profile only mode 1 is evident, as in (a), but at $F_0 = 0.56$ there is a component of mode 2, and it dominates at $F_0 = 0.45$. Note that, as in three- and four-layered systems (Figures 11.3, 11.4, 11.6 and 11.7), the presence of mode 2 upstream (for $F_0 < 0.5$) has no noticeable effect on the critical amplitude for mode 1, which occurs at the continuation of the line for $F_0 > 0.5$; the same is found for mode 2 in the presence of mode 3 upstream when $F_0 < 1/3$. For $F_0 = 0.32$ (Figure 11.13e), the upstream profile at blocking is dominated by mode 3 with a small element of mode 2.

Exploring further details of Figure 11.13 for larger H is not computationally trivial with this model. An alternative approach with modes in continuous stratification with a rigid lid is described in §11.4.3.

11.4.2 Many equal layers, with a pliant upper boundary

Compared with §11.4.1, here we have an extra mode, mode 0 (see §10.7.2), and some representative numerical results are presented. The corresponding (F_0, H)-diagram to Figure 11.12 for 64 layers with a pliant upper boundary is shown in Figure 11.14, where the computations have concentrated on the range $0.3 < F_0 < 1$, where $F_0 = \pi U/2ND$ here. The computed boundary marking the onset of critical flow (for $F_0 > 1/3$) coincides with the boundary obtained from Long's model in Figure 10.26. The contours of constant u_1/U, where u_1 is the velocity of the bottom layer, are also shown. Up to the point where the dashed line is reached, only mode 0 appears in the upstream flow. Beyond this point, mode 1 becomes subcritical upstream and modes 0 and 1 increase together with h_m/D.

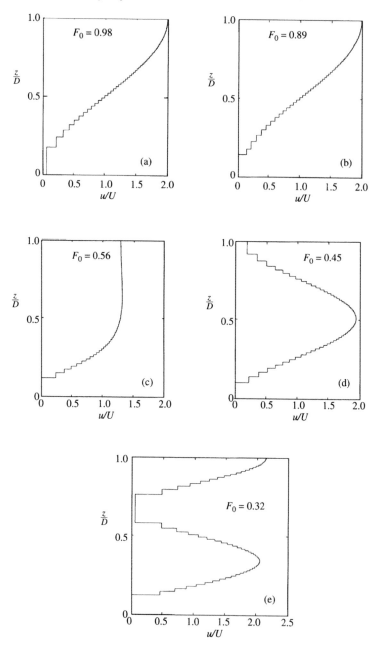

Figure 11.13 Some representative upstream velocity profiles obtained from the 64-layer model with a rigid upper boundary. (a) At the point of maximum upstream disturbance for $F_0 = 0.98$ (i.e. mode 1 is critical upstream – see Figure 11.10 for $n_1 = 64$). (b) At the point of blocking of the lowest layer for $F_0 = 0.89$; (c) for $F_0 = 0.56$; (d) for $F_0 = 0.45$; (e) for $F_0 = 0.32$. The associated obstacle heights can be inferred from Figure 11.12. (From Baines & Guest, 1988, reproduced with permission.)

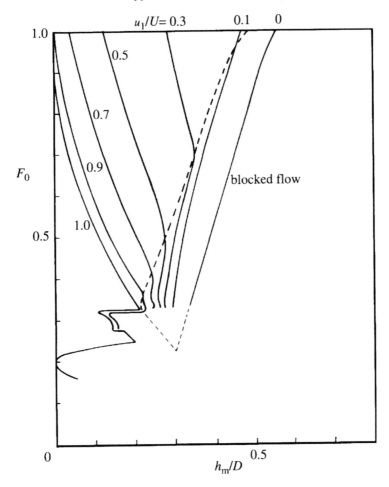

Figure 11.14 The (F_0, H_m)-diagram for the 64-layer hydraulic model with a pliant upper surface, showing the region of hydraulic transition and contours of u_1/U upstream, where u_1 is the velocity of the lowest layer (note that here $F_0 = \pi U/2ND$). The onset of critical flow coincides with the limit of the Long-model solutions shown in Figure 10.26, marked here as $u_1/U = 1.0$. The dashed line for $F_0 > 0.33$ denotes where, for larger h_m, the second mode (mode 1) begins to affect the flow. The upstream motions were calculated assuming upstream rarefactions as described in §11.2.3. Locations marked X denote parameter values for profiles shown in Figures 11.15 and 11.16.

In Figure 11.14, to the left (small h_m) of the dashed line, only mode 0 appears in the upstream flow. Beyond this point mode 1 is present but is subcritical upstream, and modes 0 and 1 increase together with $H = h_m/D$, with mode 0 critical at the crest and mode 1 critical upstream, as for the rigid upper boundary case in

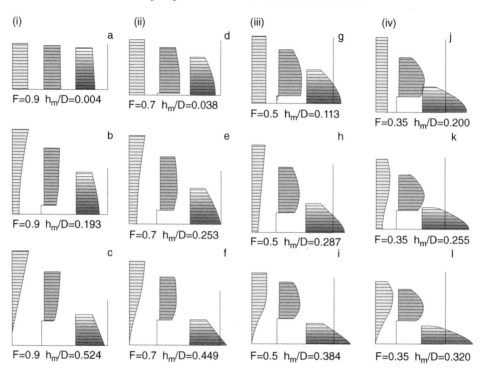

(i)

a
F=0.9 h$_m$/D=0.004

b
F=0.9 h$_m$/D=0.193

c
F=0.9 h$_m$/D=0.524

(ii)

d
F=0.7 h$_m$/D=0.038

e
F=0.7 h$_m$/D=0.253

f
F=0.7 h$_m$/D=0.449

(iii)

g
F=0.5 h$_m$/D=0.113

h
F=0.5 h$_m$/D=0.287

F=0.5 h$_m$/D=0.384

(iv)

j
F=0.35 h$_m$/D=0.200

k
F=0.35 h$_m$/D=0.255

F=0.35 h$_m$/D=0.320

Figure 11.15 Representative velocity profiles from the pliant-upper-surface 64-layer hydraulic model (§§10.7.2, 11.4.2) at locations upstream, at the obstacle crest and downstream, for (i) $F_0 = 0.9$, (ii) $F_0 = 0.7$, (iii) $F_0 = 0.5$, (iv) $F_0 = 0.35$, up to the point of upstream blocking of the lowest layer. The locations are marked with × in Figure 11.14. In each case (i)–(iv) the central profile at the obstacle crest is critical for mode 0. The bottom frames in each case (i)–(iv) show the minimum obstacle heights for which the bottom layer upstream is brought to rest. The single vertical line on the right denotes the initial depth of the stratified fluid (with no obstacle). (From Baines & Granek (1990): reproduced by permission of Kluwer Academic Publishers.)

Figure 11.12. The dynamically inert upper layer surmounting these profiles is not shown.

Figures 11.15 and 11.16 show horizontal velocity profiles upstream, at the obstacle crest and downstream, for steady-state flows for the parameter values marked X in Figure 11.14. For each value of F_0, the first three profiles show the velocity profiles upstream, at the obstacle crest and downstream for the value of h_m/D where the flow is just critical at the obstacle crest. The second and third sets of profiles (for each F_0) show flows for taller obstacles which deform the upstream flow, with the third set marking the minimum height at which the lowest layer becomes blocked.

Figure 11.16 shows two sets of profiles for $F = 0.3$, under the same conditions as

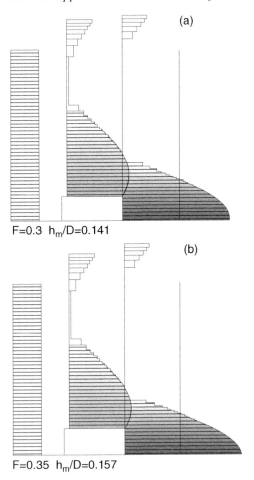

(a)

F=0.3 h_m/D=0.141

(b)

F=0.35 h_m/D=0.157

Figure 11.16 Velocity profiles from the 64-layer model with a free upper surface as for Figure 11.15, but for $F = 0.3$ with (a) $h_m/D = 0.141$, the critical height for mode 1 with the upstream flow undisturbed, and (b) $h_m/D = 0.157$, where mode 1 is critical upstream. (From Baines & Granek (1990): reproduced by permission of Kluwer Academic Publishers.)

Figure 11.15. The first denotes the profiles at the point where mode 1 is just critical at the obstacle crest, and the upstream flow is undisturbed. The second shows flow with a slightly larger h_m, where mode 1 is present upstream, but has become critical there. A further increase in obstacle height will generate mode zero upstream. The flow over and downstream of the obstacle is a combination of mode 0 and mode 1.

11.4.3 Continuously stratified flow of finite depth: long obstacles

As the analysis in §11.3.4 shows, the dynamics of continuously stratified flow over obstacles is much more complex than that of the layered flows described in §11.3.1–11.3.3. Modern numerical methods provide other approaches to solving problems of this nature. This can involve models based on values at grid points, or describing the flow as a sum of orthogonal modes. This section provides results from both approaches to the problem of inviscid incompressible two-dimensional uniformly-density-stratified hydrostatic Boussinesq flow over an obstacle, in a channel of depth D, with mean velocity U. The relevant time-dependent equations are given by (1.3)–(1.8), and may be expressed as

$$u_t + uu_x + wu_z = -\frac{1}{\rho}p_x, \qquad p_z = -g\rho, \qquad \rho_t + u\rho_x + w\rho_z = 0, \qquad (11.36)$$

and writing $u = -\psi_z$, $w = \psi_x$, we obtain

$$\psi_{zzt} + \psi_x\psi_{zzz} - \psi_z\psi_{xzz} = -\frac{1}{\rho}g\rho_x, \qquad \rho_t - \psi_z\rho_x + \psi_x\rho_z = 0. \qquad (11.37)$$

Writing

$$T = \frac{Ut}{D}, \qquad X = x/D, \qquad Z = z/D, \qquad \Psi = \psi/UD, \qquad (11.38)$$

we obtain equations for the two variables: stream function Ψ and density ρ:

$$\frac{\partial}{\partial T}\Psi_{ZZ} + \Psi_X\Psi_{XZZ} - \Psi_Z\Psi_{XZZ} = -\frac{g}{D}\left(\frac{\rho}{\rho_0}\right)_X \frac{D^2}{U^2},$$

$$\frac{\partial}{\partial T}\frac{\rho}{\rho_0} - \Psi_Z\left(\frac{\rho}{\rho_0}\right)_X + \Psi_X\left(\frac{\rho}{\rho_0}\right)_Z = 0, \qquad (11.39)$$

where ρ_0 is the mean density of the fluid. If $\Delta\rho$ denotes the range of densities and the fluid is uniformly stratified in the mean, we may write

$$\Psi(X,Z,T) = -\zeta + \sum_{n=1}^{\infty} B_n(X,T)\sin n\pi\zeta,$$

$$R(X,Z,T) = \frac{\rho - \rho_0}{\Delta\rho} = \frac{1}{2} - \zeta + \sum_{n=1}^{\infty} R_n(X,T)\sin n\pi\zeta, \qquad (11.40)$$

where

$$\zeta = \frac{z - h(x)}{D - h(x)} = \frac{Z - H(X)}{1 - H}, \qquad \text{where} \quad H(X) = h(x)/D. \qquad (11.41)$$

With

$$N^2 = \frac{g\Delta\rho}{\rho_0 D} \qquad \text{and} \quad K = \frac{1}{F_0} = \frac{ND}{\pi U}, \qquad (11.42)$$

the above equations become

$$\frac{\partial}{\partial T}\Psi_{ZZ} + \Psi_X\Psi_{ZZZ} - \Psi_Z\Psi_{XZZ} = -\pi^2 K^2 R_X,$$

$$\frac{\partial R}{\partial T} - \Psi_Z R_X + \Psi_X R_Z = 0.$$

(11.43)

In the event that the flow is steady, the lines of constant density are also streamlines, and these equations reduce to

$$\Psi_X\Psi_{ZZZ} - \Psi_Z\Psi_{XZZ} + \pi^2 K^2 \Psi_X = 0. \tag{11.44}$$

The time-dependent equations (11.43) have been integrated from a state of rest with a sudden commencement of the flow, equivalent to a sudden commencement of motion of the obstacle to constant speed U from a state of rest. The two relevant dimensionless parameters are $K = ND/\pi U$, and $H_{\mathrm{m}} = h_{\mathrm{m}}/D$, as in §10.7. The integration was carried out for a range of parameter values for sufficient time to reach a steady-state in the flow over the obstacle, avoiding any possible effects from upstream and downstream, using OpenFOAM code. Results from these integrations are shown in Figures 11.17 and 11.18.

Here the focus is on the flow over obstacles that are tall enough to partially block the upstream flow at low levels in the steady-state. Figure 11.17 shows the (K, H_{m})-diagram for the various types of uniformly stratified flow. The regions denoted "LM" are the regions where *Long-model* solutions are valid, as described in §10.7 and shown in Figure 10.25 (for hydrostatic flow over long obstacles). The curves denoted "CF" denote the conditions for which the flow is *critical* at the location of the highest point of the obstacle, in the normal hydraulic sense, for the appropriate mode. For the curves shown, mode 1 is critical in the range $1 < K < 2$, mode 2 in the range $2 < K < 3$, etc. They can be determined algebraically, but a large number of modes must be included for accurate results. In the regions between the boundary of LM and curves CF, the flow solutions are stable and there are no permanent upstream disturbances, but they are not described by Long's model.

For H-values above the CF curves, the commencement of (initially) uniform flow from rest causes upstream columnar motion (or modes) which are not transients, and a hydraulic transition to develop over the downstream side of the obstacle, rather than at the point of maximum obstacle height. For $1 < K < 2$, this motion is dominated by mode 1, and the hydraulic transition "flips the sign" of the downstream flow. As H is increased above CF, the amplitude of the upstream columnar wave increases, and when curve BF is reached, it is sufficient to bring the upstream fluid at the lowest level to rest. If H is increased beyond this curve, numerical integrations show that the upstream columnar mode remains but the depth of low-level stationary fluid upstream increases.

Some examples of these blocked flows are shown in Figure 11.18, where the

Stratified flow over two-dimensional obstacles

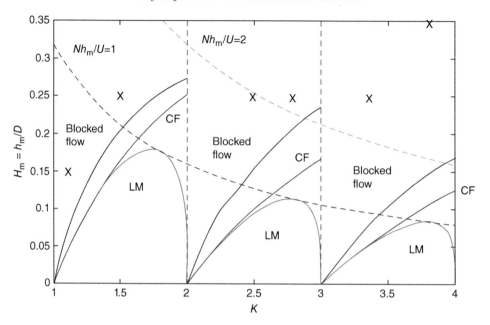

Figure 11.17 This figure shows the (K, H_m)-diagram for steady flow commenced from a state of rest over a long obstacle of finite length. The regions denoted "LM" are locations where the steady-state *Long-model* solutions are valid (as in Figure 10.25); the lines denoted "CF" are where the flow at the obstacle crest is critical, and the black lines mark the lower boundaries of the regions where low-level upstream flow is blocked. The points marked "X" indicate the parameter values for the flows shown in Figure 11.18.

flow has reached a steady state in each case. Figures 11.18a,b are in the range $1 < K < 2$, and the flow is dominated by mode 1, with a change in sign (a hydraulic transition) over the lee side of the obstacle, rather than at the point of maximum height. In 11.18b K is larger and the obstacle is higher than in 11.18a; the most conspicuous difference is that the hydraulic transition is now closer to the centre-point. In Figures 11.18c–f, $K > 2$ and the critical transition is effectively over the highest point of the obstacle in each of these flows.

In all of these blocked flows, the structure of the flow over the obstacle is dominated by mode 1, even for values of K up to 3.8. This is principally due to the low-level upstream blocking which increases the mean velocity in the overlying fluid and constrains the vertical structure. If the obstacle height is increased above the values shown in Figure 11.17, this dominant mode 1 flow structure over the obstacle continues. If the K value is increased beyond $K = 4$, the upstream blocking will increase, to the extent that the only fluid passing over the obstacle will be from the upper levels.

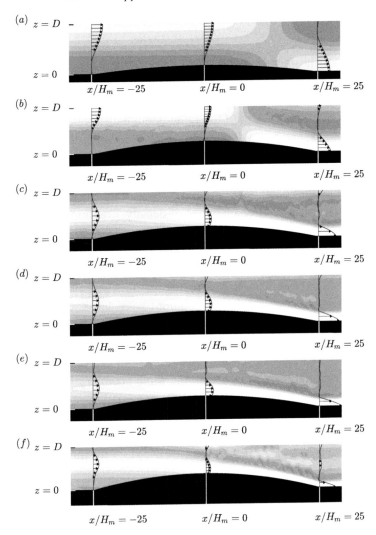

Figure 11.18 Six examples of steady-state flow over a long obstacle in uniform stratification, commencing from a state of rest. (a) $K = ND/\pi U = 1.1$, $H_m = h_m/D = 0.15$. (b) $K = 1.1$, $H_m = 0.25$. (c) $K = 2.5$, $H_m = 0.25$. (d) $K = 2.8$, $H_m = 0.25$. (e) $K = 3.3$, $H_m = 0.25$. (f) $K = 3.8$, $H_m = 0.35$. Velocity profiles immediately upstream at the obstacle peak and immediately downstream are shown, and the background shading shows the spatial flow pattern. The locations of these flows in Figure 11.17 are shown by crosses X. (Figure courtesy of L. Chan.)

11.5 Observations and numerical results for finite Nh/U in finite depth: short obstacles

We next consider the more general case of flows that are not constrained to be hydrostatic, for all values of Nh/U. We distinguish between the various observations at larger values in this light. We distinguish between the different upper boundary conditions: "infinite-depth" as in §11.6, finite-depth with a rigid upper boundary as in most experiments, and finite-depth with a reflecting upper-level critical layer, as in some numerical experiments. If they are sufficiently high, the presence of these upper boundaries may be inconsequential when the low-level flow is in hydraulic transition, and the finite- and infinite-depth cases are described together.

The first (known) observational studies of density-stratified flow past obstacles were by Browand & Winant (1972), who towed horizontal cylinders at low speeds through uniformly salt-stratified fluid. The cylinder diameters and towing speeds were both small so that the Reynolds numbers based on cylinder diameter were low, ranging from 0.42 to 11.4, implying that the flow was viscously constrained. Nh/U values (equal to $R_i^{1/2}$ in their notation) were large, ranging from 19 to 80. These ranges are in contrast to the parameter ranges of most interest here. Nonetheless, stratification effects were prominent, and triangular shaped upstream blocking was observed, with hydraulic transition to thin wakes on the downstream side. A numerical analysis of flows with the same geometry by Winters & Armi (2012) with Reynolds numbers of order 100 gave qualitatively similar but more extensive upstream flows, as expected. In both the experimental and numerical situations, the total depth of the fluid was large enough to have no impact on the nature of the observed flow near the obstacle (see §11.7 for further discussion).

11.5.1 Lee-side flows

When $Nh/U < (Nh/U)_c$ with a rigid upper boundary, numerical computations by Lamb (1994) for $1 < K < 2$ (for "Witch of Agnesi" with $a/D = 0.17$) have shown that the approach to the solution for Long's model can be extremely slow, with small departures persisting for 75 or more buoyancy periods after commencement of motion. Further, even if Nh/U is only $\sim 2/3(Nh_m/U)_c$, the transient motion contains overturning regions on the lee side, that eventually disappear or are swept downstream as time increases. Observations of such overturning regions were reported by Baines (1977), and these were taken to imply that Long's model was invalid. It now appears that, for sufficiently long obstacles, the flow approaches the solution to Long's model given sufficient time, but that the time (or obstacle length) was insufficient in these experiments. However, for flows forced by short obstacles the situation is complicated by the fact that a periodic wavetrain (and the Long

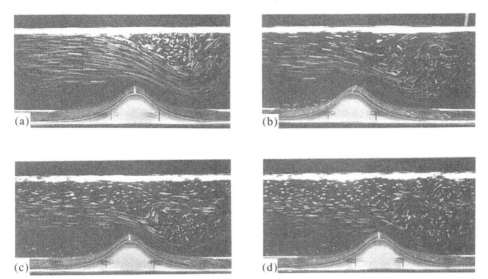

Figure 11.19 Examples of flow relative to large obstacles ("Witch of Agnesi" shape with $h_m/a = 1.26$), showing stagnant regions in finite-depth experiments with $h_m/D = 0.325$. (a) $K = 1.82$, $Nh/U = 1.86$; (b) $K = 2.1$, $Nh/U = 2.14$; (c) $K = 3.3$, $Nh/U = 3.37$; (d) $K = 4.1$, $Nh/U = 4.18$. Note the hydraulic transition character of these flows, and the similarity to the flows of Figures 10.3–10.6. Note also that upstream blocking occurs for (b), (c) and (d), where $Nh/U > 2$. (From Baines, 1977, reproduced with permission.)

model solution) may be unstable at finite amplitude (Davis & Acrivos, 1967), and significant time-dependent non-linear interactions may occur over the obstacle, as described below in connection with drag fluctuations.

If $Nh/U > (Nh/U)_c$ the development of the flow is quite different, and overturning regions appear quickly and dramatically on the lee side shortly after commencement of motion. Figure 11.19 shows some examples of near-steady-state flows with a rigid lid, for large Nh/U. As with most laboratory experiments of this type, these flows are not hydrostatic. Nevertheless, the upper stagnant regions and the hydraulic character of the main stream of flow over the obstacle are apparent, particularly for the flows in (b), (c) and (d). Lee waves of various modes are apparent in these flows, and they coexist in an untidy fashion with undular motion of the main stream that could perhaps be described in terms of an undular hydraulic jump. When $1 < K < 2$, mode 1 is apparent downstream as a stationary lee-wave mode, but when $K > 2$ the higher modes generated tend to have a forward phase tilt with height resembling the infinite-depth flows of Figures 10.3–10.6. This is attributed to the relatively larger rate of dissipation of these smaller-scale waves, as they propagate vertically and reflect from the upper boundary. Boyer & Tao (1987) describe a number of quantitative observations of lee-side properties for large Nh/U over long time intervals, but these observations must be affected by upstream motions reflected from the upstream end of the tank (these reflections are important; in the limit of very slow flows they reduce to the "squashing" phenomenon (Foster & Saffman, 1970; Castro & Snyder, 1988).

The formation of rotors in uniformly stratified flow has been simulated numerically by Sachsperger et al. (2016), obtaining results that are consistent with laboratory experiments (Baines & Hoinka, 1985) and lee-side separation as depicted in Figure 10.8. In particular, these rotors have amplitudes and scales that are consistent with linear wave theory, and with differing simulations by Doyle & Durran (2002, 2004). In flow over single mountains, the strongest rotors tend to form in flows with $Nh/U > 1$ behind mountains with vertical aspect ratios between 0.1 and 0.325. If one includes the presence of a low-level interface in the form of a boundary-layer inversion, this can introduce hydraulic effects such as hydraulic jumps, and the associated rotors become part of finite-amplitude lee waves (Sachsperger et al., 2017).

The introduction of surface friction in numerical models reduces the momentum at the lowest levels and increases the time taken for the downslope flow to develop, but does not substantially alter the final mean state over the obstacle from that of an inviscid model (Richard et al., 1989), unless the lee-side slope is small (Miller & Durran, 1991). It also restricts the downstream extent of the supercritical flow, so that increased friction has the effect of moving a stationary downstream jump closer to the obstacle.

11.5.2 Upstream motions

Steady upstream motions generated by stratified flow over an obstacle were first described by Wei et al. (1975). More detailed observations (for two-dimensional obstacles) have been described by Baines (1977, 1979b) and Castro & Snyder (1988) for a rigid upper boundary. As expected, for uniform stratification these motions consist of a sum of linear columnar modes with vertical wavenumber $n < K$. When observed at any fixed distance upstream of the obstacle, these motions take a long time to approach a steady value, as observed downstream. Generally, for any given mode there may be an initial rise and fall in the amplitude (due to "start-up"), followed by a generally steady increase. The motion may be complicated by reflections of the columnar modes from the upstream end of the tank (if it is a simple rectangle), and over a long time there may be multiple reflections from both ends. These motions are attributed to the momentum-source process (§10.2.3) if $Nh/U < (Nh/U)_c$, and to the hydraulic processes (for the most part) otherwise, but a clear-cut distinction between these two sources is not obvious from the observations

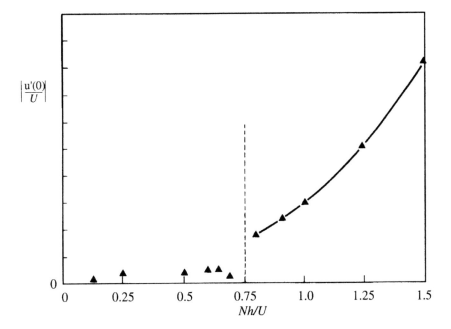

Figure 11.20 Magnitude of the steady-state upstream velocity perturbation at ground level $|u'(0)|/U$ for various Nh/U, obtained from an hydrostatic numerical study with a Gaussian-shaped obstacle. The model was approximately inviscid, with the lower boundary as a streamline. $Nh/U = 0.75$ denotes the point of overturning waves for this obstacle shape. (From Pierrehumbert & Wyman, 1985: reproduced by permission of the American Meteorological Society.)

(Garner, 1995). For instance, it is sometimes possible to choose parameters so that the momentum-source model fits the observations in the range where the source should be hydraulic.

For infinite depth, the inviscid numerical computations of Pierrehumbert & Wyman (1985) with the lower boundary as a streamline give little or no permanent upstream motion for $Nh/U < (Nh/U)_c$, but there is a small but sudden increase when this becomes an equality (Figure 11.20), due to the onset of the hydraulic process. For Nh/U above this value (0.75 for their hydrostatic Gaussian obstacle), the amplitude of the upstream motion (measured by the surface velocity) increases slowly with time to a maximum value that depends on Nh/U, as shown in Figure 11.21 for $Nh/U = 1.5$ and 2. This is consistent with the observations of §10.1 in this parameter range.

For the infinite-depth case (as described in §10.1) and the rigid upper boundary case with $K > 2$, the amplitude of these upstream motions is large enough to block

(a)

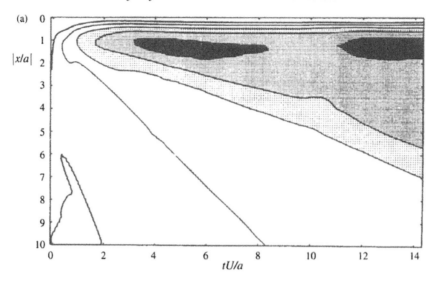

(b)

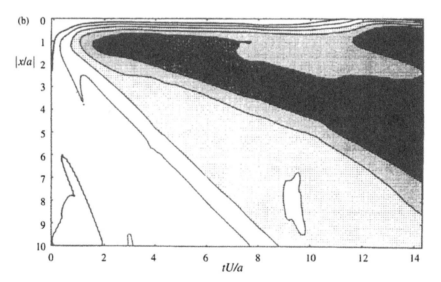

Figure 11.21 Evolution with time of the ground-level "wind field" from the model of Figure 11.20 for (a) $Nh/U = 1.5$, and (b) $Nh/U = 2$. (From Pierrehumbert & Wyman, 1985: reproduced by permission of the American Meteorological Society.)

the flow at low levels if (Baines, 1979b; Baines & Hoinka, 1985)

$$Nh/U > 2 \pm 0.2. \tag{11.45}$$

This condition has been found for a range of obstacle shapes and conditions, and was also obtained numerically by Pierrehumbert & Wyman (1985) but there may be

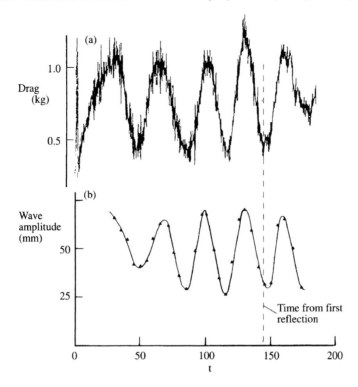

Figure 11.22 (a) The variation of the observed drag with time on a "Witch of Agnesi" obstacle ($h_m/a = 1.27$) towed at constant speed in uniformly stratified fluid of finite depth, for $K = 1.7$, $Nh/U = 1.28$. The vertical dashed line denotes the first arrival time of any wave reflected from the upstream end of the tank. (b) Amplitude of the vertical displacement in the lee wave, observed at the same time. (From Castro et al., 1990: reproduced by permission of the Royal Society.)

exceptions for strongly asymmetric shapes. When blocking occurs in this manner, the thickness of the region of blocked fluid is approximately $h_m/2$; this thickness increases with Nh/U, approaching its maximum of h_m as $Nh/U \to \infty$.

11.5.3 Drag fluctuations

One measurable quantity that gives information about the overall flow is the drag force F_D on the obstacle. Computations of the drag for flows with $Nh/U > (Nh/U)_c$ have been useful in indicating the long time taken to approach steady-state (e.g. Peltier & Clark, 1979). Direct measurements of the drag (Castro et al., 1990) for non-hydrostatic finite-depth flows with Nh/U mostly less than $(Nh/U)_c$ have shown a large-amplitude long-term periodic variation when $1.5 < K < 2$ or $2.5 < K < 3$, as shown for example in Figure 11.22. Typical ranges of observed amplitudes and

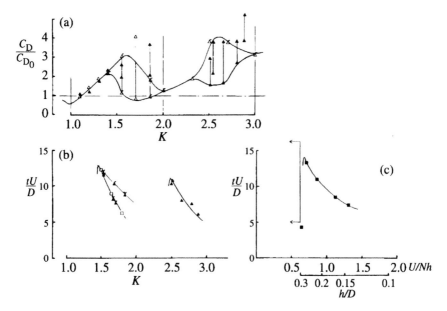

Figure 11.23 Properties of the drag variation under steady conditions shown in Figure 11.22. (a) Variation of the drag coefficient (normalised by the value for homogeneous fluid) as a function of K for △: two-dimensional fence, ▲: "Witch of Agnesi". (b) Period of the unsteadiness as a function of K for ▲: "Witch of Agnesi", ■: wide three-dimensional fence. (c) Period of unsteadiness as a function of Nh/U for $K = 1.7$, two-dimensional fence. (From Castro et al., 1990: reproduced by permission of the Royal Society.)

periods for this phenomenon are shown in Figure 11.23. Numerical simulations by Lamb (1994) and Rottman et al. (1996) have also reproduced these drag variations, with comparable periods. These show that the fluctuations also exist for $1 < K < 1.5$, that they are associated with the generation of long forced waves that propagate upstream, and in the case of $1 < K < 2$, are dependent on interactions with the forced mode 2. The period of these variations equals that of linear waves with $c_g = 0$ (see (8.36)), and tends to infinity as $K \to 1+$. This behaviour still requires suitable explanation.

11.5.4 Flows with an upper-level critical layer

Several studies (Clark & Peltier, 1984; Durran & Klemp, 1987; Bacmeister & Pierrehumbert, 1988; Scinocca & Peltier, 1991) have examined flow over obstacles with mean wind profiles where $U(z)$ decreases with height, and reverses direction at some height D. This constitutes a critical level for internal waves in steady flow. In numerical simulations, orographically generated internal waves that are not heavily dissipated (i.e. (8.105) is not satisfied) incident on these critical layers produce weakened density gradients there. These layers then tend to act as imperfect reflecting boundaries, in conformity with the discussion in Chapter 8. Flows with these layers can take a long time to approach steady state, and this time is found to increase as the Richardson number at the critical level decreases (velocity gradient increases) towards 1/4. Three examples with $Nh/U = 0.5$ and different critical-level heights are shown in Figure 11.24. These were calculated with a time-dependent layered model, that gave essentially the same results as those from a grid-point model (described in Bacmeister & Pierrehumbert, 1988).

The variation in the drag with the free-surface Froude number $F_0 = \pi U/2ND$ from these and other studies, is shown in Figure 11.25 for $Nh/U = 0.5, 0.625$ and 0.75. This drag variation may be interpreted in terms of the (F_0, H)-diagram, Figure 10.26, by tracing the change in behaviour along the lines $Nh_m/U = $ constant. In general, the "high-drag" states of Figure 11.25 correspond to the hydraulically controlled regions of Figure 10.26, and the low-drag regions to those where no hydraulic transition occurs. However, the match of these transitions is not perfect, and the Long-model solution for Figure 10.26 has zero rather than finite drag. Presumably, these differences occur because the velocity profiles of the two systems are not exactly the same, the critical layer is not a perfectly reflecting free boundary, the numerical situation is not hydrostatic in some cases, and the computations must contain some dissipation.

11.6 Application of the hydraulic model to infinite-depth flows

We now return to the discussion of flows of effectively infinite depth over topography, where the obstacle is high enough for the lee waves to be unstable $(Nh/U > (Nh/U)_c)$, and the flow undergoes a hydraulic transition, as in §10.4. The development of the preceding sections enables us to construct a *conceptual model* for infinite-depth flows over long obstacles that may be used to calculate flow properties. One does not attempt to describe the full complexity of these flows with this model, but it captures the essence of the phenomenon, and permits quantitative estimates of the important quantities in the steady state.

Flows of the "hydraulic transition" type may be realised in several ways. The basic requirement is for U and N profiles that specify a horizontal wave guide with at least one discrete vertical wave mode, and that this mode may become stationary relative to the topography. Ways of achieving this are: (i) wave-overturning to produce a reflecting upper boundary, as described above for uniform U and N; (ii) a reflecting critical level in the mean flow, as in §11.5.4 above; (iii) l^2 decreasing to zero with increasing height, as in §§8.5 and 10.5.1; and (iv) l^2 decreases but does not vanish for large z, giving a different type of hydraulic flow without wave-overturning, as described in §10.7.3. For this last case the hydraulic flow is mainly confined to the lower region, but some wave energy leaks through the upper region (Durran, 1986).

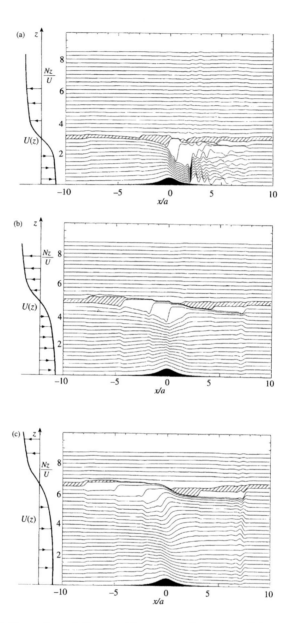

Figure 11.24 Examples of flows with a mean-flow critical layer over a long
obstacle, obtained from time integrations with a layered numerical model. The
initial velocity profile is shown on the left. The obstacle has Gaussian form
$h = h_m \exp{-x^2/a^2}$, and the critical level has initial height D. In each case
$Nh/U = 0.5$, and with F_0 defined as $F_0 = \pi U/2ND$, F_0 and the times from commencement for the three cases are: (a) $F_0 = 0.455$, $t = 18a/U$; (b) $F_0 = 0.333$,
$t = 28a/U$; (c) $F_0 = 0.238$, $t = 40a/U$. (Figures courtesy of J. Bacmeister.)

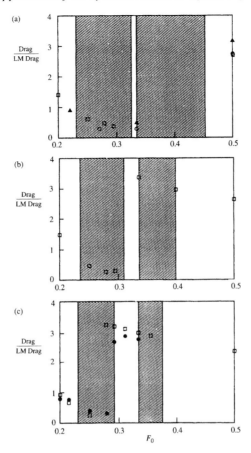

Figure 11.25 Variation of the drag (expressed in terms of the value given by Long's model for infinite depth for the same obstacle, denoted "LM Drag") for flow types such as those shown in Figure 11.24, as a function of F_0, obtained from various numerical studies. (a) $Nh/U = 0.5$, (b) $Nh/U = 0.625$, (c) $Nh/U = 0.75$. In the unshaded ranges of F_0 the flow is in a hydraulic-transition regime of Figure 11.2, and in the shaded ranges it is in a stable Long-model regime. The symbols refer to the following sources: •: Clark & Peltier (1984); ∘: Durran & Klemp (1987); ▲: Bacmeister & Pierrehumbert (1988); □: Scinocca & Peltier (1991).

Here we assume for simplicity that the values of U, N are uniform. The conceptual model is as follows. If the hydrostatic approximation is valid, Figure 10.16 shows that the value of $(Nh/U)_c$ lies in the range 0.5–1.0, with the precise value depending on the shape of the obstacle. When Nh/U exceeds this value the lee waves overturn, producing a substantial region of (relatively) homogeneous fluid over or slightly behind the obstacle. Laboratory observations (Baines & Hoinka, 1985) and numerical studies (Peltier & Clark, 1983; Pierrehumbert & Wyman, 1985; Durran, 1986; Durran & Klemp, 1987; Laprise & Peltier, 1989a, 1989b) imply that, locally at least, the lower boundary of this homogeneous region then acts as a pliant surface for disturbances in the fluid below. In other words, the overturning lee waves effectively convert the system from one of infinite depth to one of finite depth locally, with a pliant upper surface – or rather, one that may be described by such a model for present purposes (Smith, 1985). A numerical simulation of the development of such a flow is shown in Figure 10.17.

Time-dependent numerical studies (Peltier & Clark, 1983) have shown that the stagnant region is maintained by a leakage of wave energy upwards, so that this steady hydraulic model is a convenient simplification of the full picture. The depth D of the associated stratified layer is then given by Figure 10.26 for uniform initial flow, where the value of Nh/U at overturning defines a value of F_0 for the solution of Long's model in the range $0.333 \leq F_0 < 0.4$. The principal hypothesis is that, when the overturning point is reached, the appropriate solution for the steady-state flow changes from the infinite-depth Long-model solution of §10.3 to the appropriate pliant-surface Long-model solution of §11.4.2. From the conceptual viewpoint, this places the system on the (F_0, H)-diagrams of Figure 10.26 and Figure 11.14.

What happens as Nh/U increases further? For modest increases, the observational evidence described in §10.1 shows that the character of the flow remains the same. Hence, as H_m increases, the pliant-surface model with many layers may be used to calculate the response of the flow if the effective value of F_0 is known. The simplest assumption that can be made is to assume that F_0 remains unchanged as h_m increases, and there are physical reasons supporting this (Baines & Granek, 1990). Once a deep stagnant region has become established over the obstacle, further increase in h_m causes local adjustments to the flow below to keep it critical at the obstacle crest, and this change is propagated upstream by a columnar disturbance mode. By itself this process is intuitively unlikely to affect the upper-level flow.

11.7 Flows with large Nh/U: deep blocked flow, topographic drag and clear-air turbulence

Two-dimensional stratified flow over barriers becomes blocked and stagnant at low levels, if the barrier is high enough. If Nh_m/U is increased to values significantly greater than unity for a given value of h_m/D (that is not too small), the upstream columnar motions superimpose to create stationary blocked flow upstream. The depth of this blocked flow progressively increases with Nh_m/U to levels just below the obstacle height, and the presence of an upper boundary is immaterial if its level is high enough. Figure 11.26 shows two steady-state examples from numerical computations. The figure on the right ($Nh_m/U = 21.7$) shows lines of constant fluid density, with stationary flow at low levels on both sides. There is some overflow from a short distance below the obstacle peak, and an upper level non-linear wave field

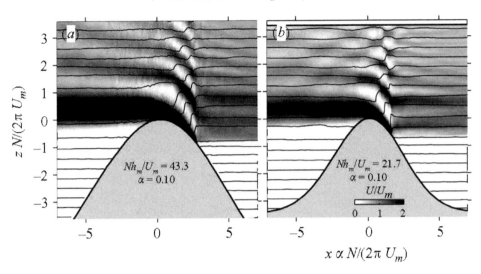

Figure 11.26 This shows two examples of steady-state flows with large Nh/U and large K. The obstacles are the same in each case, with $h_\mathrm{m}/D = 0.5$. Panel (b) on the right shows the full depth of the fluid, with $Nh/U = 21.7, K = 27.6$. Panel (a) on the left with $Nh/U = 43.3$, $K = 55.2$ only shows the central portion of the flow. Each figure is scaled with the vertical wavelength of the dominant mode. (From Klymak et al., 2010, reproduced with permission.)

similar to that shown in Figure 11.19d. The figure (a) on the left with $Nh/U = 43.3$ only shows the upper part of the obstacle, but the overall flow pattern is remarkably similar, with the waves on a smaller scale.

In such flows, the moving layer above the level of the obstacle top has a mean velocity $U_m = UD/(D - h_\mathrm{m})$, with a dominant vertical wavelength of $2\pi U_m/N$. These waves are visible in Figure 11.26, but have little impact on the overall dynamics. The overall drag is mainly determined by the difference between the upstream and downstream isopycnal levels.

Not all upstream flow profiles are dynamically possible. If the upstream flow structure is caused by flow over the obstacle, and governed by these equations, one may expect that the flow will be critical at the obstacle crest, subcritical on the upstream side and supercritical on the downstream side, and this places restrictions on possible upstream flows. If the moving upstream fluid has uniform density gradient (i.e. constant N), the simplest permissible form of continuous upstream velocity profile has a parabolic shape.

Hydraulic models of such flows have been developed by Winters & Armi (2014), and by Jagannathan et al. (2019). A conceptual two-dimensional diagram is shown in Figure 11.27. Here the flow is assumed to be hydrostatic, inviscid and Boussinesq, and the flow is confined between the upper density ρ_0 and the lower density $\rho_0 + \Delta\rho$.

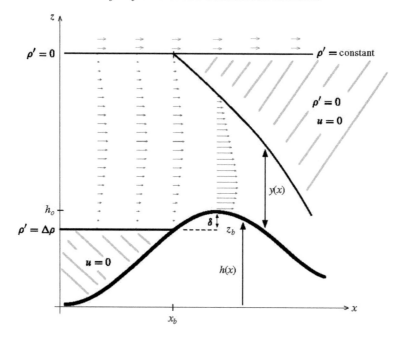

Figure 11.27 Schematic of a steady, jet-like stratified flow with uniform upstream stratification N, with large Nh_{m}/U. The flow is blocked and at rest below the upstream level $z_b = h_{\mathrm{m}} - \delta$, and accelerates over a large isolated obstacle. The uppermost streamline of the accelerating layer, which coincides with the $\rho' = 0$ isopycnal, bifurcates into two moving branches at (x_b, H). The fluid above this streamline consists of constant horizontal flow, and the two branches are separated by a stagnant homogeneous layer with $\rho' = 0$. Density and pressure perturbations, ρ' and p' respectively, are defined relative to the density of the bifurcating streamline ρ_0. (From Winters & Armi, 2014, reproduced with permission.)

The upstream fluid below the level where $\rho = \rho_0 + \Delta\rho$ is assumed to be at rest, and the fluid above the level where $\rho = \rho_0$ is at rest or in uniform motion. Each streamline in the range of intermediate densities passes over the obstacle and descends on the lee side, where the fluid above the uppermost streamline is assumed to be at rest with density ρ_0 – a configuration validated by numerical models (Winters & Armi, 2014). With each streamline being a line of constant density, the governing equations are the conservation of mass/volume, and the Bernoulli equation, which may be used to solve for the flow profiles over the topography for the given upstream conditions.

Winters & Armi identify what they call the "optimal flow", which takes the form of parabolic flow upstream, with critical flow over the obstacle peak. With the notation of Figure 11.27, the upstream flow lies between the levels z_b and $\hat{H}$, where

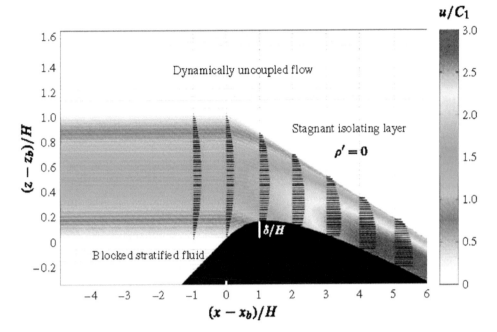

Figure 11.28 The optimal solution with an upstream, parabolic profile with $u(x_b, z) = 6C_1 \frac{z}{H}(1 - \frac{z}{H})$, with $C_1 = NH/\pi$. As in Figures 11.26 and 11.27, the flowing layer overlies blocked fluid upstream and plunges beneath a stagnant isolating layer downstream of the blocking point x_b. Above the bifurcation level $z = H$, the uncoupled flow is set equal to zero. Since the flow is hydrostatic, the solution should be viewed as a set of z profiles that only depend on the topographic height. (From Winters & Armi, 2014, reproduced with permission.)

$z_b = h_m - \delta$, and has the parabolic profile

$$u(z) = \frac{6Q}{H^3}(\hat{H} - z)(z - z_b). \tag{11.46}$$

Here $Q = NH^2/\pi$ is the total flux of fluid and $H = \hat{H} - z_b$. Time-dependent numerical simulations tend to approach this form of steady flow, an example of which is shown in Figure 11.28.

This idealised flow pattern is based on the assumption that the low-level blocked stratified fluid, and the upper-level flow (if any) has little effect on the dynamics of the concentrated flow over the mountain, and this is consistent with the observed "infinite depth" laboratory flows shown in Figures 10.3–10.6, for example.

These profiles are different from those shown in Figure 11.18, but as for most hydraulic flows controlled at the obstacle peak, the largest leeside velocity increase occurs at the lowest level, namely the fluid in contact with the ground.

Airflow past a mountain barrier of finite transverse length may become blocked and stagnant in the same manner. This blocked fluid can be maintained at constant depth in the atmosphere, and yet flow around the barrier or through deep valleys. Observations of airflow over high mountainous areas such as the European Alps (Armi & Mayr, 2007) and the Californian Sierra Nevada (Armi & Mayr, 2015) show that, in many cases, low-level air on the upstream side can be stationary (at least in the main flow direction) up to heights in excess of 2 km. This stationary air then forms the lower boundary for strong flow over the mountain barrier, with rapid descent on the lee side.

Numerical simulations by Stein (1992) with a primitive-equation hydrostatic atmospheric model have explored the flow regimes for various Nh_m/U values and obstacle heights, for uniform upstream flow and stratification in a compressible atmosphere. The main features of interest included the variation of the drag on the obstacle and the depth of the blocked flow region, as a function of Nh_m/U. Time-dependent runs with constant conditions led to (effectively) constant steady states. The nett drag force for a number of different runs (scaled with the linear form) is shown in Figure 11.29. After the initial sharp increase near $Nh_m/U = 1$ where blocking occurs, the subsequent decrease is approximately linear. The depth of the blocked region then increases approximately linearly with h_m. Integrations with upper-level critical levels in the background flow in this model also show that the drag on the obstacle may vary with the height of the critical level, consistent with results by Smith (1985).

Similar two-dimensional simulations of internal waves generated by steady flow over topography, in an atmosphere with exponentially decreasing density with height, show that the resulting increase in wave amplitude with increasing height can result in turbulence at high altitudes (Bacmeister & Schoeberl, 1989). This is due to substantial wave-breaking and overturning of initially upward-propagating lee waves, downward wave energy propagation and consequent turbulence, in the height range of 30–50 km. In most realistic situations the topography is not two-dimensional and airflow is not constant for long periods, but this process is probably a principal cause of the familiar "clear-air turbulence".

Similar phenomena to some of the above may also occur in the ocean, notably in flow through straits and channels, and a very readable summary has been presented by Legg (2012).

11.8 Details of the dynamics of downslope windstorms

Downslope flows in the atmosphere (and ocean) are potentially prone to instability and subsequent mixing. Analysis of the potential instability of various realistic downslope profiles by Jagannathan et al. (2017) show that a variety of Kelvin–Helmholz and Holmboe instabilities can exist, particularly if the overlying fluid is

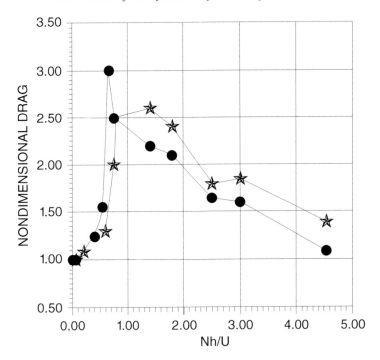

Figure 11.29 Drag force for hydrostatic flow over an obstacle as a function of Nh_m/U, scaled with the linear Boussinesq value $D_{lin} = (\pi/4)\rho_s NUh_m^2$, where ρ_s is the surface density. The parameter $S = NU/g = 0.01$, (stars), 0.034 (circles). (Adapted from Stein, 1992.)

stagnant. Such profiles are not uncommon on the lee side of mountains (Winters & Armi, 2014; Winters, 2016), and represent a possible mechanism for downslope windstorms.

"Downslope windstorms" may be defined as anomalously high surface winds that are experienced on occasions on the lee sides of mountain ranges. Such windstorms should be distinguished from "foehns" (or föhns), which is a generic meteorological term for all warm, dry downslope winds, some of which may have the same cause. Normally, windstorms are caused by the flow undergoing a hydraulic transition that takes a severe form, giving rapid supercritical flow on the lee side. This may occur in several ways, as described in §11.5. In the real world the presence of factors like upstream topography may complicate the picture, and as with most mesoscale meteorological events, good field observations of downslope windstorms are rare.

The severe form of downslope windstorms is distinguished by three significant features (Wurtele et al., 1996), as follows, though documentation is limited. First, the downslope surface wind is strong and extremely gusty, oscillating with a periodicity of several minutes between near-calm conditions. Secondly, such storms are characterised by very high drag, together with sporadic patches of clear-air turbulence throughout the depth of the troposphere, and thirdly, the disturbance is propagated rapidly up into the stratosphere.

One case that has been extensively studied and simulated is the 11th January 1972 windstorm in Boulder, Colorado, on the lee side of the Rocky mountains. For modelling and descriptive purposes this complex terrain is usually assumed to be two-dimensional on the large scale, so we also assume this here. An inferred observational section of the flow on this occasion, where the mesoscale details were obtained by instrumented aircraft, is shown in Figure 11.30. The "hydraulic transition" character of the flow is quite evident, and this is reinforced by the regions of "turbulence" at 200–300 mb and on the lee side near the ground. There are similarities with the flows with $Nh/U > 2$ in Figure 10.3 (for example), but this flow is more complex because of the extensive upstream terrain and the non-uniform wind and temperature profiles.

Surface wind observations from strong downslope flows such as those of Figure 11.31 often show variations on the time-scale of 5 to 15 minutes, in addition to the shorter period motions due to boundary-layer turbulence. An anemometer trace for the 11th January 1972 windstorm in Boulder, with a significant spectral peak in this range, is shown in Figure 11.32. Numerical studies of this situation and others by Scinocca & Peltier (1989) and Peltier & Scinocca (1990), extending earlier studies by Clark & Farley (1984) with the anelastic model of Clark (1977), have captured and described this variability, which they term "pulsations" in the wind. This model is essentially inviscid with a free-slip lower boundary, but it contains a mixing parametrisation that operates where the Richardson number falls below 1/4. This implies that stress on the downslope flow due to mixing with the overlying fluid is included, but stress on the lower boundary is not; the former is the more important in such flows (see, for example, Manins & Sawford, 1979).

These pulsations only appear after the model has been run for several hours, and the drag and downslope wind speed have increased to reach maximum, saturation levels about which they fluctuate. An example depicting horizontal wind speed for this situation is shown in Figure 11.32, and a corresponding Hovmüller diagram of the surface wind is shown in Figure 11.33. The pulsations affect the whole depth of the flow. At low levels they appear as isolated regions of wind maxima that travel downwind at near uniform speed, but slower than the wind. The mean wind speed increases with downslope distance (as expected in hydraulic flow), and so do the amplitudes of the pulsations. Their detailed properties are sensitive to the initial upstream profiles. For instance, they still appear if N and U are uniform for hydrostatic flow with $Nh/U = 1$, but at higher frequency and smaller wavelength.

These studies have shown that the pulsations only appear if $R_i < 1/4$ in the flow, and are due to growing waves caused by shear instability, centred on the

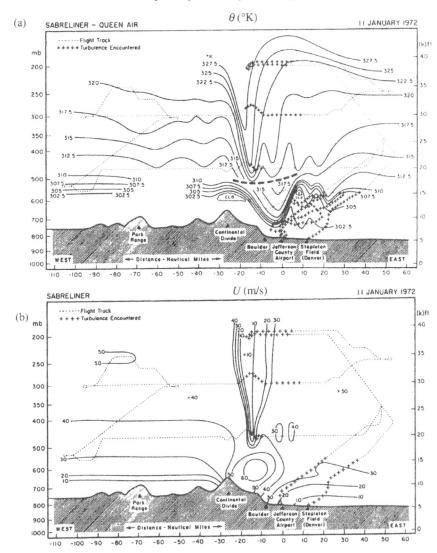

Figure 11.30 (a) Cross-section of the potential temperature field (degrees K) on an East–West line through Boulder, Colorado, as obtained from aircraft data taken between 1330 and 2000hr on 11th January 1972. Dotted lines indicate flight tracks, and crosses denote where turbulence was encountered. (b) Contours of horizontal velocity (m/s) inferred from data obtained along the same cross-section as in (a). (From Klemp & Lilly, 1975: reproduced by permission of the American Meteorological Society.)

sheared region at the top of the downslope stream. These waves are initiated over the mountain (or obstacle), slightly upstream of the crest, by some turbulence or noise in the flow. Their amplitudes then grow as they are carried downstream by the

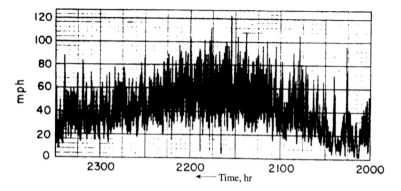

Figure 11.31 Anemometer trace (in mph) for the 11th January 1972 downslope windstorm in Boulder, Colorado from 2000 to 2300hr (time from right to left). (From Klemp & Lilly, 1975: reproduced by permission of the American Meteorological Society.)

supercritical lee-side flow to give wind pulsations that are comparable in strength with the mean. However, as waves, these disturbances propagate against this stream, so that their downstream speed is less than that of the surface wind (by a typical factor of 1/3), which passes through them. The growth of the waves is restricted by mixing with the overlying fluid. The instability mechanism may be interpreted as due to the interaction between waves on two vorticity interfaces (or regions with large U''), as described in §8.8. To date there has been no indication of these pulsations in laboratory experiments, presumably because they depend on the form of the upstream velocity and density profiles as shown (for example) in Figure 11.32; also Na/U and the Reynolds number may not be large enough. These pulsations may be contrasted with the roll waves in single-layer flow down slopes described in §2.3.4. However, numerical simulations where the disturbance is permitted to become three-dimensional (Clark & Farley, 1984) show that this two-dimensional description is probably an over-simplification of the real atmospheric phenomenon.

Another possible mechanism is the response to highly sheared upstream flow in the lower atmosphere. If the buoyancy frequency N is constant but the low-level flow is highly sheared, the flow on the downstream side of the obstacle can be very different from that for uniform flow. Linear lee-wave analysis shows that a background upstream profile of the form

$$u(z) = U_0 \tanh(z/d), \qquad N = \text{constant}, \qquad (11.47)$$

where U_0 is constant and d is somewhat larger than the obstacle height, can give lee-side surface velocities and lee waves of significantly larger amplitude than that for uniform flow (Lott, 2016). This is particularly pronounced when $J = (Nd/U_0)^2 \geq 1$, where linear solutions are suggestive of downslope windstorms.

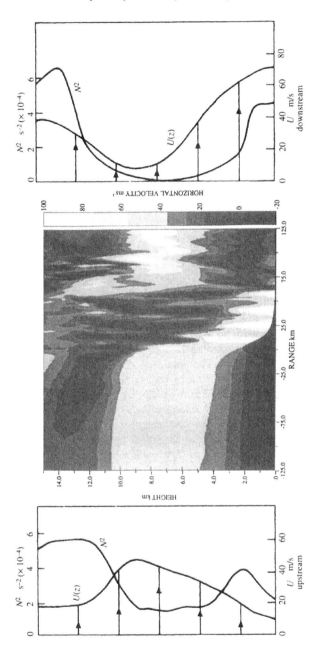

Figure 11.32 Horizontal wind speed at a particular time ($t = 275$ min) in a numerical simulation of the downslope windstorm depicted in Figures 11.30 and 11.31. A sequence of wind speed maxima on the downslope side of the obstacle is evident. Simulated wind and N^2 profiles on the upstream and downstream sides of the topography are shown. (Adapted from Scinocca & Peltier, 1989; Peltier & Scinocca, 1990.)

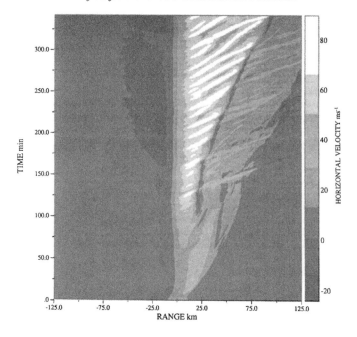

Figure 11.33 Hovmüller plot of surface wind speed for the simulation of Figure 11.31. Continuous downslope progression of the wind maxima and minima at uniform speed is evident. (From Scinocca & Peltier, 1989: reproduced by permission of the American Meteorological Society.)

11.9 Flow across valleys

In the preceding sections we have been concerned with flow over an obstacle with a single peak. We next consider flow over valleys ("valley", "depression" and "hole" are here used interchangeably), which may or may not lie between two peaks. This geometry presents a new and important question with practical relevance, namely, under what circumstances is the fluid within a valley stagnant, or completely swept out ("sweeping flow") by the flow across the topography?

The situation with hydrostatic one-layer flow over depressions was described in §2.3.2, where the flow could remain subcritical if $F_0 < 1$ or supercritical if $F_0 > 1$, regardless of the depth of the hole (the same applies to two-layer flows where $r = 0$, see (2.58)). In these situations the valley is always "swept out". For the opposite extreme of initially uniform U and N profiles, a number of different possible situations present themselves, but before discussing these it is useful to consider the situation with solutions of Long's model as a theoretical preliminary.

11.9.1 Solutions of Long's model over valleys

For these solutions to be valid, all streamlines must begin at upstream infinity where (for hydrostatic flow) N and U are uniform with height. This excludes completely periodic topography, for example. Solutions over valleys for general non-hydrostatic flow when Nh/U is small are approximately equal to the linear boundary perturbation solutions, which are the same as those for the positive topography of the same shape, but with reversed sign for the vertical displacement and stream function perturbation. Hydrostatic solutions for finite depth for valleys or "holes" are given in §10.7, and the limits to their applicability (where streamlines in the flow become vertical, so that $u = 0$ somewhere) are given in Figure 10.25 for a rigid upper boundary, and Figure 10.26 for a pliant upper boundary. For the case of infinite depth, which is our primary concern here, the relevant equations are given in §10.3. Specifically, for a valley where $h(x)$ is everywhere negative, $h(x) = h_-(x) \le 0$ say, then the corresponding function $f(x) = f_-(x)$ in (10.48) is given by

$$f_-(x) = f_+(-x), \tag{11.48}$$

where $f_+(x)$ is the f-function obtained from (10.50) with $h(x) = h_+(x) = -h_-(-x)$. This means that the displacement field over a valley is an inverted and reversed form of that over a mountain with the same shape. The streamlines may be obtained from the same diagrams by inverting them (by rotation about the y-axis), and placing the obstacle at a level where it coincides with a streamline. An example with the negative "Witch of Agnesi" is shown in Figure 11.34, which may be obtained from Figure 10.15. There is only one symmetric shape in this figure, but any of the other streamlines could be taken as the topography instead. The direction of flow in this diagram is again from the left.

The criterion for vertical streamlines in the valley solutions is the same as for flow over the same reversed mountain, and hence the limits for validity of the solutions must be the same. In particular, for hydrostatic flow over the "Witch of Agnesi" valley, the limiting value of Nh/U is $|Nh/U|_c = 0.85$, and the values for the asymmetric forms of this profile ((10.53) with h_m negative) may be obtained from Figure 10.16 if a_u/a_d is replaced by a_d/a_u. These give $0.5 < |Nh/U|_c < 1.0$, depending on the obstacle shape.

Figure 11.34 shows (from the streamline spacing) that the velocity is a maximum on the downslope and a minimum on the upslope. From (10.51) this implies a minimum pressure perturbation on the downslope and a maximum on the upslope, and hence there is a positive drag on the valley. This drag is given by (10.52), and substitution using (11.48) shows that the drag on $h(x)$ is the same as that on $h_+(x) = -h_-(-x)$.

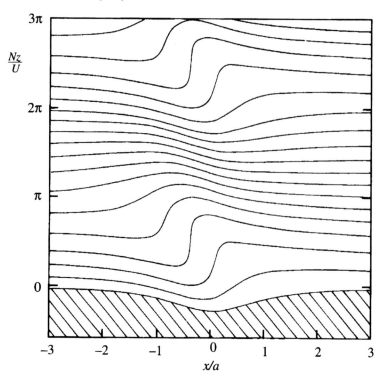

Figure 11.34 The hydrostatic Long-model solution for flow of infinite depth over a valley of "Witch of Agnesi" shape with $Nh/U = -0.85$ (cf. Figure 10.15).

11.9.2 Flow regimes for initially uniform U and N profiles

We consider flows commenced from a state of rest in the same ways as previously (§§2.1, 2.2 and 10.2.1), and assume that the flow is approximately hydrostatic unless otherwise specified. Four different valley geometries or cases may be identified. We take the simpler geometries first, so that in Cases 1 and 2 the topography is entirely negative with $h(x) \le 0$, although the experiments and numerical studies have been made with the more complex topography in Cases 3 and 4.

Case 1. Isolated depression without heavy fluid
We consider an initial state with steady horizontal flow, and allow the depth of a valley with a profile resembling that of Figure 11.34 to increase slowly with time from zero. The resulting flow may then be approximated by a succession of steady states. Under these circumstances the lower boundary remains a streamline, and the Long-model solution applies up to the point where $|Nh/U| = |Nh/U|_c$. Here the flow field has the form shown in Figure 11.34. At present there are no direct observational or numerical studies of what happens if $|Nh/U|$ is increased beyond

this point, and we proceed on an *a priori* basis. Straightforward use of the pliant-surface model of §§11.4.2 and 11.5 gives unrealistic flow patterns, and hence the model is not useful over valleys.

A guide to the way in which the flow does adjust when $|Nh/U| > |Nh/U|_c$ may be obtained from Kimura & Manins (1988), as described in more detail below. When the lee waves become steep and break at low levels, as shown in incipient fashion in Figure 11.34, most of the form drag on the fluid can no longer be carried to upper levels by the waves. In the breaking process, this drag must act via the associated pressure field to retard the lower-level fluid below the breaking region, causing stagnation on the upslope side of the valley, where the fluid velocities are smallest. This process causes fluid of the maximum density entering the valley to accrete to this stagnant region, increasing its size so that it eventually extends across the bottom of the valley. When it has formed, the mean depth of this region may be expected to be given (very roughly) by $|h_m - h_{mc}|$, where h_{mc} denotes the depth for the Long-model limit ($0.85U/N$ for the "Witch of Agnesi").

Under certain conditions this valley flow may be controlled by lee-side flow separation. If the downslope side is sufficiently steep so that $NA_d/U < \pi$, where A_d is the width of the downslope side of the valley, boundary-layer separation occurs there (see §10.1). If also $NL/U < \pi$, where L is the width of the valley, the "wake" of the downslope side will dominate the valley, and the separation region will occupy most of it. If instead $NL/U > \pi$, when separation occurs its effect will be localised within the valley. When boundary-layer separation does not occur, stagnation on the upslope side of the valley is another form of "post-wave" separation (see Figure 10.8). In summary, for Case 1 sweeping flow occurs for $|Nh/U| < |Nh/U|_c$, and otherwise some stagnant fluid is expected in the valley.

Case 2. Isolated depression with heavy fluid
We next discuss flow over the same topography as Case 1 ($h \leq 0$), but with the initial condition that the flow over the topography is commenced from rest with uniform N. The fluid initially in the valleys is therefore heavier than that at level $z = 0$. We first discuss the situation for an inviscid fluid, where the separating flow state

$$u = \begin{cases} U, & z > 0, \\ 0, & z < 0, \end{cases} \qquad (11.49)$$

is a possible steady solution for this geometry. Under what conditions is this flow situation realised? Immediately following the sudden commencement of motion, the fluid in the valley moves downstream so that its upper boundary or "interface" tilts upward in the downstream direction, and some fluid may be lost at the downstream end. The buoyancy forces become manifest after a time of $O(N^{-1})$, and if the initial heavy fluid has all been swept out of the valley by this time, the resulting flow in the valley will evolve in the same way as for Case 1. For a valley of width $2A$, the criterion for this to occur is

$$2NA/U \ll 1, \tag{11.50}$$

which also implies that boundary-layer separation should not occur. On the other hand, if $2NA/U$ is large, nearly all the heavy fluid will remain in the valley, and the flow is approximated by (11.49). Between these two extremes, where $2NA/U \sim 1$, if enough fluid is swept out of the valley initially so that the depth of penetration exceeds h_{mc} wave-overturning as in Case 1 should ensue, resulting in the accumulation of stagnant fluid on the upsloping lee side of the valley with the density of the level $z = 0$.

   In practical situations with real fluids, viscous effects, turbulence and mixing are important in these situations. An experimental example of the near-steady state of one of these flows is shown in Figure 11.35, where $2NA/U = 6.7$, and the initial heavy fluid lost is negligible. Turbulent stresses and mixing occur at the boundary between the overlying and the heavy fluid, and the latter is located more to the upwind side of the valley. Over a long period of time, these turbulent processes must continuously erode the heavy fluid and remove it from the valley, so that the flow eventually evolves toward Case 1 where the fluid at the bottom of the valley has the density of the level $z = 0$. In summary, for inviscid fluids state (11.49) is established if the flow is commenced sufficiently slowly, or $2NA/U$ is large. For real viscous fluids at large Reynolds numbers, (11.49) is approximately established in most practical circumstances, but turbulent mixing should erode and remove the heavy fluid given sufficient time, so that Case 1 applies.

Case 3. Periodic valleys

Purely sinusoidal topography may be regarded as a sequence of hills with valleys in between, or as a sequence of valleys with no hills – the choice of the reference level makes no difference. However, the choice of initial conditions is again important. If there is no heavy fluid in the valleys and the density on $z = h(x)$ is initially uniform, this system is a periodic form of Case 1, with similar properties. If instead there is heavy fluid in the valleys initially, the situation is a periodic form of Case 2, with the important difference that fluid swept out of one valley will enter the next. Hence heavy fluid is not lost from a valley. Numerical studies of this system have been made by Kimura & Manins (1988) for infinite depth (who repeated in more detail the work of Bell & Thompson (1980) for finite depth) with cosine topography with $L/h_m = 16$ where L is the wavelength, and using both inviscid and dissipative models. These studies support the heuristic interpretation given above for Cases 1 and 2. Their experiments had an initial impulsive start to the motion,

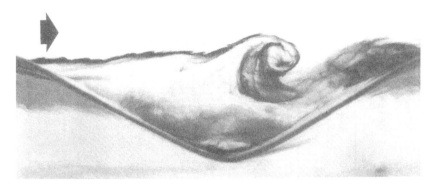

Figure 11.35 A laboratory example of "steady" flow over a valley filled with heavy fluid, with $|Nh/U| = 1.5$, $h_m/A = 0.45$. Dark fluid at the bottom of the valley denotes initial heavy fluid. A source of dye at the upwind side of the valley marks the line of separation between flow over the valley and flow inside it. Note the turbulent mixing and entrainment region over the upslope side.

and the mean motion was then allowed to decrease due to bottom drag (whereas Bell & Thompson's was not). This complicates the interpretation for our present discussion, which assumes constant U values.

The development of the flow after fairly short times, obtained from the inviscid model, is shown in Figure 11.36. In panel (a) is shown the flow state for $|Nh/U| = 0.84$, at $t = 1.3L/U$. The flow is evolving slowly with time and shows rapid downslope flow, steeply ascending flow over the central part of the valley, and weak or stagnant flow on the upsloping lee side, so that the overall pattern is similar to that of Figure 11.34. The beginning of the accumulation of stagnant flow may be seen. Figure 11.36b shows inviscid flow for $|Nh/U| = 2.61$ at $t = 0.38L/U$. The descending and ascending regions of flow have moved to the upstream side of the valley, and the development of the stagnant region is much more advanced. This flow is of "post-wave separation" type (see §10.1). The situation for much larger times is shown in Figure 11.37, with results from the dissipative model. These numerical experiments give "sweeping flow" if

$$|Nh/U| < 0.45, \text{ at} \qquad t = 3.4L/U, \qquad (11.51)$$

i.e. after the fluid has crossed $3\frac{1}{2}$ valleys. This value decreases with time to about 0.33 at $t = 6.8L/U$. After allowing for the slow run-down in mean flow speed, this change seems to be due to the natural tendency for the heavy fluid to accumulate in the valleys. Weakly stratified stagnant fluid is visible in the valley in all three cases in Figure 11.37, where $|Nh/U|$ exceeds the criterion (11.51). Except for flows where $|Nh/U|$ is close to the value for sweeping out, the depth of the stagnant fluid

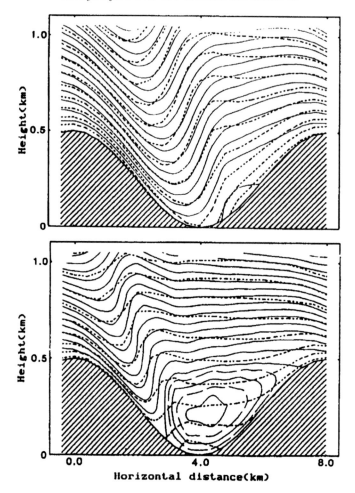

Figure 11.36 Examples of inviscid hydrostatic numerical simulations of flow over periodic sinusoidal valleys (wavelength L), shortly after commencement from rest. Both cases are producing unstable waves. Solid lines denote instantaneous streamlines, and dashed lines potential density. (a) $|Nh/U| = 0.84$, $t = 1.3L/U$; (b) $|Nh/U| = 2.61$, $t = 0.38L/U$. In the latter, stagnant fluid has already formed without involving friction or overturning directly. (From Kimura & Manins, 1988: reproduced by permission of Kluwer Academic Publishers.)

is about $0.7h_\mathrm{m}$, with a slight increase as $|Nh/U|$ increases, as these examples show. The result for the slowest flow (Figure 11.37c) resembles (11.49), as expected, and the dissipative numerical results show that the flow tends toward this limit for large times, unless $|Nh/U| < 0.33$.

Case 4. Finite number of valleys between hills
This is a truncated version of Case 3, with a finite number of periodic valleys where

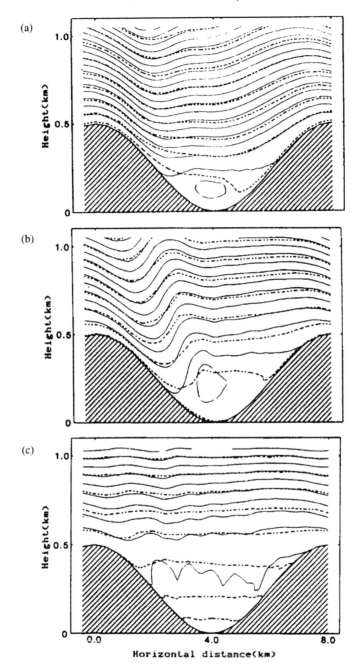

Figure 11.37 Simulations of the same flow as that in Figure 11.36 but using a model that parametrises turbulent dissipation, and at much larger times ($t = 3.4L/U$ for each). (a) $|Nh/U| = 0.69$; (b) $|Nh/U| = 1.38$; (c) $|Nh/U| = 2.44$. Stagnant fluid has formed in each case. (From Kimura & Manins, 1988: reproduced by permission of Kluwer Academic Publishers.)

the topography approaches the same horizontal level ($h = 0$) at each end. Again the flow is commenced from a state of rest with uniform stratification. There is an extra variable here, namely the position of the level $h = 0$ relative to the valleys. The behaviour when this is the level of the tops of the valleys may be inferred from the discussion for Cases 1 and 2. The situation where $h = 0$ is the level of the bottom of the valleys, with $h \to 0$ as $|x| \to \infty$, was studied experimentally by Bell & Thompson (1980) and Kimura & Manins (1988), but their observations were limited to short times and central locations so that they effectively simulated periodic topography. Here we discuss this finite "Loch Ness monster" topography, which is interesting in itself.

We first consider the case of a *single* valley between two hills with $h(x) \geq 0$. When flow over two identical hills is commenced from rest, if $Nh/U < (Nh/U)_c$ for a single hill, the combined Long-model solution will apply in the steady state; the valley will be swept out, and there will be little interaction between the flow due to the two obstacles (assuming that there is no significant lee-side flow separation). If instead $Nh/U > (Nh/U)_c$, laboratory experiments indicate that the flow in the valley develops in a manner similar to that described for Case 1: a quantity of slow-moving or stagnant fluid develops in the bottom of the valley due to the pressure field associated with the form drag on the low-level fluid. Over the second obstacle, the hydraulic transition develops in the same manner as for a single obstacle, since it is not affected by anything downstream. This hydraulic transition will generate upstream columnar modes, which will promote stagnation in the valley. This situation may be compared with that of a single layer flowing over two obstacles, as discussed in §3.4. In each of these systems, the drag due to the flow over the first obstacle reduces the overall flow force, but in a different manner: by forming stagnant fluid in the valley in the stratified case, and by producing a periodic wavetrain (possibly with breaking waves) between the obstacles in the single-layer case.

If more than two obstacles with the same height are placed in succession, for $Nh/U > |Nh/U|_c$ the flow should develop to be similar in each valley, with the flow not attaining a critical state over any obstacle crest except the last, where the hydraulic transition occurs. In summary, the principal feature of these flows is that the flow over the final obstacle is different from that over the others, in the same manner as for a single layer.

Another series of experiments by Tampieri & Hunt (1985) has examined the effect of valley width in stratified flow over two identical stationary obstacles of small height, in a "race-track" channel. These obstacles lay within the laminar viscous boundary-layer, so that the external velocity profile was the non-uniform highly sheared boundary-layer profile. The h_m/A values used were 1/2 and 2/3, so that the flow was not hydrostatic. The principal factor varied in these experiments was the distance L between the two obstacles, and although comparison with the previous results can only be qualitative the results are consistent. Generally, greater penetration into the valley was observed for $|Nh/U| < 1$ than for $|Nh/U| > 1$, but the maximum value of $|Nh/U|$ for which complete sweeping was obtained was found to increase with NL/U. Penetration increases with valley width, and is largest for wide valleys where $NL/U \gg 1$.

Case 5. Elevated valleys
Where valleys lie between two mountains (or two mountain ranges), the heights on each side may be unequal, and the flow within the valley may be subject to lee waves and the effects of low-level blocking upstream of the mountain on the upstream side. If the heights of both mountains are small so that $Nh/U < 1$ for each, the motion within the valley will be dominated by the lee waves of the upstream mountain. But if $Nh/U > 1$, non-linear and blocking effects may be important.

   If the lee-side mountain is the higher, upstream blocking by this mountain will tend to reduce the flow within the valley, and minimise the effects of the (lower) upstream mountain. But if the upstream mountain is the higher, particularly if $Nh/U > 2$, the situation is very different. Here, in an environment of (approximately) uniform N, we may expect blocked fluid upstream of this mountain, and a strong downflow on the lee side, entering the valley. However, the flow within the valley is then very dependent on the height of the second, downstream mountain.

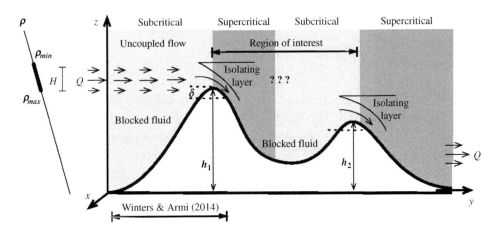

Figure 11.38 Schematic diagram of stratified flow over a pair of two-dimensional ridges. The flow is hydraulically controlled at the two crests, and so undergoes a transition from super- to subcritical flow between the ridges. The blocking heights, a distance of δ beneath the ridge crests, are indicated by the dashed lines. The nature of the internal hydraulic jump is unknown and indicated with the question marks. (From Winters, 2016, Fig. 1, reproduced with permission.)

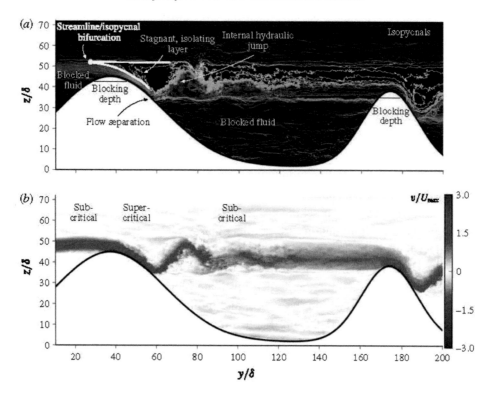

Figure 11.39 (a) Instantaneous isopycnals along the centre-line of the numerical simulation in the quasi-steady regime for $\Delta = 0.84$, where $\Delta = (h_1 - h_2)/H$, and these terms are defined in Figures 11.37, 11.38. (b) The corresponding normalised speed (positive in the main stream). (From Winters, 2016, Fig. 4, reproduced with permission.)

Figure 11.38 shows a schematic of possible quasi-steady flow over a valley with a high mountain upstream. The upstream flow is assumed to have the same form as in Figure 11.27, with a concentrated upstream flow above a stratified stationary blocked region. This flow stream flows through a hydraulic transition over the first mountain and descends in the lee, surmounted by effectively stationary, inactive fluid. However, the presence of the lower second mountain downstream controls the flow in the valley and results in a blocked quasi-stagnant region within the valley, below the second mountain height. The flow over the first mountain then passes through a transition from supercritical to subcritical flow, and continues at a reduced, approximately constant height to pass over the second mountain, where the flow is (notionally) governed by a second critical condition to supercritical flow downstream (Winters, 2016). A numerical simulation of this is shown in Figure 11.39. As far as the valley is concerned, despite all the activity above, the fluid near the ground is relatively stationary, though this does depend on the flow being two-dimensional.

These simulations show the character of the flow in some detail. Assuming that the upstream flow has the form shown in Figure 11.26, this flow pattern seems to be relatively insensitive to the strength of this upstream flow. The volume flux of this of course depends on the history of how this flow was developed in the first place.

12

Stratified flow over three-dimensional topography: linear theory

> Oh ye'll tak' the high road, and I'll tak' the low road,
> And I'll be in Scotland afore ye.
>
> ANON., *The Bonnie Banks o' Loch Lomond.*

We now proceed to the more general case of three-dimensional topography. Here the distinction between finite and infinite depth is less significant. Disturbances caused by an isolated topographic feature spread out laterally as well as vertically, so that the presence of an upper boundary or interface has a minor effect locally unless it is at a low level. Hence we concentrate on infinite-depth systems, although low-level variations in N and U may be important.

How relevant are the two-dimensional flows of the preceding chapter to flows past three-dimensional topography? We may address this question by considering the flow over a three-dimensional obstacle that is very long and approximately uniform in the direction transverse to the flow. Specifically, let this obstacle have breadth $2B$, and downstream length $2A$ where $A \ll B$, in fluid with buoyancy frequency N that is set in motion with uniform initial velocity U. This starting flow produces internal waves with a dominant vertical wavenumber-scale of $n = N/U$ (see §8.11), and if the horizontal length-scale is longer than the vertical so that the sum of the squares of the horizontal wavenumbers $k^2 + m^2 \ll N^2/U^2$, the horizontal component of the group velocity is of magnitude $N/n \approx U$.

Waves with $n < N/U$ have larger group velocity, and are able to propagate against and across the stream. In the central region of the obstacle, the initial development of the flow is the same as for two-dimensional topography, but in the regions near the ends the development is three-dimensional. Since the characteristic group velocity of the significant waves is equal to U, the influence of the ends of the obstacle reaches the centre in a time of order B/U. Up to this time, the flow in the central region is approximately two-dimensional, and afterward it evolves toward its three-dimensional form. In the atmosphere for example, if $B=150$ km and $U=10$ m/s, this time is about 4 hours.

As a general rule therefore, two-dimensional solutions are relevant to nearly two-dimensional topography for a time that depends on the location and the geometry. After this time the arrival of internal waves carrying information about the more distant topography will cause the flow to approach the fully three-dimensional steady state. For isolated three-dimensional topography (i.e. occupying a finite region), in this final steady state the flow is found to be undisturbed at locations that are sufficiently far upstream (from both observations and theory) for all Nh/U, because of the lateral spreading of disturbances. This contrasts with the two-dimensional situation (see Chapter 10) and does not apply downstream, as seen below. For the remainder of this chapter we concentrate on the nature of these steady (or approximately steady) flows.

The general character of the observed flow past three-dimensional obstacles is quite complex. For this reason we do not follow the order of the previous chapter and instead discuss first some theoretical models that have been applied to these flows. We may then use these to interpret the observations described later. As for two-dimensional obstacles, if the obstacle height is sufficiently small we may expect linear perturbation theory to be relevant, and we begin by describing its implications for three-dimensional obstacles. As used here, the terms "obstacle", "hill" and "mountain" are normally equivalent, with the particular choice based on poetic relevance.

12.1 Linear theory for small-amplitude topography, with the lower boundary as a stream surface

Linear perturbation theory for flow over three-dimensional obstacles is a natural extension of that for two-dimensional obstacles described in §10.2.1. The governing equations are given in §8.11, and are specifically (8.121)–(8.125). In this chapter the undisturbed wind velocity is uniform and aligned with the x-axis (unless otherwise specified), so that U is constant and $V = 0$. If the lower boundary is $z = h(x, y)$, the associated boundary condition is again

$$w' = U\frac{\partial h}{\partial x}, \qquad \text{at } z = 0, \quad t > 0. \tag{12.1}$$

The upper boundary condition is again a radiation condition that specifies that there is no incoming energy, as discussed in §8.11. Confining attention to steady solutions, for which the vertical displacement of a streamline η from its level in undisturbed flow is given by $w' = U\partial\eta/\partial x$, the governing equations are

$$U\frac{\partial u'}{\partial x} = -\frac{1}{\rho_0}\frac{\partial p'}{\partial x}, \quad U\frac{\partial v'}{\partial x} = -\frac{1}{\rho_0}\frac{\partial p'}{\partial y}, \quad U\frac{\partial w'}{\partial x} = -\frac{1}{\rho_0}\frac{\partial p'}{\partial z} - \frac{\rho' g}{\rho_0},$$

$$U\frac{\partial \rho'}{\partial x} = -\frac{d\rho_0}{dz}w', \quad \frac{\partial u'}{\partial x} + \frac{\partial v'}{\partial y} + \frac{\partial w'}{\partial z} = 0. \tag{12.2}$$

From these we obtain the equation for η or w'

$$\frac{\partial^2}{\partial x^2}\left(\frac{\partial^2}{\partial x^2} + \frac{\partial^2}{\partial y^2} + \frac{\partial^2}{\partial z^2}\right)\eta + \frac{N^2}{U^2}\left(\frac{\partial^2}{\partial x^2} + \frac{\partial^2}{\partial y^2}\right)\eta = 0. \tag{12.3}$$

In order to justify this linearisation of the equations, we require $|\mathbf{u}'/U| \ll 1$, as in §10.2.1, and these conditions are again given by scale analysis. If A and B are the length-scales of the topography in the x- and y-directions respectively, then $w' \sim Uh_m/A$, so that $|w'/U| \ll 1$ requires $h_m/A \ll 1$, as before. The steady-state forms of (8.121), (8.122) then give $p \sim \rho_0 U u'$, $v' \sim Au'/B$, so that (8.125) gives

$$u'/U \sim \frac{B^2}{A^2 + B^2}Nh_m/U, \qquad v'/U \sim \frac{AB}{A^2 + B^2}Nh_m/U. \tag{12.4}$$

Hence the linearisation is justified if $Nh_m/U \ll 1$ and $h_m/A \ll 1$, and there are no restrictions on the value of B. These conditions are the same as those in §10.2.1.

12.1.1 Flow over periodic topography

For periodic topography of the form

$$h(x, y) = h_m \cos kx \cdot \cos my, \tag{12.5}$$

the solution to (12.3) for the vertical displacement η is

$$\eta = \begin{cases} h_m \cos(kx + nz) \cdot \cos my, & 0 < k < N/U, \\ h_m \cos kx \cdot \cos my \cdot e^{-nz}, & k > N/U, \end{cases} \tag{12.6}$$

(cf. (10.11)) where from (8.129) with ω and V zero,

$$n = \left(1 + \frac{m^2}{k^2}\right)^{1/2}\left|\frac{N^2}{U^2} - k^2\right|^{1/2}. \tag{12.7}$$

The sign of n has been chosen so that each Fourier wave component has upward group velocity. The corresponding pressure field is given by

$$p' = -\left(\frac{N^2/U^2 - k^2}{1 + m^2/k^2}\right)^{1/2}\rho_0 U^2 h_m \sin(kx+nz)\cdot\cos my, \qquad 0 < k < N/U, \tag{12.8}$$

and with u' given by $u' = -p'/\rho_0 U$, the pressure has a maximum and $u = U + u'$ a minimum on the upstream face of each periodic obstacle, at the same location as for the two-dimensional topography (given by the present expressions with $m = 0$), but with reduced amplitude.

The horizontal drag force on the topography for inviscid flow, the "form drag" $\mathbf{F}_D$, is given by

$$\mathbf{F}_D = - \iint \nabla p h(x, y) dx dy = \iint p \nabla h dx dy, \tag{12.9}$$

where p is the pressure on the surface. For the flow represented by (12.6) this gives, to leading order and per unit horizontal area,

$$\mathbf{F}_D = \begin{cases} \frac{1}{4} \rho_0 U^2 h_m^2 k^2 (k^2 + m^2)^{-1/2} \left(\frac{N^2}{U^2} - k^2 \right)^{1/2} \hat{\mathbf{x}}, & 0 < k < N/U, \\ 0, & k > N/U, \end{cases} \tag{12.10}$$

(cf. (10.13)), where $\hat{\mathbf{x}}$ is the unit vector in the x-direction. Note that the limit $k \to 0$ is discontinuous, in that the limit of the area-average denoted by (12.9) is finite, but the drag on topography with no x-variation at all must vanish. If the topography is not symmetric about the wind direction, the drag force usually has a transverse component that may be termed "lift", copying the terminology from aerodynamics. This component may be in either direction, depending on the topographic shape. For example, if the periodic topography (12.5) is aligned at an angle α to the wind, the flow will consist of the sum of two solutions of the form (12.6), with velocities $U \cos \alpha$ along one axis of symmetry for the first and $U \sin \alpha$ along the other for the second. If the topography is written in the form

$$h = h_m \cos kx' \cos my', \tag{12.11}$$

where the x'-axis makes angle α with the x-axis (the wind direction), from (12.10) the total drag force per unit area is

$$\mathbf{F}_D = \frac{\rho_0 U^2 h_m^2}{4(k^2 + m^2)^{1/2}} \left\{ \left[k^2 \left(\frac{N^2}{U^2} - k^2 \right)^{1/2} \cos^2 \alpha + m^2 \left(\frac{N^2}{U^2} - m^2 \right)^{1/2} \sin^2 \alpha \right] \hat{\mathbf{x}} \right.$$

$$\left. + \left[k^2 \left(\frac{N^2}{U^2} - k^2 \right)^{1/2} - m^2 \left(\frac{N^2}{U^2} - m^2 \right)^{1/2} \right] \sin 2\alpha \hat{\mathbf{y}} \right\}, \tag{12.12}$$

when both $k, m < N/U$, and the form for other values of k, m may be readily inferred from (12.10). The component in the y-direction is the "lift" component.

12.1.2 *General solution*

For an obstacle of general shape, the solution to (12.3) may be obtained from a Fourier integral over x and y. If we define

$$\hat{h}(k,m) = \int_{-\infty}^{\infty} \int_{-\infty}^{\infty} h(x,y)e^{-i(kx+my)}dxdy, \qquad (12.13)$$

then

$$\eta = \frac{1}{4\pi^2} \int_{-\infty}^{\infty} \int_{-\infty}^{\infty} \hat{h}(k,m)e^{i[kx+my+n(k,m)z]}dkdm, \qquad (12.14)$$

where the radiation condition requires

$$n(k,m) = \begin{cases} (k^2 + m^2)^{1/2} \left(\frac{N^2}{U^2} - k^2\right)^{1/2} k^{-1}, & 0 < |k| < N/U, \\ i(k^2 + m^2)^{1/2} \left(1 - \frac{N^2}{k^2 U^2}\right)^{1/2}, & |k| > N/U. \end{cases} \qquad (12.15)$$

For simplicity we confine attention to obstacles that are symmetric about two perpendicular axes. If the direction of the mean wind is not aligned with one of these axes, it may be resolved along them, so that the resulting solution is the vector sum of two solutions each with a component of the mean wind along an axis, as noted above for periodic topography.

For some purposes it is necessary to consider particular topographic shapes, and we will describe some flow properties with the versatile form

$$h(x,y) = \frac{h_m}{\left[1 + \left(\frac{x}{a}\right)^2 + \left(\frac{y}{b}\right)^2\right]^v}. \qquad (12.16)$$

Here horizontal sections are ellipses and the exponent v ("upsilon") is a positive number. The obstacle half-widths A and B are given approximately by $A = 2a$, $B = 2b$. Larger v implies a more compact shape. If $v \le 1$, this topography has an infinite volume, and if $v \le 1/2$ each vertical section has infinite area. We will be mainly concerned with $v = 3/2$ and 2, which avoid these properties. Given that $v > 1$, we expect that the physical properties of flow over these obstacles will not be very sensitive to the value of v. The shapes with some values of v are more amenable to mathematical treatment than others, and the results obtained are taken to be representative of those with most v values. By expressing x and y in terms of elliptical polar coordinates s, ω in the form

$$x = sa\cos\omega, \qquad y = sb\sin\omega, \qquad (12.17)$$

the Fourier transform (12.13) of (12.16) is (Phillips, 1984)

$$\hat{h}(k,m) = \hat{h}(\kappa) = 2\pi h_m ab \left(\frac{\kappa}{2}\right)^{v-1} \frac{K_{v-1}(\kappa)}{\Gamma(v)}, \qquad \kappa = (k^2 a^2 + m^2 b^2)^{1/2}, \quad (12.18)$$

provided $\upsilon > 1/4$. Here $K_{\upsilon-1}$ is the Bessel function of imaginary argument, and Γ is the gamma function. The term $\hat{h}(0)$ is equal to the volume of the obstacle, $\mathcal{V}$. For κ large, $\hat{h}(\kappa)$ has the asymptotic form

$$\hat{h}(\kappa) \sim \pi^{3/2} h_m ab \left(\frac{\kappa}{2}\right)^{\upsilon-3/2} \frac{e^{-\kappa}}{\Gamma(\upsilon)}, \qquad \kappa \gg 1, \tag{12.19}$$

which decreases rapidly with increasing κ, particularly if $\upsilon \leq 3/2$. For the case $\upsilon = 3/2$, (12.18) has the simple form

$$\hat{h}(k,m) = \hat{h}(\kappa) = 2\pi h_m abe^{-\kappa}, \upsilon = 3/2. \tag{12.20}$$

It may be noted that in the limit $b \to \infty$, $\hat{h}(k,m)$ approaches a delta function form in m, so that $\hat{h}(k,m) \to \hat{h}(k,0)\delta(m)$, and the flow attains the two-dimensional limit. In the discussion below the integrals are expressed in polar coordinates (κ, γ), which is suitable for three-dimensional obstacles but is not convenient for relating the flow to the two-dimensional limit of large b.

12.1.3 Flow over short obstacles: stationary phase approximation for waves in the far-field

The archetypal obstacle for the situation where the obstacle length-scales $A, B \ll U/N$ is that where

$$h(x, y) = h_m \delta(x/A)\delta(y/B). \tag{12.21}$$

Here we have

$$\hat{h}(k,m) = h_m AB = \mathcal{V}, \tag{12.22}$$

where $\mathcal{V}$ may be taken to be the volume of the obstacle. Equation (12.14) then becomes

$$\eta = \frac{1}{4\pi^2} \mathcal{V} \int_{-\infty}^{\infty} \int_{-\infty}^{\infty} e^{i[kx+my+n(k,m)z]} dk dm, \tag{12.23}$$

so that for η, all horizontal wavenumbers (k,m) are forced equally. The corresponding vertical wavenumbers are given by the wavenumber surface in Figure 8.13a. In order to discuss this flow we introduce cylindrical polar coordinates (R, θ, x) and spherical coordinates (r, ϕ, θ) as shown in Figure 12.1, where

$$\begin{array}{cccc} R^2 = y^2 + z^2, & \sin\theta = z/R, & \cos\theta = y/R, & \tan\theta = z/y, \\ r^2 = x^2 + y^2 + z^2, & \sin\phi = R/r, & \cos\phi = x/r, & \tan\phi = R/x. \end{array} \tag{12.24}$$

It may be shown that for the flow represented by (12.23), the "swirl" velocity component v_θ in the θ-direction (i.e. in circles about the x-axis) vanishes everywhere (Janowitz, 1984); that is

$$v_\theta = (yw' - zv')/R = 0. \tag{12.25}$$

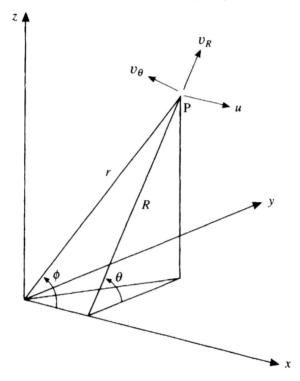

Figure 12.1 Coordinates for flow past three-dimensional obstacles. v_R, v_θ denote velocity components at point P in the directions of R, θ increasing, in the plane $x = $ constant.

In other words, steady flow over the δ-function obstacle (12.21) induces no swirling flow component about the x-axis, and the velocity vectors are confined to the planes of constant θ, which intersect the x-axis. For each of these planes the velocity is specified by u' and the component in the R-direction, v_R. The entire flow field may be conveniently described by a stream function ψ, for which

$$\psi = \frac{R^2}{z}\eta(x, y, z), \qquad (12.26)$$

and

$$u' = -\frac{1}{R}\frac{\partial \psi}{\partial R}, \quad v_R = \frac{1}{R}\frac{\partial \psi}{\partial x}, \quad v' = \frac{y}{R^2}\frac{\partial \psi}{\partial x}, \quad w' = \frac{z}{R^2}\frac{\partial \psi}{\partial x}, \qquad (12.27)$$

where all of these partial derivatives are taken in cylindrical polars.

An approximation to (12.14) when $r \gg U/N$ may be obtained from the method of stationary phase (Lighthill, 1978; Janowitz, 1984). This involves finding points in (k, m)-space where the phase $kx + my + nz$ is stationary, since these should be the regions that contribute most to the value of the Fourier integral. Elsewhere, contributions should be small due to the rapid oscillations caused by the large values of x, y and/or z. An exception to this is the region near $(k,m) = (0,0)$, since this may give a significant contribution to the integral even if x and/or y are large; this region is discussed separately below. In the present case, equating derivatives of the phase function to zero shows that there are two points of stationary phase, given by $\mathbf{k} = \pm \mathbf{k}^*$, where $\mathbf{k}^* = (k^*, m^*, n^*)$, and k^*, m^* are given by

$$k^* = \frac{N}{U} \sin\theta \cos\phi, \quad m^* = -\frac{N}{U} \frac{\sin 2\theta \cos^2 \phi}{2 \sin \phi}, \tag{12.28}$$

and n^* by (12.15). These points may range over the whole wavenumber surface of Figure 8.13a, depending on the direction of the point (r, θ, ϕ) from the origin. Expanding the phase function in a power series in k, m about (k^*, m^*) then enables the integral to be evaluated locally, giving the "stationary phase" approximation for the distant wave field, which for η we will denote by η_w. In the present case this gives

$$\eta_w \sim \frac{\mathcal{V} N}{2\pi U} \frac{z}{r^2} \frac{e^{i(k^*x+m^*y+n^*z)}}{|K_G|^{1/2}} + \text{complex conjugate}, \tag{12.29}$$

where K_G is the Gaussian curvature (e.g. Struik, 1950) of the wavenumber surface (12.15) shown in Figure 8.13a. If K_G is small, so that the surface is locally relatively cylindrical, the local contribution to the integral is large, and conversely if K_G is large it will be small. Evaluating K_G then gives (Wurtele, 1957; Crapper (1959,1962) and Janowitz (1984) obtain essentially the same expression with differing methods)

$$\eta_w = \frac{z}{R^2} \psi_w = \frac{N\mathcal{V}}{\pi U} \frac{z}{R^2} \cos\phi \cdot (1 - \sin^2\theta \cos^2\phi)^{1/2} \cos\left(\frac{Nr}{U}\sin\theta\right), \tag{12.30}$$

or in terms of x, y and z

$$\eta_w = \frac{N\mathcal{V}}{\pi U} \frac{z}{R^3} \frac{x(R^4 + x^2y^2)^{1/2}}{r^2} \cos\left(\frac{rz}{R}\right). \tag{12.31}$$

With (12.26), (12.27), this may be used to infer the velocity field and other variables. The resulting flow pattern is analogous to the "Kelvin wake" behind a ship, described briefly in §2.2.2. This expression should be contrasted with (10.20) for two-dimensional flow; there is similarity, but also some significant differences that relate to surprising properties, and these we now discuss.

Equation (12.30) consists of two factors: a phase term $\cos[(Nr/U)\sin\theta]$, so that the phase oscillates as r increases for constant θ, ϕ with wavelength $2\pi U/N\sin\theta$, and an amplitude term. We recall that the velocity field in the disturbance is confined to planes of constant θ, so that a fluid parcel initially in one of these planes will remain there. Near the x-axis the wavelength in each of these planes corresponds to the frequency $N\sin\theta$ of buoyancy oscillations in that plane, so that the fluid particles in each plane make buoyancy oscillations there, with zero group velocity. Further away from the x-axis the periods of oscillation of the particles are longer, and the group velocity is finite. Representative streamlines for the flow in these planes are shown in Figure 12.2. In physical space the scale of the pattern varies continuously with θ, and the wavelength becomes progressively longer as the planes become more horizontal ($\theta \to 0$ and π).

The structure of the amplitude function is more surprising, in that it remains finite as $x \to \infty$ for fixed y and z, where the largest amplitudes are close to (but not above) the x-axis. If we examine the structure of (12.31) in planes of constant height z, we may write

$$\eta_w = \frac{N\mathcal{V}}{\pi U} \frac{1}{z} \mathcal{A}(x/z, y/z) \cos(rz/R), \qquad (12.32)$$

and the function $\mathcal{A}$ is shown plotted in Figure 12.3. The ridge of maximum values is ∪-shaped, with a maximum of 0.5 at $x/z = 1, y = 0$, and far downstream a maximum of $2/3\sqrt{3}$ at $y/z = 1/\sqrt{2}$. The dashed lines denote representative lines of constant phase $r/R = $ constant in horizontal planes, and are hyperbolic in shape. In physical space, the flow pattern of Figure 12.3 expands and decreases in amplitude with increasing height. Hence the largest-amplitude waves generated in this flow are the buoyancy oscillations at low levels, lying close to the x-axis, and after the amplitudes reach initial maxima behind the obstacle, they decrease to finite values that persist indefinitely downstream.

When compared with flow over a two-dimensional ridge with the same δ-function profile, there is a substantial region downstream at low levels where this three-dimensional disturbance is larger than the two-dimensional one. For example, on the central plane $y = 0$ this occurs where $z < (Ux/N\pi)^{1/2}$, for x large. The comparison is particularly noticeable as $x \to \infty$, where the two-dimensional disturbance vanishes at finite z. This difference is attributed to focussing of the motion in the θ-planes where they converge on the x-axis, causing the large-amplitude motions in the wake region immediately downstream. Since these are approximately buoyancy oscillations, they propagate very slowly away from this region of fluid, and leave a residue of pure buoyancy oscillations far downstream.

If the lateral scale of the obstacle is increased so that NA/U, NB/U are no longer small, the flow loses its "swirl-free" character. However, the stationary phase approximation to (12.14) for general obstacles may be readily obtained from (12.29)

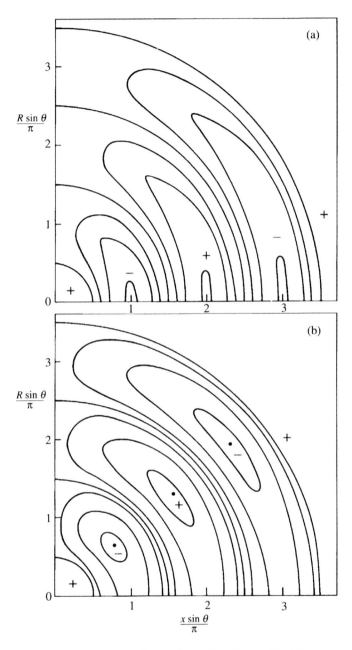

Figure 12.2 Linear perturbation theory for uniformly stratified flow past a point-like obstacle, stationary phase approximation: patterns of the perturbation stream function in planes intersecting the x-axis and inclined at angle θ to the horizontal. (a) $\theta = \pi/8$; (b) $\theta = 3\pi/8$. (Modified from Janowitz, 1984, reproduced with permission.)

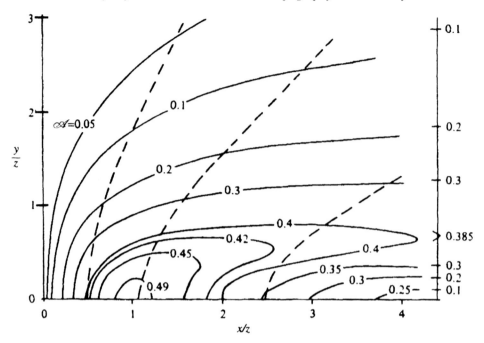

Figure 12.3 Linear perturbation theory for uniformly stratified flow past a point-like obstacle, stationary phase approximation: the amplitude function $\mathcal{A}(x/z, y/z)$ for the vertical displacement field η_w, showing the relative amplitudes in a plane of height z. The values on the right show the asymptotic values as $x/z \to \infty$. (From Janowitz, 1984, reproduced with permission.)

if the obstacle is symmetric about the x- and y-axes. All that is required is to replace $\mathcal{V}$ in (12.29) by $\hat{h}(k^*, m^*)$, which is real under these conditions.

The relation between these far-field approximations and the flow near the obstacle must be evaluated numerically. An example of a numerically computed displacement field for flow past a circular obstacle of the form (12.16) with $b = a$, and $\upsilon = 3/2$; that is,

$$h(x, y) = \frac{h_m a^3}{(x^2 + y^2 + a^2)^{3/2}}, \tag{12.33}$$

is shown in Figure 12.4. Here $Na/U = 1$, and the lines denote vertical displacements at a height $Nz/U = \pi/4$. The fluid rises on the upstream side of the obstacle, and then descends over the lee side into the first trough of a train of waves that occupy a wake region directly downstream, with a width that is comparable to the height. The wavelengths also approximately correspond to those for buoyancy oscillations. Hence this pattern is quite consistent with the stationary phase approximation, which therefore gives a useful description of these details when Na/U is as large as unity.

12.1.4 Hydrostatic flow over long obstacles

For typical atmospheric values of $N = 10^{-2}$ rad/s and $U = 10$ m/s, we have $Na/U = 1$ for $a = 1$ km. For topography on the surface of the Earth of significant height, most relevant length-scales are much longer than 1 km; hence $Na/U \gg 1$, and the flow in the atmosphere is approximately hydrostatic.

When Na/U or Nb/U is large, $\hat{h}(k^*, m^*)$ becomes small, and hence the above stationary phase term becomes small. In other words, the amplitudes of the waves of the form (12.29) become vanishingly small if the obstacle becomes long in the across-stream direction (where it approximates a two-dimensional ridge), or in the downstream direction. Under these conditions the only significant contribution to (12.14) comes from the region of (k, m)-space near the origin; the dynamical equations are hydrostatic, and n in (12.15) may be replaced by its hydrostatic form

$$n = \frac{N}{Uk}(k^2 + m^2)^{1/2}. \tag{12.34}$$

The nature of the hydrostatic flow near the obstacle may be obtained by numerical integration of (12.14) in (k, m)-space with n given by (12.34). Calculations for the axisymmetric obstacle (12.33) give a displacement field as shown in Figures 12.5, 12.6 and 12.7, which may be taken as representative of nearly axisymmetric obstacles. At low levels there is one major region of elevation on the upwind side of the mountain, and one major U-shaped region of depression on the downstream side. Rapid descent occurs over the lee side of the mountain, and between the elevated and depressed regions. As z increases, this pattern expands laterally, the amplitude of the elevation in the upwind region decreases, and the downstream region of depression moves upstream and is replaced by another region of elevation behind it. This implies a forward tilt of the phase lines, and hence an upward propagation of wave energy. Overall, the energy of this wave field is located more on the downstream side of the obstacle than it is for a two-dimensional obstacle, where in the latter case it lies directly over it.

For the obstacle (12.16) with $\upsilon = 3/2$, s and w defined by (12.17), κ by (12.18) and γ by

$$k = \frac{\kappa}{a}\cos\gamma, \qquad m = \frac{\kappa}{b}\sin\gamma, \tag{12.35}$$

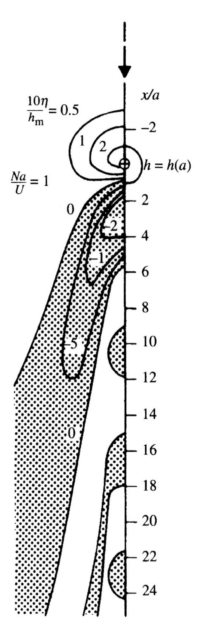

Figure 12.4 Vertical displacement field computed from linear perturbation theory for flow past a circular obstacle of the form (12.33) with $Na/U = 1$, at a height $Nz/U = \pi/4$. Shaded areas denote depressed streamlines, unshaded areas elevated ones; + denotes the centre of the obstacle. Note the lee waves extending downstream, in conformity with Figure 12.3. (From Smith, 1980: reproduced by permission of *Tellus*.)

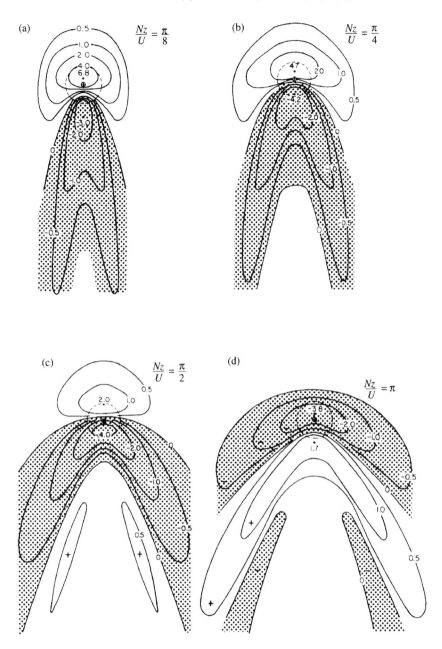

Figure 12.5 Linear perturbation theory for hydrostatic flow past a circular mountain of the form (12.33): computed vertical displacement fields at various heights as in Figure 12.4. Here the horizontal pattern is scaled with a (represented by the dashed circle centred on the origin), and the displacement amplitudes are scaled with $h_{\mathrm{m}}/10$ as in Figure 12.4. (From Smith, 1980: reproduced by permission of *Tellus.*)

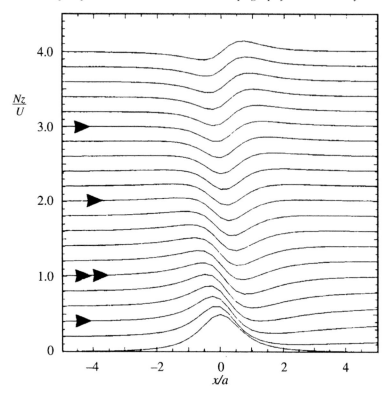

Figure 12.6 Streamlines in the plane of symmetry ($y = 0$) for the flow described in Figure 12.5, computed from (12.37) with $a = b$. The double arrow denotes the height of the stream surface shown in Figure 12.7. (Modified from Smith, 1988: reproduced by permission of the American Meteorological Society.)

we may write (12.14) as

$$\eta = \frac{h_m}{2\pi} \int_0^{2\pi} \int_0^{\infty} e^{-\kappa} \cdot e^{i\kappa s \cos(\gamma - \omega)} \cdot \exp i \left[\frac{Nz(\cos^2 \gamma + \frac{a^2}{b^2} \sin^2 \gamma)^{1/2}}{U \cos \gamma} \right] \kappa d\kappa d\gamma.$$

$$(12.36)$$

The integral with respect to κ may be evaluated exactly to give

$$\eta = \frac{h_m}{2\pi} \int_0^{2\pi} \frac{\exp i \left[\frac{Nz(\cos^2 \gamma + \frac{a^2}{b^2} \sin^2 \gamma)^{1/2}}{U \cos \gamma} \right]}{[1 - is \cos(\gamma - \omega)]^2} d\gamma. \qquad (12.37)$$

This expression reduces to the two-dimensional form (the equivalent of (10.21)) in the limit $a/b \to 0$, and simplifies when $a = b$. An analytical approximation to

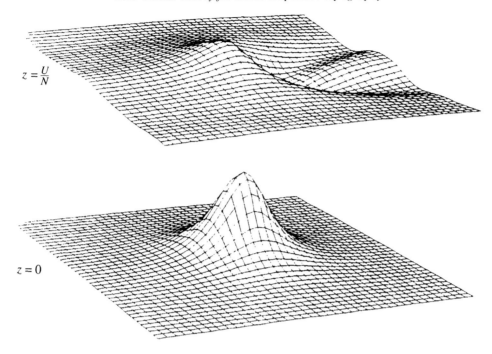

$$z = \frac{U}{N}$$

$$z = 0$$

Figure 12.7 Structure of the material surface from an initial upstream height $z = U/N$ (denoted by the double arrow in Figure 12.6 above the obstacle), for the same flow as in Figure 12.5. (From Smith, 1989b: reproduced by permission of Academic Press.)

these hydrostatic flow fields may be obtained when

$$s = \left(\frac{x^2}{a^2} + \frac{y^2}{b^2} \right)^{1/2} \tag{12.38}$$

is large. The principal contributions to this integral must then come from the ranges of γ near where

$$\gamma = \omega + \frac{\pi}{2}, \quad \omega + \frac{3\pi}{2}. \tag{12.39}$$

Evaluating and summing the contributions from these regions gives (extending Smith, 1980)

$$\eta \sim 2h_{\mathrm{m}} \frac{1}{s} |\beta e^{-\beta}| \cos \left[\frac{Nz(x^2 + y^2)^{1/2}}{Uy} \right] + O\left(\frac{1}{s} \right)^2, \tag{12.40}$$

where

$$\beta = \frac{Nzbx}{Uy^2} \left[\frac{x^2 + (ay/b)^2}{x^2 + y^2} \right]^{1/2}. \tag{12.41}$$

The factor $|\beta e^{-\beta}|$ has a maximum value of $1/e$ when $\beta = 1$, so that the expression (12.40) is small unless β is close to this value. This implies that η is small except at points in physical space close to the curve $\beta = 1$. For axisymmetric obstacles where $a = b$, this surface is a paraboloid (Figure 12.8a)

$$y^2 = \frac{Nazx}{U},$$

(12.42)

which is a "developable surface" swept out by straight lines radiating out from the origin. A section of the surface at each level z gives a parabola with its vertex above the origin (Figure 12.8b). Within the region of physical space close to (12.42), the amplitude of the disturbance decays with distance as $1/s$ (which is generally slower than the stationary phase expression (12.30), which varies as z/R^2). The waves in this paraboloidal region are described by the phase factor in (12.40), $\cos[(1 + x^2/y^2)^{1/2}Nz/U]$. This shows that in horizontal planes the wave crests and troughs are aligned along lines $y/x = $ constant, which are straight lines radiating from the point in the plane above the origin. These lines lie on surfaces of constant phase that are tilted upward and outward, giving downward and outward phase propagation relative to the fluid (see Figure 12.8b). This corresponds to upward and outward energy propagation relative to the obstacle. The waves on the surface (12.42) are represented on the wavenumber surface (Figure 8.13a) above the origin, which is shown in hydrostatic form in Figure 8.13b. For the non-axisymmetric obstacles where $b \neq a$, these general remarks are still applicable, except that the surface $\beta = 1$ is not a simple paraboloid.

At the vertex of each curve $\beta = 1$, above the obstacle, the wave crests are perpendicular to the mean flow direction as for two-dimensional flows, but (12.40) is not valid here because s is not large. Instead, a different asymptotic expression applies with $s \ll 1$ and $Nz/U \gg 1$, of the form (extending Smith, 1980)

$$\eta \sim \frac{h_m b}{a} \left(\frac{U}{\pi Nz}\right)^{1/2} \left[\left(1 - \frac{2x}{a} - \frac{x^2}{a^2}\right) \cos\left(\frac{Nz}{U}\right)\right.$$
$$\left. - \left(1 + \frac{2x}{a} - \frac{x^2}{a^2}\right) \sin\left(\frac{Nz}{U}\right)\right]\left(1 + \frac{x^2}{a^2}\right)^{-2}.$$

(12.43)

This expression is independent of y, and assumes that y/b is small. The weak vertical decay as $(Nz/U)^{1/2}$ is related to the curvature of the wavenumber surface (Figure 8.13a) on the plane of symmetry where $m = 0$. The expression has similarities to those for hydrostatic flow over two-dimensional obstacles (see §10.2.1) which do not decay with height, but is not related to them in a simple manner because of the essential three-dimensionality of this limit.

For obstacles that are long in the transverse direction, so that they may be described as two-dimensional ridges of finite length, the flow pattern (from linear

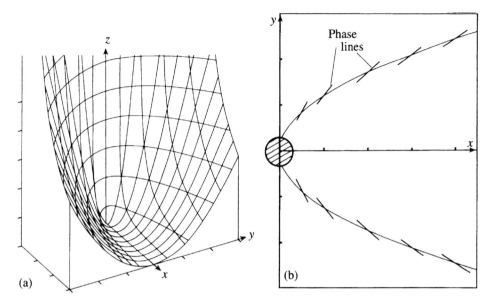

Figure 12.8 A three-dimensional representation of the surface $y^2 = Nazx/U$, the locus of the wake region for flow over the obstacle (12.33), for $Na/U = 1$ (arbitrary scale). (b) A horizontal section of the surface in (a) with length-scale equal to the height z, showing the orientation of the wave crests.

theory) over most of the ridge is very similar to that over a two-dimensional obstacle. The exception is near the ends, where it more resembles that shown in Figures 12.5–12.7. Figure 12.9 shows the displacement field for flow near the end of a long ridge at a height $Nz/U = \pi/8$, evaluated numerically from (12.14), (12.34). The region of elevation of the streamlines trails around and behind the obstacle, and a region of deep depression exists in the lee of the end of the ridge. This behaviour is due to the fact that the low-level flow is deflected around the ridge, as discussed below. This is another example in which the disturbance behind three-dimensional obstacles at low levels is larger than that behind two-dimensional obstacles, because of lateral deflection.

12.1.5 The hydrostatic flow near the ground

The surface pressure field may be evaluated by integrating the displacement field vertically downwards and using the hydrostatic relation, to give

$$p'(x,y,0) = \frac{\rho_0 N^2}{4\pi^2} \mathrm{i} \iint \frac{\hat{h}(k,m)}{n} \mathrm{e}^{\mathrm{i}(kx+my)} \mathrm{d}k\,\mathrm{d}m. \qquad (12.44)$$

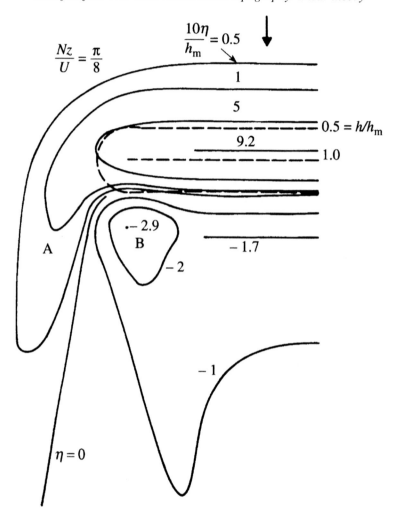

Figure 12.9 The pattern of vertical displacement in linear hydrostatic flow at a height $z = \pi U/8N$, near the end of a long ridge of Gaussian shape $h(x, y) = h_m e^{-(\bar{r}/a)^2}$, where $\bar{r}$ is the distance from the ridge crest line (shown dashed, as is the $h = h_m/2$ contour). Away from the end region the flow is two-dimensional, as if the ridge were infinitely long. Lateral deflection of the low-level flow causes the raised streamlines in region A, and depressed streamlines in region B. (From Smith, 1980: reproduced by permission of *Tellus*.)

For the axisymmetric obstacle (12.33), this expression may be evaluated analytically to give (Smith, 1980)

$$p' = -\rho N U h_m \frac{x/a}{\left[1 + (x/a)^2 + (y/a)^2\right]^{3/2}}, \qquad (12.45)$$

which shows a positive pressure perturbation on the upstream side of the obstacle, and a symmetric negative pressure perturbation on the downstream side. Figure 12.10 shows the corresponding pressure patterns computed for the obstacle (12.16) with $a = 0.5b$, $\upsilon = 2$, but with three different orientations. Note that the pressure perturbation is much larger for the obstacle elongated transversely. For each value of y, these perturbations are generally less than those for flow over a two-dimensional obstacle having the same cross-section as at that y-value. Note also that the pressure pattern for the obstacle inclined at 45° is tilted to lie between the wind direction and the major axis of the obstacle.

From (12.2) the velocity component u' is given by

$$u' = -p'/\rho_0 U, \tag{12.46}$$

so that the above pressure fields also describe the surface u' field. Clearly, $u = U+u'$ is decreased on the upstream side and increased on the downstream side, with a minimum velocity on the upstream side at the point of maximum pressure.

The transverse velocity component v' may be obtained by integrating (12.2) using the above expression for p', and for the obstacle (12.33) we obtain

$$v' = Nh_m \frac{\frac{y}{a}}{\left[1 + \left(\frac{x}{a}\right)^2 + \left(\frac{y}{a}\right)^2\right]^{3/2}}. \tag{12.47}$$

This may be used to obtain the lateral displacement $\xi(x, y)$ of a surface streamline from its upstream position by evaluating

$$\xi(x, y) = \frac{1}{U} \int_{-\infty}^{x} v'(x', y)dx', \tag{12.48}$$

assuming that this is small. For the obstacle (12.33), (12.47) and (12.48) give

$$\xi(x, y) = \frac{Nh_m}{U} \frac{y}{\left(1 + \frac{y^2}{a^2}\right)} \left\{1 + \frac{\frac{x}{a}}{[1 + \left(\frac{x}{a}\right)^2 + \left(\frac{y}{a}\right)^2]^{1/2}}\right\}. \tag{12.49}$$

This lateral displacement of the surface streamlines is shown in Figure 12.11. The displacement is always away from the x-axis (the line of symmetry) and increases to a maximum far downstream. Fluid approaching the obstacle with $y \neq 0$, is deflected away from the centre of the obstacle by the pressure field (12.45). When a fluid particle is abreast of the obstacle the pressure gradient vanishes but, from (12.47), v' is still directed away from the centre-line. On the downstream side the pressure minimum centred on the x-axis causes the streamlines to deflect back towards the x-direction, so that they resume their original direction far downstream.

This lateral displacement on the topographic surface downstream implies lateral divergence at low levels over the obstacle, and this is compensated for by

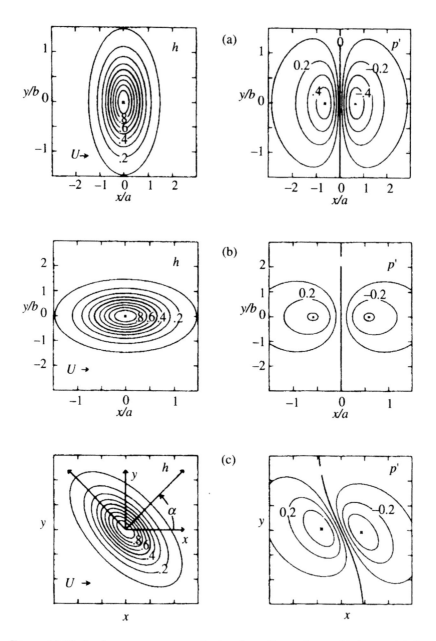

Figure 12.10 Surface pressure perturbation from linear theory (right panels) for the obstacles shown on the left ((12.16) with $v = 2$) and (a) $a/b = 0.5$, (b) $a/b = 2.0$ and (c) the same obstacle at $45°$ to the flow. The mean wind direction is from left to right, and h is in units of h_m, p' in units of $\rho_0 N U h_m$. The extreme pressure perturbations are labelled by crosses. (From Phillips, 1984: reproduced by permission of the American Meteorological Society.)

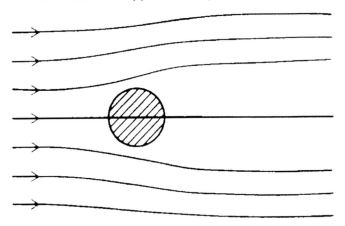

Figure 12.11 Plan view of the pattern of lateral displacement of the surface stream-lines for the obstacle (12.33). (From Smith, 1980: reproduced by permission of *Tellus*).

fluid descending from aloft, as shown in Figure 12.6. At levels above the ground, numerical evaluation shows that this lateral displacement and depression are not permanent and the streamlines eventually rise and return to their original (y, z)-locations, though the smaller z is, the longer the depression lasts. In other words, as one moves further downstream, the lateral deflection and vertical depression are confined to a progressively shallower region adjoining the ground. This lateral displacement is a hydrostatic effect, but it is still present in flows that are substantially non-hydrostatic. For an obstacle of arbitrary shape it may readily be shown (Smith, 1980) from (12.14)–(12.15) that the nett lateral displacement of a streamline on the surface may be expressed as a Fourier integral containing only the integrated cross-sectional area of the obstacle for each y, so that

$$\xi(\infty, y, 0) = -\mathrm{i}\frac{N}{U}\frac{1}{2\pi}\int_{-\infty}^{\infty}(\operatorname{sgn}m)\widehat{\mathcal{A}}(m)\mathrm{e}^{\mathrm{i}my}\,\mathrm{d}m, \tag{12.50}$$

where

$$\widehat{\mathcal{A}}(m) = \int_{-\infty}^{\infty}\mathcal{A}(y)\mathrm{e}^{-\mathrm{i}my}\,\mathrm{d}y, \qquad \mathcal{A}(y) = \frac{1}{2\pi}\int_{-\infty}^{\infty}h(x, y)\,\mathrm{d}x. \tag{12.51}$$

Hence this general effect of divergence of the low-level flow from regions where $\mathcal{A}(y)$ is large, and convergence to regions where it is small, is quite ubiquitous and independent of the small-scale dynamics.

12.1.6 The drag on an isolated obstacle: $Nh/U < 1$

The form drag on an obstacle is given by (12.9). This drag is the nett force on the obstacle by the pressure perturbation field, the latter being generally positive on the upstream side and negative on the downstream side. From linear perturbation theory we may use (12.2), (12.14) to express the pressure in terms of $\hat{h}$, and obtain an expression for this drag on the obstacle,

$$\mathbf{F_D} = \frac{\rho_0 U^2}{4\pi^2} \int_{-N/U}^{N/U} \int_{-\infty}^{\infty} |\hat{h}(k,m)|^2 \left[\frac{N^2 U^{-2} - k^2}{k^2 + m^2} \right]^{1/2} k\mathbf{k} \, dk \, dm, \qquad (12.52)$$

where here $\mathbf{k} = (k, m)$. This reduces to the one-dimensional form (10.24) in the event that h is independent of y. For hydrostatic flow, for which $\hat{h}$ is small unless $|k| \ll N/U$, (12.52) gives

$$\mathbf{F_D} = \frac{\rho_0 U N}{4\pi^2} \int_{-\infty}^{\infty} \int_{-\infty}^{\infty} \frac{k\mathbf{k}}{(k^2 + m^2)^{1/2}} |\hat{h}(k,m)|^2 \, dk \, dm. \qquad (12.53)$$

For obstacles of the general form

$$H(R) = \frac{h(x, y)}{h_m} = \frac{1}{(1 + R^2)^{\upsilon}}, \qquad \text{where} \quad R^2 = \frac{x^2}{a^2} + \frac{y^2}{b^2}, \qquad (12.54)$$

and where the axis with width scale a makes an angle α with the wind direction, the total vector drag on the obstacle is

$$\mathbf{F_D} = \rho_0 U N h_m^2 b G(D_x, D_y), \qquad (12.55)$$

where G is the steepness parameter, and D_x and D_y denote the normalised drag in the x- and y-directions, with the form

$$D_x = (B \cos^2 \alpha + C \sin^2 \alpha), \qquad D_y = (B - C) \sin \alpha \cdot \cos \alpha. \qquad (12.56)$$

Here B and C are functions only of $\gamma = a/b$, which is the departure from axisymmetry of the obstacle, where (Lott & Miller, 1997)

$$B(\gamma) = 1 - 0.18\gamma - 0.04\gamma^2, \qquad C(\gamma) = 0.48\gamma + 0.3\gamma^2. \qquad (12.57)$$

In general the drag force is not aligned with the wind direction but has a transverse component, as shown for periodic topography in §12.1.1. For obstacles of the form (12.54), the ratio of these two components is shown in Figure 12.12a. It is seen that the perpendicular component ($F_\perp$) can be much larger than the parallel ($F_=$) when the wind is nearly aligned with the long axis of a very long obstacle.

The parameter G is a function of the exponent υ only, and takes the form

$$G = 4 \int_0^{\infty} \rho^2 (\hat{H}(\rho))^2 d\rho, \qquad \text{where} \quad \hat{H}(\rho) = \int_0^{\infty} R H(R) J_0(\rho R) dR; \qquad (12.58)$$

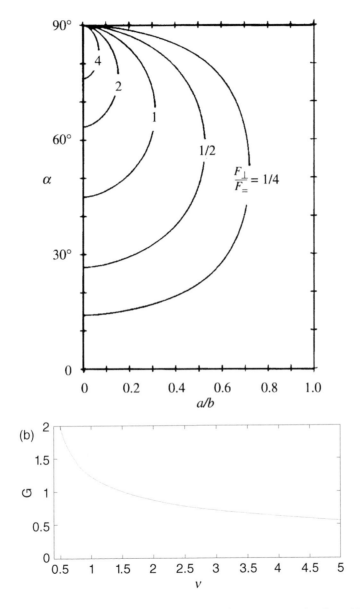

Figure 12.12 (a) Ratio of the transverse "lift" force ($F_\perp$) to the drag ($F_=$) for elliptical obstacles ((12.54), any v), as a function of a/b and the inclination α of the wind direction to the minor axis. (From Phillips, 1984: reproduced by permission of the American Meteorological Society.) (b) The parameter G as a function of the exponent v, which is a measure of the steepness, or sharpness of the obstacle in the form (12.59).

it is a measure of the steepness of the obstacle: larger values of v imply steeper slopes for fixed a and b. In fact, G may be expressed in the form

$$G = 4\Gamma \left(\frac{3}{2}\right) \frac{\Gamma(2v-1/2)}{\Gamma(2v+1)} \left[\frac{\Gamma(v+1/2)}{\Gamma(v)}\right]^2, \qquad (12.59)$$

where Γ denotes the gamma function: a plot of this dependence is shown in Figure 12.12b. The drag is clearly larger for smaller values of v, causing larger values of G and more "spread out" topography. These parameters are significant for modelling sub-grid-scale topographic effects in weather and climate models, as described in Chapter 15.

In summary, for flow with uniform U and N over obstacles of small height, various parts of the wavenumber spectrum of the topography force a range of different patterns of waves, and although we have discussed these patterns and their far-field forms separately, a typical obstacle will excite all of them to varying degrees. These "patterns" include buoyancy oscillations at low elevations directly downstream, hydrostatic waves over the obstacle with vertical wavenumber N/U, and waves on the downstream side on a three-dimensional paraboloid that expands with height. In addition, there is a lateral deflection of the streamlines at low levels near the obstacle with a compensating downward deflection of lighter fluid from above. This effect persists for a considerable distance downstream, and the deflection is permanent at ground level.

12.2 Linear theory for trapped lee waves

If N and $\mathbf{U} = (U, V)$ vary with height, the procedure of the previous section needs modification as follows. As described in Chapters 4 and 5, such inhomogeneities cause partial or total reflection of the internal waves, and this may result in trapped or leaky modes, where the vertical wavenumber spectrum is wholly or partially discrete. The relevant equations for steady flow are the steady forms of (8.121)–(8.125), and the solution may again be obtained by Fourier methods. With the lower boundary condition (12.1) and the Fourier transform of the topography given by (12.13), the solution for w' (for example) may be expressed in the form

$$w' = \frac{1}{4\pi^2} \int_{-\infty}^{\infty} \int_{-\infty}^{\infty} i(kU+mV)\hat{h}(k,m)\hat{w}(k,m,z)e^{i(kx+my)}\,dk\,dm, \qquad (12.60)$$

where $\hat{w}(k,m,z)$ satisfies (8.129) with $\omega = 0$, and the lower boundary condition

$$\hat{w} = 1, \qquad z = 0. \qquad (12.61)$$

This steady solution must also satisfy the radiation condition for large $x^2 + y^2 + z^2$, specifying that there is no incoming energy from infinity. This is achieved by ensuring that this applies to each Fourier component, $\hat{w}e^{i(kx+my)}$. For the important case where the wind is unidirectional so that $V = 0$, (8.129) with $\omega = 0$ becomes

$$\frac{d^2\hat{w}}{dz^2} + \left(l^2 + \frac{N^2m^2}{U^2k^2} - k^2 - m^2\right)\hat{w} = 0, \tag{12.62}$$

where

$$l^2 = \frac{N^2}{U^2} - \frac{1}{U}\frac{d^2U}{dz^2}. \tag{12.63}$$

If N and U vary with z, solutions for w with the boundary condition (12.61) can introduce singularities into the integrand of (12.60) in the (k,m)-plane, and these result in trapped modes in the solution for w'. The radiation condition (or its equivalent) must be applied to these modes in order to determine the contours of integration around these singularities, and hence to obtain the correct steady-state solution. Details are given in Lighthill (1978). In a situation where N and U become constant for sufficiently large z, where $N = N(\infty)$, $U = U(\infty)$ say, then in general we expect the vertical spectrum to be continuous for $|k| < l(\infty) = N(\infty)/U(\infty)$, and discrete otherwise.

A linear numerical model of this form for three-dimensional waves in the atmosphere has been described by Simard & Peltier (1982). Here we concentrate instead on simpler models that may be used to capture the essence of observed phenomena. For atmospheric flows there are two common situations that are observed to cause trapping: a strongly stratified layer or temperature inversion that may be approximated by a density interface, and secondly, progressively increasing wind speed with height. In order to illustrate these effects we consider the possible waves in three specific examples of U and N profiles: a single interface, a uniformly stratified layer with a rigid upper boundary, and uniformly stratified flow with exponentially increasing wind speed. For these we assume that the wind is unidirectional, so that $V = 0$, and U is uniform unless otherwise stated.

(i) *Density interface at height d_0, with stratified fluid above*
This simple situation provides a rough approximation to the atmospheric boundary-layer on a hot sunny day. For the density profile implied by

$$\begin{aligned}N^2(z) &= \frac{g\Delta\rho}{\rho_0}\delta(z - d_0), \\ &= N_1^2, \qquad z > d_0,\end{aligned} \tag{12.64}$$

where N_1 is constant, the solution of (12.62) with the boundary condition

$$\hat{w} = 0, \qquad z = 0, \tag{12.65}$$

gives the stationary trapped wave modes

$$\hat{w} = \begin{cases} \dfrac{\sinh(k^2+m^2)^{1/2}z}{\sinh(k^2+m^2)^{1/2}d_0}, & 0 < z < d_0, \\ e^{-(k^2+m^2)^{1/2}(1-N_1^2/k^2U^2)^{1/2}(z-d_0)}, & z > d_0, \end{cases} \qquad (12.66)$$

provided that $|k| > N_1/U$, with the dispersion relation

$$K^2 = \frac{(K^2+M^2)^{1/2}\tanh\left[\frac{(K^2+M^2)^{1/2}}{F_0^2}\right]}{1 + \left[1 - \left(\frac{N_1U}{g'}\right)^2 \frac{1}{K^2}\right]^{1/2}\tanh\left[\frac{(K^2+M^2)^{1/2}}{F_0^2}\right]}, \qquad (12.67)$$

where

$$K = \frac{kU^2}{g'}, \quad M = \frac{mU^2}{g'}, \quad F_0^2 = \frac{U^2}{g'd_0}, \quad g' = \frac{g\Delta\rho}{\rho_0}. \qquad (12.68)$$

For $|k| < N_1/U$ (i.e. $|K| < N_1U/g'$), the modes are leaky as described in §8.5, and are not relevant here. The dispersion curves are shown in Figure 12.13a,b for $N_1U/g' = 0,1$. The first set ($N_1 = 0$) is very similar to those for a single layer described in §2.2, and in particular, the wedge angle for small F_0 is the same as that for the Kelvin wake: 19.5°. For this reason we may expect the waves behind an obstacle to resemble those behind a ship, when this simple layered model of the stratification is applicable.

However, wavy wake patterns on shallow inversion layers in the atmosphere may be expected to be much more varied than those of ships for several reasons, as follows. First, the Froude number in the atmosphere may have a wide range of values, from much less than to much greater than unity, whereas for boats and ships F_0 is almost invariably small. Second, the height of the obstacle or mountain may not be small compared with the depth of the layer, as seen in the next section. Further, when upper-layer stratification is present, the wavenumber cut-off at N_1/U has the effect of decreasing the maximum wedge angle and, if N_1 is sufficiently large, removing the transverse waves with m near zero for $F_0 < 1$, as in Figure 12.13b. Fourth, if the density variation is spread out over a finite layer instead of being concentrated at an interface, there is a high wavenumber cut-off to the spectrum of waves: only waves with $|k| < N_m/U$ can propagate, where N_m is the maximum value of the buoyancy frequency within the layer. The presence of shear across the layer also has the effect of introducing a high-wavenumber cut-off, due to instability of the flow at short wavelength (see §4.2).

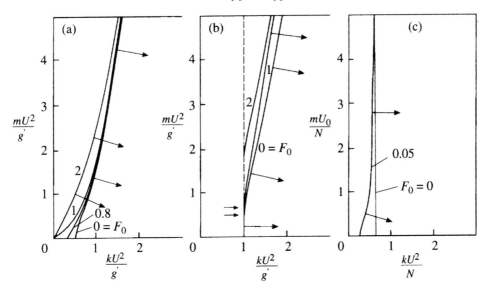

Figure 12.13 (a) Dispersion curves for stationary wave modes on an interface for various Froude numbers, calculated from (12.67), with upper-layer buoyancy frequency $N_1 = 0$. (b) As for (a) but with $N_1 U/g' = 1$. (c) Dispersion curves for stationary wave modes in flow with uniform N and exponential wind given by (12.72), with $NH/U_0 \gg 1$. The curves have been calculated from (12.75).

(ii) *Uniformly stratified layer of depth D*

With U and N constant and the boundary conditions

$$\hat{w} = 0, \qquad z = 0, D, \tag{12.69}$$

(12.62) has the solutions

$$\hat{w} = \sin j\pi z/D, \qquad j = 1, 2, 3, \ldots, \tag{12.70}$$

with the dispersion relation

$$(jF_0)^2 K^2 = (K^2 + M^2)(1 - K^2), \tag{12.71}$$

where $K = kU/N$, $M = mU/N$, and $F_0 = 1/K = U/(ND/\pi)$ is the Froude number based on the lowest mode, as in §10.5. Equation (12.71) is the same relation as that plotted in Figure 8.13a for infinite depth, if jF_0 is identified with nU/N. Hence, for a given mode j, (12.71) is a horizontal slice of the surface in Figure 8.13a, at a level depending on jF_0. This dependence on F_0 is qualitatively similar to that of (12.67), shown in Figure 12.13a. The wedge angle tends to $90°$ as $jF_0 \to 1$, but there are two important differences: the wedge angle is zero for small F_0, and K is finite ($\to 1$) for large M. Flow past a small obstacle in this system will produce a wake comprised of all the modes. Since jF_0 is the Froude number of the jth mode, those modes with $jF_0 \ll 1$ will have narrow wakes with transverse waves made up of buoyancy oscillations with wave crests perpendicular to the flow, and those with $jF_0 \gg 1$ will have wakes with narrow wedge angles containing waves propagating nearly transversely to the flow. Some examples have been described by Sharman & Wurtele (1983).

(iii) *Uniform N, exponential U*
An increase in wind speed with height causes wave-trapping by refracting the wave energy back toward the ground. If we take

$$U(z) = U_0 e^{z/H}, \qquad N = \text{constant}, \tag{12.72}$$

then (12.62) with the boundary condition (12.65) has the solution

$$\hat{w} = J_v \left[\frac{NH\,(k^2 + m^2)^{1/2}}{U_0} \frac{}{k} e^{-z/H} \right], \qquad v = (1 + k^2 H^2 + m^2 H^2)^{1/2}, \tag{12.73}$$

where J_v denotes the usual Bessel function of order v, with the lower boundary condition giving the expression for the eigenvalues

$$J_v \left[\frac{NH\,(k^2 + m^2)^{1/2}}{U} \frac{}{k} \right] = 0. \tag{12.74}$$

The zeros of J_v are all real, and each one corresponds to a discrete vertical mode. The vertical structure of each of these modes has a finite number of oscillations, with exponential decay at large z. If $NH/U_0 \gg 1$, which is a common situation for the atmosphere, we may approximate (12.74) by its asymptotic form, and defining $K = kU_0/N$, $M = mU_0/N$, we obtain

$$K \frac{U_0}{NH} \left(j - \frac{1}{4} \right) \pi \simeq (K^2 + M^2)^{1/2} - \frac{\pi}{2} K \left[\left(\frac{U_0}{NH} \right)^2 + K^2 + M^2 \right]^{1/2}, \tag{12.75}$$

where j is an integer. Here the Froude number for the jth mode is approximately

$$F_0 \simeq U_0 \Big/ \left[NH \left(j + \frac{1}{4} \right) \pi \right], \qquad j = 1, 2, 3, \dots \tag{12.76}$$

and curves from (12.75) for $F_0 = 0, 0.05$, and $j = 1$, are shown in Figure 12.13c. The similarity to (12.71) is obvious. Examples of wakes for various parameter values for this situation are described by Sharman & Wurtele (1983).

Hence the two very different forms of wave-trapping in examples (ii) and (iii) have similar dispersive properties, despite their differences. They may both be contrasted with example (i): the interface, in that at small F_0 the dispersive properties of the latter give a Kelvin-type wake with the classical angle as in Chapter 2, whereas in (ii) and (iii) the wake for small F_0 consists of buoyancy oscillations in a narrow region directly downstream of the obstacle.

12.3 Atmospheric lee waves

At this point we examine some examples of observed internal waves forced by topography in the atmosphere. These waves extend some distance from the topography, so that the above far-field analysis is applicable.

In the atmosphere, the vertical column of air at each location contains a vertical profile of water vapour concentration, which normally decreases with height. At each level this concentration has a maximum possible value, at which the air is said to be "saturated". This saturation value depends on the temperature and pressure of the air, and also decreases with height (see, for example, Wallace & Hobbs, 1977) If the air is close to saturation, and is lifted to higher levels by flow over topography or in wave crests, the saturation level can be reduced to below the local concentration. This produces water droplets (condensation), and hence forms local clouds. This condensation process occurs very rapidly, in a time-scale of a few seconds, and the associated cloud pattern is the primary means of visualising waves produced by flow over topography in the atmosphere. When the air again descends below the level for condensation, evaporation of the water droplets may occur as rapidly as their formation. However, at heights above about 2 km part or all of the water in the cloud may be converted to ice, which evaporates more slowly. Consequently, at these high elevations, on its downstream side the cloud may be depressed below its initial level on the upstream side. An illustration of this is seen in the lowest cloud in Figure 12.14, which shows rapid descent on the lee side of Mt Fuji. The two separate higher-level clouds indicate the regions of largest-amplitude waves in a non-hydrostatic flow, consistent with Figures 12.2, 12.3, although this flow pattern may well be non-linear because Nh/U may not be small.

Most of our information about three-dimensional patterns of lee waves in the atmosphere has come from satellite observations of such cloud patterns. There is now a substantial number of well-described examples (e.g. Scorer, 1986, 1990) that show great variety, depending on atmospheric conditions and terrain shape. Figures 12.15, 12.16 and 12.17 show three examples that are typical of many others, and are presented to illustrate the relevance of the preceding theoretical analysis. These examples are from isolated islands at high latitudes, where there is often a deep cold layer of nearly saturated air near the surface, so that waves may be trapped on the stratified layer or temperature inversion at its upper boundary. Sometimes cloud may be present below the inversion , and its reaction to changing elevation due to the waves can render the latter visible. Wave-trapping on such layers may cause the wave pattern to extend many wavelengths and hundreds of kilometres downstream with little visible attenuation, and the observation of such an extensive pattern is a clear indication that the wavetrain is in fact trapped.

Figure 12.15 shows waves downstream of Macquarie Island, an approximately

Figure 12.14 Wave clouds over and downstream of Mt Fuji, Japan, height 3800 m (source unknown).

two-dimensional ridge of length 30 km and mean height of 300 m, oriented approximately North–South in the Southern Ocean. This shows an initially nearly two-dimensional transverse wave pattern that spreads laterally with distance downstream. Here cloud indicates elevated fluid and denotes wave crests, as in Figure 10.12. There is some indication that these crests become more tapered back at the ends as one moves downstream, so that the wavetrain more resembles a ship wake, due to spreading effects from the ends of the island. There is also an apparent increase in wavelength after the first seven or so crests from the island, from 10–15.4 km. Balloon soundings on Macquarie Island taken before and after the image show an inversion layer at a height of approximately 1.5 km with l^2 (see §8.5) small above, and the difference between the soundings shows that the wind decreases and the inversion strength increases with time. Based on the simple interface model (i) of §12.2 with $N_1 = 0$, the Froude number is less than unity and the decreasing wavelength is consistent with the change in background conditions.

Figure 12.16 shows a diverging wake from Jan Mayen, with no transverse waves visible at all. Jan Mayen has two localised peaks of different heights (769 m and 2777 m) that are approximately aligned with the wind in this image, and the domi-

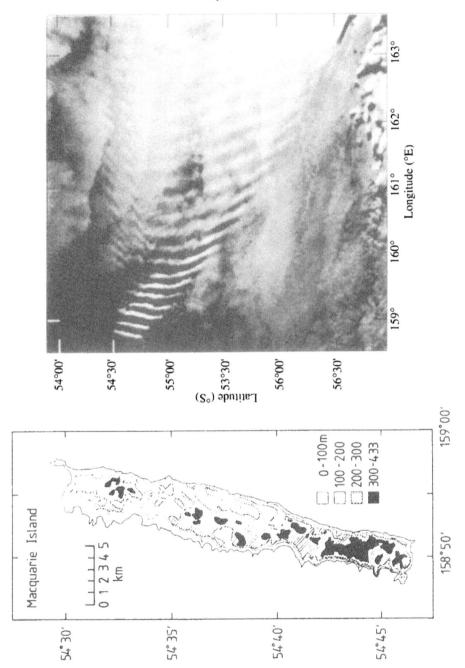

Figure 12.15 Lee-wave clouds downstream of Macquarie Island (shown in inset). Wind direction is from WNW (approx.), and the central ridge of the island coincides with the western edge of the first cloud band. Image from NOAA/AVHRR channel 2 at 2024 UT on 16th October 1985. (From Mitchell et al. (1990): reproduced by permission of the Royal Meteorological Society.)

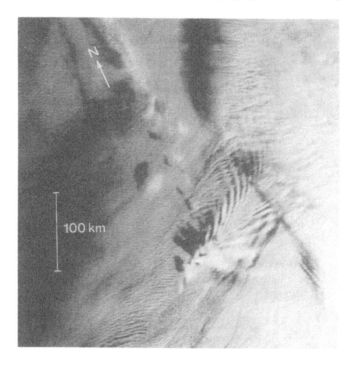

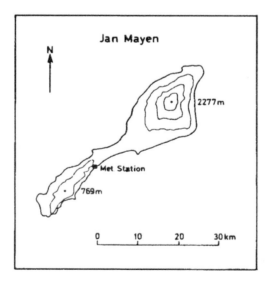

Figure 12.16 Lee-wave pattern downstream of Jan Mayen (71.0°N, 8.5°W, shown in inset). Wind direction is from WSW, so that the wave patterns from the two peaks are superimposed. The waves are made visible by a cloud layer at an inversion at a height of ∼ 1600 m. (The broad dark regions are due to shadows from higher-level cloud.) Image from NOAA5 at 1011 UT, 10th October 1978. (From Gjevik, 1980: reproduced by permission of WMO.)

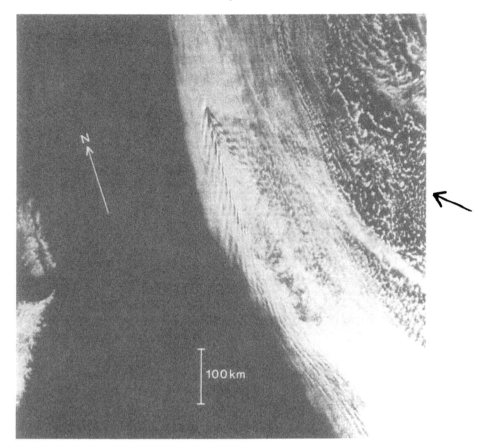

Figure 12.17 Dark, broad-angled wake (indicated by arrow) from Jan Mayen, in addition to a narrower wake similar to that in Figure 12.16. Wind direction is from NNE. Image from NOAA5/VHRR at 1122 UT on 29th December 1976. (From Gjevik & Marthinsen, 1978: reproduced by permission of the Royal Meteorological Society.)

nant part of the visible wave field could be produced by either or both. The nature of the wave clouds suggests that the waves are concentrated on a cloud-top inversion, so that model (i) of §12.2 with $N_1 \neq 0$ may be used as a guide (Figure 12.13b). The relatively narrow wake with wavenumbers $m \gg k$ suggests that $F_0 \gg 1$, which is consistent with the data from local soundings.

Figure 12.17 shows another image of flow past Jan Mayen. Here there is also a lee-wave wake resembling that of Figure 12.16, but more particularly there is a "dark band" of width 14 km extending from the island at a broad angle to the wind direction. There is an inversion (of strength 10°C at a mean height of 1 km, and the wind is from the north with speed 30 m/s. For the interface model (i) of §12.2, this gives a Froude number $F_0 = 1.6$. The probable explanation of the observed "dark band" is that it constitutes one side of a supercritical wake of the same form as those in shallow water shown in Figures 2.3, 2.4. The observed angle of 38° of the line to the wind direction is consistent with a Froude number of 1.6, and the higher peak of Jan Mayen extends above the inversion to 1.7 km, so that the flow of the lower layer is around it, as in Figure 2.4c. The width of the dark band is comparable to that of the higher peak. Such a wake has elevated fluid at the leading part of the wave and depressed fluid in the trailing (dark) part, and is composed of a spread of wavenumbers where $m \sim k$, so that the phase lines are straight and all emanate from the obstacle at the same angle that depends on F_0.

The vertical structure of lee waves in the atmosphere is not readily seen with natural tracers, and detailed observations of this structure have to be made directly, using such things as balloons and aircraft. Such experiments have been restricted to a small number of locations, usually in a vertical plane aligned with the wind, over topography that may be regarded as approximately two-dimensional. One such example is given in Figure 11.29 (Rocky Mountains, USA), and others in other locations have been described by: Hoinka (1985) – European Alps; Smith (1987) – Dinaric Alps, Yugoslavia; and Pitts & Lyons (1989) – the Darling Scarp, Perth, Western Australia. These observations mostly concern flow over a large orographic complex with a single main ridge.

A different type of situation is displayed in Figure 12.18. This shows the vertical velocity field inferred from balloon and aircraft observations on 26th November 1991 along a downwind section across the Lake District in Cumbria in north-west England. Here the mean wind speed increases at a uniform rate from 15 m/s near the ground to 40 m/s near 10 km, and the lapse rate (and hence N) is approximately uniform at 7°/km over this range of heights. The figure shows a dominant mode trapped by the wind shear below about 7 km, where the phase lines are approximately vertical with $F_0 \sim 0.05$, as in example (iii) of §12.2. Above 7 km the phase lines are tilted, implying some leakage and upward propagation of energy. The simultaneous observation of the lateral behaviour of such phenomena is not possible with current techniques and resources.

12.4 Limitations and extensions of linear theory

As Nh/U is increased, the amplitude of the perturbation solutions of §12.1 increases, and the linearisation on which they are based becomes less justifiable. As shown in §13.2, for a given obstacle shape at large Reynolds number, a limit is reached beyond which the observed flow has a different character from that of the linear solution. The nature of the linear solutions suggests three possible ways in which this breakdown may occur. These are: (a) the development of stagnation

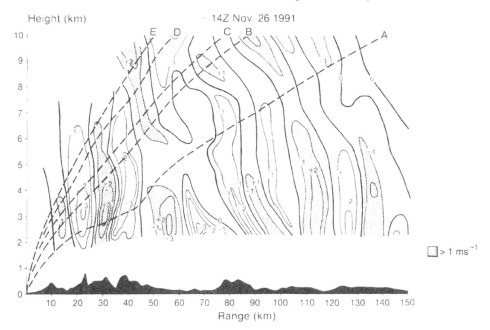

Figure 12.18 Inferred vertical velocity field from balloon and aircraft observations over Cumbria, England on 26th November 1991, in a flow from the SW with approximately uniform N and positive dU/dz. The dashed lines show the tracks of balloons with different buoyancy released at different times. The flow displays a deep mode trapped by the wind shear. (From Shutts & Broad, 1993: reproduced by permission of the Royal Meteorological Society.)

points and flow-splitting on the upstream face of the obstacle; (b) the initiation of wave-overturning aloft; and (c) "singular behaviour" on the downstream side, which may involve flow separation and a complex wake. For two-dimensional in-viscid flow with uniform U and N this question is resolved by Long's model and found to be option (b): the breakdown of the linear-like flow solutions is due to wave-breaking aloft regardless of the obstacle profile, as described in §10.3. For three-dimensional obstacles, observations show that the departure from linear-like behaviour is instead generally due to option (a): the development of a stagnation point on the upstream side and consequent flow-splitting. The details of how this occurs are described in the following sections. In particular, upper-level overturning and wake formation may affect upstream flow-splitting (see §§12.4.3, 13.3.3).

The flow fields described by Long's model are essentially non-linear, and the linear form of the equation depends on the choice of a special choice of variable, namely $\eta(x, z) = z - z_0(x, z)$, where z_0 is the elevation far upstream of the streamline passing through the point (x, z). If one makes the same assumptions and employs the same variables for three-dimensional flow, namely steady flow that is independent of z and y far upstream, from the inviscid form of (1.1), (1.3) and (1.4) one obtains the three-dimensional generalisation of (10.42) (Drazin, 1961)

$$(\nabla \times \mathbf{u}) \times \mathbf{u} + \left[U \frac{dU}{dz_0} + \frac{1}{2\rho_0} \frac{d\rho_0}{dz} (U^2 - \mathbf{u}^2) + N^2 \eta \right] \times \left(\mathbf{k} - \frac{\partial \eta}{\partial x} \mathbf{i} - \frac{\partial \eta}{\partial y} \mathbf{j} \right) = 0. \quad (12.77)$$

Here $\mathbf{i}, \mathbf{j}$ and $\mathbf{k}$ are the unit vectors in the directions of x, y and z increasing respectively, and $\mathbf{u}, \eta$ and z_0 are all functions of x, y and z. There does not appear to be any transformation that renders this complex equation into linear form, but there is scope for further analytical study.

12.4.1 Flow in isosteric coordinates

Some useful information may be obtained by expressing the equations in isosteric (constant density) coordinates (x, y, α), where $\alpha = 1/\rho$. Realistically, the application of these coordinates is restricted to situations where the vertical density gradient of the fluid is non-zero everywhere, for obvious reasons. For inviscid incompressible flow, (1.1), (1.3) and (1.4) have the isosteric form (extending Smith, 1988)

$$\frac{D_\alpha}{Dt} \mathbf{u}_H + \alpha \nabla_\alpha p + \left(1 + \frac{1}{g^2} \frac{D_\alpha^2}{Dt^2} \phi \right) \nabla_\alpha \phi = 0, \quad (12.78)$$

$$\frac{D_\alpha}{Dt} \frac{\partial \phi}{\partial \alpha} + \frac{\partial \phi}{\partial \alpha} \nabla_\alpha \cdot \mathbf{u}_H = 0, \quad (12.79)$$

$$\alpha \frac{\partial p}{\partial \alpha} + \frac{\partial \phi}{\partial \alpha} \left(1 + \frac{1}{g^2} \frac{D_\alpha^2}{Dt^2} \phi \right) = 0, \quad (12.80)$$

where $\phi = gz$ is the geopotential,

$$\mathbf{u}_H = u\mathbf{i} + v\mathbf{j}, \qquad \frac{D_\alpha}{Dt} = \frac{\partial}{\partial t} + \mathbf{u}_H \cdot \nabla_\alpha, \quad (12.81)$$

and ∇_α implies the gradient with α constant. If we make the simplifying assumption that the vertical acceleration of the fluid is much less than g, (12.78) reduces to

$$\frac{D_\alpha}{Dt} \mathbf{u}_H + \alpha \nabla_\alpha p + \nabla_\alpha \phi = 0. \quad (12.82)$$

The curl of (12.82) yields

$$\frac{D_\alpha}{Dt} \left(\frac{\nabla_\alpha \times \mathbf{u}_H}{\partial \phi / \partial \alpha} \right) = 0, \quad (12.83)$$

so that the vertical component of potential vorticity defined on isosteric surfaces is conserved. This equation is a restatement of the unforced form of (1.33). If the vertical component of vorticity is zero far upstream, it will remain zero on each isosteric surface; that is,

$$\frac{\partial u}{\partial y} = \frac{\partial v}{\partial x}, \tag{12.84}$$

where these partial derivatives have α constant. Hence we may define a velocity potential $\Phi(x, y, \alpha)$ for the horizontal motion on isosteric surfaces by $\mathbf{u}_H = \nabla_\alpha \Phi$. Note that this does not imply that the vertical component of vorticity in space coordinates (x, y, z) is zero, and in general it is not, because of tilt of the density surfaces.

For steady flow, from (12.79) we also have

$$\nabla_\alpha \cdot \left(\frac{\partial \phi}{\partial \alpha} \mathbf{u}_H \right) = 0, \tag{12.85}$$

which may be interpreted as saying that the fluid between adjacent density surfaces is conserved. Equations (12.85) may also be used to define a stream function, and to obtain the equation for Φ:

$$\frac{\partial^2 \Phi}{\partial x^2} + \frac{\partial^2 \Phi}{\partial y^2} + \frac{\partial \Phi}{\partial x} \frac{\partial}{\partial x} \ln \frac{\partial \phi}{\partial \alpha} + \frac{\partial \Phi}{\partial y} \frac{\partial}{\partial y} \ln \frac{\partial \phi}{\partial \alpha} = 0. \tag{12.86}$$

Equation (12.78) integrates to give the Bernoulli integral

$$\frac{1}{2} u_H^2 + \alpha p + \phi = \text{constant} = R, \tag{12.87}$$

along a streamline. If the flow is independent of x and y upstream, this constant is uniform on an isosteric surface, and is therefore a function of α alone, given by $R(\alpha) = \alpha p(\alpha) + \phi(\alpha) + \frac{1}{2} U(\alpha)^2$.

If we make the hydrostatic approximation, (12.80) simplifies to

$$\alpha \frac{\partial p}{\partial \alpha} + \frac{\partial \phi}{\partial \alpha} = 0, \tag{12.88}$$

whereas (12.79), (12.82) and the above deductions therefrom remain unchanged. Using (12.87), (12.88) we may then obtain

$$p = \frac{\partial C}{\partial \alpha}, \qquad \phi = C - \alpha \frac{\partial C}{\partial \alpha}, \qquad \text{where} \quad C = R(\alpha) - \frac{1}{2} \left[(\partial \Phi \partial x)^2 + (\partial \Phi \partial y)^2 \right], \tag{12.89}$$

and substituting for ϕ in (12.86) gives

$$\frac{\partial}{\partial x} \left(\frac{\partial^2 C}{\partial \alpha^2} \frac{\partial \Phi}{\partial x} \right) + \frac{\partial}{\partial y} \left(\frac{\partial^2 C}{\partial \alpha^2} \frac{\partial \Phi}{\partial y} \right) = 0, \tag{12.90}$$

which is an equation for Φ alone.

If the lower boundary at $z = h(x, y)$ coincides with a surface of constant density, which it does if Nh/U is sufficiently small, the lower boundary condition at $\alpha = \alpha_0$ say, may be expressed as

$$\phi(x, y, \alpha_0) = gh(x, y). \tag{12.91}$$

If Nh/U is large enough some of the density surfaces will be penetrated by the topography, so that some of the streamlines on the topography may have different fluid densities. For this more general situation the boundary condition takes the form

$$\frac{D_\alpha}{Dt}[\phi(x, y, \alpha) - gh(x, y)] = 0, \qquad \text{on} \quad \phi = gh. \tag{12.92}$$

This implies that a streamline on the topography remains there. However, it is possible for streamlines to "land on" and "take off from" the topography through singular points and lines where (12.92) does not apply, as will be discussed in §13.1. Such points and lines are an integral part of the flow pattern in situations where (12.91) is not applicable.

At large heights (large enough α), for flow over isolated topography the disturbance will be small and the equations effectively linear, and the radiation condition for linear theory is appropriate. Hence the upper and lower boundary conditions may be applied exactly, in the same manner as for linear equations. Unfortunately, the field equations are still inherently non-linear, for both the hydrostatic and non-hydrostatic forms, and some approximation is necessary to obtain solutions by analytical means.

12.4.2 Linear hydrostatic flow in isosteric coordinates

It appears that the closest we can get to a three-dimensional equivalent of the two-dimensional linear Long-model equation is obtained by linearising (12.82) and (12.79), with (12.88), about a mean upstream state. Writing $u = U(\alpha) + u'$, $v = v'$, $\phi = \Phi(a) + \phi'$, we may obtain an equation for ϕ'

$$\frac{\partial^4 \phi'}{\partial \alpha^2 \partial x^2} + \frac{g^2}{U^2 N^2 \alpha^2}\left(\frac{\partial^2 \phi'}{\partial x^2} + \frac{\partial^2 \phi'}{\partial y^2}\right) = 0, \tag{12.93}$$

where it is assumed that U and N are uniform upstream. This equation may be expressed in terms of the undisturbed height $z_0(\alpha)$ of a streamline and its vertical displacement from this height, $\eta = z - z_0$, which with the Boussinesq approximation has the form

$$\frac{\partial^4 \eta'}{\partial x^2 \partial z_0^2} + \frac{N^2}{U^2}\left(\frac{\partial^2 \eta}{\partial x^2} + \frac{\partial^2 \eta}{\partial y^2}\right) = 0, \tag{12.94}$$

where the partial derivatives here relate to the coordinates (x, y, z_0). The lower boundary condition may then be expressed in the exact form

$$\eta(x, y, z_0 = 0) = h(x, y). \tag{12.95}$$

Equation (12.94) is identical to the hydrostatic form of (12.3), except that z has been replaced by z_0 and the lower boundary is a streamline so that the boundary condition is satisfied exactly. Consequently, the general solution for arbitrary topography has the form (12.14), and for obstacles of the form (12.54) with $\upsilon = 3/2$, namely

$$h(x, y) = \frac{h_{\mathrm{m}}}{\left[1 + \left(\frac{x}{a}\right)^2 + \left(\frac{y}{b}\right)^2\right]^{3/2}}, \tag{12.96}$$

the solution is given by (12.37), so that

$$\eta = \frac{h_{\mathrm{m}}}{2\pi} \int_0^{2\pi} \frac{\exp i \left[\frac{N z_0 (\cos^2 \gamma + \frac{a^2}{b^2} \sin^2 \gamma)^{1/2}}{U \cos \gamma}\right]}{[1 - is \cos(\gamma - \omega)]^2} d\gamma, \tag{12.97}$$

where

$$s = \left(\frac{x^2}{a^2} + \frac{y^2}{b^2}\right)^{1/2}, \qquad \tan \omega = \frac{bx}{ay}. \tag{12.98}$$

Figure 12.6 shows the displacement field in the plane $y = 0$ computed from these equations for the case $a = b$, $Nh/U = 0.5$. Computations show that the range of applicability of these solutions is limited by a convergence, or "collapse", of the streamlines on the lee side of the obstacle, which occurs when $Nh/U \geq 0.5$. This is an artefact of the linearisation of the equations. Essentially, it means that the streamlines overlap or intersect within a region above the lee side, and this region is found to broaden in the x-, y- and z-directions as Nh/U increases. If $b/a \ll 1$, one may deduce from (12.97) that this lee-side streamline collapse on $z_0 = 0$ occurs for $Nh_{\mathrm{m}}/U = O(b/a)$, but for two-dimensional obstacles where $b/a \to \infty$, on the other hand, it occurs for $Nh/U \approx 2$, implying a monotonic increase with b/a. Hence the linear solution (12.97) is not valid in this region on the lee side, but it may still be a useful guide to the flow outside this region.

Aside from this lee-side streamline collapse, the linear solution (12.97) predicts that for $a = b$, a stagnation point occurs on the upstream side of the obstacle when $Nh_{\mathrm{m}}/U \approx 1.3$, and at larger values for $b/a < 1$, smaller for $b/a > 1$. For $a = b$ it also predicts that an upper level stagnation point will occur at approximately the same value. Which one occurs for the lower value of Nh/U in practice must depend primarily on non-linear factors, to which we now turn.

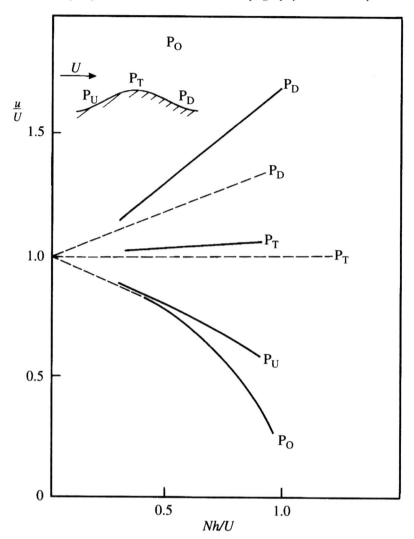

Figure 12.19 Flow speeds at the four points P_O, P_U, P_T and P_D for hydrostatic flow past the Gaussian hill (12.96), as a function of $Nh/U < 1$, using an operational mesoscale numerical forecasting model (modified from Smith & Grønås, 1993), compared with the dashed-line trends from linear theory.

12.4.3 Numerical computations of hydrostatic nearly linear behaviour

Numerical studies of flow past three-dimensional topography are described in more detail in §13.3. Our objectives here are limited to inferences about the limits to linear-like behaviour. If the hydrostatic inviscid forms of the equations (1.1)–(1.4) are scaled in the manner suggested by linear theory in §12.1, where x and y are scaled with a, z with U/N, u' and v' with Nh_m, and w' with Uh_m/a, the resulting non-linear equations and boundary conditions contain only one dimensionless parameter: Nh/U. The development with time of the approach to steady state for flow over a Gaussian hill of the form

$$h(x, y) = h_{\mathrm{m}} e^{-(x^2+y^2)/a^2}, \tag{12.99}$$

has been described by Smith & Grønås (1993) using an operational sigma (pressure)-coordinate model. In a number of runs, Nh/U was progressively increased by slowly decreasing U, and the wind speeds at four critical points of the flow were monitored. These points were: the upper level point of minimum u above the lee slope (P_O), the point of minimum u on the upwind slope (P_U), the point of maximum windspeed on the lee slope (P_D), and the point on the mountain top (P_T).

The wind speeds at these four points are shown in Figure 12.19, with the dashed lines showing the trends for linear theory. The results implied that a single steady state exists for $Nh/U < 1$, with the values represented by the solid curves. For this obstacle shape, the velocities at points P_O and P_U initially decrease together as Nh/U increases, but eventually that at P_O decreases faster, and u there becomes zero at a value of about 1.1. This would suggest that upper-level overturning occurs before stagnation on the upwind surface for hydrostatic flow past axisymmetric obstacles. But since the latter is observed to occur when $Nh/U > 1.1$ (see §13.2), one may infer instead that upper-level overturning helps its occurrence. For $Nh/U > 1$ the time-dependent numerical integrations behaved differently without appearing to approach a steady state, implying that the flow is in a different regime. We return to numerical studies of this parameter range in §13.3.

Before we discuss any further theoretical aspects it is appropriate to examine the observed properties of flows past topography over the whole range of Nh/U. Many of these properties are compatible with the linear theory of the preceding sections, but other flows have a different topological structure, and we return to look at these from a theoretical viewpoint in §13.3.

12.4.4 The effect of critical levels on 3D orographic flows

If the velocity of unidirectional density-stratified flow over an obstacle decreases with height, the flow may contain a critical level. Flow over the obstacle generates lee waves in the usual manner which propagate upwards, and encounter the critical level as described in §8.9. An analysis of this phenomenon with linear theory and numerical simulations (Grubišić & Smolarkiewicz, 1997) with Richardson number $R_i > 1$ shows that the main features of the two-dimensional analysis of critical layers (§8.9) are also prominent in the 3D wave field. Here the critical level lies above the maximum height of the obstacle, and linear theory shows that a 3D wave field resembling that for uniform flow (such as those depicted in Figures 12.2–12.5) is generated beneath it. The asymptotic far-field solutions yield waves confined to paraboloidal envelopes that widen quickly with height. Hence the wave fronts tend to become normal to the mean flow as the critical level is approached (rather than as $z \rightarrow \infty$). On passing through the critical level the waves are strongly attenuated, but those more elongated with the flow are attenuated more severely than those perpendicular to it. As a result, the wave field above the critical level has reduced amplitude and is primarily oriented normal to the flow direction. These results are generally confirmed by non-linear numerical simulations, but they depend on the obstacle height being of modest amplitude.

13

Three-dimensional stratified flow over finite obstacles

Fire, brimstone and pits bottomless swallow you all alive,
in case you do not firmly believe all that I shall relate unto you in this present
chronicle.

FRANÇOIS RABELAIS, *Works, Prologue to Book II.*

13.1 The topology of the flow field on the surface of an obstacle

Flow patterns over three-dimensional obstacles can be much more varied and complex than those for two-dimensional obstacles. Before describing observations in detail, it is useful to discuss the possible structures of flow fields on the surface of a solid obstacle. Such patterns are readily observed, and they provide a helpful framework for interpreting the character of the flow pattern as a whole. Knowledge of these surface flows reveals the conditions under which elevated fluid from upstream may come in contact with (i.e. become attached to) the surface of the obstacle, and when fluid on the surface may separate from it. Such separation or attachment may occur at lines or at isolated points on the surface, and a possible structure for separation is also one for attachment, with the flow directions reversed. These considerations are based on the topology of vector fields, and are independent of the dynamics involved in producing them.

We suppose that at any point on the surface we may define a curvilinear coordinate system (x_1, x_2, x_3) that is locally Cartesian, in which the surface of the obstacle is specified by $x_3 = 0$, and neighbouring points on the surface are specified by values of x_1 and x_2. Unit vectors in the directions of x_1, x_2 and x_3 increasing are denoted by $\hat{\mathbf{x}}_1$, $\hat{\mathbf{x}}_2$ and $\hat{\mathbf{x}}_3$ respectively, so that the normal to the surface at $(x_1, x_2, 0)$ is given by $\hat{\mathbf{x}}_3$. For inviscid flow, at a given time t the motion on the surface of the obstacle may be described by a velocity field $\mathbf{v}(x_1, x_2)$. For viscous flow, the fluid velocity is zero on the surface, and the surface flow is characterised instead by a surface stress

$\tau(x_1, x_2)$, given by

$$\tau = \mu \frac{\partial \mathbf{v}}{\partial x_3}(x_1, x_2, 0) = (\tau_1, \tau_2),$$

where τ_1, τ_2 denote the components in the x_1- and x_2-directions respectively. Since all experiments are conducted with viscous fluids, the motion of tracers on or very close to the surface usually occurs within a thin boundary-layer, and their streamlines may be taken to give the local direction of τ if the variation of the flow with time is not too rapid (Hunt et al., 1978). Such observations may be compared with results from an inviscid theoretical or numerical model by comparing the direction of τ with that of $\mathbf{v}$; these directions should coincide if the inviscid model surface velocity describes the flow just outside the boundary-layer.

The relative velocity field near a point P on a two-dimensional surface may be described locally as a pure straining motion plus a rigid rotation (Batchelor, 1967), and this may be used to interpret possible flow patterns on the surfaces of obstacles. We assume here that the flow is steady, which is adequate for subsequent applications. We define the surface flow in terms of τ (for the inviscid case one may substitute $\mathbf{v}$ for τ in what follows). Near point P on the surface the stress field relative to that at P is given by, to leading order,

$$\begin{bmatrix} \tau_1 \\ \tau_2 \end{bmatrix} = \begin{bmatrix} \partial\tau_1/\partial x_1 & \partial\tau_1/\partial x_2 \\ \partial\tau_2/\partial x_1 & \partial\tau_2/\partial x_2 \end{bmatrix} \begin{bmatrix} x_1 \\ x_2 \end{bmatrix}, \tag{13.1}$$

where the gradients in the matrix are evaluated at the point P. From the stress gradient matrix we may define three scalar quantities: J, Δ and Ω, that are independent of the choice of axes:

$$J = \det \begin{bmatrix} \partial\tau_1/\partial x_1 & \partial\tau_1/\partial x_2 \\ \partial\tau_2/\partial x_1 & \partial\tau_2/\partial x_2 \end{bmatrix} \qquad \begin{aligned} \Delta &= (\partial\tau_1/\partial x_1) + (\partial\tau_2/\partial x_2), \\ \Omega &= (\partial\tau_2/\partial x_1) - (\partial\tau_1/\partial x_2). \end{aligned} \tag{13.2}$$

One may rotate the axes to new coordinates y_1, y_2 with stresses τ_1', τ_2', so that (13.1) becomes

$$\begin{bmatrix} \tau_1' \\ \tau_2' \end{bmatrix} = \begin{bmatrix} T_{11} & \Omega \\ -\Omega & T_{22} \end{bmatrix} \begin{bmatrix} y_1 \\ y_2 \end{bmatrix}, \tag{13.3}$$

where $T_{11} = \partial\tau_1'/\partial y_1$, $T_{22} = \partial\tau_2'/\partial y_2$ are the principal stress gradients. The four components of the matrix of (13.2) have been reduced to the three variables T_{11}, T_{22} and Ω, and we have

$$J = T_{11}T_{22} + \Omega^2, \quad \Delta = T_{11} + T_{22}. \tag{13.4}$$

It is customary to use J, Ω and Δ to specify the local properties because they are axes-independent; T_{11} and T_{22} may then be obtained from (13.4). One of these may be taken as a scale factor, so that the number of parameters determining the local behaviour is reduced to two: J/Δ^2 and Ω/Δ, for example, if $\Delta \neq 0$ and its sign is chosen.

For inviscid flow, τ is interpreted as a velocity field, Δ and Ω are the divergence and vorticity on the surface, and T_{11} and T_{22} are the principal rates of strain. Since the normal velocity $v_3 = 0$ on the surface, near the point P we have $v_3 = -\Delta.x_3$. Hence if $\Delta \neq 0$ there is motion towards (or away from) the surface close to point P. However, integrating this relation for constant Δ shows that a fluid particle would take a logarithmically infinite time to reach (or leave) the surface from (or to) a finite distance above it. Hence, for inviscid fluids, even if $\Delta \neq 0$ over a finite region of the surface of an obstacle, this would preclude separation and attachment, except for points where $\mathbf{v} = 0$. Fluid particles on streamlines intersecting the latter points may reach the surface because they have infinite time to do so in steady flow.

For viscous flows described in terms of τ, near point P we have $v_3 = -C\Delta x_3$, where C is a positive constant. The definition of whether or not a fluid particle makes contact with the surface is less clear-cut. For practical purposes the relevant criterion for separation or attachment may be taken to be whether or not a fluid particle enters or leaves a boundary-layer. In this case separation or attachment may occur if $\Delta \neq 0$ along the path of a fluid particle for a finite time. We return to this possibility below.

For the above reasons, points on the surface where $\tau = 0$ have special importance and are termed *singular* points (or *stagnation* points when relating to $\mathbf{v}$) for the vector field, because more than one vector (or streamline) may pass through them. The nature of the flow on the surface in the neighbourhood of such points is specified by the values of J, Δ and Ω, as shown in Figure 13.1 (see, for example, Coddington & Levinson, 1955; Lighthill, 1963; Perry & Fairlie, 1974). If $J < 0$, the point is a *saddle point*, with two intersecting surface flow lines called *separatrices*, one of which has its vectors directed toward the singular point, and the other away from it. All other neighbouring flow lines on the surface are asymptotic to these two lines, one having the character of a line of separation, and the other of a line of attachment. If $\Delta > 0$, the separatrix of attachment extends off the surface as a *plane of attachment*, containing vectors directed toward the saddle point. If $\Delta < 0$, on the other hand, the other separatrix extends from the surface as a *plane of separation*, containing vectors directed away from the saddle point, as shown in Figure 13.2. With a saddle point, therefore, fluid from a range of directions and locations within a plane becomes attached to the surface, or contains fluid that has separated from it.

If $J > 0$, the singular points are termed *nodes*, where every neighbourhood surface stress line is directed into ($\Delta < 0$) or out of ($\Delta > 0$) the point. If $0 < J < (\Delta/2)^2$, all of these lines are tangent to one line at the point, with one exception. Hence there is one prominent line to which the others converge, or diverge from. If instead $J > (\Delta/2)^2$, every line is an infinite spiral into ($\Delta > 0$) or out of ($\Delta < 0$) the

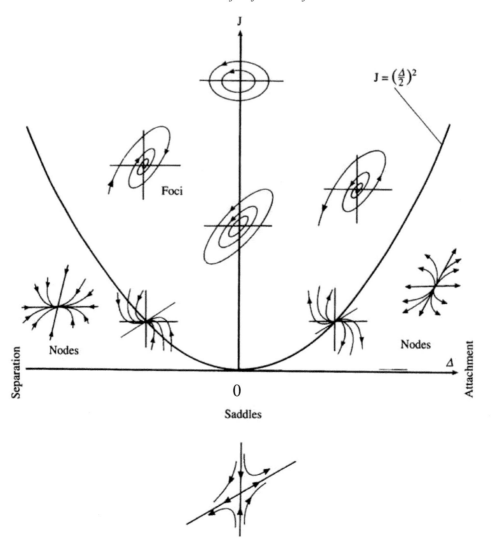

Figure 13.1 The (J, Δ)-plane, showing examples of the structure of the inviscid surface flow at singular points in various regions. To define the structure uniquely, J, Δ and Ω must all be specified. Note the three flow types: foci, nodes and saddles. For viscous flow, these flow patterns exist a short distance above the no-slip surface. Δ and J are both zero at the origin. Ω is positive in some cases, negative in others.

point, with the direction of rotation depending on the sign of Ω. Such nodal points are termed *foci*; they occur when Ω is large or the boundary is rotating, but are not important for our discussion here. For all nodes, there is a single streamline or trajectory from off the surface that passes through the singular point, directed inward for $\Delta > 0$ and outward for $\Delta < 0$ (see Figure 13.2). Saddle points and nodes

$\Delta > 0$ $\Delta < 0$

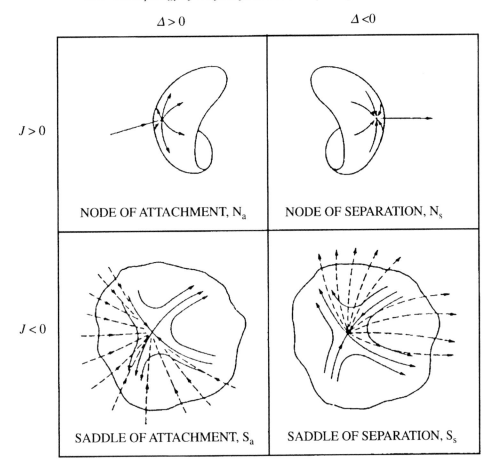

	$\Delta > 0$	$\Delta < 0$
$J > 0$	NODE OF ATTACHMENT, N_a	NODE OF SEPARATION, N_s
$J < 0$	SADDLE OF ATTACHMENT, S_a	SADDLE OF SEPARATION, S_s

Figure 13.2 The three-dimensional structure of the flow at nodes and saddles of attachment and separation, drawn for cases where $\Omega = 0$. (Adapted from Chapman & Yates, 1991: reproduced by permission of ASME.)

of attachment and separation may be conveniently denoted by Sa, Ss, Na and Ns respectively.

A wide variety of these singular points may coexist on any given surface, but there are restrictions of a book-keeping nature that relate the numbers of each type. On an isolated obstacle such as a sphere or its topological equivalent, from the Poincarè–Bendixson (or "hairy-sphere" theorem), one may deduce that the number of nodes Σ_N and the number of saddles Σ_S are connected by (Lighthill, 1963)

$$\Sigma_N - \Sigma_S = 2. \qquad (13.5)$$

For flow past an isolated solid obstacle situated on an infinite plane surface, which is our primary concern here, (13.5) may be modified to give (Hunt et al., 1978)

$$\Sigma_N - \Sigma_S = 0, \tag{13.6}$$

where the two missing nodes have moved out to upstream and downstream infinity.

This classification of singular points on the surface of the topography or obstacle may be extended to other two-dimensional sections of the flow field, such as a vertical section through an axis of symmetry, for instance. Ignoring the velocity component normal to this section, there may be nodes and saddles in the flow in the section, remote from the topography, and nodes and saddles on the surface of the topography. These singular points on the surface are only half-present, since the other half would lie within the obstacle. Consequently they are only counted as half-nodes and half-saddles in the book-keeping accounts. For a vertical section above a rigid boundary the equivalent of (13.6) is then

$$\Sigma_N + \frac{1}{2}\Sigma_N' - (\Sigma_S + \frac{1}{2}\Sigma_S') = 0, \tag{13.7}$$

where Σ_N' and Σ_S' respectively denote the number of half-nodes and half-saddles on the obstacle surface. If instead a horizontal section is taken, which slices through an obstacle so that there is a single isolated body in the section, the sum in (13.7) totals -1 rather than zero. For more complex shapes the reader is referred to Hunt et al. (1978).

The location and type of the singular points on the surface of an obstacle and on suitably chosen sections provides a framework for the description and classification of flows past most shapes, and seven basic types have been proposed by Chapman & Yates (1991) on this basis. Not all of these are relevant here, and instead we restrict consideration to the forms that flow separation takes in stratified flows. From various observations, some of which are described below, we may identify five types of flow separation, some of which are in common with those of Chapman & Yates.

The first type (type I) of separation (or attachment) is the *simple node*, where one streamline in the flow impinges on the surface and connects with an infinite number of surface flow (shear stress) lines. The second type (type II) is the *simple saddle point*, where an infinite number of streamlines within a plane impinge on the surface and connect with two lines along the *separatrices*, or lines of separation. For separation, all the fluid within the plane emanates from these two directions on the surface, and they are depicted in Figure 13.2. In both of these types, separation or attachment only occurs at a single point on the surface. The third type (type III) involves a singular line of points (along which $\tau = 0$), so that this is effectively a three-dimensional extension of the two-dimensional case. This may occur on obstacles that are axisymmetric about the flow direction in homogeneous flows.

The fourth type (type IV) occurs on lines that originate from a saddle point and extend to a node, where τ or $\mathbf{v}$ is directed along the line (in either direction). This separation is not described by the inviscid equations and depends on the broader definition for viscous flows, namely whether or not the fluid leaves the boundary-layer. It stems from the plane of the separatrix of the saddle point, and depends on sufficiently large values of $|\Delta|$ along the line. The location of such lines is an important part of the overall flow. In the case of attachment, the impinging fluid comes from the same plane as for type II, but avoids passing through the saddle point and takes shorter routes to the surface. Hence the overall effect is similar to that of type II, but the fluid paths are "rounded off".

The fifth type (type V) of separation occurs along lines that are not associated with singular points on the surface of the obstacle, but are associated with *half-singular points* on the obstacle in sections through the flow. In this case, separation occurs as a result of the gradual convergence of stress lines (or streamlines) on a smooth surface, or toward a sharp edge, where $|\Delta| \neq 0$, as above. The associated flow topology in the cross-flow plane (i.e. the plane perpendicular to the main flow) shows a *half-saddle point* on the obstacle at the point of convergence, and this has been termed "cross-flow separation" by Chapman & Yates.

Stratified flows provide some good examples of such topological structure, as described in the next section.

13.2 Observations of the flow past three-dimensional obstacles

Virtually all observational studies of flow past three-dimensional obstacles have been carried out in flows of finite depth D, with $D/h_{\mathrm{m}} \geq 4$. This situation may be assumed to give a reasonable approximation to infinite depth provided that the obstacle is not too long, in contrast with two-dimensional situations where it does not. These studies are of flow past obstacles mounted on a plane horizontal surface at large Reynolds number, where these are either towed through stationary fluid or placed in a stratified stream. Experiments on flow past obstacles isolated in space are also relevant to this geometry, and descriptions of flow past a sphere are included (§13.2.4).

As described in Chapter 9, for steady external conditions two-dimensional flows past topography tend to reach an approximate steady state near the obstacle in many cases, but for some this could take a very long time, and for others there is an inherent periodic variability. The situation for flow past three-dimensional obstacles is similar, but the sources of the variability are different, and are due to periodicity in the wake and overturning on the upstream side. Observational studies have concentrated on the flow on the surface of the obstacle and in the central vertical plane (usually of symmetry). Techniques used have included dye released upstream and on the surface, shadowgraph, Schlieren, and laser-illuminated slices of the flow containing tracers.

13.2.1 More theoretical preliminaries

Because of the variety of three-dimensional shapes, there are more parameters involved than for two-dimensional flow. If the obstacle has half-widths A and B in the along-stream and across-stream directions respectively ($A \approx 2a$, $B \approx 2b$ for obstacle shapes like (12.33)), and is symmetric about these two directions, the major dimensionless parameters are: Nh/U, h/A and A/B. If the obstacle is not so symmetric, other details such as orientation and shape, and the values of A and B on the upstream and downstream or left and right sides (A_u, A_d, B_l, B_r) may be significant. For the most part we discuss doubly symmetric obstacles here, where the mean flow is directed along one axis.

We consider the effect of different values of B/A on the extent of upstream influence of the obstacle, for the interpretation of some of the observations described below. From §8.11, the equation for steady linear wave motion in uniform unidirectional flow in terms of the vertical displacement η is

$$(\eta_{xx} + \eta_{yy} + \eta_{zz})_{xx} + \frac{N^2}{U^2}(\eta_{xx} + \eta_{yy}) = 0. \tag{13.8}$$

If we consider approximately columnar modes that have vertical wavenumber n so that $\eta_{zz} = -n^2\eta$, the motion is effectively hydrostatic and satisfies

$$\left[1 - \left(\frac{nU}{N}\right)^2\right]\eta_{xx} + \eta_{yy} = 0. \tag{13.9}$$

If $|n| < N/U$, the criterion for waves to be able to propagate upstream, (13.9) is Laplace's equation if x is replaced by $X = x/[1 - (nU/N)^2]^{1/2}$.

In stratified flow over two-dimensional barriers, if Nh/U is large enough the flow has a hydraulic transition over the obstacle (see §11.4.3), which acts as a source for upstream waves. This may also occur for three-dimensional obstacles, where the wave-source now has finite crossflow length. If we model this source simply by specifying a forcing at $x = 0$ that is periodic in y

$$\eta = 1 + \cos my, \quad \text{at} \quad x = 0, \tag{13.10}$$

the solution to (13.9) on the upstream side ($x < 0$) is

$$\eta = 1 + e^{mX} \cos my, \qquad x < 0. \tag{13.11}$$

This implies that the effects of the obstacle extend a distance $[1-(nU/N)^2]^{1/2}/m$ upstream, plus an x-independent component that corresponds to the two-dimensional effect. For an isolated obstacle of width $2B$, this steady component vanishes but the distance of upstream influence is the same, namely $[1 - (nU/N)^2]^{1/2}B$, although the decay with $|x|$ is now algebraic rather than exponential.

If Nh/U is large enough, low-level flow may be forced to pass around an isolated obstacle in an approximately two-dimensional fashion. This implies approximately potential flow (see §13.3), and solutions of flow past obstacles of elliptical shape (see, for example, Milne-Thomson, 1960) show that the effect of the obstacle is felt on a scale of distance B upstream, where $2B$ is again the breadth of the obstacle transverse to the stream. The existence of flow separation on the lee side does not affect this scale. Hence, for both columnar waves and lateral flow, the upstream effect produced by an isolated obstacle has the scale of the obstacle half-width.

In the following description of flow past three-dimensional obstacles we start with small values of Nh/U and proceed to progressively larger values, so that the flow properties may be related to the neutral ($N = 0$) case, and to the linear perturbation solutions of the previous sections. From these observations we note three particular flow properties that help to characterise the overall flow. These are:

(i) the nature and position of upstream stagnation points and separation, where these occur, and the origin of the fluid that makes contact with the obstacle surface;

(ii) the presence or otherwise of upper-level wave-overturning (as in two-dimensional flows); and

(iii) the presence and nature of a lee-side separation region and associated wake.

There are some important observable quantities that relate to item (i) that will be used below as descriptors. These are the heights $z(N_a)$, z_r, z_s (i.e. $z_{stagnation}$) and z_t, (i.e. z_{top}) which are shown in the inset in Figure 13.10a, and defined as follows. First, $z(N_a)$ is the height of a node of attachment N_a on the upstream face of the obstacle, where such a node exists. Next, z_t is the height far upstream of the streamline that touches the crest of the obstacle ($h = h_m$). Third, z_s is the height far upstream of the streamline that impinges directly on the nodal point N_a (see §13.1), and z_r is the height of the same streamline immediately upstream of the obstacle, near the point $x = -A$, say. Finally z_r is often not well-defined and will be used only when it is, namely when $B/A \gg 1$. Depending on obstacle shape, z_s may be approximately equal to z_t or z_r.

Some important inferences may be obtained from Bernoulli's equation, which for a streamline in steady flow is

$$p + \frac{1}{2}\rho u^2 + \rho g z = p(z_0) + \frac{1}{2}\rho U^2 + \rho g z_0, \qquad (13.12)$$

where $z_0(x, y, z)$ is the elevation of the streamline far upstream as in §10.3. For present purposes we assume that the flow is hydrostatic, that the Boussinesq approximation is valid, and U and N are uniform upstream. If we write $\eta(x, y, z) = z - z_0$, the difference between the pressure $p(x, y, z)$ and its value far upstream at the same level $p_{-\infty}(z)$ may be expressed as

$$p^*(x, y, z) = p(x, y, z) - p_{-\infty}(z) = \rho(z_0)N^2 \int_z^\infty \eta\, dz, \qquad (13.13)$$

which, with a change of variable from z to z_0, may be written (Smith, 1988) as

$$p^* = \rho(z_0)N^2 \left[\int_{z_0}^\infty \eta(x, y, z_0')\,dz_0' - \frac{1}{2}\eta^2 \right]. \qquad (13.14)$$

Substituting this into (13.12) gives

$$\mathbf{u}^2 = U^2 - \frac{2p^*}{\rho(z_0)} - N^2\eta^2 = U^2 - 2N^2 \int_{z_0}^\infty \eta\, dz_0'. \qquad (13.15)$$

We may differentiate this equation with respect to z_0 to obtain

$$\frac{\partial \mathbf{u}^2}{\partial z_0} = 2\mathbf{u} \cdot \frac{\partial \mathbf{u}}{\partial z_0} = 2N^2\eta. \qquad (13.16)$$

At a node of attachment on the obstacle surface where $\mathbf{u}$ vanishes, (13.16) implies that $\eta = 0$. Hence, for the streamline impinging on a node of attachment on the upstream face of the obstacle, we have

$$z_s = z(N_a), \qquad (13.17)$$

for steady hydrostatic flow, and the excess pressure $p^* = \frac{1}{2}\rho U^2$, at N_a. Similarly, the elevation of an upper-level stagnation point (where $\mathbf{u} = 0$) or local velocity minimum (where $\partial \mathbf{u}/\partial z_0 = 0$) will equal the upstream elevation of the streamline passing through it.

13.2.2 *Flow of homogeneous fluid* ($N = 0$)

The flow of homogeneous fluid (i.e. $N = 0$) past obstacles at large Reynolds number is a vast subject, for which only a brief summary is appropriate here. Many of the details are still the subject of current research, particularly for bluff obstacles, because of the difficulties of observing the various properties of complex time-dependent three-dimensional flow fields. For inviscid flows where U is uniform upstream, the flow is irrotational so that $\nabla \times \mathbf{u} = 0$, and the total velocity field $\mathbf{u}$ may be expressed as $\mathbf{u} = \nabla \phi$, where $\phi(x, y, z)$ is a velocity potential. Conservation of mass ($\nabla \mathbf{u} = 0$) then implies that $\nabla^2 \phi = 0$, which with the boundary condition that the normal component of $\mathbf{u}$ at the boundary must vanish, specifies "potential flow". In a general sense, this provides the smoothest and simplest flow pattern over a complex obstacle shape that is possible. For an incompressible fluid, this flow pattern adjusts instantaneously to changes in U if the latter varies with time.

For flow of *viscous* fluids with uniform U upstream past an obstacle at large Reynolds number R_e, the boundary condition is now $\mathbf{u} = 0$ on the surface, and the flow pattern departs from potential flow because of boundary-layers on the surface and separation of the flow from it. Such separation may occur in various places depending on the obstacle shape: on the upstream face of the obstacle if it is sufficiently steep, or over the obstacle, or downstream in an extensive wake region. This implies that there are substantial volumes of fluid where the flow pattern is affected by boundary processes and singular points. In a general sense, the flow outside these regions is closely approximated by potential flow around them, but this is of little practical use since the shape of these regions and the interchange of external fluid with them must be known before ϕ can be determined. In general, a separated wake behind an isolated obstacle at large R_e does not re-attach and form a closed bubble, but is instead made up of a complex turbulent region that trails downstream. A wide variety of topological flow patterns have been inferred from observations. These depend on obstacle shape, orientation and Reynolds number, and I will only briefly describe two representative examples, shown in Figures 13.3 and 13.4.

Figure 13.3a,b show the flow of homogeneous fluid on the surface and in the centre-plane of symmetry past a smooth axisymmetric obstacle with maximum slope of unity and $h/A \approx 0.5$, situated on a flat plate and towed through stationary fluid at large Reynolds number ($\approx 10^4$). A boundary-layer forms on the surface, beginning at the leading edge of the plate. Separation first occurs on the centre-line before the summit of the "hill" is reached, but near the crest the structure is complicated with three singular points rather than just one: two saddle points of separation denoted S_s, and a node of attachment, N_a (these symbols are used to denote the points themselves as well as their type, if there is no ambiguity). Two symmetrical lines of separation or attachment stem from each of these points, one on each side and trailing downstream. Fluid away from the surface leaves or joins the surface layer on these lines. The nett effect of these three closely spaced singular points and lines is that of a single saddle and its symmetrical lines of separation.

On the downstream side of the obstacle there are two symmetrically placed nodes of separation N_s, and a saddle point of attachment S_a on the centre-line, with an associated line of attachment. These singular points satisfy (13.6), although the ends of the attachment and separation lines are not specified, presumably because the associated motions are too unsteady. The observed flow pattern is complex and the description is incomplete, but these lines of separation appear to be of type IV of §13.1.

The inferred overall mean flow pattern, in simplified form, is shown in Fig-

(a)

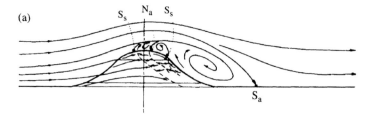

(b)

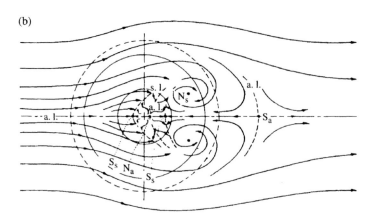

(c)

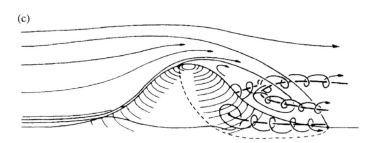

Figure 13.3 (a) Side-view of the mean surface shear stress pattern and streamlines on the centre-plane of symmetry for homogeneous flow at large Reynolds number with a turbulent boundary-layer, over an obstacle of the form (termed "polynomial hill" by Snyder et al. (1980): $z = h_m \left(\dfrac{1.04}{1 + \bar{r}^4} - \dfrac{0.083}{1 + (a\bar{r} - r_1)^2/a_1^2} - 0.03 \right)$, where $\bar{r} = (x^2 + y^2)^{1/2}/a$, $h_m = a = 22.9\,\text{cm}$, $r_1 = 20.3\,\text{cm}$, $a_1 = 7.6\,\text{cm}$. N_a, N_s, S_a, S_s denote nodes and saddle points of attachment and separation (see §13.1). (b) As for (a) showing a plan view of the pattern of surface stress; a.l. and s.l. denote lines of attachment and separation respectively. (c) An inferred picture of the three-dimensional flow pattern of (a) and (b). An instantaneous flow may deviate considerably from this mean, particularly in the wake region. (From Hunt & Snyder, 1980, reproduced with permission.)

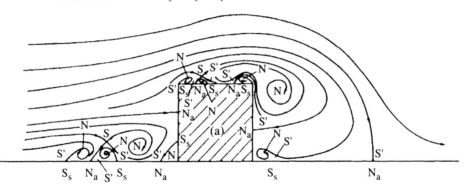

Figure 13.4 Flow on the central plane of symmetry of homogeneous fluid past a cuboid (volume $h_m ab$) with $h_m = 6.5\,\text{cm}$, $a = 3.25\,\text{cm}$, $b = 10\,\text{cm}$ at large Reynolds number with a turbulent boundary-layer in a wind tunnel. This shows singular points on the surface flow (S_s, N_a) and in the flow in the central plane (S, N, S'), as interpreted by Hunt et al. (1978), reproduced with permission.

ure 13.3c. The wake region downstream of the horseshoe-shaped separation line contains two distinct counter-rotating horseshoe vortices. One of these connects with the downstream vortex in the centre-plane shown in Figure 13.3a, and the second emanates from the two nodes on the surface of the obstacle (so that the apex of this horseshoe is inside the obstacle). The vorticity in the first (uppermost) of these horseshoe vortices comes from the flow in the boundary-layer upstream of the first separation line. On reaching the latter the fluid leaves the vicinity of the surface, and then rolls up to form the vortex. The vorticity in the second (lower) vortex of Figure 13.3c, of opposite sign to the first, comes from the flow in the boundary-layer on the downstream side of the obstacle, where the fluid leaves the vicinity of the surface at the two nodes of separation, N_s in Figure 13.3b. In practice there is considerable unsteadiness and variability in the flow about this mean, particularly in the wake where substantial lateral oscillations and asymmetries may occur.

Figure 13.4 shows the flow pattern on the centre-plane of flow incident on a cuboid (with $b/a = 3.1$), oriented normal to the flow in a wind tunnel. Here the upstream flow is turbulent, with the logarithmic wind profile in the mean. Some similarities to Figure 13.3 are evident, but there is substantial additional complexity due to the steep sides and the sharp corners. Saddle points of separation (S_s) and nodes of attachment (N_a) in the surface flow pattern alternate on this central plane as in Figure 13.3, but now there are seven of each. The last of these is now a node of attachment N_a rather than a saddle S_a. Further, in the flow in this plane there are nodes N and saddles S in the flow, and every node or saddle in the surface flow becomes a half-saddle S' in the central plane. The sum total of these points satisfies (13.7).

In practice, for sufficiently large R_e there may well be more detail than shown on a figure such as this, but if present, such additional motions are of weaker velocity and smaller scale. Because of the bluff forward face of the cuboid and the vorticity in the upstream boundary-layer, there is upstream separation and a number of low-level vortices, one of which is dominant. The associated separation lines and vortices extend around the obstacle on each side. There is also some separation on the flat upper surface, and a major vortex on the lee side, paralleling that shown in Figure 13.3a. Substantial changes to this pattern may occur if the obstacle is inclined to the flow direction or the shape is varied (Hunt et al., 1978).

13.2.3 Flow at finite Nh/U past obstacles with circular horizontal cross-section

Here we have (half-length) $B = A$ (half-width), so that the important parameters are Nh/U and h/A. For practical reasons, most experimental studies have used obstacles with maximum slopes of order unity ($h/A \lesssim 1$), so that linear perturbation theory of §12.1 is not formally valid even when Nh/U is small. Fortunately, it nevertheless appears to be quite robust and useful beyond its expected range of validity. A comparison between this theory and flow over a towed axisymmetric obstacle of a similar form to that of Figure 13.3, but with $h_m/a = 0.4$, has been described by Thompson et al. (1991) for a range of Nh/U values. Basing their comparisons on computed and observed streamline patterns, they concluded that linear theory gave an accurate description of the streamline paths provided $Nh/U \leq 0.5$. As Nh/U increased above this value the agreement deteriorated, and it was generally poor for $Nh/U \geq 1$. This gives confidence in linear theory at the lower end of the Nh/U range, even for topography with moderate slopes.

As stated above, most experimental topographic slopes are not small, and this certainly applies to the "polynomial hill" of Figure 13.3 ($h_m/a = 1$, $h_m/A \approx 0.5$), one of the shapes for which detailed studies have been made. These observations may be taken to be representative, and form the basis for the description given here, supplemented by numerous observations made by the author.

Figure 13.5 shows the flow on the surface and in the centre-plane for flow over this obstacle for $Nh/U = 0.59$. This flow pattern shows many of the same features as that for $Nh/U = 0$, but modified by the effects of stratification in ways that may be expected from the linear perturbation theory of §§12.1 and 12.4. Reynolds numbers for the experiments of Figures 13.3, 13.5–13.7 vary, but all are somewhat larger than 10 000, so that the boundary-layer on the obstacle is turbulent. The flow in Figure 13.5 separates on the lee side with a similar topological structure to that shown in Figure 13.3, implying that the vortical structure in the wake is also similar. The nodes and saddle points observed in homogeneous flow are also found for $Nh/U = 0.59$, in a similar pattern but different locations. The initial saddle point

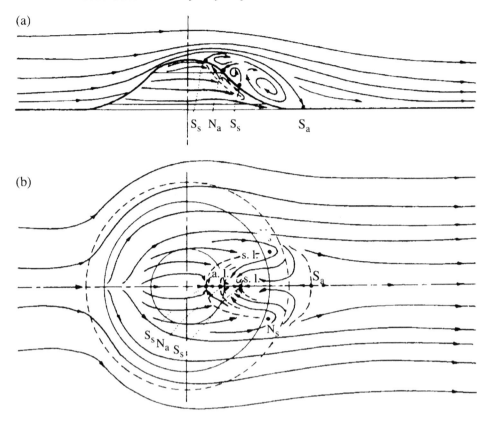

Figure 13.5 As for (a) and (b) of Figure 13.3, but for $Nh/U = 0.59$. (From Hunt & Snyder, 1980, reproduced with permission.)

of separation on the centre-line is now downstream of the peak (rather than slightly upstream), and the trio of singular points S_s, N_a, S_s and associated separation and attachment lines are now centred on the steepest part of the downstream side of the obstacle.

The downstream saddle point of attachment S_a is now closer to the obstacle, so that the separated vortical wake region is smaller and narrower, although it apparently still trails downstream in a horseshoe pattern. This comparison with Figure 13.3 is consistent with linear theory, from which one would expect that descending motion on the lee side due to stratification would reduce and compress the separated wake region.

The surface shear-stress pattern shown in Figure 13.5b is also consistent with expectations from §12.1.5, in that substantial and permanent lateral deflection of the surface streamlines occurs on passing the obstacle. The surface streamlines tend to follow the horizontal topographic contours over the lower part of the obstacle on

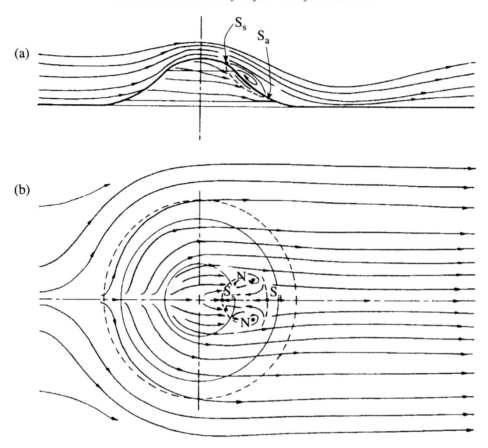

Figure 13.6 As for Figure 13.5, but for $Nh/U = 1.0$. (From Hunt & Snyder, 1980, reproduced with permission.)

the upstream side, and then become approximately aligned with the flow direction on the downstream side. This lateral divergence of the low-level flow is also seen in the vertical section (Figure 13.5a), where the downstream streamlines in the figure have lower elevations than on the upstream side. It is possible that this vertical displacement is partly associated with a lee wave, but Figure 13.5b and the long length-scale imply that it is mostly hydrostatic.

Figure 13.6 shows the corresponding patterns for $Nh/U = 1$. Here the lee-side separation region again has similar topology but has contracted still further in size, and the former trio of singular points on the lee side is now only resolvable as a saddle point of separation. The lateral divergence in the surface flow at lower levels now extends over most of the upstream side of the obstacle, excepting only the region near the crest, and a lee wave with wavelength approximately $6a = 6h_m$ is apparent on the lee side. The effects of stratification are again generally consistent with the linear-theory expectations: descending motion on the lee side that suppresses separation, lee waves in the centre-plane of wavelength $\sim 2\pi U/N$, and permanent lateral deflection of the surface flow pattern. It should be noted that for $Nh/U = 0$, upstream fluid at ground level from a range of y-values (centred on $y = 0$) flows over the surface of the obstacle until it separates, but as Nh/U increases, this range of y-values decreases, so that at $Nh/U = 1$, the fluid in contact with the obstacle surface has (until separation) all originated from the central upstream streamline, or very close to it.

A change occurs in the flow pattern when Nh/U increases beyond a value of about 1.05. This change is associated with the appearance of stagnation points on the centre-line on the upstream side when Nh/U reaches this value, as suggested by the trend of linear theory. Figure 13.7 shows the flow pattern for the "polynomial hill" for $Nh/U = 2.5$, which may be taken as representative of the new flow pattern. Two stagnation points are now seen on the centre-line on the upstream side: a saddle point of separation S_s, near the foot of the obstacle, and a node of attachment N_a at a height of $\sim 0.4h_m$, joined by a line of separation along which there is weak flow directed from the latter down to the former.

Whereas for $Nh/U \leq 1$ the stream surface of maximum density appears to cover the surface of the obstacle up to the first line of separation on the lee side, in Figure 13.7 this stream (or density) surface splits at this initial saddle point S_s, and flows around the base of the obstacle. The two parts of the divided central surface streamline do not meet up again in the flow shown. The elevated streamline incident on the node N_a also divides, as do all the constant-density surfaces and centre-line streamlines beneath it, so that this attachment appears to be of type IV as described in §13.1. This phenomenon is known as "flow-splitting". These divided streamlines first descend slightly on the upstream side, and then flow around the obstacle in approximately horizontal paths, until they meet the separation line on the downstream side. The descending flow in this region close to the upstream surface is inherently unsteady, and the pattern shown in Figure 13.7 represents the mean flow over time. On the upper part of the obstacle above the level of the node N_a, the fluid on the surface all comes from the streamlines impacting on or close to the node, and we denote the density of this fluid as $\rho(N_a)$. The flow of this fluid is similar to that for the flow over the whole obstacle when $Nh/U = 1$, as shown in Figure 13.6.

On the downstream side of Figure 13.7, there is the same number and pattern of singular points on the surface as in Figure 13.6 – saddle points S_s and S_a on the centre-line, and two nodes of separation N_s (probably), one on each side of it – but the flow pattern is different and the scale has increased. The fluid with density close to $\rho(N_a)$ that passes over the crest and descends on the surface on the lee side of the obstacle encounters a line of separation that emanates from the saddle

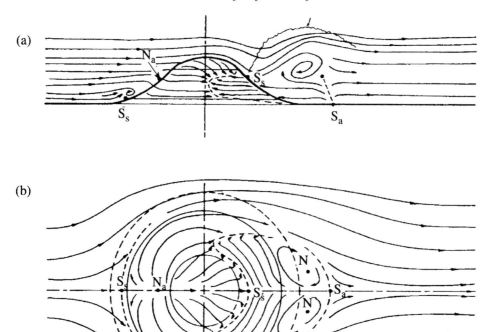

Figure 13.7 As for Figure 13.5, but for $Nh/U = 2.5$. (From Hunt & Snyder, 1980, reproduced with permission.)

point S_s. This separation line is approximately horizontal, at near the same height as N_a on the upstream side. It meets the uppermost of the dividing surfaces that are split at the line S_s–N_a at points that are slightly downstream of the transverse plane of symmetry of the obstacle. Here the separation line curves back towards the downstream direction and descends, where the streamlines impinging on it have increasing density from the split density surfaces. Each side of this separation line appears to join (approximately) with another separation line emanating from the downstream saddle point S_a, so that they enclose a separated wake region on the surface, with fluid recirculating around the two symmetrical nodes N_s. As for the homogeneous case, this wake region is inherently unsteady at these Reynolds numbers. This motion appears to be approximately horizontal up to the level of N_a, with fluid of different density at each level. Above this region, the downstream vortex evident in Figure 13.7a suggests an upper horse-shoe vortex trailing downstream, similar to those seen for $Nh/U \leq 1$, but of smaller scale. This corresponds to

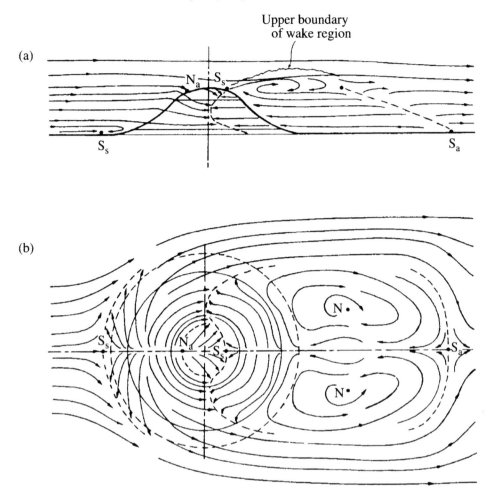

Figure 13.8 As for Figure 13.5, but with $Nh/U = 5.0$. (From Hunt & Snyder, 1980, reproduced with permission.)

the "cowhorn eddy" described by Brighton (1978). It is located below the first downstream lee-wave crest, and appears to be due to shear flow instability between the wake fluid and the rapidly moving stream above it.

If Nh/U is increased further this flow pattern persists, with the conspicuous change that the upstream node N_a moves close to the obstacle crest; hence the region of approximately horizontal flow around the obstacle expands vertically and the region of flow of fluid of density $\rho(N_a)$ over the crest contracts accordingly. Also, the separated wake region expands in the downstream direction. These properties are seen in Figure 13.8a,b, which shows the flow for $Nh/U = 5$.

The upper-level stagnant region that was so prominent in two-dimensional flows

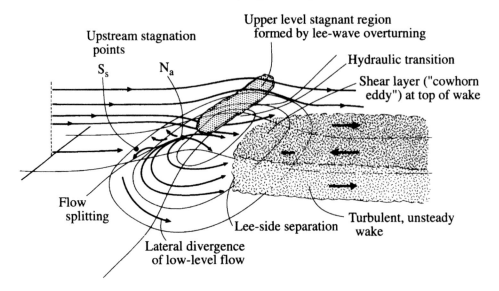

Figure 13.9 Schematic diagram of the flow past a symmetrical obstacle where $Nh/U \gg 1$. Lines denote streamlines on the surface of the obstacle, and in the central plane of symmetry.

is observed with axisymmetric obstacles if Nh/U is somewhat greater than unity (depending on obstacle shape), and the Reynolds number is large enough. This is discussed in more detail in §13.2.5.

Figure 13.9 shows in schematic form the inferred three-dimensional picture of this flow pattern for Nh/U large, and this may be contrasted with Figure 13.3c. The figure has been drawn for $B/A > 1$, and anticipates some of the results described in §13.2.5. It shows the upstream stagnation points and flow-splitting, lee-side separation and the wake with large horizontal eddies, the upper-level stagnant region, and the flow over the obstacle with the "cowhorn eddy". The latter is present for a variety of obstacle shapes, but its size contracts as Nh/U increases. The observations suggest that the flow pattern above $z = z_s$ is approximately similar for all $Nh/U > 1$, contracting in scale as Nh/U increases, provided that the Reynolds number $U(h_m - z_s)/\nu$ remains large.

It should be remembered that these flow patterns are mean flow patterns, and that there is considerable unsteadiness, particularly in the separated wake region. For $Nh/U > 1$, in particular, the wake may be asymmetrical and be dominated by a single eddy at a given time. This eddy may then separate and be replaced by another of opposite sign, to give a periodic behaviour that is similar to a Karman vortex street. This behaviour is particularly noticeable for non-symmetric obstacles (see §13.2.5). A second complicating property of these flows when $Nh/U > 1$ is that they involve flow separation of fluid of varying density along lines on the surface. Such a process is bound to be untidy, and to involve some mixing of the fluid, which increases turbulence. The rough line above the wake in Figures 13.7a and 13.8a represents the upper boundary of this behaviour.

An additional feature when $Nh/U > 1$ is that the weak downward motion on the upstream side from N_a to S_s is associated with some small-scale overturning close to the surface. Although it is not conspicuous in experiments on this scale, this motion is intermittent, and it introduces overturning and a vortical component to the fluid motion close to the divided streamlines that are advected around the obstacle. We return to this property in §13.3.3.

The values of z_s, $z(N_a)$ and z_t provide quantitative measures of these flows when $Nh/U > 1$. It was shown in §13.2.1 that z_s and $z(N_a)$ are the same, assuming hydrostatic flow. For obstacles with $A = B$, z_t is observed to be very slightly greater than z_s because the divergence of the flow on the surface above $z(N_a)$ is small. The streamline from height z_t has been termed "the dividing streamline" by Snyder et al. (1985). However, it is more appropriate to regard the streamline from height z_s impinging on N_a as identifying the "dividing stream-surface", because the density surface on which it lies separates the flow into a lower part that mostly flows around the obstacle, and an upper part that mostly flows upward and over it. For obstacles where $B \geq A$, z_t lies only slightly above z_s.

Figure 13.10a shows observations of the $z(N_a)$, for an obstacle of the form (12.33) with $h_m/a = 0.38$, (with $h/A \sim 0.2$), for which the flow is approximately hydrostatic. Two points from Figures 13.8 and 13.9 are also included. Since zero values of $z(N_a)$ denote that no stagnation point is present, this figure shows the very dramatic change in flow type as Nh/U increases beyond 1.05. There is some scatter in the data, but in general it is approximately described by

$$z(N_a)/h = 1 - 0.7U/Nh, \qquad Nh/U > 1.05. \qquad (13.18)$$

The sudden appearance of stagnation at $Nh/U \approx 1.05$ at a finite height is initially surprising, but is what would be expected from an extrapolation from linear theory (§12.1), since the latter gives a maximum in the pressure perturbation and minimum in velocity u at $z \approx h_m/2$ (see Figure 12.10).

Snyder et al. (1980, 1985) have described observations of z_t for a range of obstacle shapes where $h/A \geq 0.5$, implying non-hydrostatic flow. They found that z_t follows (13.18) (with z_t in place of $z(N_a)$) in some cases, but for most of them it is described by

$$z_t/h_m = 1 - U/Nh, \qquad (13.19)$$

as shown in Figure 13.10b. Since $z_s \leq z_t$, and $z_s = z(N_a)$ for hydrostatic flow, (13.18) and (13.19) together suggest that $z_s \leq z(N_a)$ for non-hydrostatic flow. But

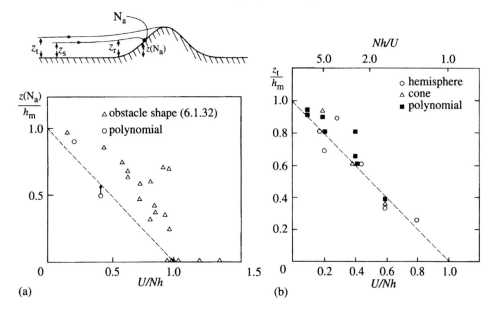

Figure 13.10 (a) Observed height $z(N_a)$ of the stagnation point N_a on the upstream centre-line, as a function of Nh/U. We denote by z_s the upstream height of the streamline intersecting N_a, by z_r the height of this streamline immediately upstream of the obstacle, and by z_t the upstream height of the streamline intersecting the obstacle peak. △: obstacle shape (12.33); ○: the "polynomial hill" of Figure 13.3. A value of zero implies that no stagnation point exists. The dashed line denotes the relation $z(N_a)/h_m = 1 - U/Nh_m$. (From Baines & Smith, 1993.) (b) The observed height z_t far upstream of the lowest streamline that reaches the crest of the obstacle, in terms of Nh/U, for a range of axisymmetric obstacle shapes. △: cone, $h/A = 0.66$; ■: "polynomial hill" of Figure 13.3, $h/A = 0.88$; ○: hemisphere, $h/A = 1$. (From Snyder et al., 1980.)

the differences between the data of Figure 13.11a and b are not large except near $Nh/U = 1$, and apart from the differing h/A values they may also be attributed to the effects of differing obstacle profiles and the vagaries of measurement. Further careful observational studies are required to resolve these details.

The empirical relationship (13.19) is the expression that one obtains if one naively equates the kinetic energy of the upstream fluid (namely $\frac{1}{2}\rho_0 U^2$) to the difference in potential energy [namely $\frac{1}{2}N^2(h_m - z_t)^2$] due to the mean stratification between z_t and the top of the obstacle, h_m. This is the "Sheppard criterion" (Sheppard, 1956), but there is no theoretical justification for this relationship because this "derivation" effectively assumes that $p^* = 0$ in (13.15), and that $\mathbf{u} = 0$ at the obstacle crest; each of these assumptions is demonstrably false. Nonetheless, Snyder et al. (1985) have reported that it provides a good description of observations of z_t for a variety of obstacle shapes and density profiles, in the generalised Boussinesq form of (13.19):

$$\frac{1}{2}U(z_t)^2 = \int_{z_t}^{h_m} (h_m - z)N^2(z)\mathrm{d}z, \qquad (13.20)$$

although there is little data near $Nh/U = 1$. When (13.19) and (13.20) are valid, we may infer that at the crest of these obstacles the $\mathbf{u}^2$ and p^* terms approximately cancel, so that $p^* \approx -\frac{1}{2}\rho\mathbf{u}^2$ there. This behaviour near the crest is quite different from that of linear theory, and may be attributed to the hydraulic-transition-like character of the flow.

Numerical simulations of the stratified flow past axisymmetric obstacles described in this section have been made by Ding et al. (2003), with good agreement in all main respects.

The above results imply that when no flow-splitting occurs on the upstream side ($Nh/U < 1$), the fluid that makes contact with the surface of the upstream side of the obstacle comes from close to ground level far upstream. However, for symmetric obstacles, if flow-splitting occurs (i.e. $Nh/U > 1$) it mainly comes instead from upstream fluid within or close to a vertical T-shaped region. The vertical stem of the "T" lies in the central vertical plane below z_s, and the cross-piece lies in the horizontal plane just above z_s. The length of this cross-piece depends on the obstacle shape. On the downstream side, upstream fluid from a range of lateral displacements and heights may enter the turbulent separation region and wake, thereby making contact with the surface there. In the atmosphere, these distinctions between the fluid that does and does not make contact with the surface will of course be made "fuzzy" by the presence of small-scale turbulence.

13.2.4 Flow past a sphere at finite Nh/U

The sphere is a special obstacle, and its flow properties have been the most studied of all three-dimensional shapes. Stratified flows are no exception. Observational studies have covered a comprehensive range of Nh/U values and Reynolds numbers, and a brief summary is provided here.

If the flow past an isolated sphere is symmetrical about a horizontal plane surface through the centre, as these flows mostly are (the exception being flow in periodic and turbulent wakes), they may be equated with the flow past a hemisphere situated on the plane, with free-slip on the plane surface. Flow past this topography has the topological characteristic that it has a singular point of attachment at the upstream point, and for the sphere as a whole the number of nodes and saddles satisfies (13.5) rather than (13.6). This provides some interesting contrasts with the flow shown in Figures 13.5–13.8. We here define Nh/U by Na/U, where a is the radius of the sphere, and the Reynolds number by $R_e = Ua/\nu$ (this is to maintain consistency with previous definitions; R_e for a sphere is usually defined in terms of the diameter). For the experiments of interest, a has values ranging from 0.5–2.5 cm, and R_e covers the range from 5 to 5000. This is somewhat smaller than most of the experiments described in §13.2.3, so that the boundary-layer on the sphere is mostly laminar. Given the small obstacle sizes, most attention has concentrated on the properties of the wake, interpreted from observations in horizontal and vertical planes.

From laboratory experiments with spheres towed through salt-stratified water, Lin et al. (1992) identified eight flow regimes based on classifying different wake types, and these are shown in terms of Na/U and R_e in Figure 13.11. For $Na/U < 1.1$, which from §13.2.3 may be identified as the non-flow-splitting regime, there are four of these flow types depending on the value of R_e. These are shown in Figure 13.11b, and in order of increasing Reynolds number they are: lee waves with attached vortices (WV), symmetric vortex shedding (SVS), asymmetric vortex shedding (AVS), and turbulent wake (T). Given the steep nature of the "topography", one cannot expect linear theory to give a good description of the flow, and the flow type that compares with it best appears to be the category WV. Observed wavelengths of the lee waves are slightly less than $2\pi U/N$, as expected. As R_e increases with constant Na/U, the bluffness of the spherical shape becomes more significant, and the wake becomes progressively more disordered and eventually turbulent. The most conspicuous effect of the stratification is then to suppress vertical motion in the wake, so that far downstream the turbulent region is spread horizontally, and dominated by large-scale horizontal motions (Lin & Pao, 1979).

If Na/U is increased above the approximate value of 1.1, which is independent of Reynolds number, a new classification (VW) of the wake is seen. Here the lee-side eddies are longer and of greater vertical extent, the lee waves are smaller, and overturning is seen in the first wave crests. This change signifies the onset of flow-splitting, with flow around the sphere at the central levels, as described in the preceding section for "polynomial hill". For $Na/U < 1.1$, the only singular point on the upstream side is a node of attachment N_a at the leading central point. For $Na/U > 1.1$, this has split to become a trio of singular points N_a, S_a, N_a: a saddle point of attachment replacing the node at the centre, and two nodes of attachment on the centre-line above and below (conserving (13.5) locally). The vortices observed at the crest of the first lee wave obtain most of their vorticity from the boundary-layer on the surface of the obstacle before it separates, and it corresponds to the upper horseshoe vortex, or "cowhorn eddy" described in §13.2.3. This is the only one of these wake classifications where this eddy was conspicuous, and its relative weakness, compared with that for the surface-mounted obstacles of §12.4.2, may be attributed to the smaller Reynolds numbers and the relatively limited scope for the development of the boundary-layer on the surface of the sphere.

As Na/U is increased beyond values ranging from 2 to 5 (depending on R_e),

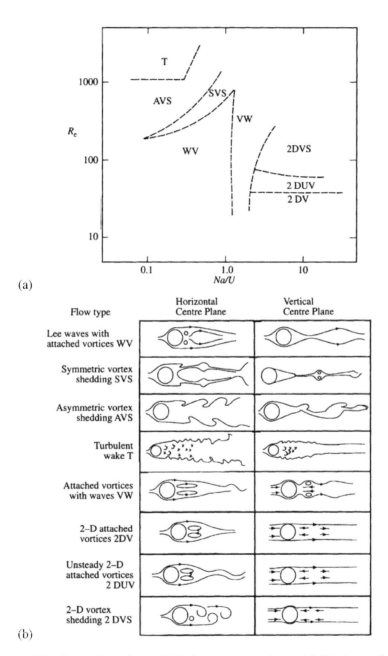

(a)

(b)

Figure 13.11 Properties of stratified flow past a sphere. (a) Regions of the $(Na/U, R_e)$-plane with different flow regimes, as specified by the behaviour of the wake. (b) Descriptions of the flow types in (a) in terms of flow in the vertical and horizontal planes through the x-axis. The first four types are for $Na/U < 1.1$, and the second four types for $Na/U > 1.1$. (Adapted from Lin et al., 1992.)

three more categories may be identified. These depend on Reynolds number, and comprise flows that are predominantly around the sphere with weak vertical motions at their upper and lower parts, so that they are a natural extension of class VW above. In order of R_e increasing, they are: steady two-dimensional attached vortices (2DV), unsteady two-dimensional attached vortices (2DUV), and two-dimensional vortex shedding (2DVS). The behaviour for yet larger Reynolds number (> 500) is expected to be the same as in this last category. All of these vortices – steady, unsteady or shed – are vertically coherent. Whereas the shedding frequency for the three-dimensional vortices in the categories SVS and AVS above is sensitive to the values of both Na/U and R_e, the (one-sided) shedding frequency ω_s for the two-dimensional vortices in category 2DVS is independent of both of these parameters and is given by $\omega_s = 0.1U/a$ (Strouhal number $2a\omega_s/U = 0.2$), the same as for vortices behind a vertical circular cylinder (Lin et al., 1992).

The shape and topology of the separation line on the lee side of a sphere has an interesting structure that varies in a complex way with both Na/U and R_e if the latter is not too large (Chomaz et al., 1992; Lin et al., 1992), and detailed studies of the turbulent wake have been made by Chomazetal et al. (1993) and Bonneton et al. (1993), but these details are peripheral to our purpose here.

Numerical results by Hanazaki (1988) for $R_e = 100$ (which lie across the centre of Figure 13.11a) confirm the interpretation of the upstream stagnation points, and show the flow-splitting and the connecting line of attachment for the fluid incident between the two nodes of attachment, when $Na/U > 1.1$. Figure 13.12 shows side views of the computed flow patterns in the centre-plane and just above the surface, for various Na/U. These may be compared with the much higher Reynolds number flows sketched in Figures 13.5a–13.8a; apart from the wake effects, most details agree well. One may note from Figure 13.12c,d and e that there is an isopycnal line crossing the upper part of the surface of the sphere, due to the divergence of the flow on the centre-line. This implies that the density on this part of the surface of the sphere is not quite uniform, so that the isopycnal with density $\rho(N_a)$ does not reach the top of the sphere, and z_s is slightly less than z_t (defined in Figure 13.10) as before. The observational and numerical data for z_s for a sphere generally lie between (13.18) and (13.19). As Figure 13.12 also shows, no upper lee-wave-overturning of the type common in two-dimensional flows is observed for a sphere for $R_e = 100$.

13.2.5 *Flow at finite Nh/U past elongated obstacles*

We next examine the differences in the properties of the flow past obstacles that do not have circular cross-section from those that do. We concentrate on shapes that are symmetrical about the flow direction and have a height maximum at a single

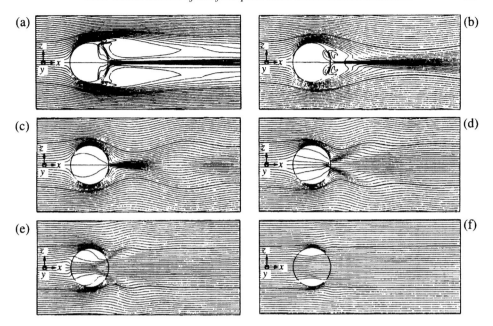

Figure 13.12 Isopycnal lines in the vertical plane of symmetry and on the obstacle surface for flow past a sphere, from numerical computations at $R_e = Ua/v = 100$. These may be compared with Figures 13.3a, 13.5a–13.8a. (a) $Na/U = 0.005$; (b) $Na/U = 0.5$; (c) $Na/U = 1$; (d) $Na/U = 1.43$; (e) $Na/U = 2$; (f) $Na/U = 4$. (From Hanazaki, 1988, reproduced with permission.)

point, so that the important parameters are Nh/U, h/A and B/A. There are two main situations, $A > B$ and $A < B$, which are treated in turn.

Symmetric obstacles with $B/A < 1$

If B/A $(= b/a)$ is decreased so that the obstacle is made progressively thinner, linear hydrostatic theory predicts that the surface pressure maximum (and pressure gradient) on the upstream side also decreases (see §6.1), implying smaller-amplitude disturbances overall. It is also intuitively obvious that a stratified fluid will have a greater tendency to pass around such an obstacle rather than over it, implying a greater degree of divergence and flow-splitting along the upstream centre-line.

Figure 13.13 shows the flow pattern on the surface of an obstacle of the form (12.16) with $v = 3/2$ and $b/a = 0.38$, for $Nh/U = 4.2$. This may be compared with Figure 13.8. The overall pattern of singular points in the two figures is similar, with the same two stagnation points (S_s, N_a) on the upstream side, the lower at the base of the obstacle. The observed heights $z(N_a)$ for obstacles with $b/a = 0.32, 0.38$, and z_t for $b/a = 0.32$ are shown in Figure 13.14, as a function of Nh/U. This pattern is quite different from that of Figure 13.10; the stagnation points are only present

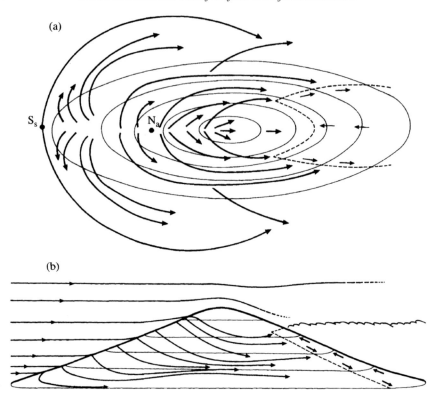

Figure 13.13 Sketches of the flow pattern on the surface and on the vertical plane of symmetry for flow past an obstacle elongated in the downstream direction of the form (12.16) with $v = 3/2$ and $b/a = 0.38$, for $Nh/U = 4.2$, $R_e = 430$. (a) Plan view of flow on the surface; (b) side view (cf. Figures 13.7, 13.8).

for $Nh/U > 1.43$, (i.e. $U/Nh < 0.7$), and the height $z(N_a)$ increases linearly from zero as U/Nh decreases below 0.6.

Below the level of $z(N_a)$, flow on the centre-line descends and splits, continuing around the obstacle as before, and above $z(N_a)$, it rises. However, this does not mean that this fluid passes over the obstacle crest. Divergence on the centre-line above N_a is so large that the fluid with density $\rho(N_a)$ only rises a small distance before it is split to the extent that it is effectively non-existent on the centre-line. Fluid from a slightly higher upstream level that makes contact with the surface on the centre-line, is also split, and so on. The centre-line above N_a is therefore a line of attachment with rising motion, in the same way that below N_a it is a line of attachment with descending motion. Both of these lines of attachment are of type IV as described in §13.1. In consequence, the upstream height z_t of the streamline that reaches the obstacle crest lies well above the line of (13.19).

This flow-splitting behaviour above N_a was noted in the numerical solutions for

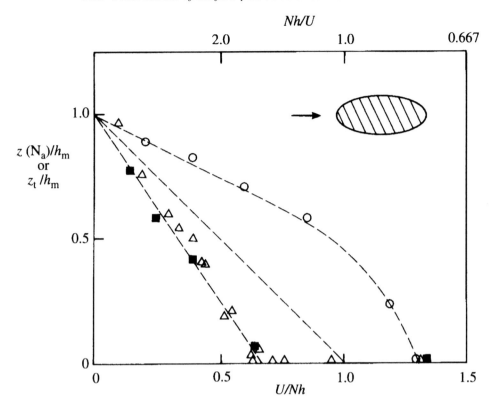

Figure 13.14 Observed values of $z(N_a)$ and z_t for obstacles elongated in the downstream direction. ■: $z(N_a)/h_m$ for the obstacle of Figure 13.13 ($b/a = 0.38$); △: $z(N_a)/h_m$ and ○: z_t/h_m for a similar obstacle with $b/a = 0.32$. The obstacle is shown schematically in the inset. (Modified from Baines & Smith, 1993.)

flow over a sphere in §13.2.4, but there the effect was weak, whereas here it is prominent over most of the upstream centre-line. In general for $B/A < 1$, as B/A becomes progressively smaller the divergence on the leading edge increases, and z_t for the streamline that actually reaches the obstacle crest will rise. For the theoretical limit of a knife-edge where $B/A \to 0$, we may expect that $z_t/h_m \to 1$ for all Nh/U.

Lee waves were visible over the downstream side on the centre-plane of these obstacles with $b/a = 0.32, 0.38$, but their slopes were modest and no upper-level overturning was observed. Hence low-level flow-splitting occurs without any upper-level overturning, for these obstacles. When $Nh/U > 1$ the observed downstream wake for these conditions tended to consist of two stationary eddies, elongated in the downstream direction.

For obstacles as shown in Figure 13.13, the fluid making contact with the obstacle surface comes from an upstream region in the form of a cross †, rather than a "T".

The vertical extension above the cross-piece is due to the flow-splitting above N_a, and for given A, the cross-piece has shorter arms than those of the "T" for $B = A$.

For obstacles that have the form of ridges elongated in the flow direction where $h = h_m$ along a line rather than at a single point, the surface flow pattern has some important differences. Some experiments by the author have shown that if the upstream face of the obstacle has circular cross-sections (such as a ridge with a rounded end), the flow pattern on the upstream side resembles that for an obstacle with $A = B$. This is seen in the values for z_t, which are generally consistent with (13.19). However, for $Nh/U > 1$, the streamlines that rise to h_m may subsequently split and descend on each side of the ridge to near their upstream levels, so that the ridge crest line can be a line of flow-splitting and attachment, in the same manner as the forward face above N_a.

Symmetric obstacles with $B/A > 1$

Figure 13.15 shows the surface streamlines on an obstacle of the form (12.16) with $B/A = 2$ and $h/a = 0.38$, for $Nh/U = 3.42$. A comparison with Figures 13.7 and 13.13 shows that the topology of the surface flow is similar in all three cases. A side view for similar conditions ($Nh/U = 2.5$) in Figure 13.16 shows the flow in the central plane of symmetry, where the upper-level overturning is conspicuous. The important features of Figures 13.15 and 13.16 are depicted in Figure 13.9.

Figure 13.17 shows the observed height $z(N_a)$ for this obstacle. The variation with Nh/U is essentially the same as for the obstacles with $A = B$ (Figure 13.10a) and (13.18), with a sudden increase or discontinuity as Nh/U increases above ~ 1.05. For obstacles with $B/A > 1$, divergence on the surface above N_a is small, so that $z_t \simeq z_s$. Observations of z_t for triangular ridges where $h/A = 2$ and $1 \le B/A \le 8$, and a sinusoidal ridge where $h/A = 0.54$, $B/A = 10.8$ (Snyder et al., 1985), give a similar dependence to (13.19). These quantitative results indicate an essential similarity in these flows for all B/A, with the scale of the horizontal low-level potential-like flow around the obstacle increasing as B increases.

Small-scale overturning on the upstream side below N_a when $Nh/U > 1$ is again observed here. For vertical barriers and steep triangular ridges where h/A is large, this may be enhanced by bluff-body effects as in Figure 13.4, and an upstream vortex may form that is up to half the obstacle height (Castro et al., 1983).

The occurrence of upper-level overturning has been investigated in a series of experiments by Castro & Snyder (1993). Figure 13.18 summarises the results and shows the bounding curves for upper-level overturning as a function of Nh/U and A/h for $B/A = 1, 2$ and 4. It also shows the corresponding curve for two-dimensional elliptical shapes from Huppert & Miles (1969); see §10.3. Corresponding curves for $B/A < 1$ presumably lie within the first of these, implying overturning within a narrower range of Nh/U if A/h is large enough. As described above, for $Nh/U > 1$

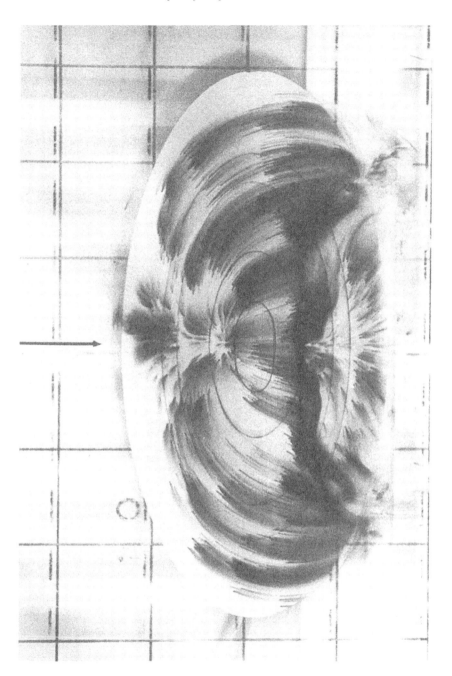

Figure 13.15 Photograph of the surface flow pattern for an obstacle of the form (12.16) with $v = 3/2$, $b/a = 2$ and $Nh/U = 3.45$, visualised with dye emanating from small grains of potassium permanganate placed on the surface. Note the upstream stagnation points and flow-splitting, and the downstream wake.

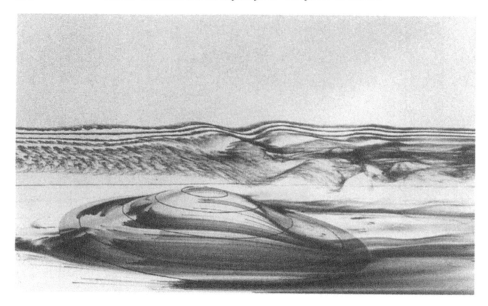

Figure 13.16 The observed flow in the central plane of the obstacle of Figure 13.15, for $Nh/U = 2.5$. The flow is visualised by dye from a vertical rake placed upstream in the plane of symmetry, in addition to the dye released on the surface of the obstacle as in Figure 13.15.

the effective height of the obstacle for wave generation is $\sim U/N$, and since the obstacles used all have (nearly) flat tops, the effective value of A/h should increase as U/N decreases.

The experimental curves have an upper limit for Nh/U for overturning, which is not present in the inviscid model, and may be attributed to Reynolds number effects. An overturning region has a length-scale of U/N, so that a realistic Reynolds number $R_e = U^2/N\nu$, which has values of order 100 in these experiments for $U/Nh = 0.1$: small enough to constrain turbulence. Hence the lower parts of these curves are not relevant to very large Reynolds number situations, such as in the atmosphere, and the dashed extensions of the upper parts of the three curves are drawn in Figure 13.18 as approximate inviscid extrapolations of them. It should be noted that it is the *mean* slope (h/A) of the obstacle that is important here, rather than the maximum slope of the obstacle or its profile.

In many experiments where the obstacle is towed over a finite distance the wake is observed to consist of two symmetrical eddies, as for narrower obstacles, even though the effective Reynolds number is much greater than 40. However, in some cases a periodic wake results if the experiment is run for a sufficiently long time, and this may be attributed to an instability of the symmetric wake (e.g. Schär & Smith, 1993b).

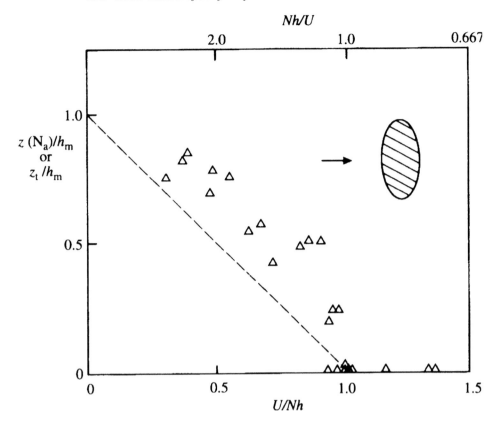

Figure 13.17 Observed values of $z(N_a)$ for the obstacle of Figure 13.15.

The fluid that makes contact with the surface of the obstacle when $Nh/U > 1$ again comes from a T-shaped region upstream, as for $A = B$, but the cross-piece lengthens with increasing B.

If B is large so that $B/A \gg 1$, shortly after the flow has commenced (i.e. $Nt \geq \pi$) the flow near each part of the obstacle is locally two-dimensional (i.e. in the (x, z)-plane), except near the ends. Information about the presence of the ends is then propagated in towards the centre of the obstacle by internal gravity waves, enabling the final steady three-dimensional pattern to be attained. These waves have the vertical scale U/N, which for hydrostatic waves have the group velocity U. For an obstacle or mountain range of length L this gives a time-scale of L/U, so that flow over longer obstacles takes a longer time to reach steady state. The initial upstream columnar motions generated along the length of the obstacle (if $Nh/U \sim 1$) spread out horizontally, so that their nett amplitude decreases with distance upstream in the final steady state, in the manner described in §12.1.1. They cause substantial modification to the velocity and density profiles near the obstacle within a distance of

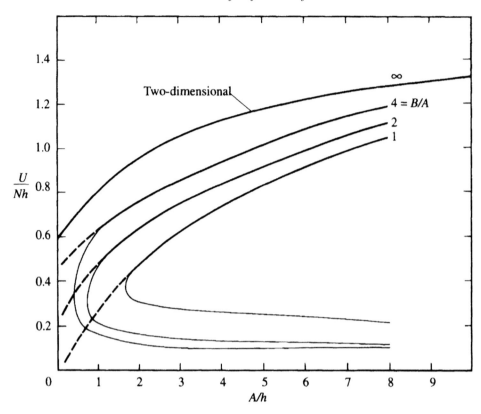

Figure 13.18 Observed criteria for the onset of upper-level overturning as functions of U/Nh and A/h, for obstacles with various values of B/A. The topmost
curve is the inviscid theoretical result of Huppert & Miles (1969) for semi-elliptical
obstacles. The (lighter) solid lower part of these curves is apparently due to
Reynolds number effects (see text). The dashed curves give the expected extrapolations for flows with large R_e, so that overturning occurs below these curves.
(Modified from Castro & Snyder, 1993.)

scale B. The low-level flow diverted around the barrier has approximately potential
flow structure as described in §13.2.1, and as B is increased the horizontal scale of
this flow pattern expands with it, but the pattern remains essentially similar. The
upper-level flow over the barrier (which is mainly in the (x, z)-plane) also remains
unchanged, so that the broad features of the flow pattern are independent of the
scale of the obstacle.

When Nh/U is small, the lateral deflection of the fluid by the obstacle is also
small, and flow over the central part of a long three-dimensional (3D) barrier is
similar to that over a two-dimensional (2D) one. As Nh/U increases, so do the
upstream disturbances and the lateral deflection, and the difference between the 2D
and the central 3D flow. This is reflected in the criteria for lee-side flow separation

(shown in Figure 10.8), which mostly apply when $Nh/U < 1$ and are the same for both 2D and 3D cases, and in the criteria for upstream blocking or stagnation, which are distinctly different ($Nh/U = 2$ for 2D, and 1 for 3D). We return to the connection with the two-dimensional case in §13.2.6.

Obstacles that are not symmetric about the flow direction

Here there is too much variety to give comprehensive results, but some general statements may be made. Firstly, sinusoidal ridges of finite length inclined at various angles to the flow give results for z_t that are consistent with (13.20) (Snyder et al., 1985). Secondly, asymmetric obstacles are often observed to have unsteady, periodic vortex-street-type wakes when $Nh/U > 1$, containing eddies shed alternately from lines of separation on each side of the obstacle. These eddies have heights less than h_m, and are generally coherent over this depth, although the flow patterns may differ somewhat with height. They are composed of fluid that has acquired vorticity from the boundary-layers upstream of their separation lines, and these shed vortices may be unequal because of lack of symmetry at separation. An example of these properties is shown in Figure 13.19, where the flow is time-dependent and the shed vortices are unequal because of this lack of symmetry.

Vortex-street-like wakes downstream of islands are sometimes observed by satellites, when stable layers of air exist at low levels. The strong or layered stratification causes the low-level fluid to flow around the obstacle and separate in an approximately two-dimensional manner. An example that may be attributable to the asymmetry of the island is shown in Figure 13.20. Island vortex wakes have been observed to have Strouhal numbers $2\omega_s B/U$ in the range 0.2–0.5 (Gjevik, 1980). These values are quite consistent with shedding frequencies for flow past a sphere (see §13.2.4), and as observed in the laboratory, so that the dynamics seem to be essentially the same in both cases.

The reason why vortex-street wakes of the form shown in Figure 13.20 are stable over many wavelengths and great distances, whereas the corresponding two-dimensional wake behind a cylinder in a wind tunnel when $R_e > 200$ is not, is due to the imposed two-dimensionality in the (x, y)-plane. In the wind tunnel, for $R_e > 200$ the initially two-dimensional vortices behind a cylinder become unstable and break up because of the growth of disturbances that vary in the third (z) dimension along the cylinder (Wei & Smith, 1986). The process of three-dimensional vortex stretching then results in a turbulent wake. Such disturbances are suppressed in the atmosphere, so that the vortices in the island wakes maintain their identity for arbitrarily large Reynolds numbers.

To date, the only results from quantitative measurements in the field appear to be those reported by Leo et al. (2015), which were made as part of the MATER-HORN project (see §14.2.7). Nocturnal observations were made of flow upstream

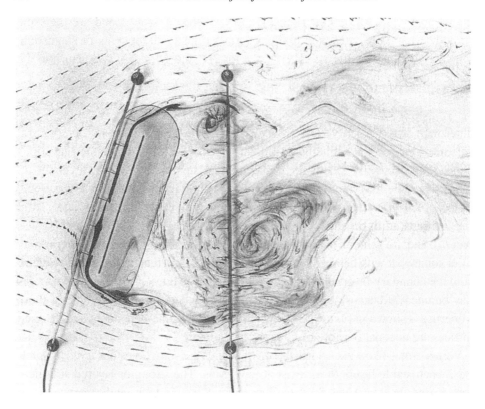

Figure 13.19 The observed horizontal flow pattern at a height $z = h_m/3$ for uniformly stratified flow past a towed barrier inclined at 70° to the flow direction. $U/Nh = 0.1$, $R_e = UL/\nu = 2990$, where L is the obstacle length. Flow visualisation is by neutrally buoyant dye released from three horizontal rakes: one upstream, one on the surface of the obstacle and one downstream. The dye released on the obstacle surface marks the vorticity produced there, and horizontal Kelvin–Helmholtz billows are evident after this dyeline separates. The vortices from these separated shear layers then become concentrated in lee-side vortices. The wake region is unsteady, with these unequal vortices being shed alternately. (From Baines, 1990: reproduced by permission of ASCE.)

of isolated Granite Mountain, irregularly shaped with a nominal height of 60 metres. Under these conditions the upstream flow was approximately uniformly stratified, with a logarithmic vertical velocity profile, consistent with Monin–Obukhov similarity theory. The Sheppard criterion for these conditions was evaluated, and although only two sets of observations were realised, the streamline impinging on the appropriate height of the mountain was close to the values calculated with the Sheppard criterion for these conditions.

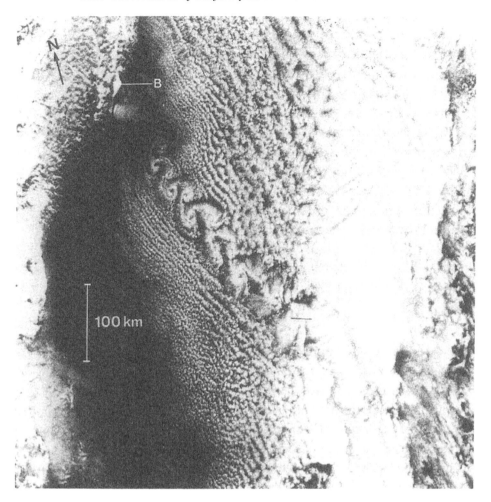

Figure 13.20 Satellite photograph of a vortex street downstream of Jan Mayen (see Figure 12.16) at 1115 UT, 14th April 1977, from NOAA-5. The mountain peak (Beerenberg) is denoted by B. Note that the scale is not uniform, and is reduced in the direction perpendicular to the scale on the picture by a factor of 0.7. (From Gjevik, 1980: reproduced by permission of WMO.)

13.2.6 Flow at finite Nh/U past two-dimensional barriers with gaps

The relation between the flow fields past two- and three-dimensional obstacles may be seen by considering the flow past two-dimensional barriers with gaps in them. This situation also has obvious practical relevance. We consider a barrier (obstacle) that has (mostly) uniform height h_m of length L (where $L = 2B$), oriented perpendicularly to the flow, and is repeated periodically along this axis at intervals W, where necessarily $L/W \leq 1$. In the two-dimensional limit $L/W \to 1$, and in the three-dimensional limit of the isolated obstacle $L/W \to 0$. If we assume for simplicity that the obstacle is symmetric about its centre-point, this periodicity is equivalent to the obstacle being in the centre of a channel of width W, and we will describe the flow in these terms. The parameter L/W is additional to Nh/U, h/A and B/A, and we now examine the changes in flow properties as L/W increases from zero.

If $Nh/U > 1$ so that the flow over the barrier generates columnar motions that propagate against the stream, the upstream disturbances that spread out laterally when $L/W = 0$ are now confined within the channel. Hence, when steady state is achieved they superimpose to give a small but non-zero amplitude at upstream infinity, proportional to L/W. This will increase with L/W to give the two-dimensional value when $L/W = 1$. In other words, for $0 < L/W < 1$, for given Nh/U the velocity and density profiles far upstream that are seen by the obstacle are different from those for the three-dimensional case, by an amount (roughly) proportional to L/W. If Nh/U is increased by a small amount (in the spirit of the hydraulic calculations of §11.2), these upstream differences cause the flow to respond differently from the situation where $L/W = 0$. Consequently, there can be a substantial difference between the flow properties in the two-dimensional and three-dimensional cases in the steady state, even close to the obstacle, as described for the simpler case of a single layer in §3.5. As L/W increases, for $Nh/U > 1$ the low-level (nearly) potential flow around the barrier must pass through a progressively narrower gap. From the theory of potential flow around an isolated barrier, for upstream velocity U the maximum velocity at the end of the barrier is $U(1 + B/A)$, in a stream with a width-scale of A^2/B, based on the local gradient at the barrier. The presence of the channel walls therefore has little effect until the gap width decreases to near this value; that is,

$$W(1 - L/W) \lesssim 2A^2/B, \tag{13.21}$$

where the factor "2" appears because the high-velocity regions appear on each side of the gap. We may define a gap parameter G by $G \equiv$ (area of the gap)$/Wh_m$, which is approximately equal to $1 - L/W$, given that the definition of L is imprecise. If WG is decreased below the scale of the value of (13.21) and if the total flux through the gap is to remain the same the velocity there must increase. In potential flow this is achieved by a local decrease in pressure, but in practice various physical factors such as drag and entrainment in the gap will constrain the flow and limit this increase in the local velocity. Consequently, the approach to two-dimensionality may not be linear with L/W.

Experiments on flow past uniform barriers with gaps, towed through stationary fluid, have been described by Baines (1979b) and Weil et al. (1981). A striking feature of these flows when $Nh/U > 2$ is the sharp distinction between low-level flow around the barrier, and higher-level flow over it that has relatively little sideways deflection. The boundary between them may be identified as the same dividing stream-surface described for isolated obstacles earlier in this section. Figure 13.21 shows an example, of flow past a ridge of "Witch of Agnesi" shape (10.14) with $h_m = 6.26$ cm and $a = 4.95$ cm (i.e. $h/A = 0.63$), with a rounded end having the same profile and elliptical height contours, with $G = 1/16$. These experiments modelled half the obstacle and gap, employing symmetry to simulate the flow past a barrier with periodic gaps of twice their actual width. They were conducted in fluid of finite depth where $K > 3$ (mostly) and upper-level overturning was present, so that the effect of the upper boundary was not important. In this example $Nh/U = 5.9$, and the flow field is essentially the same in all four frames. In the two left-hand frames, approximately neutrally buoyant dye is released from a horizontal upstream rake at a level just below the dividing level, and for the right-hand frames the rake has been elevated 0.9 cm so that the fluid is released just above it. The presence of an approximately horizontal dividing surface between the lower-level flow in horizontal planes and the upper-level flow in near-vertical planes is apparent. On the downstream side for $G \ll 1$, fluid passing through the gap continues along the tank, whereas the low-level fluid downstream of the barrier below the hydraulic stream is relatively quiescent, as for two-dimensional flow with "post-wave" lee-side separation (see §5.1).

The height immediately upstream, z_r (as defined in §13.2.1, Figure 13.10a), of this dividing stream-surface for the experimental configuration of Figure 13.21 was found to vary with G in the manner shown in Figure 13.22. For given values of G and Nh/U, all upstream fluid at heights below the line for z_r would eventually pass around the obstacle through the gap, whereas at heights above it, the fluid passed over the barrier with relatively little lateral deflection. The measured quantity z_r becomes z_s when the obstacle is effectively three-dimensional with $G > 0.5$.

The reasons for the variations of z_r in Figure 13.22 are as follows. For $G = 0$ the fluid below z_r is totally blocked, as described in Chapter 5. z_r is given approximately by the line

$$z_r/h_m = \begin{cases} 0, & Nh/U < 2, \\ 1 - U/Nh, & Nh/U > 2, \end{cases} \tag{13.22}$$

although the observations show considerable scatter (see Figure 10.7b for infinite depth, Baines (1979b) for finite depth), and because of this the curve is shown dashed in Figure 13.21. If G is increased very slightly, some of this blocked fluid is able to "leak" through the gap; z_r begins to decrease and the flux is controlled by local drag and mixing as described above. But as G is increased further these physical factors become less important and the level z_r of the surface descends

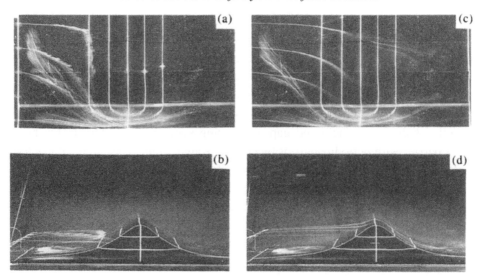

Figure 13.21 Flow past a two-dimensional barrier with a gap at one end, seen from above (upper two frames) and from the side (lower two frames). The obstacle has "Witch of Agnesi" cross-section with $a = 4.95\,\text{cm}$, $h_\text{m} = 6.26\,\text{cm}$, tank width $W/2 = 23\,\text{cm}$ and gap parameter $G = 1/16$. The flow is the same in all four frames, and is in approximately steady state locally, with $Nh/U = 5.9$. In the two left-hand frames the dye is released from a rake below the dividing stream-surface height z_r and on the right it is released 0.9 cm higher, above it. (Adapted from Baines, 1979b.)

rapidly toward the line

$$z_r/h_\text{m} = 1 - 2U/Nh, \qquad (13.23)$$

which it attains when $G = 1/16$. Here (13.21) is satisfied. The flow is still largely governed by two-dimensional hydraulic control over the barrier, which generates the upstream disturbances, but the increased gap width has enabled the blocked fluid to leak freely around the barrier, reducing its depth on the upstream side. The surface level z_r is controlled by a balance between this leakage and the generation of the upstream disturbances by the barrier. As G is increased further, $W - L$ increases to the point where the tank walls no longer compress the flow through the gap, and (13.21) no longer applies. For the obstacles of Figures 13.21, 13.22, this occurs at $G \approx 1/4$. The upstream disturbance also weakens because of decreasing L/W, and z_r approaches the form (13.18) or (13.19) for isolated three-dimensional barriers, as Figure 13.22 shows.

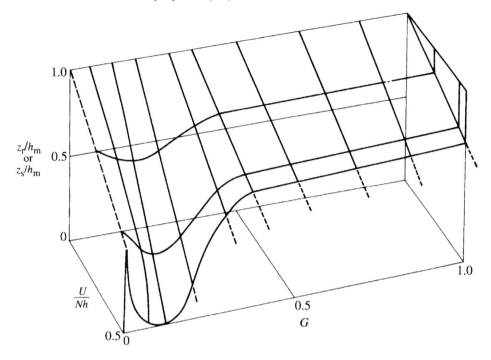

Figure 13.22 The approximate observed upstream height of the dividing stream surface z_r, or z_s (i.e. the height of the highest streamline that does not pass over the barrier), observed upstream of the obstacle as in Figure 13.21, as a function of Nh/U and G. The figure is based on data with one obstacle shape, and should be seen as "schematic" until more data is available. For $G > 0.5$ the surface is depicted following (13.18). The surface is only shown for $Nh/U > 2$, with extensions indicated by dashed lines.

13.3 Flow properties for finite Nh/U: theoretical aspects

We now return to theoretical studies, specifically those that are applicable to situations where Nh/U is not small, with a view to understanding some of the non-linear phenomena described in the previous section. We first examine an analytic approach based on a perturbation expansion, and then proceed to corresponding numerical studies.

13.3.1 Perturbation solution for Nh/U large

We begin with the non-linear inviscid incompressible form of the equations of motion from §1.1, with the Boussinesq approximation, and write the pressure and density in the form

$$p = p_0(z) + p', \qquad \rho = \rho_0(z) + \rho', \tag{13.24}$$

where p_0 and ρ_0 are related by hydrostatic balance (1.10). Defining $\sigma = g\rho'/\bar{\rho}$, where $\bar{\rho}$ is the mean value of $\rho_0(z)$, (1.1), (1.3) and (1.4) become

$$\frac{\partial \mathbf{u}}{\partial t} + \mathbf{u} \cdot \nabla \mathbf{u} + \frac{1}{\bar{\rho}}\nabla p' + \sigma \hat{\mathbf{z}} = 0, \tag{13.25}$$

$$\frac{\partial \sigma}{\partial t} + \mathbf{u} \cdot \nabla \sigma - N^2 w = 0, \tag{13.26}$$

$$\nabla \cdot \mathbf{u} = 0. \tag{13.27}$$

For flow past an isolated obstacle of the form $z = h(x, y)$, which has height h_{m} and horizontal length-scale A, the boundary conditions are

$$\mathbf{u} \to [U(z), 0, 0], \qquad p', \sigma' \to 0, \quad \text{as } x^2 + y^2 + z^2 \to \infty, \tag{13.28}$$

$$\mathbf{u} \cdot \nabla[z - h(x, y)] = 0, \quad \text{on} \quad z = h(x, y). \tag{13.29}$$

The steady-state form of these equations may be integrated to give

$$\sigma = N^2(z - z_0), \tag{13.30}$$

where z_0 is the height of the streamline at (x, y, z) far upstream, and the Bernoulli integral

$$\frac{1}{2}\mathbf{u}^2 + \frac{p'}{\bar{\rho}} + \frac{1}{2}N^2(z - z_0)^2 = \frac{1}{2}U(z)^2. \tag{13.31}$$

If the vertical displacements are of the order of the obstacle height, we have $\sigma \sim N^2 h$ (here, as previously, h denotes h_{m}). If we denote a characteristic value of $U(z)$ by U, and concentrate on situations where Nh/U is large, we have $p' \sim \bar{\rho}N^2 h^2$. If we then define dimensionless variables u^*, v^*, etc. by

$$(u, v) = U(u^*, v^*), \quad w = Uhw^*/A, \quad (x, y) = A(x^*, y^*),$$
$$z = hz^*, \quad \sigma = N^2 h\sigma^*, \quad p' = \bar{\rho}N^2 h^2 p^*, \tag{13.32}$$

we obtain the dimensionless equations

$$\epsilon \mathbf{u}^* \cdot \nabla \mathbf{u}_H^* + \nabla_H^* p^* = 0, \qquad \epsilon \frac{h^2}{A^2}\mathbf{u}^* \cdot \nabla^* w^* + \frac{\partial p^*}{\partial z^*} + \sigma^* = 0,$$
$$\mathbf{u}^* \cdot \nabla \sigma^* + w^* = 0, \qquad \nabla^* \cdot \mathbf{u}^* = 0, \tag{13.33}$$

where $\epsilon = (U/Nh)^2$, and the subscript "H" denotes the two horizontal dimensions only. If the starred quantities all have the same order of magnitude, the dimensional quantities have the magnitudes given by (13.32). If $h/A \lesssim 1$ and ϵ is small, we may look for a solution as an expansion in powers of ϵ (Drazin, 1961), namely

$$(\sigma^*, p^*, \mathbf{u}^*) = (\sigma_0^*, p_0^*, \mathbf{u}_0^*) + \epsilon(\sigma_1^*, p_1^*, \mathbf{u}_1^*) + O(\epsilon^2), \tag{13.34}$$

where the zeroth-order solution $(\sigma_0^*, p_0^*, \mathbf{u}_0^*)$ satisfies (13.33) with $\epsilon = 0$. This is generally known as *Drazin's model*, or *solution*. Equation (13.33) then gives

$$\nabla_H^* p_0^* = 0, \tag{13.35}$$

which with the upstream conditions implies that

$$\sigma_0^* = p_0^* = w_0^* = 0, \tag{13.36}$$

everywhere, and that

$$\nabla^* \cdot \mathbf{u}_{H0}^* = 0. \tag{13.37}$$

Since the upstream flow is assumed to be uniform in the horizontal, the flow at each level is irrotational, with $\nabla \times \mathbf{u}_H = 0$. Hence, at this zeroth order, the flow consists of horizontal potential flow at each level, around the obstacle shape at that level.

For the solution at the next order in powers of ϵ, we have

$$\nabla_H^* p_1^* = -\mathbf{u}_0^* \cdot \nabla^* \mathbf{u}_0^*, \tag{13.38}$$

so that p_1^*, σ_1^* and w_1^* are given by

$$p_1^* = \frac{1}{2}\left[\frac{U(z)^2}{U^2} - \mathbf{u}_0^2\right], \quad \sigma_1^* = -\partial p_1^*/\partial z^*, \quad w_1^* = -\mathbf{u}_0^* \cdot \nabla^* \sigma_1^*. \tag{13.39}$$

The pressure field at each level is therefore due to the zeroth-order flow pattern, and vertical displacements and velocities depend on the vertical gradient of this pressure field. u_1^*, v_1^* and higher-order fields may also be deduced, but the details become progressively more complex and less relevant.

In dimensional form, therefore, the approximate solution given by this model may be expressed in terms of a stream function $\psi(x, y, z)$ at each level defined by

$$u = U - \frac{\partial \psi}{\partial y}, \quad v = \frac{\partial \psi}{\partial x}, \tag{13.40}$$

where ψ satisfies

$$\nabla_H^2 \psi = 0, \tag{13.41}$$

with

$$\begin{array}{ll} \psi \to 0 & \text{as} \quad x^2 + y^2 \to \infty, \\ \psi - Uy = 0, & \text{on the obstacle.} \end{array} \tag{13.42}$$

(Here zero circulation around the obstacle has been assumed – see the discussion below.) The quantities p', w, σ and the vertical displacement η are then given in terms of ψ by

$$p' = \frac{1}{2}\bar{\rho}(U^2 - u^2 - v^2), \quad \eta = \frac{\sigma}{N^2} = -\frac{1}{\bar{\rho}N^2}\frac{\partial p'}{\partial z}, \quad w = \mathbf{u}_H \cdot \nabla\eta. \tag{13.43}$$

Note that this solution also applies for variable $N(z)$ (Brighton, 1978), and that this horizontal flow field is independent of Nh/U (being that for $U/Nh = 0$).

The theory of potential flow around obstacles is highly developed and well known. Figure 13.23 shows an example of potential flow around an elliptical obstacle satisfying (13.41), (13.42), which immediately raises the following questions about the application of this inviscid theory to stratified flows at large Reynolds number. First, as noted in the experiments described in §13.2, lee-side separation is a common occurrence in stratified flow around obstacles at large Nh/U. If potential flow theory is to be applied in such cases, some modification of the shape of the obstacle to take account of the separation region is required. Secondly, the shape of the obstacle may induce circulation around it, particularly if it is streamlined in some way so that separation is absent. For example, the upstream or downstream stagnation points may be moved to nearby sharp corners, as in flow around an aerofoil. It is necessary to know or be able to infer what this circulation is before the flow solution can be obtained. If circulation around the obstacle is present, in general there is a transverse force on the obstacle (corresponding to "lift" on an aerofoil), which exerts a corresponding opposite transverse force on the fluid as it passes the obstacle.

Flow around obstacles that are symmetrical about the flow direction induces no nett circulation, and the solution (13.40)–(13.43) for flow past an obstacle with circular contours that have radius $a(z)$ at height z, with polar coordinates r, θ defined by $x = r \cos \theta$, $y = r \sin \theta$, is given by

$$\psi = U(z)\frac{a^2}{r}\sin\theta, \tag{13.44}$$

$$\eta = -\frac{dU}{dz}\frac{Ua^2}{N^2 r^2}\left(2\cos 2\theta - \frac{a^2}{r^2}\right) - \frac{da}{dz}\frac{2U^2 a}{N^2 r^2}\left(\cos 2\theta - \frac{a^2}{r^2}\right). \tag{13.45}$$

In the case where $dU/dz = 0$ and $da/dz < 0$, the variation of η is shown in Figure 13.24. There is upward motion that raises the streamlines slightly as the fluid particles approach the obstacle, up to a horizontal distance of $\sim (\sqrt{2} - 1)a$ from the surface. Closer to the surface the velocity is downward, and the streamlines return to their upstream level at $x = -a$. The vertical scale of the region where this descending motion next to the boundary occurs is $[\sqrt{2} - 1]a(da/dx)$. Note that this solution does not predict any reversed flow (at least at this lowest order in the expansion), since u is never negative. Numerical results by Ding et al. (2003) show that Drazin's model gives poor agreement for $Nh/U > 0.4$.

As the fluid flows around the obstacle, for $da/dz < 0$ the streamlines are displaced downward, because the pressure above is greater at the same radius r. Comparisons between (13.45) and observations show that the general pattern of descent of flow incident below the upper upstream stagnation point is consistent, and Hunt & Snyder

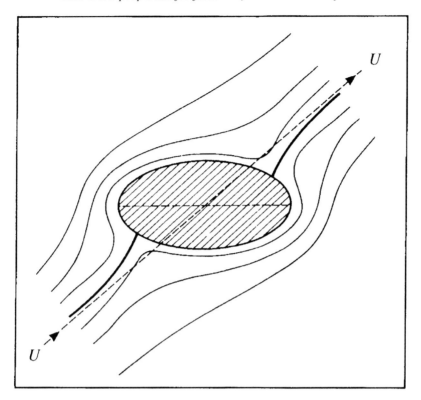

Figure 13.23 Streamlines for two-dimensional potential flow past an ellipse inclined at an angle to the mean flow.

(1980) found reasonable quantitative agreement when $Nh/U \gtrsim 5$ for the vertical
displacements at mid-levels, as exemplified in Figure 13.8.

From the nature of solution (13.44), (13.45), it is clear that it does not apply close
to the top of an obstacle where $a(z) \to 0$, and the solution becomes singular, or at
least discontinuous with the flow above. Both scaling and observation indicate that
this range of non-applicability extends a distance of $O(U/N)$ below the obstacle top,
and the governing equations are non-linear and three-dimensional, being similar
to those for the flow over a whole obstacle for which $Nh/U = 1$. Also, if the
obstacle has a "skirt" so that $da/dz \to \infty$ at the base where $z \to 0$, the solution
again becomes singular with a range of non-applicability that is of $O(U/N)$, and
the governing equations in this "verge region" are again non-linear and three-
dimensional (Greenslade, 1992).

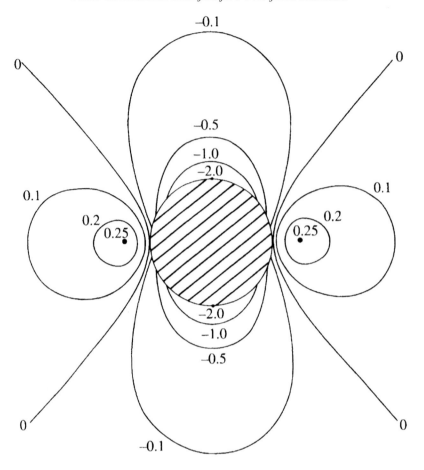

Figure 13.24 Interface displacement for flow around a circular obstacle with $dU/dz = 0$, $da/dz < 0$, from Drazin's model for small U/Nh, in units given by (13.45).

13.3.2 The momentum-source model

Stratified flow around obstacles at large R_e is likely to separate on the lee side and develop a turbulent wake region, as described in §13.2. This phenomenon introduces additional effects on the flow outside the wake, because of the drag associated with separation. For the two-dimensional (x, z) case in Chapter 10, these effects were described by representing the wake by a momentum source in the form of a body forcing term within the fluid. This may also be done for three-dimensional obstacles. Such models cannot describe the total effect of the topography on the flow, and are generally not formally justifiable, but they do represent the wake effects realistically and are useful in describing some effects in the far-field. Since these effects are mostly hydrostatic, we again assume that the flow is linear and hydrostatic for simplicity.

Hence if the effect of the topography is represented by a horizontal body force $\mathcal{B} = (\mathcal{B}_1, \mathcal{B}_2, 0)$ in (13.25) we obtain a forced hydrostatic form of (12.3) for η (cf. (10.28))

$$L(\eta) \equiv \frac{\partial^4 \eta}{\partial x^2 \partial z^2} + \frac{N^2}{U^2}\left(\frac{\partial^2 \eta}{\partial x^2} + \frac{\partial^2 \eta}{\partial y^2}\right) = -\frac{1}{\rho_0}\frac{\partial}{\partial z}\nabla.\mathcal{B}, \qquad (13.46)$$

and equivalently for the transverse velocity v'

$$L(v') = -\frac{N^2}{\rho_0 U^3}\frac{\partial \mathcal{B}_1}{\partial y} + \frac{1}{\rho_0 U}\frac{\partial}{\partial x}\left(\frac{\partial^2}{\partial z^2} + \frac{N^2}{U^2}\right)\mathcal{B}_2. \qquad (13.47)$$

The boundary conditions are then $\eta = 0$ at $z = 0$, with a radiation condition at infinity. These equations may be solved by Fourier transforms for chosen forms of $\mathcal{B}_1$ and $\mathcal{B}_2$. Using the same type of representation as in §10.2.3, the simplest form for $\mathcal{B}$ is to take $\mathcal{B}_2 = 0$, and $\mathcal{B}_1$ localised and uniform, so that

$$\mathcal{B}_1 = \begin{cases} -\dfrac{1}{2}\rho_0 C_\mathrm{D}\dfrac{U^2}{A}\delta\left(\dfrac{x}{A}\right)\delta\left(\dfrac{y}{B}\right), & 0 < z < z_\mathrm{m}, \\ 0, & z > z_\mathrm{m}. \end{cases} \qquad (13.48)$$

Here the function $C_\mathrm{D}(z)$ is a drag coefficient of order unity, A and B are the usual horizontal length-scales of the obstacle, and z_m is the height over which the drag force acts. We assume that C_D is constant in this discussion, and both C_D and z_m depend on Nh/U. The resulting flow is representative of that produced by a localised obstacle, except for very close to the origin. If $Nh/U = 0$, from the drag on a sphere (for example) we expect $C_\mathrm{D} < 0.5$ (for $A = B$) and $z_\mathrm{m} < h_\mathrm{m}$. As Nh/U increases, from §13.2 stratification suppresses the lee-side separation and wake to a maximum extent when $Nh/U \sim 1$, so that C_D and z_m should decrease. For $Nh/U > 1$, fluid at levels below $h_\mathrm{m} - U/N$ flows around the obstacle and then separates, so that we expect $C_\mathrm{D} \approx 1$ and $z_\mathrm{m} = h_\mathrm{m} - U/N$.

For forcing of the form (13.48), from our experience with the unforced form of (13.26) we may expect the solution to have a simple horizontal structure, except close to the origin. Regardless of the values of C_D and z_m, the solution is given by

$$\eta(x, y, z) = \frac{C_\mathrm{D} BU}{2\pi^2 N}\int_0^\infty \int_0^\infty dk\,dm \frac{\cos my}{(k^2 + m^2)^{1/2}} \times \begin{cases} \sin(kx + nz_\mathrm{m})\sin nz & z < z_\mathrm{m} \\ \sin(kx + nz)\sin nz_\mathrm{m} & z > z_\mathrm{m}, \end{cases} \qquad (13.49)$$

where n is given by (12.34). On the x- and y-axes this reduces to

$$\eta(x, 0, z) = \frac{C_D B U}{8\pi N x}[J_0(Z_+) + J_0(Z_-)], \qquad z < z_m,$$

$$\eta(x, 0, z) = \frac{C_D B U}{8\pi N x}[J_0(Z_+) - J_0(Z_-)], \qquad z > z_m, \tag{13.50}$$

$$\eta(0, y, z) = 0,$$

where

$$Z_+ = \frac{N}{U}(z + z_m), \quad Z_- = \frac{N}{U}(z - z_m). \tag{13.51}$$

This indicates that η has a simple horizontal pattern at each level, being depressed on the upstream side and elevated on the downstream side (or vice versa, depending on Z_+, Z_-) with a more complex vertical structure and a discontinuity at $z = z_m$. The lateral velocity v' is more significant, and has the form

$$v'(x, y, z) = \tag{13.52}$$

$$\frac{C_D B U}{2\pi^2} \int_0^\infty \int_0^\infty dk\,dm \frac{m \sin my}{(k^2 + m^2)} \times \begin{cases} \cos kx - \cos(kx + nz_m)\cos nz & z < z_m \\ \sin(kx + nz)\sin nz_m & z > z_m. \end{cases}$$

On the axes this solution reduces to

$$v'(x, 0, z) = 0,$$

$$v'(0, y, z) = \frac{U}{8\pi}\frac{C_D B}{y}\int_{|Z_-|}^{Z_+} J_0(x')dx', \tag{13.53}$$

which shows a diverging pattern with sideways deflection around the obstacle away from the x-axis, in the same manner as for linear theory on the plane $z = 0$. For the solution (13.52), however, we have $\partial v'/\partial z = 0$ at $z = 0$, so that the pattern of lateral deflection is extended upwards into the fluid. This deflection decays with height, and for $N z_m/U \gg 1$ it is reduced by $1/2$ at $z \approx z_m$.

For $N z_m/U \gg 1$, from the stationary phase approximation the z-dependent term for $z < z_m$ is $O[(U/N z_m)^{1/2}]$, so that v' is approximately given by the first term, which is

$$v' = \frac{U C_D B y}{4\pi(x^2 + y^2)}. \tag{13.54}$$

For each y this is symmetric about $x = 0$, and it may be integrated to give the lateral displacement $\xi(x)$ for each x-value of a streamline originating from a point $(-\infty, y_s)$ far upstream, which is

$$\xi(x) = \frac{C_D B}{4\pi}\,\mathrm{sgn}(y_s)\left[\frac{\pi}{2} + \arctan\left(\frac{x}{|y_s|}\right)\right], \tag{13.55}$$

where $\mathrm{sgn}(y_s)$ denotes the sign of y_s. This has obvious similarities to (12.49). Hence the momentum-source model implies that when lee-side separation occurs, it causes the fluid at the same levels to be permanently displaced laterally, away from the obstacle.

As for two-dimensional obstacles, we may heuristically add the two forms of linearised solution – the linear perturbation theory of §12.1 and the momentum-source model – to obtain a realistic qualitative description of the nett effect of the obstacle on the far-field flow, when $Nh/U < 0.5$, and there is lee-side separation. Quantitative tests of the efficacy of such a superposition have, to present knowledge, yet to be made.

13.3.3 Numerical studies

The preceding sections show that the information obtainable from analytical studies when Nh/U is not small is limited, but the nature of such flows (at least for simple obstacles) is reasonably clear from laboratory experiments. More specific information is available from numerical studies, notably by Suzuki & Kuwahara (1992), Ding et al. (2003) and Smolarkiewicz & Rotunno (1989, 1990). The latter have examined the flow past an isolated obstacle using the anelastic non-hydrostatic model of Clark (1977) (previously invoked in §§10.4 and 11.8) with some numerical improvements, including incorporation of the interactive nesting scheme of Clark & Farley (1984). This model simulated an essentially inviscid fluid of infinite depth, with a free-slip lower surface boundary condition, and permitted "zooming in" on two smaller scales. However, such a model does not simulate the flow separation in large R_e experiments realistically, and the results must be interpreted with care. The obstacles studied had the form

$$h(x, y) = \frac{h_m}{\left[1 + \left(\frac{x}{a}\right)^2 + \left(\frac{y}{b}\right)^2\right]^{3/2}}, \tag{13.56}$$

(i.e. (12.16) with $\upsilon = 3/2$), with $h_m/a = 0.12$, a smaller slope than for any of the laboratory experiments. For $Nh/U < 0.5$ and $b = a$, the numerical solutions were close to the hydrostatic analytic linear solutions as described in §§12.1.4 and 12.1.5. As Nh/U increased further, the flow pattern remained similar to that of linear theory but the disturbance amplitude was larger. In particular, the velocity minimum on the upstream face of the obstacle became more pronounced, and increased to become a stagnation point when $Nh/U = 1.6$ (i.e. $U/Nh = 0.63$), at approximately the same location as that given by linear theory. A small further increase in Nh/U caused this stagnation point to spread vertically and form a line of attachment where the flow divided and passed around the obstacle.

This upstream flow-splitting has the same character as that of the observations described in §13.2, except that it occurs for $Nh/U > 1.6$ rather than 1.1. This difference may be affected by the obstacle shape, but more probably, it is due to the absence of lee-side separation in the numerical simulation. Lee-side separation is controlled by boundary-layer processes, and occurs over the whole range of Nh/U values shown in Figures 13.3, 13.5–13.8. Its effect on the rest of the flow is described by the momentum-source model of §13.3.2, where it is shown that $\eta < 0$ on the upstream side for $z < z_m$, a perturbation that promotes flow-splitting. This effect is absent from the numerical simulations for $Nh/U < 2$. In §12.4.3 it was argued from the numerical results of Smith & Grønås (1993) that upper-level overturning promotes flow-splitting, and these present results imply that lee-side separation does also. However, flow-splitting does not *depend* on either, as the observations of §13.2 show.

The cause of this upstream stagnation and flow-splitting may be seen from the Bernoulli equation (13.31), which may be expressed in the form

$$\mathbf{u}^2 = U^2 - N^2(z - z_0)^2 - 2 \int_z^\infty \frac{g\rho'}{\bar{\rho}} dz, \qquad (13.57)$$

if the pressure is given by its hydrostatic value. The numerical solutions show that the fluid column up to a considerable height above the face of the obstacle is lifted, increasing the hydrostatic pressure on the surface, and this is the main reason for the fluid on the surface stagnating and splitting at $z/h_m \lesssim 0.5$.

When flow-splitting occurs, the fluid descends on and close to the centre-line, so that the flow becomes statically unstable in a region that is narrower (in the x-direction) than the region of descent given by Drazin's solution $(0.4x(dh/dx))$. In the high-resolution numerical simulations, this gives a broad region on the upstream face where "spotty" convective instability occurs, very close to the surface. This is concentrated at the lower part of the flow-splitting range, as in Figures 13.7, 13.8, where it extends laterally around the height contours. The largest of these over-turnings occurs near the lower stagnation point S_s, where there is a semi-persistent "ripple".

As Nh/U is increased above 1.67, the ripple (and the stagnation point) moves closer to the obstacle, and its amplitude decreases. Its character is similar to that of the "horseshoe vortex" due to separation of the frictional boundary-layer, observed at the base of steep obstacles in homogeneous fluid (as shown in Figure 13.4, for example), but here it is due to the pressure field caused by the stratification. The horseshoe vortices due to steep slopes are also observed in stratified fluids (Castro et al., 1983), and which mechanism dominates will depend on the obstacle slope and flow geometry. The upstream ripple itself is due to the three-dimensional geometry but it has an approximately two-dimensional structure, and it causes a disturbance to the fluid overhead. This has the form of a weak but conspicuous vertically propagating internal wave, as from flow over a small obstacle. This shallow unstable region has significant consequences for flow in the atmosphere, as discussed in Chapter 14.

If B/A decreases below unity (a narrower obstacle, spanwise), the numerical studies imply that the value of Nh/U for the onset of upstream stagnation also increases, to values greater than those for the occurrence of lee-side eddies. If B/A increases above unity, the pattern of upstream stagnation and flow-splitting persists, and the position of the upstream "ripple" (for values of Nh/U for which it occurs) moves upstream. However, the picture is now complicated by the presence of upper-level overturning and the generation of upstream disturbances by the hydraulic-transition mechanism, as for two-dimensional flows. Smith (1989a) has shown that linear perturbation theory (i.e. small Nh/U as in §12.1) suggests that for $B/A < 1$, upstream stagnation occurs at smaller Nh/U than does upper-level overturning (if the latter occurs at all), but that for $B/A > 1$ it suggests the reverse. This expectation is generally borne out in the laboratory and numerical experiments, except that for $B/A > 1$ if there is upper-level overturning there is also likely to be upstream stagnation. For obstacles that vary slowly in the y-direction, all models and observations show that the flow over the obstacle crest varies little with y except near the ends, and upper-level overturning, if it occurs, extends along the length of the obstacle. Concomitant with this, both observations (Baines, 1979b) and the numerical results (Smolarkiewicz & Rotunno, 1990) show that there is a well-defined height above which the fluid flows over the obstacle in near two-dimensional (x, z) fashion, and below which it flows around it in near horizontal fashion.

When upstream flow-splitting occurs in laboratory experiments, the flow around the obstacle separates from the surface and this results in a wake that consists of a stationary vortex pair, with near vertical axes, or a periodic vortex street. The vorticity in these lee-side eddies stems from the boundary-layer that has separated from the surface of the obstacle. It is important to note that viscosity may play a part in determining where separation occurs, but that it does not affect the total amount of vorticity in the boundary-layer, or the amount that enters the body of the fluid when it separates (Morton, 1984). The integral of the vorticity across a boundary-layer is equal to the fluid velocity just outside it, for a no-slip surface, even in the limit $v \rightarrow 0$.

The numerical experiments of Smolarkiewicz & Rotunno contain a wake and stationary eddies that have much the same appearance (as in the experiments), but there is an important difference, in that the flow in their model is essentially inviscid with a free-slip boundary condition. Here, each isosteric surface that is "split" at the upstream line of attachment is constrained to maintain contact with the surface of the topography, once contact is established, so that no separation occurs.

Nonetheless, the fluid assumes a form that is superficially similar to that seen in the large-Reynolds-number experiments. It is helpful to regard this numerically simulated wake as a "dissipative structure" in the flow, in the sense of Prigogine (see, for example, Prigogine, 1980; Prigogine & Stengers, 1984). It is important in the energy balance, in an analogous manner to that of a hydraulic jump in the flow of a single layer around a tall, gently sloping obstacle, as described in §3.5.

From the preceding observations, theoretical and numerical studies, we may summarise the situation for flow-splitting, upper-level overturning and lee-side separation, for stratified flow over single isolated obstacles at large R_e, as follows. For an obstacle of given slope, as Nh/U increases the principal phenomenon that alters the flow pattern is upstream flow-splitting, which occurs for $Nh/U \geq 1$, depending on shape. This is due to the upstream pressure field causing stagnation on the upstream obstacle surface, at a location given approximately by linear perturbation theory. If upper-level overturning or lee-side separation forming a wake occur, both tend to promote flow-splitting (i.e. cause it to occur at smaller Nh/U than it would otherwise) by increasing the pressure on the upstream slope, but they are not necessary for its occurrence. Conversely, flow-splitting tends to promote lee-side separation and reduce upper-level overturning, since it involves more lateral flow around the obstacle. The application of this type of study in representing the effects of topography in weather-forecasting models is described in §15.3.

Subsequently to the above, there have been several other numerical studies of the development with time of uniformly stratified flow over single obstacles of varying width and differing surface boundary conditions. These include Olafsson & Bougeault (1996), Schär & Durran (1997), Bauer et al. (2000), Epifanio & Durran (2001) and Epifanio & Rotunno (2005), and collectively these studies provide a picture of the variety of these three-dimensional flows.

More recently, Jagannathan et al. (2019) have shown that for stratified flow past long ridges of finite length with large Nh/U (greater than 6), the flow primarily consists of upper-level two-dimensional flow over the ridge with a hydraulic transition, above blocked lower-level lateral potential flow around the ridge.

13.4 The drag force on isolated obstacles: $Nh/U > 1$

The drag on obstacles where Nh/U is small so that the flow is effectively linear is described in §12.1.6. For large obstacles such as those described in this chapter, the same processes still apply, with drag due to lee-wave generation and pressure difference across the obstacle. Here this is increased because of slower flow on the upstream side (relative to the mean) and faster downslope flow on the lee side. But as is evident from the flow shown in Figures 13.6–13.19, when $Nh/U > 1$, low-level flow passes horizontally around the obstacle, and this results in a much larger total drag on the obstacle.

Hence, there are two components to the drag on a single symmetric obstacle, of height h_m. In dimensional terms, the first component is the "wave drag" which has the form

$$\tau_w = C_w \rho_0 N (h_m - h_0)^2 \mathbf{U}, \qquad (13.58)$$

where C_w is a constant that depends on the geometry, and h_0 is the depth of the lower-level blocked fluid. This drag is manifested in the generation of vertically propagating internal gravity waves.

The second component is the "blocked flow" drag due to low-level separated flow around the obstacle, which has the familiar form for drag due to flow past obstacles:

$$\tau_b = 0.5 C_D \rho_0 b h_0 |U| \mathbf{U}, \qquad (13.59)$$

where b is a measure of the breadth of the obstacle, and C_D is an appropriate drag coefficient. This drag is manifested at the levels of flow separation.

The nett drag on the obstacle is the sum of these two components. How this drag affects the flow of the atmosphere (or the overlying stratified fluid) is described in §15.3.

14

Flow over complex and realistic terrain in the atmosphere and ocean

Out of intense complexities, intense simplicities emerge.

WINSTON CHURCHILL

14.1 Flow over complex terrain

So far we have concentrated attention on isolated obstacles of various shapes, with a single maximum in height at a point, or along a line. If two or more such obstacles are in close proximity, the disturbance field of any one may affect the external flow past the others. Shorter obstacles have less controlling effect on the flow than taller obstacles. For present purposes we assume that all obstacles in a complex region have comparable heights, so that they also have comparable Nh/U values. If $Nh/U < 1$, the disturbance field of each obstacle is approximately linear, so that the interaction is minimal. If instead $Nh/U > 1$ so that flow-splitting, vortex wakes and lee-wave overturning occur, interaction may be significant, particularly if one obstacle lies downstream of another. The effect of all of these factors depends on the flow geometry, but upper-level wave-overturning is not expected to affect downstream flows greatly unless the topography is nearly two-dimensional, and this type of system has been discussed in §11.9. The most important interaction for present purposes concerns the flow that has split and passed around the obstacles, resulting in substantial wakes, and this discussion concentrates on the effects of this behaviour.

If the obstacles are relatively isolated and spread out so that $A/S \ll 1$, where A is the typical obstacle dimension and S is the typical spacing between them, the interaction is negligible. If the nett effect is weak it may be modelled as a linear disturbance, and we may then obtain the effect of a whole region of such obstacles by taking the sum of the linear superposition of the nett effects of each obstacle and its wake. This may be represented by a number of isolated momentum sources, or more simply in a smoothed form by a horizontal drag force spread over the whole topographic region. The region therefore acts as a momentum source (or more accurately, sink) for the flow at levels where the flow is around the obstacles, and the long horizontal length-scale implies that this disturbance will be hydrostatic.

The relevant equations for this form of model are given in §13.3.2. We assume that the body force $\mathcal{B}$ varies continuously in the horizontal, and as an example we take the form

$$
\begin{aligned}
\mathcal{B}_1 &= -\frac{1}{2}\rho_0 U^2 C_D \frac{a^2}{(x^2 + y^2 + a^2)^{3/2}}, & z &< z_m \\
\mathcal{B}_2 &= 0, & z &< z_m \\
\mathcal{B}_1 &= \mathcal{B}_2 = 0, & z &> z_m.
\end{aligned}
\right\} \quad (14.1)
$$

This gives a smoothed version of the solution for δ-function forcing (13.48), to which it is qualitatively similar. In particular, for $z < z_m$ and $N z_m/U > 1$, the flow diverges laterally around the whole region in the same manner as described by (13.54). Experiments on stratified flow past a localised bed of vertical nails (Newley et al., 1991) with $A/S \sim 0.3$ have been compared with results using (14.1) where C_D was estimated from the drag of individual nails, giving qualitative agreement for the vertical streamline displacements.

In complex terrain in general, however, the hills and mountains (here termed obstacles) are not so thinly spaced. Distances between nearby obstacles are often comparable with their widths ($A/S \sim 1$), so that it is common for an obstacle to lie within the wake of a close upstream neighbour. The flow through such a region of randomly arranged and oriented complex terrain where $Nh/U > 1$ may be expected to consist of a superposition of wakes, in which the nett momentum is small. In fact, the first approximation to the nett flow through such a region is to assume that it is zero, and this may be made for parametrisation schemes as described in Chapter 15. Effectively, this would imply that there was no flow through the region at heights below $h_m - U/N$, so that the external flow passes around it entirely. The region effectively behaves as a single, broad obstacle with a drag coefficient of unity.

Laboratory observations of flow past arrays of obstacles where $A/S < 1$ show that this picture is valid in general, but there are differences in detail, which may be expected to increase with A/S. Two features in particular may be noted. The first is variability. The low-level wakes behind typical obstacles are unsteady (at large R_e), consisting of vortices being shed alternately from each side. The effect of such wakes is to weaken the incident flow on obstacles downstream, and to cause its direction to vary. This increases the variability of the wake of the second obstacle and the overall flow. Some experiments with a clump of six obstacles towed at uniform speed give what appear to be completely aperiodic flow at low levels in the region between the obstacles, varying in a random way on time-scales greater than the eddy-shedding time-scale ($\sim 0.2A/U$). Figure 14.1 shows an example of the flow at two different times for $Nh/U \sim 5$, showing this inherent variability.

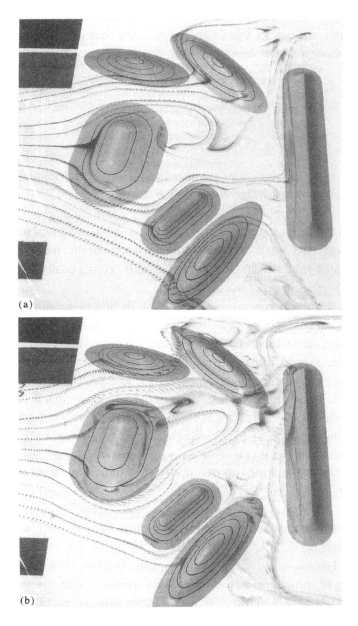

Figure 14.1 Two examples of near-horizontal flow through a towed clump of obstacles representing complex terrain, with $Nh/U = 5$. The second photograph was taken at a later time in the same run to show the varying flow pattern. The flow has been visualised by neutrally buoyant dye released from a rake placed horizontally upstream at a level well below h_m.

The second noteworthy feature of flow through complex terrain is the existence of concentrated streams or "jets", which originate from high-velocity regions at the sides of obstacles where $B/A > 1$. In particular, as described in §13.2, potential flow around an elliptically shaped obstacle (see Figure 13.23) with semi-axes of length A and B, perpendicular to the stream, has maximum velocities at the sides of $U(1 + B/A)$ (more general expressions may be derived for obstacles that are not perpendicular). The width of this region is given by $|u/(\partial u/\partial y)|$ at $(x, y) = (0, B)$, which is A^2/B, containing a flux of $\sim U(A + B)$ per unit height. If $B/A \gg 1$, this gives a narrow region with large velocity at the sides. This flow is prone to separate, so that this fluid leaves the surface of the obstacle in the form of a concentrated stream or "jet", that may then thread its way between the obstacles downstream. Such jets are apparent in Figure 14.1, where their pattern varies with time in a random fashion. In a sense, the pattern of these jets defines the flow, because the remainder of the fluid between the obstacles is relatively stagnant. The vertical structure of this low-level horizontal flow pattern is not exactly uniform, because the separation points on obstacles whose widths vary with height cannot be the same (in x, y terms). However, the flow is observed to be broadly coherent in the vertical, with lee vortices and concentrated jets occurring in approximately the same locations at different heights.

If $A/S \gtrsim 1$, so that the obstacles are closely spaced and steep-sided with $B/A > 1$ and $NA/U < 1$, separation in the vertical plane may become more important than in the horizontal plane. Under these circumstances where typically $A<2\,\mathrm{km}$, the obstacles may act more like roughness elements in a mechanically generated turbulent boundary-layer over a rough flat surface than like obstacles in stratified flow (Grant & Mason, 1990). Turbulent layer thicknesses up to approximately $2h_m$ have been observed, and the flow may be described with roughness lengths z_0 of up to $10\,\mathrm{m}$ for values of h_m of several hundred metres. Here the overlying stratification may have little or no effect on the flow below the maximum topographic height. This picture is more complex if a buoyancy flux due to solar heating or cooling is present. In nocturnal conditions, surface cooling may resuscitate the low-level stratification and restore the stratified horizontal flow regime, where appropriate, but such phenomena are generally specific to the region concerned.

Oscillating stratified fluids can have quite complex flows, particularly in closed containers. Some examples are described in Baines (1997).

14.2 Some atmospheric examples in the troposphere

14.2.1 Wakes of airflow past islands

The large island of Hawaii is situated in the persistent northeasterly trade winds, and it presents a good example of an isolated obstacle (with two peaks over 4 km) where the flow around it normally has $Nh/U > 2$. This flow shows many of the features described in the preceding sections, and these have ramifications for the weather and climate of the island. Figure 14.2 shows a Space Shuttle photograph of Hawaii. Apart from the relatively familiar orographic clouds near the summit, there is a band of cloud wrapped around the upwind eastern side of the island. These cloud bands, known as Hilo's band clouds, form 10–50 km east of Hilo (on the northeastern coast) nearly every day and in the same location, with a tendency to move onshore after sunrise with possible onshore rain. Modelling studies by Smolarkiewicz et al. (1988) and Rasmussen et al. (1989) have shown that these clouds are not due to surface-heating effects as was previously thought, but are caused by the orographically induced overturning and "ripple" on the upstream side of the island, as described in §§13.2.3 and 13.3.3.

The numerical model used in this study was the same as in §13.3.3, with the additions that it modelled liquid water, vapour and rain, and surface sensible and latent-heat flux (the details of which are omitted here). Nesting on three scales was used. Figure 14.3 shows results from numerical simulations of flow past Hawaii, initialised with realistic wind from due east, temperature and water-vapour profiles, and horizontal potential flow around the island at each level. These give $Nh/U = 5$, a typical value for Hawaii. The integration was begun at sunset, and after three hours the transients had largely propagated away and the surface flow pattern was as shown in Figure 14.3a. Here the solid contours denote liquid water content at a height of 750 m, representing clouds. Except for the wake region, this pattern remained approximately constant until sunrise in the model. The clouds in the model are formed by the upstream "ripple", and are identified with the band clouds seen in Figure 14.2. Figure 14.3b shows flow on the northeastern side in the "zoomed-in" integration with the same conditions, just before sunrise. Here the contours represent vertical velocity, which is concentrated in the cloud-band region, where upward motion coincides with liquid water.

After sunrise, surface heating over the island reverses the orographically induced downslope flow of the "ripple". As a result, the vertical motion in the cloud-band region in the simulations is reduced, but it is still present at the same location. The orographic dynamics are therefore seen to be the stronger factor here, compared with surface heating and cooling, and the latter play only a secondary, modifying role.

On the lee side these simulations show a time-dependent vortex wake, with eddies shed alternately in the usual manner for a vortex street. However, field observations (Smith & Grubišić, 1993) in July and August 1990 of the wake of Hawaii with wind from ENE show a symmetric wake consisting of two extended eddies (as in the pattern shown in Figure 13.8) that is apparently stationary over a period of at least five days. The absence of the oscillating vortex wake is attributed to the effect

Figure 14.2 Space Shuttle photograph of Hawaii showing a typical band cloud (denoted by two arrows). Time: 0840 HST, 30th August 1985. (From Smolarkiewicz et al., 1988: reproduced by permission of the American Meteorological Society.)

of bottom friction due to the drag on the surface of the island and the surrounding ocean (Grubišić et al., 1995).

Observations of the wake of the island of St. Vincent – one of the Windward Islands of the Caribbean – present a very different picture from the flow over the big island of Hawaii. These islands are smaller, and St. Vincent is representative, with a diameter of about 20 km and height of about 1 km, and its surface is relatively steepsided. The island is observed to have a turbulent wake with no conspicuous eddies or reversed flow, and typically extends a distance of over 300 km downstream, with an approximately constant width of 20 km (Smith et al., 1997).

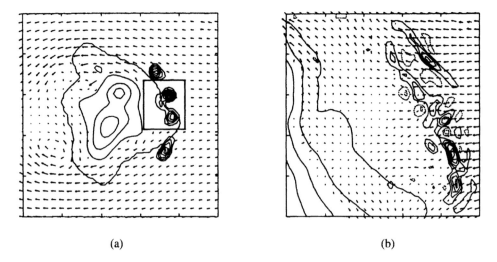

(a)   (b)

Figure 14.3 Numerical model results for flow around the island of Hawaii by Smolarkiewicz et al. (1988) for representative flow conditions with a free-slip lower boundary and $Nh/U = 5$. (a) Velocity vectors on the surface of the topography, and contours of liquid water content (denoting clouds) at 750 m above the surface (small solid features), 3 hours after commencement. Topographic contours are at intervals of 941 m. (b) The "zoom-in" to the boxed region on the northeastern corner of the island in (a), with the higher-resolution nesting, but at a later time (7 hours). Arrows again show the surface flow, and the closed contours denote the vertical velocity field at a height of 250 m (solid denotes up, dashed down). Topographic contour interval here is 171 m. (Reproduced by permission of the American Meteorological Society.)

   These and other observations (Schär & Smith, 1993a, 1993b) have suggested a classification of island wakes according to two parameters, as shown in Figure 14.4. Four different flow regimes are identified, based on parameters for a single layer as in Chapter 2. The positions of the boundaries are indicative, and are largely unknown. Low-level islands have simple or zero wakes, and tall islands may have steady wakes as for Hawaii, or oscillating or turbulent ones if the parameter R_e is large enough. The airflow incident on islands is not always uniform. The effect of a large-scale cyclone, represented as a large vortex, can have a significant effect on the island wake, as described by Smith & Smith (1995).

14.2.2 *The Olympic mountains*

The Olympic mountains of the northwestern United States constitute an isolated, roughly circular complex of 50 km in diameter with heights ranging from near sea-level to 1800 m in the centre. The whole may be crudely approximated for theoretical purposes by a single smooth circular obstacle with this diameter and height. The

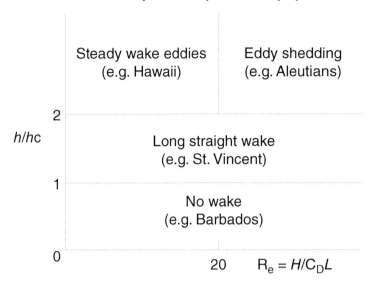

Figure 14.4 This shows a classification of island wake-types according to two dimensionless parameters: the appropriate height h of the island, and R_e, a form of Reynolds number. h is scaled with h_c, a critical height for wave-breaking if the flow is considered as a single layer of thickness H; also, C_D is a drag coefficient for flow over the lower boundary, and L is an appropriate horizontal scale (the island width) (Smith et al., 1997).

region is exposed to the prevailing westerly winds from the Pacific Ocean. In a field experiment carried out in 1988, the existing meteorological network was supplemented by a number of sensitive microbarographs, giving a description of the surface-pressure variations on an almost continuous basis for nearly six months (Mass & Ferber, 1990). The resulting pressure pattern was systematically compared with the gross predictions from linear theory for a single obstacle. The value of U used was centred on 850 mb, N was averaged over a range of low-level heights, and $h_m = 1800$ m. The data analysis and results were largely statistical, with some 591 identified cases.

These observations showed that there was a high-pressure perturbation on the upstream side and a low on the downstream side, as predicted (see Figure 11.10). However, the downstream perturbation was generally smaller than the upstream one, and this difference increased with Nh/U. Overall, the pressure perturbations agreed better with linear theory when $Nh/U < 1$ than when $Nh/U > 1$, as would be expected. This general agreement between field observation and theory gives confidence in the applicability of the theories described above, in spite of the approximations made and the vagaries of measurement.

14.2.3 Tasmania

Southeastern Australia is one of the regions of the world that is susceptible to wildland ("bush") fires in summer. This is because of the combination of high temperatures, strong winds and ample dry fuel due to growth in winter and subsequent drying out. The island Australian state of Tasmania lies 250 km to the south of the mainland and is separated from it by Bass Strait. Of the whole southeastern region, the southern portion of Tasmania (latitude $43°$S), while being most attractive country, is perhaps the most bushfire prone of all, as evidenced by the fires that threatened Hobart in 1967. In contrast, the northern part of the state is relatively fire-free, and it is of interest to see why this situation exists.

Fires are most likely to occur in southeast Australia when hot dry northerly winds from the centre of the continent blow across the region. These winds are due to the varying synoptic flow pattern, and occur in summer on an almost weekly basis. The depth of this hot dry air may be up to 4 km, due to the extensive convection in the central deserts. When these winds reach Bass Strait, the air at the surface is cooled by the sea so that a relatively cool and moist boundary-layer forms. This boundary-layer grows in thickness as the air moves southward, and when it reaches the coast of northern Tasmania it has a thickness of about 300m and is colder by about 5–10°C.

The central region of Tasmania consists mainly of an elevated plateau of height 1.5 km, and this cold low-level air is not able to surmount this barrier (even assuming uniform stratification with $U = 10$ m/s, $N = 10^{-2}$/s and $h_m = 1500$ m, so $Nh/U = 1.5$). In consequence, the cold low-level air is directed around the central plateau, and the air passing over it is hot and dry. This air originates from an elevated level over the mainland so that it is, if anything, hotter and dryer than the mainland air at ground level. On reaching the southern region of Tasmania, this air descends to ground level in the region around Hobart. Here it dries out the fuel bed, and provides ideal fire conditions because of its relative strength and temperature. Local foresters and weather forecasters refer to dangerous fire conditions of "superdry air" where the humidity is effectively zero, and the reasons for this are seen to be due to the Tasmanian topography and the presence of the Bass Strait.

14.2.4 The Antarctic Peninsula and island of South Georgia

The phenomenon described in §14.2.3 occurs in slightly different forms in many other locations on a variety of scales, which include the Antarctic Peninsula and the South Atlantic island of South Georgia. The mountainous region of South Georgia has a length of $\sim$ 150 km, and lies across the mean westerly wind in mid-ocean. The Antarctic Peninsula is about 10 times longer, and much broader, and the flow there is affected by the Earth's rotation, but many of the important flow features are the same and these include the dominant eastward wind.

In particular, both regions are subject to the "foehn effect", which can cause significant weather events downstream of mountain ranges. In the classic form of this effect, as air ascends the upstream side of a mountain or mountain range, the water vapour in this air condenses. This forms clouds that cause rain, which removes (some of) the water. The latent heat resulting from this condensation causes this air to be warmer and drier, as it rises to pass over the peak of the mountain (or range) to descend down the lee side. As it descends, the temperature of this air is increased by adiabatic compression giving (relatively) hot, dry air at ground level. The same effect can be obtained without clouds if the upstream flow is partially blocked, and becomes stationary at low levels, or is deflected around the mountain away from the downstream side. When this happens, the air that passes over the mountain and descends on the downstream side comes from higher upstream levels (where it is generally drier, with higher potential temperature than air at ground level), as described in §13.2.6. Both of these processes operate over the Antarctic Peninsula and South Georgia (Orr et al., 2008; Bannister & King, 2015).

Since the prevailing westerly winds are very common in these locations, the environment on the downstream side of these ranges is much warmer, drier and generally more benign than on the upstream side (at these latitudes), though the dry downslope winds may be strong at times. Further, with climate trends over the past 100 years the foehn winds have become stronger, resulting in a decrease in glaciers and other forms of ice on the lee side of both ranges.

14.2.5 The PYREX experiment in the Pyrenees

The Pyrenees/PYREX experiment was a joint observational program by the French and Spanish weather services, also involving research institutes from France, Spain and Germany. The Pyrenees constitutes a massive mountain range extending from the Atlantic Ocean to the Mediterranean over a distance of nearly 400 km, with a maximum height of over 3000 m in the central part, and a width varying from 30–70 km. The main aim of the experiment was to establish an observational data base that can be used in improving the representation of mountains of this scale in weather forecasting models.

The experiment took place in October and November 1990 with 10 intense observing periods (IOPs) lasting 2–3 days. A wide range of observations were made during these periods, including observations from four instrumented aircraft, balloons for soundings and at constant levels, sodars (wind measurement), radar and lidar, and with an array of surface observational networks. The main objectives were to improve the knowledge of the wind distribution all around the range, and

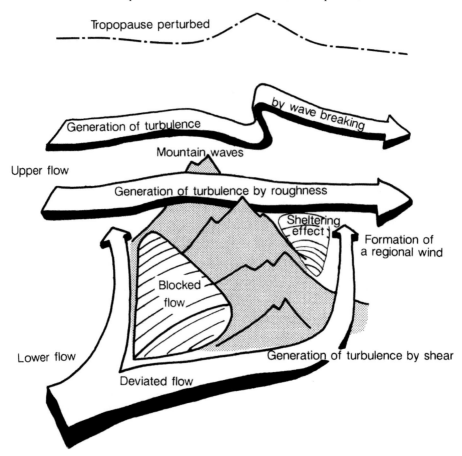

Figure 14.5 A schematic of the main features of flow around the Pyrenees. (From
Bougeault et al., 1990.)

more specifically to measure the mountain pressure drag, the wave momentum flux
and turbulence associated with mountain waves.

The observational programme was also linked with the purpose of improving
the representation of topographic effects in numerical weather forecasting models.
A schematic of the main features of the flow, identified in advance, is shown in
Figure 14.5. These are generally consistent with the experimental flow patterns
described above, but obviously on a much larger scale.

A summary of the main results is described by Bougeault et al. (1997). Some
salient observational results are as follows: measurement of the drag on the atmosphere by the mountains, and ways of parametrising it; measurement of mountain
waves, which were generally consistent with (appropriate) linear theory; measurement of low-level winds (the Tramontana on the Mediterranean side) around the
mountains, with the result that surface winds around both sides of the mountains are

very well correlated, and the measurement of boundary-layer turbulence in these flows. There was also a high level of interaction between the numerical modellers and the observationalists, leading to improved representations of topographic effects in numerical forecasting models, as described in more detail in the next chapter.

Although the data set is now nearly 30 years old, it is still the most complete set of observations of airflow over a large isolated mountain range.

14.2.6 T-REX and The Sierra Rotors Project

The Owens Valley on the lee side of the Sierra Nevada range in California has been well known as the location for large amplitude lee waves and "rotors" for many years. Rotors comprise circulation of the fluid about a horizontal axis beneath a lee-wave crest (or form of a hydraulic jump) that is driven by downslope flows from steep topography, with possible support from topography on the other side of the valley. In particular, part of the flow is often turbulent, and they constitute a significant threat to local aviation. The valley was the location of the Sierra Wave Project, which was a well-publicised observational project in 1951–1952 and 1955, using mostly sailplanes (gliders) and surface observations with occasional soundings (Grubišić & Lewis, 2004). This resulted in some excellent photographs, useful information for gliders and detailed reports to UCLA and the US Weather Bureau, but little by way of public information in the scientific literature.

More recent information with modern equipment has been gathered by the Sierra Rotors Project (SRP) in 2004 and the Terrain-induced Rotor Experiment (T-REX) in 2006 in the same location (Grubišić et al., 2008). The SRP embodied a Special Observing Period during March and April 2004 in the Owens Valley, using a variety of ground-based instrumentation, and observed nine significant wave and/or rotor events during the two months. The T-REX experiment was focused on a Special Observing Period for the same two months in 2006, using a broader range of ground-based instruments, plus three instrumented aircraft making observations over three different ranges of elevation, from near ground level to 14 km. The main observations were made in 15 intense observing periods (IOPs). A typical example of these remarkable events was observed on 9–10 April 2006, in which the descending air was sufficiently cold to drive a massive internal hydraulic jump that effectively filled the whole valley, with significant turbulent upper level clouds on the lee side (Armi & Mayr, 2011).

The full story of rotors is (of course) more complex than the original conceptions. The results from T-REX together with associated numerical modelling studies have now been presented in a substantial number of publications, and a useful introduction to them is presented in Straus et al. (2016).

14.2.7 The MATERHORN project

The MATERHORN project was a multi-disciplinary observational programme of flow over and around complex terrain on the mesoscale, with participants from 12 institutions. The experiment was carried out in the US Army Dugway Proving Ground in Nevada, around a central peak named Granite Mountain (maximum height 800 m above local terrain), with the background objective of improving weather forecasts in complex terrain. Observations were made using a wide variety of ground-based instrumentation (with some items on masts) and with small aircraft and some UAVs (unmanned aero vehicles), during 9 intense observing periods (IOPs) in the Autumn (Fall) of 2012 and 10 IOPs in the Spring of 2013 (Fernando et al., 2015). Numerical modelling of the airflow was also a significant part of the whole project. Comparisons between observations of the flow field and model simulations with 0.5 km resolution showed how each data set can be used to interpret the other (Silver et al., 2020), and what are the limitations of both. The data set from this project is likely to be studied for some time to come.

14.2.8 Atmospheric internal wave generation by volcanic eruptions

Volcanic eruptions can cause a variety of disturbances in the atmosphere. This is often manifested as a source of buoyant material in the form of lava, as from the Soufriére Hills volcano in Montserrat, in the Caribbean, or in the form of steam, as from White Island, New Zealand. These eruptions may be sudden and explosive, or be steady sources for a significant period of time. They may also cause significant amounts of volcanic ash to rise to heights of several kilometres in the atmosphere, which can constitute a significant risk to human health and aviation that may extend to several thousand kilometres from the volcanic source.

Since most volcanic eruptions are not predictable, detailed observational studies are rare, with the exception of those on Montserrat. Here, after a period of several hundred years of inactivity, the Soufriére Hills volcano began a 15-year period of activity in 1995, with unpredictable eruptions separated by periods of several months. The details of many of these events were recorded, and the results published in the literature (see Voight & Sparks (2010) and the 27 articles that follow it, and the references therein).

Some of the eruptions on Montserrat were explosive, and others lasted for much longer periods. Microbarograph recordings of explosive eruptions on Montserrat, with time scales of the order of 1 minute, have shown signals with periods of several minutes and amplitudes of about 1 mb, which imply the generation of atmospheric internal waves. These events have been modelled with the linearised compressible equations for stratified flow, with the sudden addition of mass, and separately of energy (Baines & Sacks, 2014; Ripepe et al., 2016). The nett results resemble the effect of "throwing a stone" into the atmosphere, producing transient waves that radiate radially away from the source. Near the source, these waves have frequency of about $0.7N$–$0.8N$ initially, with N the buoyancy frequency, which then increases toward N and decreases in amplitude with time. The results suggest that forcing due to the injection of mass (mostly solid particles) into the atmosphere is the principal factor in forcing the observed internal waves. The addition of thermal energy (heat) produces waves that have frequencies closer to N, and persist for much longer periods than those observed at the stations on Montserrat.

For volcanic eruptions that are not explosive but consist of a continuous source of buoyant fluid into the atmosphere, it is more appropriate to represent the flow as a turbulent buoyant plume that entrains fluid as it rises, and then spreads laterally at a neutrally buoyant level (e.g. Woods, 1988; Woods, 2010; Baines, 2013; Rooney & Devenish, 2014). If the fluid reaches a maximum height z_M, entrainment occurs up to heights $0.7z_M$, above which it spreads radially (Baines, 2013). This may be at levels above or below the tropopause. The spreading process generates internal waves, some of which propagate down to ground level and may produce a measurable signal (Baines & Sacks, 2017). Comparison with observations shows that this dynamical model can produce results that are representative of short-lived explosive eruptions also (see Figure 14.6).

Very large volcanic eruptions are fortunately rare (the largest eruption of the past 10 000 years was Tambora, in 1815!). But their dynamics are of interest for geologists and geophysicists, in particular. Large rotating ash clouds can form in the stratosphere, and their dynamics are dependent on the latitude of the source. Details have been described by Baines & Sparks (2005), and Baines et al. (2008).

14.3 Internal waves in the upper atmosphere

Internal waves generated in the troposphere by (for example) air flow over mountains can propagate upwards in the atmosphere to great heights, some into the stratosphere and mesosphere, to heights in excess of 100 km. In particular, they can cause irregularities in the ionospheric D, E and F layers that reflect radio signals and enable global radio communication in the 80–200 km height range. That internal waves were the probable cause of such fluctuations was formally proposed in a seminal paper by Hines (1960), using observations based on deformations of meteor trains, which are particular tracks through the atmosphere left by falling meteors.

The meteor tracks were initially approximately linear, but in the time-scale of several minutes large deformations became evident, on vertical scales of 10–20 km, and less. Further, these deformations tended to have larger amplitudes at greater heights, and showed oscillatory behaviour with frequency less than that of the local

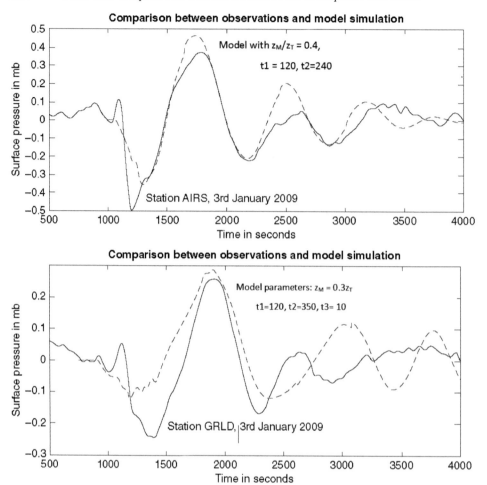

Figure 14.6 Comparisons between observations of surface pressure variations (solid line) and a model of turbulent plumes (dashed line), of the eruption on Montserrat on 3rd January 2009 at (upper figure) 5 km from the source, and (lower figure) 10 km from the source. The results imply that this eruption lasted for 4 minutes, and reached a height of 5.6 km. (From Baines & Sacks, 2017.)

buoyancy frequency, consistent with internal waves. These waves showed substantial horizontal propagation, but were always propagating upward. This vertical propagation is generally consistent with the conservation of wave action $\mathcal{A}$ in §8.7; as the wave propagates upwards the frequency and wavelength are conserved, but the fluid velocity within it increases as the density of the air decreases. As the height increases and the background gases become more rarefied, viscous dissipation increases and the waves become attenuated, to the extent that there is little observed wave activity above heights of about 200km.

Mathematical modelling studies have also demonstrated that airflow past large-amplitude topography can generate large-amplitude internal waves that reach well into the stratosphere. A study by Leutbecher & Volkert (2000) of airflow over southern Greenland, with topographic heights of 2 km above sea level, showed internal waves reaching heights in excess of 26 km, with the details confirmed by observations from aircraft. Substantial wave amplitudes were observed/modelled with wind from all directions, with amplitude generally increasing with height, which is not surprising given the shape of the environment.

In general, gravity waves generated in the troposphere can play a dominant role in the dynamics of the middle and upper atmosphere, even to the extent of influencing the quasi-biennial oscillation, but observations on the required scale are still lacking (Fritts & Alexander, 2003).

14.3.1 The DEEPWAVE project

More recently, observations of atmospheric internal waves have been made by a variety of techniques: radiosondes, stratospheric balloons, radar and lidars, plus others. But apart from the fact that vertically propagating internal waves in the equatorial wave guide are the principal cause of the quasi-biennial oscillation in the upper stratosphere (see §8.7), there seems little evidence that these internal waves affect large-scale motion in the troposphere, and studies of them have not had high priority. The Deep Propagating Gravity Wave Experiment (DEEPWAVE) in June and July 2014 was the first comprehensive measurement programme that aimed to describe the generation, propagation and ultimate dissipation of internal waves from ground level to heights of ~100 km (Fritts et al., 2016).

The DEEPWAVE programme was based in Christchurch, New Zealand, and aimed to observe and describe internal waves generated over the two main islands of New Zealand, and Tasmania, Australia, but concentrating on the NZ South Island. Participants came from a wide range of institutions, mostly in the USA but also Germany, Austria, Australia and New Zealand. Observations were made from two instrumented aircraft and a broad range of ground-based instruments with a wide range of observing techniques, with 16 intense observing periods. The observed gravity waves were generated by flow over orography, and by jet streams, frontal systems and deep convection.

A large amount of data was accumulated, and processing is continuing, but some results to date are as follows. As expected, the most mountainous areas, the Mount Cook and Mount Aspiring regions (in NZ South Island) were strong sources of internal waves, but the energy fluxes were highly variable, and were sensitive to active weather. Other observations included ship wave patterns due to flow over small isolated islands, that generated waves on small scales that reached large amplitudes at heights above the stratosphere. Waves generated over small mountains could still penetrate to very high altitudes with large amplitudes and momentum fluxes. And large-scale gravity waves are often refracted into the polar vortex, which is a cyclonic circulation centered over the South Pole.

Internal waves are also observed to be prominent in the stratosphere and thermosphere over Antarctica (Chu et al., 2018), particularly in winter. The sources are currently unresolved, but these observations are not surprising given the generally steep coastal topography and the strong winter winds around the continent.

14.3.2 New observations in the upper atmosphere

Upper atmosphere observations using new techniques such as lidar and satellite nightglow imagery are beginning to give a quantitative picture of wave motion in the upper atmosphere. Observations by lidar have shown that internal waves are prominent in the stratosphere and thermosphere over Antarctica (Chu et al., 2018), particularly in winter. And nightglow imagery can record the presence of internal waves in the upper atmosphere generated by jet streams and volcanic eruptions, and by thunderstorms and flow past mountainous regions below (Miller et al., 2015).

14.4 Internal waves in the deep ocean

Internal waves are prominent in the (density-stratified) oceans at all frequencies between N and the Coriolis frequency f. Internal waves in mid-ocean are mostly initially generated at long wavelengths and low frequencies by tides (in coastal regions) and by surface wind stress. Over time, this wave energy spreads through non-linear wave–wave interactions toward shorter wavelengths and higher frequencies, reaching a form of equilibrium described by the Garrett–Munk spectrum (Garrett & Munk, 1972).

Internal waves generated in mid-ocean can propagate onto continental shelves and slopes. Also, tidal motion in coastal environments generates internal waves of tidal frequency that mostly propagate offshore. For an overview of internal waves in mid-ocean and their relation to turbulence, the reader is referred to the treatise by Thorpe (2005), and for information on internal tides in the deep sea and coastal environments, to the treatise by Vlasenko et al. (2005).

14.5 Topographic effects in coastal oceanography

Mid-ocean internal waves propagating into coastal areas can result in a wide variety of consequences depending on the form of the local topography. Theoretical studies in this topic have tended to concentrate on idealised situations, and observational studies are focused on particular locations. Not surprisingly, each location is different: observations of the dynamics of a particular region can provide understanding, but may only apply to the local topographic geometry. The number of detailed observational studies is limited. Nonetheless, some general properties can be identified, as follows.

Internal waves propagating shorewards (in the x-direction), normal to a uniform coastal environment (in the y-direction), can show a variety of outcomes depending on the local geometry, stratification, and the possibility of coastal currents. Waves propagating shorewards from any direction over a continental slope tend to refract towards a normal direction when they approach the edge of the continental shelf. Large amplitude internal solitary waves are not uncommon (Lamb, 2014). Generally speaking, such waves are not reflected, and tend to become unstable and break, causing local mixing over the shelf and upper parts of the continental slope. These regions are energy sinks for the ocean internal wave field.

Continental shelves and slopes often contain across-shore bottom roughness and larger variations such as submarine canyons and ridges. The effect of these indentations on alongshore coastal currents is to generate local lee waves (Thorpe 1992, 1996; MacCready & Pawlak, 2001). These waves may initially be stationary, but over time they tend to propagate and refract upslope, carrying their energy and momentum upward to shallower depths on the continental shelf where the stratification permits. A general survey of the impact of internal waves on the coastal environment is provided by Woodson (2018).

Continental shelves are often the source of dense water, primarily because of the heat loss to the atmosphere in winter, through cooling, evaporation and freezing (ice formation), and their relatively shallow depth. This cold (relatively) dense water can then cascade down the adjacent continental slope (Shapiro et al., 2003). A similar effect may occur in summer, due to the increase in salinity because of evaporation. These processes have a significant impact on the formation of major water masses in the world ocean, such as the Antarctic Intermediate Water, and Antarctic Bottom Water.

Observations indicate that these cascades can proceed in four stages, as follows. First a *preconditioning stage* where dense water accumulates on the continental shelf, and a density front is formed. This is followed by an *active stage*, where the leading edge of the dense water accelerates downslope, followed by a *main stage* with quasi-steady downslope flow. In the *final stage*, the descended water spreads at its level of neutral density.

Locations around the globe where such downslope flows have been observed have been described by Ivanov et al. (2004), who have carried out a "world-wide traul of oceanic data bases as well as scientific publications". The dynamics involved is generally as described in Chapter 7, with the additional factor of the Coriolis force due to the Earth's rotation, which generally implies that the direction of the flow is not directly downslope.

One location where downslope flow is particularly strong and persistent is around Antarctica, especially in winter. This gives rise to the formation of Antarctic Bottom Water, and also a phenomenon known as the "Antarctic Slope Front". The latter involves a structure of coastal currents governed by quasi-geostrophic dynamics, essentially driven by the downslope flow, and appears to be a semi-permanent feature situated above or near long sections of the shelf break of the Antarctic continental shelf (Baines, 2009).

14.6 Oscillating ocean flows and tides

One of the most significant examples of density-stratified flows over topography is seen in the ocean tides. These are due to the gravitational attraction of the Moon and the Sun, whose effects are superimposed, forcing oscillating barotropic motion in the global ocean with periods of approximately 12 hours. The presence of bottom topography and density-stratification in the deep ocean, and at continental boundaries, results in internal or baroclinic tides. The nett forcing of this motion, (as seen by the height of the free surface) is well-observed by satellite altimetry, but the resulting subsurface motion is not. The bottom topography is well-known, but there are variations in the density stratification which imply that the subsurface tidal motion is more variable than the height of the free surface. Seamounts, in particular, are common topographic features in mid-ocean, and local tidal motion can be quite significant (Baines, 2007).

This subsurface motion is substantial and ubiquitous, and there is an extensive body of research on this topic, for which the recommended reference is by Vlasenko et al. (2005), with updated summaries by Garrett & Kunze (2007), and more recently by Sarkar & Scotti (2017).

15

Applications to practical modelling of flow over complex terrain

The main point with numerical models is to find out what's wrong with them.

ANONYMOUS

The information described in the preceding chapters may be applied in two practical ways to infer the properties of flow over realistic terrain. The first way concerns laboratory models. One may construct physical models of the terrain which may be used to model the flow around them in a dynamically realistic fashion in the laboratory. For sufficiently complex or extensive terrain, some distortion of the vertical scale in the model may be necessary. The second way concerns numerical models. Flow over reasonably simple terrain may be modelled directly, and several examples have been given in preceding chapters. However, when modelling flow over extensive regions of complex terrain it is usually not possible to resolve all features of the topography that may be important. Hence it is necessary to represent or parametrise them in some way, and this must be based on the physical properties described previously.

15.1 Laboratory modelling

The most common technique for modelling stratified flows involves salt-stratified water, although wind tunnels with heated or cooled air are also used. The applicability of experiments with water to flow of the atmosphere has been discussed in Chapter 1. There are two types of such water-based experiments: towing tank experiments; and "race-track" experiments. In the latter, stratified fluid is circulated around a closed circuit, a part of which is the working section containing the obstacle of interest. The common method of driving the mean motion is by a set of pairs of rotating discs with vertical axes. Discs at different heights may have different radii, and choosing these radii determines the velocity profile. However, facilities of this type that are in existence are generally too small to accommodate the study of a large region of complex terrain.

Most experimental studies of flow over topography have been made with towing tanks, and the principles have been described by Baines & Manins (1989). These experiments are normally limited to a uniform external flow, where the model is towed through stationary fluid. A given piece of complex terrain has a range of length-scales, but it is often possible to subjectively identify a representative height-scale h, an along-flow length-scale A and a transverse (breadth)-scale B. In fluid of depth D, mean buoyancy frequency N and mean flow speed U, the relevant dimensionless quantities are then the familiar ones:

$$\frac{Nh}{U}; \quad \frac{h}{A}, \quad \frac{B}{A}, \quad K = \frac{ND}{\pi U} \quad \text{and} \quad R_{\text{e}} = \frac{2UB}{\nu}, \tag{15.1}$$

where $1/K$ is the Froude number. Ideally, all of these parameters would have the same values in the experiment as in the situation being modelled. Normally this is not possible, so that some compromises must be made. The Reynolds number in the atmosphere is usually much larger than can be realised in the experiment. Normally one makes it as large as possible, and this should be adequate if $R_{\text{e}} \geq 10^3$. Also, here one is attempting to represent an infinite-depth system by a finite-depth one. For an isolated three-dimensional obstacle, the finite depth does not affect the flow near the obstacle provided K is sufficiently large, since the vertically propagating wave energy is small (see §12.1). For a region of complex terrain containing ridge-like components this is more problematical, and the best one can do is to make D and K as large as is convenient, with $K > 3$ at least, and preferably larger.

Ideally, the model would have the same ratio of scales h/A and B/A as the modelled flow. It is obviously essential that each horizontal section of the model be similar to that of the modelled situation, so that B/A is the same, but it may not be possible to have the same vertical scale if a large horizontal region is to be modelled. If Nh/U is large, most of the flow at low levels passes around the terrain rather than over it (see §13.2), with flow over occurring only near the crests. Accordingly, if the vertical scale of the model is increased, so that h/A is made larger, the distortion of the flow around the terrain is small. This stretching has the following advantages: it permits greater detail to be seen in the vertical structure, it helps to reduce the over-emphasis of boundary-layers in the model due to the smaller value of R_{e}, and it enables the modelling of larger horizontal areas of terrain than would be possible if h/A must have the same value. However, it is essential that Nh/U have the same value in the model and modelled situations.

Stretching the scale of the model in the vertical is therefore permissible dynamically if one is primarily concerned with the flow at lower levels, in valleys and so forth, where the flow is around the topography. If details of the flow over the tops and associated lee waves are required, models with the correct ratio of horizontal and vertical scales must be used. The biggest problem in the modelling of flow over complex terrain is in making the model. To date, this has been done by extracting heights and contours from maps by hand, smoothing subjectively in the process, and converting this to a plaster or plastic model. This is a time-consuming and consequently expensive process, but the use of computerised databases and automated machine tools promise to make this task easier in the future. The first known use of this technique with complex terrain was the investigation of the Melbourne eddy (Baines & Manins, 1989), a low-level nocturnal eddy over the city of Melbourne, Australia that is largely due to the separation of stable air from an upstream plateau (Mt Baw Baw). In fact, the study of stable nocturnal flows is one of the main potential uses of this technique.

Two further meteorological examples are presented.

15.1.1 Airflow around Antarctica

One of the most instructive applications of this technique was to large-scale flow around Antarctica (Baines & Fraedrich, 1989). A topographic model of Antarctica was made with a maximum height of 8 cm and diameter of 65 cm. This model was placed in a circular tank filled with fresh or stratified water, which was rotated clockwise and spun up to constant speed, so that the whole was in a state of rigid rotation. The topographic model was then rotated slowly in an anti-clockwise sense to simulate a mean barotropic westerly wind around Antarctica. After about 10 tank rotation periods, a dominant wavenumber-three structure with cyclonic eddies in the Ross and Weddell seas and Prydz Bay was observed as an approximately steady state. Flow over the topography was relatively stagnant, with weak anticyclonic circulation. These flows show remarkable similarity to the observed mean 700 mb height wind fields around Antarctica. This strongly suggests that the same dynamical factors are operating, namely conservation of potential vorticity and strong coupling in the vertical, so that these motions are virtually barotropic. Hence the dominant wavenumber-three pattern of large cyclonic eddies over the Southern Ocean is primarily forced by flow separation from prominent coastal irregularities such as the Antarctic Peninsula.

15.1.2 Airflow in mountain valleys in Bulgaria

Another example has been the study of flow in the central region of Bulgaria by Tucker (1989). Figure 15.1 shows the model topography of eastern central Bulgaria, including the Sofia valley. This is an extremely complex region but is typical of mesoscale topography. In stable conditions where $Nh/U > 1$, the flow in the valley

Figure 15.1 A laboratory model of central Bulgaria, including the Sophia valley (upper centre), viewed looking in the NNE direction. The maximum height of the model is 12.5 cm. The vertical scale of the model has been stretched so that 1 cm in the model corresponds to 200 m, and horizontally the model length of 60 cm corresponds to 150 km. (Model made by George Scott.)

is difficult to predict a priori, and is dominated by interacting wake effects and channelling by the various extensive topographic features.

The model was a square of side 60 cm, which represented a square area of side 150 km. Since the focus of this study was on flow in valleys under stable conditions, the scale of the model was stretched in the vertical so that 1 cm would correspond to 200 m. The model was towed to represent mean flow from various directions, and flow in the lower part of the valley was visualised by releasing dyed fluid from eight small tubes. Figure 15.2 shows typical flow patterns observed when the model is towed in two directions, and gives some indication of their complexity. At these low levels, no difference was observed between the flow behaviour when $Nh/U = 3.4$ and 6.9.

While such studies are not an answer for everything, they are useful in revealing many aspects of a complex situation in a way that is physically sound and readily visualised. For all their virtues, numerical models can have trouble in coping with such complexity, and have difficulty in representing flow separation and its consequences.

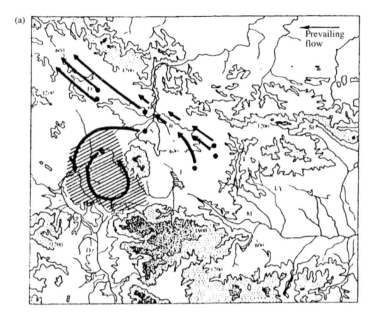

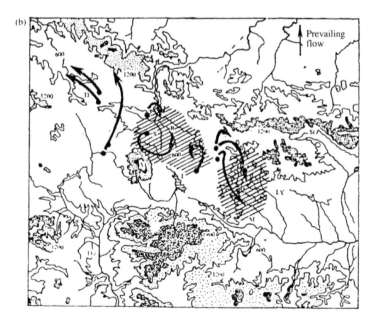

Figure 15.2 Representative flow patterns at low levels in the valleys in the model of Figure 15.1, caused by towing the model through uniformly stratified fluid. The motion is observed with dye released from a number of relevant locations. (a) Easterly (westward) external flow; (b) southerly (northward) external flow. The shaded regions denote stagnant areas. (From Tucker, 1989.)

15.2 The natural ventilation of buildings

This is a relatively new application of density-stratified fluid dynamics that has grown substantially over the past 30 years or so. The subject is too extensive to permit a detailed description here, though §6.3 on exchange flows through doorways and windows gives an example of the nature of the dynamics involved. Laboratory experiments with sources and sinks of density-stratified water have been used to simulate the effects of sources of heat in rooms, and inflow and outflow at various different levels in buildings. Numerical models of internal flows are also extensively used. A useful summary of the fundamentals is provided by Linden (1999).

The results of these studies have become essential information for architects of modern complex buildings, with the broad-scale objectives of creating pleasant internal environments that can be maintained at minimum cost. An example of the work involved is presented by the recent article by Song et al. (2018).

15.3 Parametrisation of the effects of sub-grid-scale orography in large-scale numerical models

A major concern in numerical models of weather and climate is the representation of the effects of mountainous terrain. In general, such models do not resolve all of the significant topography because there is an inevitable limit to the resolution or minimum grid size of any regional or global general circulation model (GCM) of the atmosphere. This means that scales of topography below a certain size cannot be represented explicitly in the model. The effect of the drag of sub-grid-scale (sgs) orography on the motion on large scales, and on the overall circulation of the atmosphere is substantial (Palmer et al., 1986; McFarlane, 1987), so that it is important that its effects be included. The increase in available computer power in recent years has caused a decrease in the minimum resolved scale, but for global models on the most powerful computers it is still about 25 km, and for many models in current use it is much more than this. This applies particularly to models used for climate studies, where long runs over many model years are required.

The nature of the drag force on a single obstacle due to atmospheric flow over it was summarised in §12.1.6 for obstacles where $Nh_\mathrm{m}/U < 1$, and in §13.4 for obstacles where $Nh_\mathrm{m}/U > 1$, where h_m is the maximum obstacle height. The nett drag on the ground of the flow past the sgs topography over a given region is a horizontal vector, $\mathbf{F}$, which is a function of time. The nett drag of this topography on the atmosphere is therefore $-\mathbf{F}$, but the question arises as to where and at what heights in the atmosphere it is felt. In both cases, if the obstacle has sufficiently large horizontal length-scales (more than 5 km), some of this drag will be manifested in internal gravity waves, and the associated drag is felt where these waves are dissipated, which may be spread over substantial heights in the atmosphere. The other forms of drag are due to flow separation around the obstacle, and possible hydraulic transition (from locally sub- to supercritical flow) over it, when $Nh_m/U > 1$. The effects of flow over smaller unresolved topographic features can also be represented as frictional drag. Here the drag is felt at the levels of such flows. These processes are shown schematically in Figure 13.9.

Since the obvious solution of increasing the resolution of the models is limited, it is necessary to parametrise the sub-grid-scale orographic effects in some way. This means, typically, that the relevant aspects of the topography of the Earth over an area that is many kilometers square must be represented by values at a single grid point near the centre, and the effect of this topography on the atmosphere must be represented at the grid points above this one. This is an inexact process, involving a number of assumptions. There are several different approaches that may be made, and it is necessary to find a balance between the limitations on numerical capacity and the need to contain sufficient physics and dynamics to represent the nett effect realistically. I outline here one such approach that incorporates the information contained in the preceding chapters, and has been implemented in global weather forecasting models, based on the methodology employed at the European Weather Centre (ECMWF). The following sections are essentially an update on Section 7.2 of the *first* edition of this volume.

15.3.1 *Representation of the topography*

Numerical models of the atmosphere typically represent the surface of the earth by a grid of fixed points. Each point represents a (approximately rectangular) region of the earth's surface, designated as the gridpoint-region (GPR) for that point. Over this region, the elevation of the topography is denoted by $h(x, y)$, with a mean value $\bar{h}$. Here $\bar{h}$ is the height of the surface at the grid point in the model and represents the resolved topography, so the sub-grid-scale orography is given by $h(x, y) - \bar{h}$. In the early versions of GCMs, sgs orography was then parametrised (in effect) by surface frictional drag, with suitably chosen roughness. As noted in Chapter 14, stratification affects motion in the atmosphere on scales greater than about 5 km ($2\pi U/N$, with $U \sim 10$ m/s, $N \sim 10^{-2}$). Steep topography on scales smaller than this may act as roughness elements in a turbulent boundary-layer (Grant & Mason, 1990), and hence may be parametrised as such (Beljaars et al., 2004; Garratt, 1992). However, this does not work for the larger length-scales where stratification is important. This has been shown on the small scale by the studies described in the preceding chapters, and on the large scale by the detection of systematic errors in GCMs that are associated with topography (Wallace et al., 1983; Lott, 1995).

For these purposes the topography at a given grid point may be represented by four parameters. The first is the variance,

$$\mu^2 = \frac{1}{S} \iint (h - \bar{h})^2 dS, \qquad (15.2)$$

where μ is the standard deviation and S is the area of the grid-point region. In general, the level 2μ roughly corresponds to the level of the tops of the mountains. As shown in Chapters 12 and 13, for some effects it is the height of the mountains or mountain barrier that is important, and in mountainous regions this is much greater than $\bar{h}$. For this reason, one might expect that increasing the mean topographic height in these regions in the GCM would improve the model results. One form of doing this was proposed by Wallace et al. (1983), who defined the *envelope orography* h_e by

$$h_e = \bar{h} + \epsilon\mu, \qquad (15.3)$$

where ϵ is an empirically chosen constant in the range $0 < \epsilon < 2$. The name suggests that h_e is equivalent to that of a membrane stretched over a range of mountains and attached to the peaks, but in operational versions (15.3) used values somewhat less than this. The model results with the ground level taken at h_e were often better than those with $\bar{h}$, particularly in winter. However, the procedure is essentially empirical. It was used by ECMWF for several years up until 1995, but was not generally adopted elsewhere.

The other three parameters specifying the topography stem from the topographic gradient correlation tensor

$$H_{ij} = \overline{\frac{\partial h}{\partial x_i} \frac{\partial h}{\partial x_j}}, \qquad (15.4)$$

where $x_1 = x$, $x_2 = y$, and the overbar denotes the mean over the whole area S of the GPR. For a given set of axes, the components of this tensor may be calculated from a digitised data set with (preferably) a resolution of 5 km or less. This tensor may be diagonalised to find the principal axes and the degree of anisotropy of the topographic variations in the GPR. If

$$\mathcal{K} = \frac{1}{2}\left[\overline{\left(\frac{\partial h}{\partial x}\right)^2} + \overline{\left(\frac{\partial h}{\partial y}\right)^2}\right], \quad \mathcal{L} = \frac{1}{2}\left[\overline{\left(\frac{\partial h}{\partial x}\right)^2} - \overline{\left(\frac{\partial h}{\partial y}\right)^2}\right], \quad \mathcal{M} = \overline{\frac{\partial h}{\partial x}\frac{\partial h}{\partial y}}, \quad (15.5)$$

the principal axis of H_{ij} is oriented at an angle α to the x-axis, where α is given by

$$\alpha = \frac{1}{2}\arctan(\mathcal{M}/\mathcal{L}). \qquad (15.6)$$

This gives the direction where the topographic variations, as measured by the mean square gradient, are largest, and the direction for minimum variation is perpendicular to this. If we change coordinates to x', y' which are oriented along the principal axes of the topography $(x' = x \cos \alpha + y \sin \alpha, \ y' = y \cos \alpha - x \sin \alpha)$, the new values of $\mathcal{K}, \mathcal{L}$ and $\mathcal{M}$ relative to these axes, and denoted $\mathcal{K}', \mathcal{L}'$ and $\mathcal{M}'$, are given by

$$\mathcal{K}' = \mathcal{K}, \qquad \mathcal{L}' = (\mathcal{L}^2 + \mathcal{M}^2)^{1/2}, \qquad \mathcal{M}' = 0. \tag{15.7}$$

We may now specify the anisotropy of the orography by the "aspect ratio" γ, defined by

$$\gamma^2 = \frac{\overline{(\partial h/\partial y')^2}}{\overline{(\partial h/\partial x')^2}} = \frac{\mathcal{K}' - \mathcal{L}'}{\mathcal{K}' + \mathcal{L}'} \leq 1. \tag{15.8}$$

From these we may define the mean square slope σ along the principal axis, and associated length-scales A and B by

$$\sigma^2 = \overline{\left(\frac{\partial h}{\partial x}\right)^2}, \qquad (A/2)^2 = (\mu/\sigma)^2, \qquad \gamma = B/A. \tag{15.9}$$

We therefore have the quantities $\overline{h}$, μ, σ, γ and α to describe any given region of complex terrain. The significance of these quantities is clear from the following. If h consists of periodic hills and valleys of the form

$$h(x', y') = h_0(1 + \cos k x' \cos m y'), \tag{15.10}$$

where necessarily $k \geq m$, then

$$\overline{h} = h_0, \qquad \mu = h_0/2, \qquad \sigma = k h_0/2, \qquad \gamma = m/k, \tag{15.11}$$

and $A = \pi/k$, $B = \pi/m$, for scales A and B as in (15.1).

There is an additional parameter G that comes from considering the general shape

$$h(x, y) = \frac{h_{\mathrm{m}}}{\left[1 + \left(\frac{x}{a}\right)^2 + \left(\frac{y}{b}\right)^2\right]^v}, \tag{15.12}$$

described in §12.1.6. This parameter G is a measure of the steepness, or sharpness of the topography, and is exclusively a function of the exponent v in equation (15.12). Its values are given in (12.59) and Figure 12.12b. (There is some overlap in the significance of G and σ.)

In what follows, we assume that the topography in this GPR is approximately uniform in a statistical sense, in that if the same statistical quantities are calculated for any (sufficiently large) subregion of the GPR, the same values of μ, σ, γ, α and G would be obtained. If this is not the case, so that the GPR contains mountains in one part and ocean in another part for example, some modification of this treatment may be necessary.

From μ, σ, γ and α, we may identify a representative obstacle with height 2μ and known shape (length A, width B), orientation (γ with $\gamma = A/B$, in general terms), and steepness G, that characterises the topography of the GPR. The actual topography within the GPR may then be conceptually replaced with repetitions of the representative obstacle, as many times as can be accommodated within the area. An approximate solution to the problem of specifying the effect of the complex terrain on the atmosphere can thus be reduced to specifying the effect of this single representative obstacle. One may note, however, that as models become increasingly sophisticated, the representation of each GPR could become more specialised.

15.3.2 Parametrisation of turbulent orographic drag on small scales

Atmospheric flow over topographic features with length-scales less than 5 km does not generate atmospheric internal waves to any significant extent. As a consequence, for this spectral range, all the drag on the atmosphere is local, and it may be treated as "rough terrain". For a region of terrain represented by values at a single datum point, the results from a succession of studies by several authors, summarised by Beljaars et al. (2004), resulted in the equation

$$\frac{\partial}{\partial z}\left(\frac{\tau_0}{\rho}\right) = -6.47.10^{-4} C_{md}\sigma_{flt}^2 |U(z)|U(z)e^{-(z/1500)^{1.5}}/z^{1.2}, \qquad (15.13)$$

where vertical distance z is in metres. Here C_{md} is a drag coefficient with the value of 0.005, $U(z)$ is the mean velocity over the grid-point region, and σ_{flt} is the standard deviation of the topography of the grid-point region, filtered to exclude all wavelengths greater than 5 km. This procedure (or modifications of it) has been employed at ECMWF since 2004.

15.3.3 Parametrisation of turbulent orographic drag on larger scales

In 1994, output from the then-existing version of the ECMWF forecasting model (T213, L31 – triangular truncation of 213 waves in the horizontal, 31 levels in the vertical) was compared with observations of flow over and around the Pyrenees made through the PYREX experiment in 1990 (§14.2.5), by Lott, 1995. The model at the time used "envelope orography", and various shortcomings were identified and described. As a result, a new parametrisation scheme for sub-grid-scale orographic drag (consistent with Section 7.2 of the first edition of this volume) was developed by Lott & Miller (1997). Along with other models, the resolution of the ECMWF forecasting model has increased significantly since this time, but the essentials of this parametrisation scheme are still employed, and are outlined here.

We consider the topography in the grid-point region, described by the five parameters μ, σ, γ, α and G of §15.3.1 above, but computed only using the topographic spectral components with horizontal length-scales greater than 5 km. The atmospheric drag scheme may be divided into two groups. The first is where $2N\mu/U < 1$, so that linear hydrostatic theory of §12.1.6 is applicable for obstacles with length-scales greater than 5 km, and the second is where this does not apply, and these cases are discussed in turn.

Atmospheric drag when $2N\mu/U < 1$

For flow over an idealised obstacle satisfying this condition, one may assume that linear theory of Chapter 12, and particularly §12.1.6, is applicable. The magnitude and direction of the drag force $\mathbf{F_A}$ applied to the atmosphere at ground level, due to hydrostatic flow over an obstacle/mountain of the form (15.12), is then given by

$$\mathbf{F_A} = -\mathbf{F_D} = -\rho_0 U_L N_L h_m^2 bG(B\cos^2\alpha + C\sin^2\alpha, (B-C)\sin\alpha.\cos\alpha). \quad (15.14)$$

Here the axis with width scale a makes an angle α with the wind direction, and

$$B(\gamma) = 1 - 0.18\gamma - 0.04\gamma^2, \qquad C(\gamma) = 0.48\gamma + 0.3\gamma^2, \qquad \gamma = a/b. \quad (15.15)$$

Here U_L and N_L denote low level values of U and N, and plots of the ratio are shown in Figure 12.12a. The value of G depends on the exponent v in (15.12), and is given in (12.59). A plot of values of G as a function of v is given in Figure 12.12b. This is the drag force on the atmosphere from a single representative obstacle within the GPR. The total drag on the atmosphere from topography within the GPR is represented by the sum of the drag force from each of the number of such mountains that can be accommodated within it. If L^2 denotes the area of the GPR, it can accommodate (approximately) $L^2/4ab$ obstacles, so that the total drag on the atmosphere from the GPR is $L^2\mathbf{F_A}/4ab$.

From Chapters 10 and 12, for topography with sufficiently long wavelength (more than 5 km), when linear theory is applicable the drag of topography on the atmosphere is manifested in a momentum flux $-\rho_0\overline{u'w'}$ of vertically propagating internal gravity waves, and the drag force is manifested at heights where this momentum flux has non-zero divergence. It has been shown by Palmer et al. (1986), McFarlane (1987), that this process may be parametrised in GCMs in a realistic way that corrects systematic errors in the model outputs. The essence of this *gravity-wave drag* scheme is as follows.

In a three-dimensional wave field forced by orography, the vertical momentum flux is a vector in the direction of $\mathbf{F_D}$, and if the horizontal velocity component in this direction with coordinate X is $u = \overline{u} + u'$, in the usual notation, then we have

$$\frac{\partial\overline{u}}{\partial t} = -\frac{\partial\overline{u'w'}}{\partial z}. \quad (15.16)$$

It was shown in §8.7–8.9 that this term is zero for steady linear inviscid wave fields unless the waves are dissipated in some way, or the mean flow velocity vanishes.

The simplest possible representation of a wave field with this momentum flux is a single monochromatic wave, propagating in this plane, and we use this wave as a surrogate for the whole wave field. We further assume that this wave is hydrostatic and that its vertical structure is governed by the slowly-varying approximation of §8.8. This is a good approximation most of the time (Laprise, 1993), exceptions being situations where reflections from inhomogeneities in the lower troposphere are significant. The vertical displacement η of this wave may then be denoted by

$$\eta = \eta_0(z)\cos(kx + Nz/U), \qquad (15.17)$$

where the amplitude $\eta_0 > 0$, and is directly related to the stress at height z by

$$\tau = -\rho_0\overline{u'w'} = C\rho_0 N U \eta_0^2, \qquad (15.18)$$

where C is a variable that decreases with height, depending on the absorpion of wave momentum. In the atmosphere, ρ_0 decreases exponentially with height, which implies that η_0 increases with height as $1/\rho_0^{1/2}$, if τ remains constant.

If the Richardson number for the mean flow at time t has the form $R_i(z)$, then if the wave is added, the minimum value (over the phase of the wave) of the nett Richardson number at height z, $\widehat{R}_i$, has the form

$$\widehat{R}_i = R_i \frac{1 - \dfrac{N\eta_0}{U}}{\left(1 + \dfrac{R_i^{1/2}N\eta_0}{U}\right)^2}. \qquad (15.19)$$

Hence, if $R_i > 1/4$ everywhere, the presence of the wave may produce much lower local values of $\widehat{R}_i$, which imply instability and overturning if $\widehat{R}_i < 1/4$. One may then invoke the saturation hypothesis of Lindzen (1981), described in §8.10, that this instability causes mixing to the point where R_i maintains values $\geq 1/4$ in the mean. The mean profile then determines the distribution of the gravity wave forcing on the mean flow, as described in §§8.7 and 8.8. If the gravity wave encounters a critical level, the assumption is made that the approach is gradual so that the critical layer is absorbing rather than reflecting, so that the stress τ is zero at this level and the wave and the associated momentum flux is absorbed below it.

This treatment of gravity wave propagation is obviously simplistic and it omits such factors as non-hydrostatic trapped modes, but it contains most of the essential physical processes, and it provides sensible and realistic results in GCMs (Miller et al., 1989; Lott & Miller, 1997).

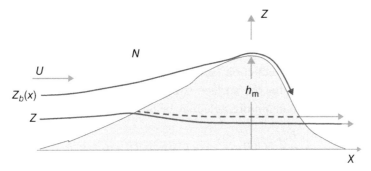

Figure 15.3 Definition sketch for flow past a representative symmetric tall obstacle/mountain where $Nh_m/U > 1$. The obstacle height is h_m, and z_b denotes the upstream elevation of the streamline that just reaches the mountain top. The incoming flow below this level is split, and these streamlines are deflected around the obstacle and contribute to the "blocked flow drag". The flow immediately above level z_b does not split, passes over the obstacle and descends to low levels on the lee side before returning to near its upstream level. This flow generates internal waves that may propagate to great heights.

Atmospheric drag when $2N\mu/U > 1$

When $2N\mu/U > 1$, linear perturbation theory alone is not applicable, and in general the drag caused by flow past the obstacle is spread amongst two different processes in the fluid flow. These are: (i) gravity-wave drag, in a similar manner to that for $2N\mu/U < 1$; and (ii) aerodynamic or "blocked flow" drag, due to the horizontal flow around the obstacle. (Hydraulic drag, previously discussed in the previous edition of this volume, is here subsumed into the blocked flow drag.) These processes are illustrated in Figures 13.5–13.9, and schematically in Figure 15.3, which specifies the notation for this section. Here at low levels the flow is nearly horizontal, flows around both sides of the obstacle, and largely remains horizontally separated on the lee side. At higher levels, within a height of U/N below the crest of the topography, the flow resembles that when $2N\mu/U < 1$. The drag on the atmosphere therefore consists primarily of "blocked flow" drag at low levels where the flow is primarily horizontal, and gravity-wave drag at upper levels. Hydraulic jumps do not seem to be common features in the low-level atmosphere, mainly because of the relatively uniform stratification, and are not a part of the schemes described here.

(i) The gravity-wave drag
Compared with the case where $2N\mu/U < 1$, the process of gravity-wave drag is similar except for two main respects: first, the effective height of the obstacles for the flow over them is now U_H/N_H (where subscript H denotes values in the upper levels of flow below h_m) rather than 2μ, measured downwards from the peak, since

the fluid at lower levels flows around them (Figure 15.3). Second, this implies that the wave-generating area is reduced in proportion (to the topography above the level $h_\mathrm{m} - U_H/N_H$), and if the reduced effective obstacle shape is similar to that of the total obstacle, the reduction factor for the drag is $(U_H/2N_H\mu)^2$. Hence an approximate expression for the surface stress (drag force) that generates gravity waves from a single mountain/obstacle may be taken to be

$$\mathbf{F_A} = -\mathbf{F_D} = -\rho_H \frac{U_H^3}{N_H} bG(B\cos^2\alpha + C\sin^2\alpha, (B-C)\sin\alpha.\cos\alpha), \quad (15.20)$$

where $B = 1 - 0.18\gamma - 0.04\gamma^2$ and $C = 0.48\gamma + 0.3\gamma^2$. Again, the total drag on the atmosphere from the GPR is $L^2\mathbf{F_A}/4ab$. The distribution of this stress in the vertical is governed by the same considerations as for $2N\mu/U < 1$.

(ii) The "blocked flow" drag

This applies for the lowest levels $0 < z < z_b = 2\mu - U_L/N_L$, where z_b denotes the upstream level of the streamline that just reaches the mountain top, and the terms U_L and N_L apply to streamlines whose upstream levels are less than z_b. From Figure 15.3, one can see that all streamlines with upstream levels less than z_b pass around the obstacle, and make no direct contribution to the gravity-wave drag. From the observations of the flow in complex terrain described in §14.1, we may note that the principal effect of the topography on the flow in this depth range is to produce a large wake region in which the mean flow velocity is substantially reduced, implying associated drag. This may be represented in any one of several ways, such as a drag coefficient of order unity, or more simply a large-body force in the GPR at these levels that brings the flow to zero. In the drag coefficient form, the vertical profile of the drag force $D_b(z)$ exerted on the flow at these levels may be expressed as

$$D_b(z) = -0.5\rho_0 C_d l(z) U_L |U_L|, \quad (15.21)$$

where the drag coefficient C_d has a value of order unity and $l(z)$ denotes the transverse width of the obstacle at height z. The mountains are assumed to have an elliptical shape (15.12), which means that, if the incident flow is perpendicular to the side with width $2b$, at height z, the width of the obstacle is given by

$$l(z) = 2b \left(\frac{z_b - z}{z} \right)^{1/2}. \quad (15.22)$$

The streamline denoted z_b marks the upper boundary of the "blocked flow" region, and integrating (15.21) from $z = 0$ to $z = z_b$ one obtains the total "blocked flow" drag force, in the form

$$\tau_b \approx 0.5 C_d \pi b \rho_0 z_b |U_L| U_L. \quad (15.23)$$

Incident streamlines at levels immediately above height z_b descend to low levels on the downstream side and contribute to the gravity-wave drag as described above, but do not directly contribute to the "hydraulic" blocked flow drag.

Unlike the gravity-wave drag, the "blocked flow" drag is aligned with the direction of the mean flow, and only operates at low levels. After some algebra, accommodating the presence of valleys below the mean level and simplifying assumptions, for practical applications the above expression may be reduced to the following (Lott & Miller, 1997)

$$\mathbf{D}_b(z) = -\rho(z)C_d l(z)\left(1 - \frac{1}{2\lambda}\right)\frac{\sigma}{2\mu}\left(\frac{z_b - z}{z + \mu}\right)^{1/2}(B(\gamma)\cos^2\alpha + C(\gamma)\sin^2\alpha)|U|U,$$
(15.24)

where B and C depend on $\gamma = a/b$ as in (12.57), μ denotes the mean depth of the mountain valleys below the level $z = 0$, and

$$\lambda = \frac{\cos^2\alpha + \gamma\sin^2\alpha}{\gamma\cos^2\alpha + \sin^2\alpha}.$$
(15.25)

The process of improving on the above schemes in forecasting models continues to this day. Tests between model results and observations have been made on isolated topography such as South Georgia and New Zealand (Vosper et al., 2016). A recent survey by Sandu et al. (2016) highlighted the most significant current problems, and Vosper et al. (2018) examined how these schemes relate to turbulent foehn flow, atmospheric rotors and the interaction between gravity waves and the stable boundary-layer.

References

Abramowitz, M. & Stegun, I.A. (1968). *Handbook of Mathematical Functions*. Dover, 1045 pp. [p. 216.]

Alford, M.H., Girton, J.B., Voet, G., Carter, G.S., Mickett, B. & Klymak, J.M. (2013). Turbulent mixing and hydraulic control of abyssal water in the Samoan Passage. *Geophys. Res. Lett.* **40**, 4668–4674. [p. 194.]

Akers, B. & Bokhove, O. (2008). Hydraulic flow through a channel contraction: multiple steady states. *Phys. Fluids* **20**, 056601. [p. 48.]

Akylas, T.R. (1984). On the excitation of long nonlinear water waves by a moving pressure distribution. *J. Fluid Mech.* **141**, 455–66. [pp. 73, 94.]

Andrews, D.G. (1980). On the mean motion induced by transient inertio-gravity waves. *Pageoph.* **118**, 177–88. [p. 226.]

Armi, L. (1986). The hydraulics of two flowing layers of different densities. *J. Fluid Mech.* **163**, 27–58. [pp. 46, 137, 137, 138, 139.]

Armi, L. & Farmer, D.M. (1986). Maximal two-layer exchange through a contraction with barotropic net flow. *J. Fluid Mech.* **164**, 27–51. [pp. 156, 161, 162.]

Armi, L. & Farmer, D.M. (1987). A generalization of the concept of maximal exchange in a strait. *J. Geophys. Res.* **92**, 14679–80. [p. 162.]

Armi, L. & Farmer, D.M. (1988). The flow of Mediterranean water through the Strait of Gibraltar. *Prog. Oceanog.* **21**, 1–105. [p. 149.]

Armi, L. & Mayr, G.J. (2007). Continuously stratified flow across an Alpine crest with a pass: shallow and deep föhn. *Quart. J. Roy. Met. Soc.* **133**, 459–477. [p. 372.]

Armi, L. & Mayr, G.J. (2011). The descending stratified flow and internal hydraulic jump in the lee of the Sierras. *J. Appl. Met. Clim.* **50**, 1995–2011. [p. 497.]

Armi, L. & Mayr, G.J. (2015). Virtual and real topography for flows across mountain ranges. *J. Appl. Met. & Climatol.* **54**, 723–730. [p. 372.]

Armi, L. & Riemenschneider, U. (2008). Two-layer hydraulics for a co-located crest and narrows. *J. Fluid Mech.* **615**, 169–184. [p. 140.]

Armi, L. & Williams, R. (1993). The hydraulics of a stratified fluid flowing through a contraction. *J. Fluid Mech.* **251**, 355–375. [pp. 146, 147, 147.]

Bacmeister, J. (1987). *Nonlinearity in transient, two-dimensional flow over topography.* Ph.D. thesis, Princeton University, 187 pp. [p. 293.]

Bacmeister, J.T. & Pierrehumbert, R.T. (1988). On high-drag states of non-linear stratified flow over an obstacle. *J. Atmos. Sci.* **45**, 63–80. [pp. 236, 364, 365, 367.]

Bacmeister, J.T. & Schoeberl, M.R. (1989). Breakdown of vertically propagating two-dimensional gravity waves forced by orography. *J. Atmos. Sci.* **46**, 2109–2134. [p. 372.]

Baines, P.G. (1971a). The reflection of internal/inertial waves from bumpy surfaces. *J. Fluid Mech.* **46**, 273–91. [p. 217.]

Baines, P.G. (1971b). The reflection of internal/inertial waves from bumpy surfaces. Part 2. Split reflection and diffraction. *J. Fluid Mech.* **49**, 113–131. [p. 217.]

Baines, P.G. (1977). Upstream influence and Long's model in stratified flows. *J. Fluid Mech.* **82**, 147–59. [pp. 314, 358, 359, 360, 360.]

Baines, P.G. (1979a). Observations of stratified flow over two-dimensional obstacles in fluid of finite depth. *Tellus* **31**, 351–71. [pp. 305, 306, 309, 310, 311, 312, 313.]

Baines, P.G. (1979b). Observations of stratified flow past three-dimensional barriers. *J. Geophys. Res.* **84**, 7834–8. [pp. 360, 362, 470, 471, 472, 483.]

Baines, P.G. (1984) A unified description of two-layer flow over topography. *J. Fluid Mech.* **146**, 127–67. [pp. 109, 116, 117, 127.]

Baines, P.G. (1988). A general method for determining upstream effects in stratified flow of finite depth over long two-dimensional obstacles. *J. Fluid Mech.* **188**, 1–22. [p. 330, 332, 334, 334, 336.]

Baines, P.G. (1990). Upstream blocking and stratified flow hydraulics. In *Stratified Flows. Proc. 3rd Int. Symp. on Stratified Flows*, E.J. List & G.H. Jirka (eds), ASCE, 113–22. [p. 341, 468.]

Baines, P.G. (1994). Mechanisms for upstream effects in two-dimensional stratified flow. In *Stably Stratified Flows: Flow and Dispersion over Topography*. I. Castro & N. Rockliff (eds), Oxford University Press, 1–14. [p. 265.]

Baines, P.G. (1997). A fractal world of cloistered waves. *Nature* **388**, 518–9. [p. 489.]

Baines, P.G. (2001). Mixing in flows down gentle slopes into stratified environments. *J. Fluid Mech.* **443**, 237–270. [pp. 182, 183, 184, 184.]

Baines, P.G. (2002). Two-dimensional plumes in stratified environments. *J. Fluid Mech.* **471**, 315–337. [p. 184.]

Baines, P.G. (2005). Mixing regimes for the flow of dense fluid down slopes into stratified environments. *J. Fluid Mech.* **538**, 245–267. [pp. 181, 184, 185, 186, 187, 188.]

Baines, P.G. (2007). Internal tide generation by seamounts. *Deep-Sea Research Part I* **54**(9), 1486–1508. [p. 504.]

Baines, P.G. (2008). Mixing in downslope flows in the ocean – plumes versus gravity currents. *Atmosphere–Ocean*, **46**(4), 405–419, doi:10.3137/AO925.2008. [pp. 188, 189.]

Baines, P.G. (2009). A model for the structure of the Antarctic slope front. *Deep-Sea Research* II, **56**, 859–873. [p. 504.]

Baines, P.G. (2013). The dynamics of intrusions into a density-stratified crossflow. *Phys. Fluids* **25**, 076601, 1–30. [p. 499, 499.]

Baines, P.G. (2016). Internal hydraulic jumps in two-layer systems. *J. Fluid Mech.* **787**, 1–15. [pp. 104, 107, 108, 108, 109, 115, 330.]

Baines, P.G. & Davies, P.A. (1980). Laboratory studies of topographic effects in rotating and/or stratified fluids. In *Orographic Effects in Planetary Flows*, GARP publication no. 23, WMO / ICSU, 233–99. [p. 41.]

Baines, P.G. & Fraedrich, K. (1989). Topographic effects on the mean tropospheric flow patterns around Antarctica. *J. Atmos. Sci.* **46**, 3401–3415. [p. 507.]

Baines, P.G. & Granek, H. (1990). Hydraulic models of deep stratified flows over topography. In *The Physical Oceanography of Sea Straits*, L. Pratt (ed), Kluwer, 245–69. [pp. 292, 293, 318, 343, 352, 353, 368.]

Baines, P.G. & Grimshaw, R.H.J. (1979). Stratified flow over finite obstacles with weak stratification. *Geophys. Astrophys. Fluid Dyn.* **13**, 317–34. [p. 282.]

Baines, P.G. & Guest, F. (1988). The nature of upstream blocking in uniformly stratified flow over long obstacles. *J. Fluid Mech.* **188**, 23–45. [pp. 299, 316, 338, 339, 340, 342, 345, 346, 347, 348, 350.]

Baines, P.G. & Hoinka, K.P. (1985). Stratified flow over two-dimensional topography in fluid of infinite depth: a laboratory simulation. *J. Atmos. Sci.* **42**, 1614–30. [pp. 264, 268, 270, 271, 272, 275, 294, 360, 362, 367.]

Baines, P.G. & Johnson, E.R. (2016). Nonlinear topographic effects in two-layer flows. *Front. Earth Sci.* **4**, 9. doi:10.3389/feart.2016.00009. [pp. 100, 109, 115, 116, 116.]

Baines, P.G. & Leonard, B.P. (1989). The effects of rotation on flow of a single layer over a ridge. *Quart. J. Roy. Met. Soc.* **115**, 293–308. [p. 54.]

Baines, P.G. & Manins, P.C. (1989). The principles of laboratory modelling of atmospheric flows over complex terrain. *J. Appl. Met.* **28**, 1213–25. [pp. 506, 507.]

Baines, P.G. & Mitsudera, H. (1994). On the mechanism of shear flow instabilities. *J. Fluid Mech.* **276**, 327–42. [pp. 245, 248.]

Baines, P.G. & Sacks, S. (2014). Atmospheric internal waves generated by explosive volcanic eruptions. *Geol. Soc. London, Memoirs* **39**, 153–168. [p. 499.]

Baines, P.G. & Sacks, S. (2017). The generation and propagation of atmospheric internal waves caused by volcanic eruptions. *Atmosphere* **8**, 60, 24 pages. [p. 499, 500.]

Baines, P.G. & Smith, R.B. (1993). Upstream stagnation points in stratified flow past obstacles. *Dyn. Atmos. Oceans*, **18**, 105–13. [pp. 454, 461.]

Baines, P.G. & Sparks, R.S.J. (2005). Dynamics of giant volcanic ash clouds from supervolcanic eruptions. *Geophys. Res. Lett.* **32**, L24808, doi:10.1029/2005GL024597. [p. 499.]

Baines, P.G. & Whitehead, J.A. (2003). On multiple states in single-layer flows. *Phys. Fluids* **15**, 298–307. [p. 41.]

Baines, P.G., Majumdar, S.J. & Mitsudera, H. (1996). The mechanics of the Tollmien–Schlichting wave. *J. Fluid Mech.* **312**, 107–124. [pp. 255, 258, 259, 260.]

Baines, P.G., Jones, M.J. & Sparks, R.S.J. (2008). The variation of large-magnitude volcanic ash cloud formation with source latitude. *J. Geophys. Res.* **113**, D21204, doi:10.1029/2007JD009568. [p. 499.]

Bannister, D. & King, J. (2015). Föhn winds on South Georgia and their impact on regional climate. *Weather* **70**, 324–329. [p. 495.]

Batchelor, G.K. (1967). *An Introduction to Fluid Dynamics*. Cambridge University Press, 615 pp. [pp. 3, 5, 84, 247, 266, 283, 434.]

Bauer, M.H., Mayr, G.J., Vergeiner, I. & Pichler, H. (2000). Strongly nonlinear flow over and around a three-dimensional mountain as a function of the horizontal aspect ratio. *J. Atmos. Sci.* **57**, 3971–3991. [p. 484.]

Beljaars, A.C.M., Brown, A.R. & Wood, N. (2004). A new parametrization of turbulent orographic form drag. *Quart. J. Roy. Met. Soc.* **130**, 1327–47. [pp. 511, 514.]

Bell, R.C. & Thompson, R.O.R.Y. (1980). Valley ventilation by cross winds. *J. Fluid Mech.* **96**, 757–67. [pp. 382, 386.]

Bell, T.H. (1974). Effects of shear on the properties of internal gravity wave modes. *Deutsche Hydrograph. Zeitschr.* **21**, 57–62. [pp. 208, 209.]

Benjamin, T.B. (1966). Internal waves of finite amplitude and permanent form. *J. Fluid Mech.* **25**, 241–70. [pp. 122, 127.]

Benjamin, T.B. (1967). Internal waves of permanent form in fluids of great depth. *J. Fluid Mech.* **29**, 559–92. [p. 120.]

Benjamin, T.B. (1968). Gravity currents and related phenomena. *J. Fluid Mech.* **31**, 209–48. [pp. 173, 175.]

Benjamin, T.B. (1981). Steady flows drawn from a stably stratified reservoir. *J. Fluid Mech.* **106**, 245–60. [p. 144, 147.]

Benjamin, T.B. & Lighthill, M.J. (1954). On cnoidal waves and bores. *Proc. Roy. Soc. A,* **224**, 448–60. [pp. 62, 64.]

Benney, D.J. (1966). Long non-linear waves in fluid flows. *J. Math. Phys.* **45**, 52–63. [p. 120.]

Binnie, A.M. (1972). Hugoniot's method applied to stratified flow through a contraction. *J. Mech. Eng. Sci.* **14**, 72–3. [pp. 130, 144.]

Binnie, A.M. & Orkney, J.C. (1955). Experiments on the flow of water from a reservoir through an open horizontal channel. II. The formation of hydraulic jumps. *Proc. Roy. Soc. A,* **230**, 237–46. [p. 70.]

Blumen, W. (1965). A random model of momentum flux by mountain waves. *Geophys. Publ.* **26**(2), 33pp. [p. 280.]

Blumen, W. (ed) (1990). *Atmospheric Processes over Complex Terrain.* Amer. Met. Soc. Met. Monographs. Vol. 23, No. 45, 323 pp. [p. 2.]

Bona, J.L., Pritchard, W.G. & Scott, L.R. (1981). An evaluation of a model equation for water waves. *Phil. Trans. Roy. Soc.* **302**, 457–510. [p. 59.]

Bonneton, P., Chomaz, J.-M. & Hopfinger, E.J. (1993). Internal waves produced by the turbulent wake of a sphere moving horizontally in a stratified fluid. *J. Fluid Mech.* **254**, 23–40. [p. 458.]

Booker, J.R. & Bretherton, F.P. (1967). The critical layer for internal gravity waves in a shear flow. *J. Fluid Mech.* **27**, 513–39. [pp. 230, 248.]

Borden, Z. & Meiburg, E. (2013a) Circulation-based models for Boussinesq internal bores. *J. Fluid Mech.* **726**, R1, doi:10.1017/jfm.2013.239. [pp. 104, 105.]

Borden, Z. & Meiburg, E. (2013b). Circulation based models for Boussinesq gravity currents. *Phys. Fluids* **25**, 101301; doi: 10.1063/1.4825035. [p. 173.]

Bougeault, P., Jansa Clar, A., Benech, B., Carissimo, B., Pelon, J., & Richard, E. (1990). Momentum budget over the Pyrenees: the PYREX experiment. *Bull. Amer. Met. Soc.* **71**, 806–818. [p. 496.]

Bougeault, P., Benech, B., Bessemoulin, P., Carissimo, B., Jansa Clar, A., Pelon, J., Petitdidier, M. & Richard, E. (1997). PYREX: a summary of findings. *Bull. Amer. Met. Soc.* **78**, 637–650. [p. 496.]

Boyer, D.L. & Tao, L. (1987). Impulsively started, linearly stratified flow over long ridges. *J. Atmos. Sci.* **44**, 23–42. [p. 359.]

Boyer, D.L., Davies, P.A., Fernando, H.J.S. & Zhang, X. (1989). Linearly stratified flow past a horizontal circular cylinder. *Phil. Trans. Roy. Soc. A* **328**, 501. [pp. 267, 274.]

Bretherton, F.P. (1966). The propagation of groups of internal gravity waves in a shear flow. *Quart. J. Roy. Met. Soc.* **92**, 466–80. [pp. 7, 222, 226.]

Brighton, P.W.M. (1978). Strongly stratified flow past three-dimensional obstacles. *Quart. J. Roy. Met. Soc.* **104**, 289–307. [pp. 451, 476.]

Britter, R.E. & Linden, P.F. (1980). The motion of the front of a gravity current travelling down an incline. *J. Fluid Mech.* **99**, 531–543. [p. 180.]

Broad, A.S., Porter, D. & Sewell, M.J. (1993). Shallow flow over general topography with applications to monotonic mountains. In *Stably-Stratified Flows: Flow and Dispersion over Topography.* I. Castro & N. Rockliff (eds), Oxford University Press, 133–8. [p. 31.]

Browand, F.K. & Winant, C.D. (1972) Blocking ahead of a cylinder moving in a stratified fluid: an experiment. *Geophys. Fluid Dyn.* **4**, 29–53. [p. 358.]

Burkill, J.C. (1956). *The Theory of Ordinary Differential Equations.* Oliver & Boyd, 102 pp. [p. 205.]

Byatt-Smith, J.G.B. (1971). The effects of laminar viscosity on the solution of the undular bore. *J. Fluid Mech.* **48**, 33–40. [p. 71.]

Cairns, R.A. (1979). The role of negative energy waves in some instabilities of parallel flows. *J. Fluid Mech.* **92**, 1–14. [p. 247.]

Carpenter, J.R., Balmforth, J. & Lawrence, G.A. (2010). Identifying unstable modes in stratified shear layers. *Phys. Fluids* **22**, 054104–1:3. [p. 245.]

Carpenter, J.R., Tedford, E.W., Heifetz, E. & Lawrence, G.A. (2011). Instability in stratified shear flow: review of a physical interpretation based on interacting waves. *Appl. Mech. Rev.* **64**, 060801–1:17. [p. 245.]

Case, K.M. (1960). Stability of an idealised atmosphere. I. Discussion of results. *Phys. Fluids* **3**, 149–54. [pp. 204, 208.]

Castro, I.P. & Snyder, W.H. (1988). Upstream motions in stratified flow. *J. Fluid Mech.* **187**, 487–506. [pp. 307, 360, 360.]

Castro, I.P. & Snyder, W.H. (1993). Experiments on wave-breaking in stratified flow over obstacles. *J. Fluid Mech.* **255**, 195–211. [pp. 462, 466.]

Castro, I.P., Snyder, W.H. & Marsh, G.L. (1983). Stratified flow over three-dimensional ridges. *J. Fluid Mech.* **135**, 261–82. [pp. 462, 482.]

Castro, I.P., Snyder, W.H. & Baines, P.G. (1990). Obstacle drag in stratified flow. *Proc. Roy. Soc. A* **429**, 119–40. [pp. 282, 283, 363, 363, 364.]

Caulfield, C.-C.P. (1994). Multiple linear instability of layered stratified shear flow. *J. Fluid Mech.* **258**, 255–285. [p. 245.]

Chandrasekhar, S. (1961). *Hydrodynamic and Hydromagnetic Stability*. Oxford: Clarendon Press. [p. 243.]

Chapman, G.T. & Yates, L.A. (1991). Topology of flow separation on three-dimensional bodies. *Appl. Mech. Rev.* **44**, 329–45. [pp. 437, 438.]

Chassignet, E.P., Cenedese, C., & Verron, J. (eds). *Buoyancy-Driven Flows*, Cambridge University Press, 436 pp. [p. 3.]

Chomaz, J.-M., Bonneton, P., Butet, A. & Perrier, M. (1992). Froude number dependence of the flow separation line on a sphere towed in a stratified fluid. *Phys. Fluids A* **4**(2), 254–8. [p. 458.]

Chomaz, J.-M., Bonneton, P. & Hopfinger, E.J. (1993). The structure of the near wake of a sphere moving horizontally in a stratified fluid. *J. Fluid Mech.* **254**, 1–21. [p. 458.]

Chu, V.H. & Baddour, R.E. (1977). Surges, waves and mixing in two-layer density stratified flow. In *Proc. 17th Int'l. Assn. Hydraul. Res.*, vol. 1, 303–10. [p. 104.]

Chu, X., Zhao, J., Lu, X., Harvey, V.L., Jones, M.R., Becker, E., Chen, C., Fong, W., Yu, A.Z., Roberts, B.R. and Dörnbrack, A. (2018). Lidar observations of stratospheric gravity waves from 2011 to 2015 at McMurdo (77.84°S, 166.69°E), Antarctica: 2. Potential energy densities, lognormal distributions, and seasonal variations. *J. Geophys. Res. Atmos.* **123**(15), 7910–7934. doi:10.1029/2017JD027386. [pp. 502.]

Clark, T.L. (1977). A small scale dynamical model using a terrain following coordinate transformation. *J. Comput. Phys.* **24**, 136–215. [pp. 293, 294, 374, 481.]

Clark, T.L. & Farley, R. D. (1984). Severe downslope windstorm calculations in two and three spatial dimensions using anelastic interactive grid nesting: A possible mechanism for gustiness. *J. Atmos. Sci.* **41**, 329–50. [pp. 374, 376, 481.]

Clark, T.L. & Peltier, W.R. (1977). On the evolution and stability of finite-amplitude mountain waves. *J. Atmos. Sci.* **34**, 1715–30. [p. 296.]

Clark, T.L. & Peltier, W.R. (1984). Critical level reflection and the resonant growth of non-linear mountain waves. *J. Atmos. Sci.* **41**, 3122–34. [pp. 364, 367.]

Clarke, S.R. & Grimshaw, R.H.J. (1994). Resonantly generated internal waves in a contraction. *J. Fluid Mech.* **274**, 139–161. [pp. 141, 141.]

Coddington, E.A. & Levinson, N. (1955). *Theory of Ordinary Differential Equations*. McGraw–Hill, 429 pp. [p. 435.]

Cokelet, E.D. (1977). Steep gravity waves in water of uniform depth. *Phil. Trans. Roy. Soc. A*, **286**, 183–230. [pp. 64, 67.]

Cole, S.L. (1985). Transient waves produced by flow past a bump. *Wave Motion* **1**, 579–87. [p. 72.]

Cornish, V. (1934). *Ocean Waves and Kindred Geophysical Phenomena*. Cambridge University Press, 164 pp. [p. 50.]

Cortes, A., Wells, M.G., Fringer, O.B., Arthur, R.S. & Ruenda, F.J. (2015). Numerical investigation of split flows by gravity currents into two-layered stratified water bodies. *J. Geophys. Res. Oceans*, **120**, 5254–5271, doi:10.1002/2015JC010722. [p. 180.]

Craik, A.D.D. (1985). *Wave Interactions and Fluid Flows*. Cambridge University Press, 333 pp. [pp. 228, 229, 247, 253.]

Crapper, G.D. (1959). A three-dimensional solution for waves in the lee of mountains. *J. Fluid Mech.* **6**, 51–76. [p. 397.]

Crapper, G.D. (1962). Waves in the lee of a mountain with elliptical contours. *Phil. Trans. Roy. Soc. A* **254**, 601–24. [p. 397.]

Dalziel, S.B. (1991). Two-layer hydraulics: a functional approach. *J. Fluid Mech.* **223**, 135–63. [p. 164.]

Dalziel, S.B. (1992) Maximal exchange in channels with nonrectangular cross-sections. *J. Phys. Oceanog.* **22**, 1188–1206. [p. 163.]

Dauxois, T., Joubaud, S., Odier, P. & Vennaille, A. (2018). Instabilities of internal wave beams. *Ann. Rev. Fluid Mech.* **50**, 131–156. [p. 260.]

Davis, R.E. (1969). The two-dimensional flow of a stratified fluid over an obstacle. *J. Fluid Mech.* **36**, 127–43. [p. 309.]

Davis, R.E. & Acrivos, A. (1967). The stability of oscillatory internal waves. *J. Fluid Mech.* **30**, 723–36. [p. 359.]

Ding, L., Calhoun, R.J. & Street, R.L. (2003). Numerical simulation of strongly stratified flow over a three-dimensional hill. *Boundary-layer Meteorol.* **107**, 81–114. [pp. 455, 476, 481.]

Doyle, J.D. & Durran, D.R. (2002). The dynamics of mountain-wave induced rotors. *J. Atmos. Sci.* **59**, 186–201. [pp. 274, 360.]

Doyle, J.D. & Durran, D.R. (2004). Recent developments in the theory of atmospheric rotors. *Bull. Amer. Met. Soc.* **85**, 337–342. [p. 274, 360.]

Drazin, P.G. (1961). On the steady flow of a fluid of variable density past an obstacle. *Tellus* **13**, 239–51. [pp. 426, 474.]

Drazin, P.G. & Reid, W.H. (1981). *Hydrodynamic Stability*. Cambridge University Press, 525 pp. [pp. 50, 243, 254.]

Drazin, P.G. & Reid, W.H. (2004). *Hydrodynamic Stability*, Second Edition. Cambridge University Press, 628 pp. [p. 243.]

Drazin, P.G. & Su, C.H. (1975). A note on long-wave theory of airflow over a mountain. *J. Atmos. Sci.* **32**, 437–9. [p. 279.]

Dressler, R.F. (1949). Mathematical solution of the problem of roll-waves in inclined open channels. *Comm. Pure Appl. Math.* **2**, 149–94. [pp. 50, 51.]

Durran, D.R. (1986). Another look at downslope windstorms. Part 1: The development of analogs to supercritical flow in an infinitely deep continuously stratified fluid. *J. Atmos. Sci.* **43**, 2527–2543. [p. 365, 367.]

Durran, D.R. (1992). Two-layer solutions to Long's equation for vertically propagating mountain waves: how good is linear theory? *Quart. J. Roy. Met. Soc.* **118**, 415–33. [p. 319.]

Durran, D.R. & Klemp, J.B. (1987). Another look at downslope winds. Part II: Nonlinear amplification beneath wave-overturning critical layers. *J. Atmos. Sci.* **44**, 3402–12. [pp. 364, 367, 367.]

Eliassen, A. & Palm, E. (1961). On the transfer of energy in stationary mountain waves. *Geophys. Publik.* **22**(3), 23 pp. [p. 223.]

Ellison, T.H. & Turner, J.S. (1959). Turbulent entrainment in stratified flows. *J. Fluid Mech.* **6**, 423–448. [p. 180.]

Engqvist, A. (1996). Self-similar multi-layer exchange flow through a contraction. *J. Fluid Mech.* **328**, 49–66. [pp. 166, 167, 168.]

Epifanio, C.C. & Durran, D.R. (2001). Three-dimensional effects in high-drag-state flows over long ridges. *J. Atmos. Sci.* **58**, 1051–1064. [p. 484.]

Epifanio, C.C. & Rotunno, R. (2005). The dynamics of orographic wake formation in flows with upstream blocking. *J. Atmos. Sci.* **62**, 3127–3150. [p. 484.]

Farmer, D.M. & Armi, L. (1986). Maximal two-layer exchange over a sill and through a combination of a sill and contraction with barotropic net flow. *J. Fluid Mech.* **164**, 53–76. [pp. 149, 163.]

Fernando, H.J.S. (1991). Turbulent mixing in stratified fluids. *Ann. Rev. FLuid Mech.* **23**, 455–493. [p. 184.]

Fernando, H.J.S. & 49 others. (2015). The MATERHORN – unravelling the intricacies of mountain weather. *Bull. Amer. Met. Soc.* **96**, 1945–1967. doi.org/10.1175/BAMS-D-13-00131.1 [p. 498.]

Finnigan, J.J. (1988). Air flow over complex terrain. In *Flow and Transport in the Natural Environment: Advances and Applications*, W.L. Steffen & T. Denmead (eds), Springer-Verlag, 183–229. [p. 266.]

Fornberg, B. & Whitham, G.B. (1978). A numerical and theoretical study of certain non-linear wave phenomena. *Phil. Trans. Roy. Soc. A* **289**, 373–404. [p. 66.]

Foster, M.R. & Saffman, P.G. (1970). The drag of a body moving transversely in a confined stratified fluid. *J. Fluid Mech.* **43**, 407–18. [p. 360.]

Fritts, D.C. (1982). The transient critical level interaction in a Boussinesq fluid. *J. Geophys. Res.* **87**, 7997–8016. [pp. 232, 232, 238.]

Fritts, D.C. & Alexander, J. (2003). Gravity wave dynamics and effects in the middle atmosphere. *Rev. Geophys.*, **41**, 1-64. See also *Rev. Geophys.* **50**, 3004, (2012). [p. 501.]

Fritts, D.C. & Geller, M.A. (1976). Viscous stabilization of gravity wave-critical level flows. *J. Atmos. Sci.* **33**, 2276–84. [p. 232.]

Fritts, D.C. & 36 others (2016). The deep propagating gravity wave experiment (DEEP-WAVE). *Bull. Amer. Met. Soc.* **97**, 425–453. doi.org/10.1175/BAMS-D-14-00269.1 [pp. 501.]

Garner, S.T. (1995). Permanent and transient upstream effects in nonlinear stratified flow over a ridge. *J. Atmos. Sci.* **52**, 227–246. [p. 361.]

Garratt, J.R. (1992). *The Atmospheric Boundary Layer*. Cambridge University Press, 316 pp. [p. 511.]

Garrett, C. (2004). Frictional processes in straits. *Deep-Sea Research II*, **51**, 393–410. [p. 47.]

Garrett, C. & Kunze E. (2007). Internal tide generation in the deep ocean. *Ann. Rev. Fluid Mech.* **39**, 57–87. [p. 504.]

Garrett, C.J.R. & Gerdes, F. (2003). Hydraulic control of homogeneous shear flows. *J. Fluid Mech.* **475**, 163–172. [p. 47.]

Garrett, C.J.R. & Munk, W. (1972). Space-time scales of internal waves in the open ocean. *Geophys. Fluid Dyn.* **3**(3), 225–254. [p. 502.]

Garrett, C.J.R., Bormans, M. & Thompson, K. (1990). Is the exchange through the Strait of Gibraltar maximal or submaximal? In *The Physical Oceanography of Sea Straits*, L.J. Pratt (ed), NATO ASI Series C, Vol. 318, 271–94. [p. 163.]

Gerdes, F., Garrett, C., & Farmer, D. (2002). On internal hydraulics with entrainment. *J. Phys. Oceanog.* **32**, 1106–1111. [p. 48.]

Gill, A.E. (1982). *Atmosphere–Ocean Dynamics*. Academic Press, 662 pp. [pp. 4, 226, 277, 280.]

Girton, J.B. & Sanford, T.B. (2003). Descent and modification of the overflow plume in the Denmark Strait. *J. Phys. Oceanog.* **33**, 1351–1364. [p. 189.]

Gjevik, B.J. (1980). Orographic effects revealed by satellite pictures: mesoscale flow phenomena. In *Orographic Effects in Planetary Flows*, WMO/ICSU, 301–16. [pp. 422, 467, 469.]

Gjevik, B.J. & Marthinsen, T. (1978). Three-dimensional lee wave pattern. *Quart. J. Roy. Met. Soc.* **104**, 947–57. [p. 423.]

Gordon, A.L., Zambianchi, E., Orsi, A., Visbeck, M., Giulivi, C.F., Whitworth III, T., & Spezie, G. (2004). Energetic plumes over the western Ross Sea continental slope. *Geophys. Res. Lett.* **31**, I.21302. [p. 189.]

Gossard, E.E. & Hooke, W.H. (1975). *Waves in the Atmosphere*. Elsevier, 456 pp. [p. 217.]

Grace, W. (1991). Hydraulic jump in a fog bank. *Aust. Met. Mag.* **39**, 205–9. [p. 140.]

Grant, A.L.M. & Mason, P.J. (1990). Observations of boundary-layer structure over complex terrain. *Quart. J. Roy. Met. Soc.* **116**, 159–86. [pp. 489, 511.]

Greenslade, M.D. (1992). *Strongly Stratified Airflow Over and Around Mountains*. Ph.D. thesis, University of Leeds, 185 pp. [p. 477.]

Grimshaw, R.H.J. (1974). Internal gravity waves in a slowly varying, dissipative medium. *Geophys. Fluid Dyn.* **6**, 131–48. [pp. 222, 226.]

Grimshaw, R.H.J. (2010). Transcritical flow past an obstacle. *ANZIAM. J.* **52**, 1–25. [p. 77.]

Grimshaw, R.H.J. & Helfrich, K.R. (2018). Internal solitary wave generation by tidal flow over topography. *J. Fluid Mech.* **839**, 387–407. [p. 77.]

Grimshaw, R.H.J. & Smyth, N. (1986). Resonant flow of a stratified fluid over topography. *J. Fluid Mech.* **169**, 429–64. [pp. 33, 73, 73, 75, 322.]

Grimshaw, R.H.J. & Yi, Z. (1991). Resonant generation of finite-amplitude waves by the flow of a uniformly stratified fluid over topography. *J. Fluid Mech.* **229**, 603–28. [pp. 323, 324.]

Grubišić, V. & Billings, B. (2008) Summary of the Sierra Rotors Project wave and rotor events. *Atmos. Sci. Let.* **9**, 176–181. [p. 274.]

Grubišić, V. & Lewis, J.M. (2004). Sierra Wave Project revisited: 50 years later. *Bull. Amer. Meteor. Soc.* **85**, 1127–1142. [p. 497.]

Grubišić, V. & Smolarkiewicz, P.K. (1997). The effect of critical levels on orographic flows: linear regime. *J. Atmos. Sci.* **54**, 1943–1960. [p. 431.]

Grubišić, V., Smith, R.B. & Schär, C. (1995). The effect of bottom friction on shallow-water flow past an isolated obstacle. *J. Atmos. Sci.* **52**, 1985–2005. [pp. 87, 491.]

Grubišić, V., & coauthors. (2008). The Terrain-Induced Rotor Experiment: an overview of the field campaign and some highlights of special observations. *Bull. Amer. Meteor. Soc.* **89**, 1513–1533. [p. 497.]

Gu, L., & Lawrence, G.A. (2005). Analytical solution for maximal frictional two-layer exchange flow. *J. Fluid Mech.* **543**, 1–17. [p. 171.]

Hammack, J., Scheffner, N. & Segur, H. (1989). Two-dimensional periodic waves in shallow water. *J. Fluid Mech.* **209**, 567–589. [pp. 94, 94.]

Hanazaki, H. (1988). A numerical study of three-dimensional stratified flow past a sphere. *J. Fluid Mech.* **192**, 393–419. [pp. 458, 459.]

Hanazaki, H. (1989). Upstream advancing columnar disturbances in two-dimensional stratified flow of finite depth. *Phys. Fluids A* **1**, 1976–87. [p. 309.]

Hanazaki, H. (1994). On the three-dimensional internal waves excited by topography in the
flow of a stratified fluid. *J. Fluid Mech.* **263**, 293–318. [p. 125.]

Härtel, C., Meiburg, E. & Necker, F. (2000). Analysis and direct numerical simulation of
the flow at a gravity current head. Part I. Flow topology and front speed for slip and
non-slip boundaries. *J. Fluid Mech.* **418**, 189–212. [p. 173.]

Hartman, R.J. (1975). Wave propagation in a stratified shear flow. *J. Fluid Mech.* **71**,
89–104. [p. 231, 232.]

Hayes, W.D. & Probstein, R.F. (1966). *Hypersonic Flow Theory*, 2nd edition. Academic
Press, 602 pp. [p. 79.]

Haynes, P.H. & McIntyre, M.E. (1987). On the evolution of vorticity and potential vorticity
in the presence of diabatic heating and frictional or other forces. *J. Atmos. Sci.* **44**,
828–41. [p. 11.]

Hazel, P. (1967). The effect of viscosity and heat conduction on internal gravity waves at a
critical level. *J. Fluid Mech.* **30**, 775–83. [p. 232.]

Henderson, F.M. (1966). *Open Channel Hydraulics*. Macmillan, 522 pp. [pp. 71, 89.]

Hines, C.O. (1960). Internal atmospheric gravity waves at ionospheric heights. *Can. J.
Phys.* **38**, 1441–1481. [p. 499.]

Hocut, C.M., Liberzon, D. & Fernando, H.J.S. (2015). Separation of upslope flow over a
uniform slope. *J. Fluid Mech.* **775**, 266–287. [pp. 194.]

Hogg, A.J., Nasr-Azadani, M.M., Ungarish, M. & Meiburg, E. (2016). Sustained gravity
currents in a channel. *J. Fluid Mech.* **798**, 853–888. [p. 175, 179.]

Hogg, A.McC. & Hughes, G.O. (2006). Shear flow and viscosity in single-layer hydraulics.
J. Fluid Mech. **548**, 431–443. [p. 47.]

Hogg, A.McC. & Killworth, P.D. (2004). Continuously stratified exchange flow through a
contraction in a channel. *J. Fluid Mech.* **499**, 257–276. [p. 167, 169, 170.]

Hogg, A.M., Ivey, G.N. and Winters, K.B. (2001a). Hydraulics and mixing in controlled
exchange flows. *J. Geophys. Res.* **106**(C1), 959–972. [p. 163.]

Hogg, A.McC., Winters, K.B. & Ivey, G.N. (2001b). Linear internal waves and the control
of stratified exchange flows. *J. Fluid Mech.* **447**, 357–375. [p. 163.]

Hoinka, K.P. (1985). Observations of the airflow over the Alps during a foehn event. *Quart.
J. Roy. Met. Soc.* **111**, 199–224. [p. 424.]

Holmboe, J. (1962). On the behaviour of symmetric waves in stratified shear layers. *Geofys.
Publik.* **24**, 67–113. [p. 248.]

Holyer, J.Y. & Huppert, H.E. (1980). Gravity currents entering a two-layer fluid. *J. Fluid
Mech.* **100**, 739–767. [p. 177.]

Hornung, H.G. (1986). Regular and Mach reflection of shock waves. *Ann. Rev. Fluid Mech.*
18, 33–58. [p. 81.]

Houghton, D.D. & Isaacson, E. (1970). Mountain winds. *Stud. Numer. Anal.* **2**, 21–52.
[p. 110.]

Houghton, D.D. & Kasahara, A. (1968). Non-linear shallow fluid flow over an isolated
ridge. *Comm. Pure Appl. Math.* **21**, 1–23. [pp. 41, 41.]

Howard, L.N. (1961). Note on a paper of John W. Miles. *J. Fluid Mech.* **10**, 509–12. [p. 244.]

Huang, D.B., Sibul, O.J., Webster, W.C., Wehausen, J.V., Wu, D.M. & Wu, T.Y. (1982).
Ships moving in the transcritical range. In *Proc. Conf. on Behaviour of Ships in
Restricted Waters, Vol. 2, (Varna, Bulgaria)*, 26.1–26.9. [pp. 71, 92.]

Hunt, J.C.R. & Snyder, W.H. (1980). Experiments on stably and neutrally stratified shear
flow over a model three-dimensional hill. *J. Fluid Mech.* **96**, 671–704. [pp. 444, 447,
448, 450, 451, 477.]

Hunt, J.C.R., Abell, C.J., Peterka, J.A. & Woo, H. (1978). Kinematical studies of the flows around free or surface-mounted obstacles: applying topology to flow visualisation. *J. Fluid Mech.* **86**, 179–200. [pp. 434, 438, 438, 445, 446.]

Hunt, J.C.R., Richards, K.J. & Brighton, P.W.M. (1988). Stably stratified shear flow over low hills. *Quart. J. Roy. Met. Soc.* **114**, 859–86. [p. 263.]

Hunt, J.C.R., Fernando, H.J.S. & Princevac, M. (2003). Unsteady thermally driven flows on gentle slopes. *J. Atmos. Sci.* **60**, 2169–2182. [p. 194.]

Huppert, H.E. & Britter, R.E. (1982). The separation of hydraulic flow over topography, *J. Hydraulics Div. ASCE* **108**, 1532–1539. [p. 48.]

Huppert, H.E. & Miles, J.W. (1969). Lee waves in a stratified flow. Part 3. Semi-elliptical obstacle. *J. Fluid Mech.* **55**, 481–96. [pp. 289, 290, 291, 462, 466.]

Huppert, H.E. & Simpson, J.E. (1980). The slumping of gravity currents. *J. Fluid Mech.* **99**, 785–99. [pp. 173, 176.]

Imberger, J. & Patterson, J.C. (1990). Physical limnology. *Advances in Applied Mech.* **27**, 303–475. [p. 285.]

Ince, E.L. (1926). *Ordinary Differential Equations.* Reprinted by Dover (1956), 558 pp. [p. 208.]

Ippen, A.T. (1951). Mechanics of supercritical flow. *Trans. Amer. Soc. Civ. Eng.* **116**, 268–95. [p. 91.]

Ivanov, V.V., Shapiro, G.I., Huthnance, J.M., Aleynik, D.L. & Golovin, P.N. (2004). Cascades of dense water around the world ocean. *Progress in Oceanography* **60**, 47–98. [p. 503.]

Jagannathan, A., Winters, K.B. & Armi, L. (2017). Stability of stratified downslope flows with an overlying stagnant layer. *J. Fluid Mech.* **810**, 392–411. [p. 372.]

Jagannathan, A., Winters, K.B. & Armi, L. (2019). Stratified flows over and around long dynamically tall mountain ridges. *J. Atmos. Sci.* **76**, 1265–1287. [p. 369, 484.]

Janowitz, G.S. (1981). Stratified flow over a bounded obstacle in a channel of finite height. *J. Fluid Mech.* **110**, 161–170. [pp. 284, 304, 307.]

Janowitz, G.S. (1984). Lee waves in three-dimensional stratified flow. *J. Fluid Mech.* **148**, 97–108. [pp. 395, 396, 397, 399, 400.]

Jeffreys, H. & Jeffreys, B. (1962). *Methods of Mathematical Physics.* 3rd edition. Cambridge University Press, 716 pp. [p. 278.]

Jiang, Q. & Smith, R.B. (2000). V-waves, bow shocks, and wakes in supercritical hydrostatic flow. *J. Fluid Mech.* **406**, 27–53. [pp. 21, 23.]

Joseph, R.I. (1977). Solitary waves in a finite depth fluid. *J. Phys. A: Math. Gen.* **10**, L225–227. [p. 121.]

Kadomtsev, B.B. & Petviashvili, V.I. (1970). On the stability of solitary waves in weakly dispersing media. *Sov. Phys. Dokl.* **15**, 539–541. [p. 92.]

Karyampudi, V.M., Bacmeister, J.T., Koch, S.E., Rottman, J.W. & Kaplan, M.L. (1991). Generation of an undular bore by a downslope windstorm on the leeside of the Rockies. In *Proc. 5th Conf. on Mesoscale Processes*, Amer. Met. Soc., 17–22. [p. 190.]

Katsis, C. & Akylas, T.R. (1987). On the excitation of long non-linear water waves by a moving pressure distribution. Part 2. Three-dimensional effects. *J. Fluid Mech.* **177**, 49–65. [p. 93.]

Kelly, R.E. & Maslowe, S.A. (1970). The non-linear critical layer in a slightly stratified shear flow. *Stud. Appl. Math.* **49**, 301–26. [p. 237.]

Killworth, P.D. (1992). On hydraulic control in a stratified fluid. *J. Fluid Mech.* **237**, 605–626. [p. 325.]

Kimura, F. & Manins, P.C. (1988). Blocking in periodic valleys. *Boundary-layer Met.* **44**, 137–69. [pp. 381, 382, 384, 385, 386.]

Klemp, J.B. & Lilly, D.K. (1975). The dynamics of wave-induced downslope winds. *J. Atmos. Sci.* **32**, 320–39. [pp. 375, 376.]

Klemp, J.B., Rotunno, R. & Skamarock, W. (1997). On the propagation of internal bores. *J. Fluid Mech.* **331**, 81–106. [p. 104.]

Klymak, J.M., Legg, S.M. & Pinkel, R. (2010). High-mode stationary waves in stratified flow over large obstacles. *J. Fluid Mech.* **644**, 321–336. [p. 369.]

Koop, C.G. (1981). A preliminary investigation of the interaction of internal gravity wave potential vorticity in the presence of diabatic heating and frictional or other forces. *J. Atmos. Sci.* **44**, 828–41. [p. 232.]

Koop, C.G. & Butler, G. (1981). An investigation of internal solitary waves in a two-fluid system. *J. Fluid Mech.* **112**, 225–51. [p. 121.]

Koop, C.G. & McGee, B. (1986). Measurements of internal gravity waves in a continuously stratified shear flow. *J. Fluid Mech.* **172**, 453–80. [p. 232, 233, 234.]

Kubota, T., Ko, D.R.S. & Dobbs, L. (1978). Weakly non-linear, long internal gravity waves in fluid of infinite depth. *AIAA J. Hydronautics* **12**, 157–65. [p. 121.]

Lamb, K.G. (1994). Numerical simulations of stratified inviscid flow over a smooth obstacle. *J. Fluid Mech.* **260**, 1–22. [pp. 307, 324, 358, 364.]

Lamb, K.G. (2014). Internal wave breaking and dissipation mechanisms on the continental slope/shelf. *Ann. Rev. Fluid Mech.* **46**, 231–254. [p. 503.]

Lamb, K.G. & Wan, B. (1998). Conjugate flows and flat solitary waves for a continuously stratified fluid. *Phys. Fluids* **10**, 2061–2079. [p. 107.]

Lamb, V.R. & Britter, R.E. (1984). Shallow flow over an isolated obstacle. *J. Fluid Mech.* **147**, 291–313. [p. 82, 84.]

Lane-Serff, G.F. (1989). *Heat Flow and Air Movement in Buildings.* Ph.D. thesis, Cambridge University, 146 pp. [p. 163.]

Lane-Serff, G.F., Beal, L.M. & Hadfield, T.D. (1995). Gravity currents over obstacles. *J. Fluid Mech.* **292**, 39–53. [p. 175.]

Lane-Serff, G.F., Smeed, D.A. & Postlethwaite, C.R. (2000) Multi-layer hydraulic exchange flows. *J. Fluid Mech.* **416**, 269–296. [p. 164, 165.]

Laprise, R. (1993). An assessment of the WKBJ approximation to the vertical structure of linear mountain waves: implications for gravity wave drag parametrization. *J. Atmos. Sci.* **50**, 1469–87. [p. 516.]

Laprise, R. & Peltier, W.R. (1989a). The linear stability of nonlinear mountain waves: implications for the understanding of severe downslope windstorms. *J. Atmos. Sci.* **46**, 545–64. [pp. 289, 294, 367.]

Laprise, R. & Peltier, W.R. (1989b). The structure and energetics of transient eddies in a numerical simulation of breaking mountain waves. *J. Atmos. Sci.* **46**, 565–85. [pp. 294, 295, 367.]

Lawrence, G.A. (1985). *The hydraulics and mixing of two-layer flow over an obstacle.* Univ. Calif. Berkeley Hydr. Eng. Lab. Report No. UCB/HEL-85/02. [pp. 127, 127.]

Lawrence, G.A. (1993). The hydraulics of steady two-layer flow over a fixed obstacle. *J. Fluid Mech.* **254**, 605–33. [pp. 99, 127, 128, 137.]

Lawrence, G.A., Lasheras, J.C. & Browand, F.K. (1990). Shear instabilities in stratified flow. In *Stratified Flows: Proc. 3rd Int'l Symp. on Stratified Flows*, E.J. List & G.H. Jirka (eds), ASCE, 15–27. [p. 253.]

Lee, S-J., Yates, G.T. & Wu, T.Y. (1989). Experiments and analyses of upstream-advancing solitary waves generated by moving disturbances. *J. Fluid Mech.* **199**, 569–93. [p. 76.]

Legg, S. (2012). Overflows and convectively driven flows. In *Buoyancy-Driven Flows*, E.P. Chassignet, C. Cenedese & J. Verron (eds), Cambridge University Press, 203–239. [p. 372.]

Leo, L.S., Thompson, M.Y., Di Sabatino, S. & Fernando, H.J.S. (2015). Stratified flow past a hill: dividing streamline concept revisited. *Boundary-layer Meteorol.* DOI 10.1007/s10546-015-0101-1. [p. 467.]

Leutbecher, M. & Volkert, H. (2000). The propagation of mountain waves into the stratosphere: quantitative evaluation of three-dimensional simulations. *J. Atmos. Sci.* **57**, 3090–3108. [p. 501.]

Lied, N.T. (1964). Stationary hydraulic jumps in a katabatic flow near Davis, Antarctica, 1961. *Aust. Met. Mag.* **47**, 40–51. [p. 190.]

Lighthill, M.J. (1963). Introduction: Boundary layer theory. In *Laminar Boundary Layers*, L. Rosenhead (ed), Oxford University Press, 46–113. [pp. 435, 437.]

Lighthill, M.J. (1965). Group velocity. *J. Inst. Maths. Applics.* **1**, 1–28. [pp. 211, 284.]

Lighthill, M.J. (1978). *Waves in Fluids.* Cambridge University Press, 504 pp. [pp. 26, 26, 28, 29, 30, 31, 60, 214, 278, 396, 415.]

Lighthill, M.J. & Whitham, G.B. (1955). On kinematic waves. I. Flood movement in long rivers. *Proc. Roy. Soc. A*, **229**, 281–316. [p. 49.]

Lilly, D.K. & Klemp, J.B. (1979). The effects of terrain shape on nonlinear hydrostatic mountain waves. *J. Fluid Mech.* **95**, 241–61. [pp. 290, 290, 292.]

Lin, C.C. (1955). *The Theory of Hydrodynamic Stability.* Cambridge University Press, 155 pp. [p. 243.]

Lin, J.T. & Pao, Y.H. (1979). Wakes in stratified fluids: a review. *Ann. Rev. Fluid Mech.* **11**, 317–38. [p. 456.]

Lin, Q., Lindberg, W.R., Boyer, D.L. & Fernando, H.J.S. (1992). Stratified flow past a sphere. *J. Fluid Mech.* **240**, 315–54. [pp. 456, 457, 458, 458.]

Linden, P.F. (1999). The fluid mechanics of natural ventilation. *Ann. Rev. Fluid Mech.* **31**, 201–38. [p. 510.]

Linden, P. (2012). Gravity currents – theory and laboratory experiments. In *Buoyancy-Driven Flows*, E.P. Chassignet, C. Cenedese & J. Verron (eds), Cambridge University Press, 13–51. [p. 172.]

Lindzen, R.S. (1974). Stability of a Helmholtz velocity profile in a continuously stratified, infinite Boussinesq fluid – applications to clear air turbulence. *J. Atmos. Sci.* **31**, 1507–14. [p. 253.]

Lindzen, R.S. (1981). Turbulence and stress due to gravity wave and tidal breakdown. *J. Geophys. Res.* **86**, 9707–14. [pp. 238, 516.]

Long, R.R. (1953). Some aspects of the flow of stratified fluids. I. A theoretical investigation. *Tellus* **5**, 42–58. [p. 287.]

Long, R.R. (1954). Some aspects of the flow of stratified fluids. II. Experiments with a two-fluid system. *Tellus* **6**, 97–115. [pp. 41, 127.]

Long, R.R. (1955). Some aspects of the flow of stratified fluids. III. Continuous density gradient. *Tellus* **7**, 341–57. [p. 315.]

Long, R.R. (1970). Blocking effects in flow over obstacles. *Tellus* **22**, 471–80. [pp. 39, 41.]

Long, R.R. (1972). Finite amplitude disturbances in the flow of inviscid rotating and stratified fluids over obstacles. *Ann. Rev. Fluid Mech.* **4**, 69–92. [p. 39.]

Long, R.R. (1974). Some experimental observations of upstream disturbances in a two-fluid system. *Tellus* **26**, 313–17. [p. 117.]

Lott, F. (1995). Comparison between the orographic response of the ECMWF model and the PYREX 1990 data. *Quart. J. Roy. Meteor. Soc.* **121**, 1323–1348. [pp. 511, 514.]

Lott, F. (2016). A new theory for downslope windstorms and trapped mountain waves. *J. Atmos. Sci.* **73**, 3585–3597. [p. 376.]

Lott, F. & Miller, M. (1997). A new subgrid-scale orographic drag parameterization: its formulation and nesting. *Quart. J. Roy. Meteor. Soc.* **123**, 101–127. [pp. 412, 514, 516, 519.]

MacCready, P. & Pawlak, G. (2001). Stratified flow along a corrugated slope: separation drag and wave drag. *J. Phys. Oceanog.* **31**, 2824–39. [p. 503.]

Manins, P.C. & Sawford, B.L. (1979). Katabatic winds: a field case study. *Quart. J. Roy. Met. Soc.* **105**, 1011–25. [pp. 180, 374.]

Marques, G.M., Wells, M.G., Padman, L. & Ozgokmen, T.M. (2017). Flow splitting in numerical simulations of oceanic dense-water outflows. *Ocean Modelling* **113**, 66–84. [p. 188.]

Maslowe, S.A. & Redekopp, L.G. (1980). Long nonlinear waves in stratified shear flows. *J. Fluid Mech.* **101**, 321–48. [p. 237.]

Mass, C.F. & Ferber, G.K. (1990). Surface pressure perturbations produced by an isolated mesoscale topographic barrier. Part I: General characteristics and dynamics. *Mon. Wea. Rev.* **118**, 2579–96. [p. 493.]

Maxworthy, T., Leilich, J., Simpson, J.E. & Meiburg, E.H. (2002). The propagation of a gravity current into a linearly stratified fluid. *J. Fluid Mech.* **453**, 371–394. [pp. 176, 178.]

Mayer, F.T. & Fringer, O.B. (2017). An umabiguous definition of the Froude number for lee waves in the deep ocean. *J. Fluid Mech.* **831**, R3, doi:10.1017/jfm.2017.701. [pp. 14, 267, 276.]

McEwan, A.D. (1971). Degeneration of resonantly excited standing internal gravity waves. *J. Fluid Mech.* **50**, 431–448. [p. 260.]

McEwan, A.D. (1973). Interactions between internal gravity waves and their traumatic effect on a continuous stratification. *Boundary-layer Met.*, **5**, 159–75. [pp. 238, 260.]

McEwan, A.D. & Baines, P.G. (1974). Shear fronts and an experimental stratified shear flow. *J. Fluid Mech.* **63**, 257–72. [pp. 210, 269.]

McFarlane, N.A. (1987). The effect of orographically excited gravity wave drag on the general circulation of the lower stratosphere and troposphere. *J. Atmos. Sci.* **44**, 1775–800. [pp. 510, 515.]

McIntyre, M.E. (1972). On Long's hypothesis of no upstream influence in uniformly stratified or rotating flow. *J. Fluid Mech.* **52**, 209–43. [pp. 276, 281, 298, 321.]

McIntyre, M.E. (1980). An introduction to the generalised Lagrangian-mean description of wave, mean-flow interaction. *Pageoph.* **118**, 153–76. [p. 226.]

Melville, W.K. & Helfrich, K.R. (1987). Transcritical two-layer flow over topography. *J. Fluid Mech.* **178**, 31–52. [pp. 121, 124, 124, 125, 126.]

Mied, R.P. & Dugan, J.P. (1975). Internal wave reflection by a velocity shear and density anomaly. *J. Phys. Oceanog.* **5**, 279–87. [p. 215.]

Miles, J.W. (1961). On the stability of heterogeneous shear flows. *J. Fluid Mech.* **10**, 496–508. [pp. 208, 244.]

Miles, J.W. (1968). Lee waves in a stratified flow. Part 2. Semi-circular obstacle. *J. Fluid Mech.* **33**, 803–14. [pp. 287, 288, 289, 314.]

Miles, J.W. (1969). Waves and wave drag in stratified flows. *Proc. 12th Int. Cong. Appl. Mech., Stanford, Calif.*, M. Hetenyi & W.G. Vincenti (eds), Springer-Verlag. [p. 15.]

Miles, J.W. (1979). On internal solitary waves. *Tellus* **31**, 456–62. [p. 121.]

Miles, J.W. (1980). Solitary waves. *Ann. Rev. Fluid Mech.* **12**, 11–43. [pp. 60, 81.]

Miles, J.W. (1981). On internal solitary waves. II. *Tellus* **33**, 397–401. [p. 124.]

Milewski, P.A. & Tabak, E.G. (2015). Conservation law modelling of entrainment in layered hydrostatic flows. *J. Fluid Mech.* **772**, 272–94. [p. 109.]

Miller, P.P. & Durran, D.D. (1991). On the sensitivity of downslope windstorms to the asymmetry of the mountain profile. *J. Atmos. Sci.* **48**, 1457–73. [p. 360.]

Miller, M.J., Palmer, T.N. & Swinbank, R. (1989). Parametrisation and influence of sub-gridscale orography in general circulation and numerical weather prediction models. *Meteorol. Atmos. Phys.* **40**, 84–109. [p. 516.]

Miller, S.D., Straka III, W.C., Yue, J., Smith, S.M., Alexander, M.J., Hoffman, L., Setvak, M. and Partain, P.T. (2015). Upper atmospheric gravity wave details revealed in nightglow satellite imagery. *Proc. Nat. Acad. Sci.* **112**, 6728–6735. [p. 502.]

Milne-Thomson, L.M. (1960). *Theoretical Hydrodynamics* 4th edition. Macmillan, 660 pp. [p. 441.]

Mitchell, R.M., Cechet, R.P., Turner, P.J. & Elsum, C.C. (1990). Observation and interpretation of wave clouds over Macquarie Island. *Quart. J. Roy. Met. Soc.* **116**, 741–52. [p. 421.]

Monaghan, J.J. (2007). Gravity current interaction with interfaces. *Annu. Rev. Fluid Mech.* **39**, 245–261. [p. 180.]

Morse, P.M. & Feshbach, H. (1953). *Methods of Theoretical Physics, Parts I & II*. McGraw-Hill, 1978 pp. [pp. 20, 301.]

Morton, B.R. (1984). The generation and decay of vorticity. *Geophys. Astrophys. Fluid Dyn.* **28**, 277–308. [p. 483.]

Morton, B.R., Taylor, G.I. & Turner, J.S. (1956). Turbulent gravitational convection from maintained and instantaneous sources. *Proc. Roy. Soc. A* **234**, 1–23. [pp. 180, 184, 185.]

Needham, D.J. & Merkin, J.H. (1984). On roll waves down an open inclined channel. *Proc. Roy. Soc. A*, **394**, 259–78. [p. 50.]

Newley, T.M.J., Pearson, H.J. & Hunt, J.C.R. (1991). Stably stratified flow through a group of obstacles. *Geophys. Astrophys. Fluid Dyn.* **58**, 147–71. [p. 487.]

Nielsen, M.H., Pratt, L. & Helfrich, K. (2004). Mixing and entrainment in hydraulically driven stratified sill flows. *J. Fluid Mech.* **515**, 415–443. [pp. 47, 48, 336.]

Ogden, K.A. & Helfrich, K.R. (2016). Internal hydraulic jumps in two-layer flows with upstream shear. *J. Fluid Mech.* **789**, 64–92. [p. 109.]

Ogura, Y. & Phillips, N.A. (1962). Scale analysis of deep and shallow convection in the atmosphere. *J. Atmos. Sci.* **19**, 173–9. [p. 6.]

Olafsson, H. & Bougeault, P. (1996). Nonlinear flow past an elliptic mountain ridge. *J. Atmos. Sci.* **53**, 2465–2489. [p. 484.]

Ono, H. (1975). Algebraic solitary waves in stratified fluids. *J. Phys. Soc. Japan* **39**, 1082–91. [p. 120.]

Orlanski, I. & Bryan, K. (1969). Formation of thermocline step structure by large-amplitude internal gravity waves. *J. Geophys. Res.* **74**, 6975–83. [p. 238.]

Orlanski, I. & Ross, B.B. (1977). The circulation associated with a cold front. Part I: Dry case. *J. Atmos. Sci.* **34**, 1619–33. [p. 293.]

Orr, A., Marshall, G.J., Hunt, J.C.R., Sommeria, J., Wang, C-G., van Lipzig, N.P.M., Cresswell, D. & King, J.C. (2008). Characteristics of summer airflow over the Antarctic Peninsula in response to recent strengthening of westerly circumpolar winds. *J. Atmos. Sci.* **65**, 1396–1413. [p. 495.]

Palmer, T.N., Shutts, G.J. & Swinbank, R. (1986). Alleviation of a systematic westerly bias in general circulation and numerical prediction models through an orographic gravity wave drag parametrisation. *Quart. J. Roy. Met. Soc.* **112**, 1011–39. [pp. 2, 510, 515.]

Peltier, W.R. & Clark, T.L. (1979). The evolution and stability of finite-amplitude mountain waves. Part II: Surface wave drag and severe downslope windstorms. *J. Atmos. Sci.* **36**, 1498–529. [pp. 296, 363.]

Peltier W.R. & Clark, T.L. (1983). Nonlinear mountain waves in two and three spatial dimensions. *Quart. J. Roy. Met. Soc.* **109**, 527–48. [p. 367, 368.]

Peltier, W.R. & Scinocca, J.F. (1990). The origin of severe downslope windstorm pulsations. *J. Atmos. Sci.* **47**, 2853–70. [pp. 374, 377.]

Peregrine, D.H. (1966). Calculations of the development of an undular bore. *J. Fluid Mech.* **25**, 321–30. [pp. 66, 68.]

Peregrine, D.H. (1968). Long waves in a uniform channel of arbitrary cross-section. *J. Fluid Mech.* **32**, 353–65. [p. 59.]

Perry, A.E. & Fairlie, B.D. (1974). Critical points in flow patterns. *Adv. Geophys. B* **18**, 299–315. [p. 435.]

Peters, H., Johns, W.E., Bowers, A.S. & Fratantoni, D.M. (2005). Mixing and entrainment in the Red Sea outflow plume. Part 1: plume structure. *J. Phys. Oceanog.* **35**(5), 569–583. [p. 189.]

Pettre, P. & Andre, J.-C. (1991). Surface pressure change through Loewe's phenomena and katabatic flow jumps: study of two cases in Adelie Land, Antarctica. *J. Atmos. Sci.* **48**, 557–71. [p. 190.]

Phillips, D.S. (1984). Analytical surface pressure and drag for linear hydrostatic flow over three-dimensional elliptical mountains. *J. Atmos. Sci.* **41**, 1073–84. [pp. 394, 410, 413.]

Phillips, O.M. (1966). *The Dynamics of the Upper Ocean*. Cambridge University Press, 261 pp. [p. 232, 260.]

Phillips, O.M. (1977). *The Dynamics of the Upper Ocean, 2nd edition*. Cambridge University Press, 336 pp. [pp. 12, 27, 238.]

Pierrehumbert, R.T. & Wyman B. (1985). Upstream effects of mesoscale mountains. *J. Atmos. Sci.* **42**, 977–1003. [pp. 293, 296, 361, 361, 362, 362, 367.]

Pitts, R.O. & Lyons, T.J. (1989). Airflow over a two-dimensional escarpment. 1: Observations. *Quart. J. Roy. Met. Soc.* **115**, 965–81. [p. 424.]

Plumb, R.A. (1977). The interaction of two internal waves with the mean flow: implications for the theory of the quasi-biennial oscillation. *J. Atmos. Sci.* **34**, 1847–1858. [p. 226.]

Plumb, R.A. & McEwan, A.D. (1978). The instability of a forced standing wave in a viscous stratified fluid: a laboratory analogue of the quasi-biennial oscillation. *J. Atmos. Sci.* **35**, 1827–39. [pp. 226, 233.]

Pratt, L.J. (1983). A note on non-linear flow over obstacles. *Geophys. Astrophys. Fluid Dyn.* **24**, 63–8. [p. 41.]

Pratt, L.J. (1984). Nonlinear flow with multiple obstructions. *J. Atmos. Sci.* **41**, 1214–25. [pp. 77, 78.]

Pratt, L.J. (1986). Hydraulic control of sill flow with bottom friction. *J. Phys. Oceanog.* **16**, 1970–80. [p. 47.]

Pratt, L. & Helfrich, K. (2005). Generalized conditions for hydraulic criticality of oceanic overflows. *J. Phys. Oceanog.* **35**, 1782–1800. [p. 325.]

Pratt, L.J. & Whitehead, J.A. (2008). *Rotating Hydraulics: Nonlinear Topographic Effects in the Ocean and Atmosphere*. Springer, 589 pp. [pp. 2, 48.]

Pratt, L.J., Deese, H.E., Murray, S.P. & Johns, W. (2000). Continuous dynamical modes having arbitrary cross-sections, with application to the Bab al Mandab. *J. Phys. Oceanog.* **30**, 2515–2534. [p. 166.]

Prigogine, I. (1980). *From Being to Becoming: Time and Complexity in the Physical Sciences*. Freeman, 272 pp. [p. 484.]

Prigogine, I. & Stengers, I. (1984). *Order out of Chaos – Man's New Dialogue with Nature*. Bantam Books, 349 pp. [p. 484.]

Queney, P. (1948). The problem of airflow over mountains: a summary of theoretical studies. *Bull. Amer. Met. Soc.* **29**, 16–26. [pp. 279, 281.]

Rasmussen, R.M., Smolarkiewicz, P. & Warner, J. (1989). On the dynamics of Hawaiian cloud bands: comparison of model results with observations and island climatology. *J. Atmos. Sci.* **46**, 1589–608. [p. 490.]

Rayleigh, Lord, (1896). *The Theory of Sound, Vol. II.* Reproduced by Dover (1948), 504 pp. [p. 246.]

Richard, E., Mascart, P. & Nickerson, E. (1989). The role of surface friction in downslope windstorms. *J. Applied Met.* **28**, 241–51. [p. 360.]

Ripepe, M., Barfucci, G., De Angelis, S., Delle Donne, D., Lacanna, G. & Marchetti, E. (2016). Modeling volcanic eruption parameters by near-source internal gravity waves. *Sci. Rep.* **6**, 36727; doi: 10.1038/srep36727. [p. 499.]

Rooney, G.G. & Devenish, B.J. (2014). Plume rise and spread in a linearly stratified environment. *Geophys. Astrophys. Fluid Dyn.* **108**, 168–190. [p. 499.]

Rottman, J.W. & Simpson, J.E. (1983). Gravity currents produced by instantaneous releases of a heavy fluid in a rectangular channel. *J. Fluid Mech.* **135**, 95–110. [p. 176.]

Rottman, J.W. & Simpson, J.E. (1989). The formation of internal bores in the atmosphere: a laboratory model. *Quart. J. Roy. Meteorol. Soc.* **115**, 941–963. [p. 176.]

Rottman, J.W., Broutman, D. & Grimshaw, R. (1996). Numerical simulations of uniformly stratified fluid flow over topography. *J. Fluid Mech.* **306**, 1–30. [p. 364.]

Rouse, H. & Ince, S. (1957). *History of Hydraulics*. Iowa Institute of Hydraulic Research, State Univ. Iowa, 269 pp. [p. 14.]

Sachsperger, J., Serafin, S. & Grubišić, V. (2016). Dynamics of rotor formation in uniformly stratified two-dimensional flow over a mountain. *Quart. J. Roy. Meteorol. Soc.* **142** 1201–1212. [p. 360.]

Sachsberger, J., Serafin, S., Grubišić, V., Stiperski, I. & Paci. A. (2017). The amplitude of lee waves on the boundary-layer inversion. *Quart. J. Roy. Meteorol. Soc.* **143**, 27–36. [p. 360.]

Sandu, I., Zadra, A. & Wedi, N. (2016). Impact of orographic drag on forecast skill. *ECMWF Newsletter 150*, 18–24. [p. 519.]

Sarkar, S. & Scotti, A. (2017). From topographic internal waves to turbulence. *Ann. Rev. Fluid Mech.* **49**, 195–220. [p. 504.]

Schär, C. (1993). A generalisation of Bernoulli's theorem. *J. Atmos. Sci.* **50**, 1437–43. [p. 11.]

Schär, C. & Durran, D.R. (1997). Vortex formation and vortex shedding in continuously stratified flows past isolated topography. *J. Atmos. Sci.* **54**, 534–554. [p. 484.]

Schär, C. & Smith, R.B. (1993a). Shallow water flow past isolated topography. Part 1: Vorticity production and wake formation. *J. Atmos. Sci.* **50**, 1373–400. [pp. 82, 83, 84, 85, 492.]

Schär, C. & Smith, R.B. (1993b). Shallow water flow past isolated topography. Part II: Transition to vortex shedding. *J. Atmos. Sci.* **50**, 1401–12. [pp. 84, 464, 492.]

Schlichting, H. (1968). *Boundary-Layer Theory*. McGraw–Hill, 817 pp. [p. 259.]

Schlichting, H. & Gersten, K. (2003). *Boundary-Layer Theory*. Springer-Verlag, 801 pp. [p. 254.]

Schubauer, G.B. & Skramstad, H.K. (1947). Laminar boundary-layer oscillations and transition on a flat plate. *J. Res. Nat. Bur. Stand.* **38**, 251–292. [p. 254.]

Schwartz, L.W. & Fenton, J.D. (1982). Strongly nonlinear waves. *Ann. Rev. Fluid Mech.* **14**, 39–60. [p. 64.]

Scinocca, J.F. & Peltier, W.R. (1989). Pulsating downslope windstorms. *J. Atmos. Sci.* **46**, 2885–914. [pp. 374, 377, 378.]

Scinocca, J.F. & Peltier, W.R. (1991). On the Richardson number dependence of nonlinear critical-layer flow over localised topography. *J. Atmos. Sci.* **48**, 1560–72. [pp. 364, 367.]

Scorer, R.S. (1949). Theory of waves in the lee of mountains. *Quart. J. Roy. Met. Soc.* **75**, 41–56. [p. 217.]

Scorer, R.S. (1986). *Cloud Investigation by Satellite*. Ellis Horwood, 312 pp. [p. 419.]

Scorer, R.S. (1990). *The Satellite as Microscope*. Ellis Horwood, 266 pp. [p. 419.]

Segur, H. & Hammack, J.L. (1982). Soliton models of long internal waves. *J. Fluid Mech.* **118**, 285–304. [p. 121.]

Shapiro, G.I., Huthnance, J.M. & Ivanov, V.V. (2003). Dense water cascading off the continental shelf. *J. Geophysical Res.*, **108**, C12:art no. 3390. [p. 503.]

Sharman, R.D. & Wurtele, M.G. (1983). Ship waves and lee waves. *J. Atmos. Sci.* **40**, 396–427. [pp. 418, 418.]

Sheppard, P.A. (1956). Airflow over mountains. *Quart. J. Roy. Met. Soc.* **82**, 528–9. [p. 454.]

Shutts, G. & Broad, A. (1993). A case study of lee waves over the Lake District, Northern England. *Quart. J. Roy. Met. Soc.* **119**, 377–408. [p. 425.]

Siddall, M., Smeed, D.A., Matthieson, S. & Rohling, E.J. (2002). Modelling the seasonal cycle of the exchange flow in Bab El Mandab (Red Sea). *Deep-Sea Res. I* **49**, 1551–1569. [p. 166.]

Silver, Z., Dimitrova, R., Zsedrovits, T., Baines, P.G. and Fernando, H.J.S. (2020). Simulation of stably stratified flow in complex terrain: flow structures and dividing streamline. *Environmental Fluid Mech.* **20**, 1281–1311, doi:10.1007/s10652-018-9648-y. [p. 498.]

Simard, A. & Peltier, W.R. (1982). Ship waves in the lee of isolated topography. *J. Atmos. Sci.* **39**, 587–609. [p. 415.]

Simpson, J.E. (1997). *Gravity Currents: in the Environment and the Laboratory*. Ellis Horwood, 244 pp. [pp. 172, 173, 173, 176.]

Simpson, J.E. & Britter, R.E. (1979). The dynamics of a head of a gravity current advancing over a horizontal surface. *J. Fluid Mech.* **94**, 477–95. [pp. 174, 174, 175.]

Smeed, D.A. (2000). Hydraulic control of three-layer exchange flows: application to the Bab al Mandab. *J. Phys. Oceanog.* **30**, 2574–2588. [p. 166.]

Smeed, D.A. (2004). Exchange through the Bab el Mandab. *Deep-Sea Research II* **51**, 455–474. [p. 166.]

Smith, R.B. (1976). The generation of lee waves by the Blue Ridge. *J. Atmos. Sci.* **33**, 507–19. [p. 127.]

Smith, R.B. (1980). Linear theory of stratified hydrostatic flow past an isolated mountain. *Tellus* **32**, 348–64. [pp. 402, 403, 405, 406, 408, 408, 411, 411.]

Smith, R.B. (1985). On severe downslope winds. *J. Atmos. Sci.* **42**, 2597–603. [pp. 317, 319, 319, 368, 372.]

Smith, R.B. (1987). Aerial observations of the Yugoslavian Bora. *J. Atmos. Sci.* **44**, 269–97. [p. 424.]

Smith, R.B. (1988). Linear theory of stratified flow past an isolated mountain in isosteric coordinates. *J. Atmos. Sci.* **45**, 3889–96. [pp. 404, 426, 442.]

Smith, R.B. (1989a). Mountain-induced stagnation points in hydrostatic flow. *Tellus* **41A**, 270–4. [p. 483.]

Smith, R.B. (1989b). Hydrostatic airflow over mountains. *Advances in Geophysics* **31**, 1–41. [p. 405.]

Smith, R.B. & Grønås, A. (1993). Stagnation points and bifurcation in 3D mountain airflow. *Tellus* **45A**, 28–43. [pp. 264, 430, 431, 482.]

Smith, R.B. & Grubišić, V. (1993). Aerial observations of Hawaii's wake. *J. Atmos. Sci.* **50**, 3728–50. [p. 490.]

Smith, R.B. & Smith, D. (1995). Pseudo-inviscid wake formation by mountains in shallow water flow with a drifting vortex. *J. Atmos. Sci.* **52**, 436–454. [p. 492.]

Smith, R.B., Geason, A.C., Gluhosky, P.A. & Grubišić, V. (1997). The wake of St. Vincent. *J. Atmos. Sci.* **54**, 606–623. [p. 491, 493.]

Smith, R.K., Crook, N. & Roff, G. (1982). The Morning Glory: an extraordinary atmospheric undular bore. *Quart. J. Roy. Met. Soc.* **108**, 937–56. [p. 190.]

Smolarkiewicz, P.K. & Rotunno, R. (1989). Low Froude number flow past three-dimensional obstacles. Part I: Baroclinically generated lee vortices. *J. Atmos. Sci.* **46**, 1154–64. [p. 481.]

Smolarkiewicz, P.K. & Rotunno, R. (1990). Low Froude number flow past three-dimensional obstacles. Part II: Upwind flow reversal zone. *J. Atmos. Sci.* **47**, 1498–511. [pp. 481, 483.]

Smolarkiewicz, P.K., Rasmussen, R.M. & Clark, T.L. (1988). On the dynamics of Hawaiian cloud bands: island forcing. *J. Atmos. Sci.* **45**, 1872–905. [pp. 490, 491, 492.]

Smyth, N.F. (1987). Modulation theory solution for resonant flow over topography. *Proc. Roy. Soc. Lond. A* **409**, 79–97. [p. 73.]

Smyth, N.F. (1988). Dissipative effects on the resonant flow of a stratified fluid over topography. *J. Fluid Mech.* **192**, 287–312. [p. 76.]

Smyth, W.D. & Peltier, W.R. (1989). The transition between Kelvin–Helmholtz and Holmboe instability: an investigation of the over-reflection hypothesis. *J. Atmos. Sci.* **46**, 3698–720. [p. 253.]

Smyth, W.D., Klaassen, G.P. & Peltier, W.R. (1988). Finite amplitude Holmboe waves. *Geophys. Astrophys. Fluid Dyn.* **43**, 181–222. [p. 253.]

Snyder, W.H., Britter, R.E. & Hunt, J.C.R. (1980). A fluid modelling study of the flow structure and plume impingement on a three-dimensional hill in stably stratified flow. In *Proc. 5th Int'l Conf. on Wind Engng*, J.E. Cermak (ed), Pergamon, pp. 319–29. [pp. 444, 453, 454.]

Snyder, W.H., Thompson, R.S., Eskridge, R.E., Lawson, R.E., Castro, I.P., Lee, J.T., Hunt, J.C.R. & Ogawa, Y. (1985). The structure of strongly stratified flow over hills: dividing streamline concept. *J. Fluid Mech.* **152**, 249–88. [pp. 453, 453, 454, 462, 467.]

Song, Jiyun, Fan, S., Lin, W., Mottet, L., Woodward, H., Davies Wykes, M., Arcucci, R., Xiao, D., Debay, J.-E., ApSimon, H., Aristodemou, E., Birch, D., Carpentieri, M., Fang, F., Herzog, M., Hunt, G.R., Jones, R.L., Pain, C., Pavlidis, D., Robins, A.G., Short, C.A. & Linden, P.F. (2018). Natural ventilation in cities: the implications of fluid mechanics. *Build. Res. Inf.* **46**, 809–828. [p. 510.]

Staquet, C. & Sommeria, J. (2002). Internal gravity waves: from instabilities to turbulence. *Ann. Rev. Fluid Mech.* **34**, 559–93. [p. 260.]

Stein, J. (1992). Investigation of the regime diagram of hydrostatic flow over a mountain with a primitive equation model. Part I: Two-dimensional flows. *Mon. Wea. Rev.* **120**, 2962–2976. [pp. 372, 373.]

Stigebrandt, A. (1990). On the response of the horizontal mean vertical density distribution in a fjord to low-frequency density fluctuations in the coastal water. *Tellus* **42A**, 605–614. [p. 166.]

Stoker, J.J. (1953). Unsteady waves on a running stream. *Comm. Pure Appl. Math.* **6**, 471–81. [p. 26.]

Stoker, J.J. (1957). *Water Waves*. InterScience, 567 pp. [p. 31.]

Strang, E.J. & Fernando, H.J.S. (2001). Vertical mixing and transports through a stratified shear layer. *J. Phys. Oceanog.* **31**, 2026–2048. [p. 184.]

Straus, L., Serafin, S. & Grubišić, V. (2016). Atmospheric rotors and severe turbulence in a long deep valley. *J. Atmos. Sci.* **73**, 1481–1506. [p. 497.]

Struik, D.J. (1950). *Lectures on Classical Differential Geometry.* Addison–Wesley, 221 pp. [p. 397.]

Sturtevant, B. (1965). Implications of experiments on the weak undular bore. *Phys. Fluids* **8**, 1052–55. [p. 71.]

Su, C.H. (1976). Hydraulic jumps in an incompressible stratified fluid. *J. Fluid Mech.* **73**, 33–47. [p. 329.]

Suzuki, M. & Kuwahara, K. (1992). Stratified flow past a bell-shaped hill. *Fluid Dyn. Res.* **9**, 1–18. [p. 481.]

Swanson, C.A. (1968). *Comparison and Oscillation Theory of Linear Differential Equations.* Academic Press, 227 pp. [p. 209.]

Tampieri, F. & Hunt, J.C.R. (1985). Two-dimensional stratified fluid flow over valleys: linear theory and a laboratory investigation. *Boundary-layer Met.* **32**, 257–79. [p. 386.]

Tanaka, M. (1993). Mach reflection of a large-amplitude solitary wave. *J. Fluid Mech.* **248**, 637–61. [p. 81.]

Taylor, P.A. (1988). Turbulent wakes in the atmospheric boundary layer. In *Flow and Transport in the Natural Environment: Advances and Applications*, W.L. Steffen & O.T. Denmead (eds), Springer-Verlag, 270–92. [p. 266.]

Thompson, R.S., Shipman, M.S. & Rottman, J.W. (1991). Moderately stable flow over a three-dimensional hill – a comparison of linear theory with laboratory measurements. *Tellus* **43A**, 49–63. [p. 446.]

Thorpe, S.A. (1968). A method of producing a shear flow in a stratified fluid. *J. Fluid Mech.* **32**, 693–704. [p. 247.]

Thorpe, S.A. (1981). An experimental study of critical layers. *J. Fluid Mech.* **103**, 321–44. [p. 232.]

Thorpe, S.A. (1987). Transitional phenomena and the development of turbulence in stratified fluids: A review. *J. Geophys. Res.* **92**, 5231–48. [p. 253.]

Thorpe, S.A. (1992). The generation of internal waves by flow over the rough topography of a continental slope. *Proc. Roy. Soc.* **439**, 115–130. [p. 503.]

Thorpe, S.A. (1996). The cross-slope transport of momentum by internal waves generated by alongslope currents over topography. *J. Phys. Oceanog.* **26**, 191–204. [p. 503.]

Thorpe, S.A. (2005). *The Turbulent Ocean.* Cambridge University Press, 439 pp. [p. 502.]

Thorpe, S.A. (2010). Turbulent hydraulic jumps in a stratified shear flow. *J. Fluid Mech.* **654**, 305–350. [p. 190.]

Thorpe, S.A. & Li, L. (2014). Turbulent hydraulic jumps in a stratified shear flow. Part 2. *J. Fluid Mech.* **758**, 94–120. [pp. 190, 191, 192.]

Thorpe, S.A., Malarkey, J., Voet, G., Alford, M.H., Girton, J.B. & Carter, G.S. (2018). Application of a model of internal hydraulic jumps. *J. Fluid Mech.* **834**, 125–148. [pp. 190, 192, 193, 194.]

Tomasson, G.C. & Melville, W.K. (1991). Flow past a constriction in a channel: A modal description. *J. Fluid Mech.* **232**, 21–45. [p. 141.]

Trefethen, L.N., Trefethen, A.E., Reddy, S.C. & Driscoll, T.A. (1993). Hydrodynamic stability without eigenvalues. *Science* **261**, 578–84. [p. 205.]

Trustrum, K. (1971). An Oseen model of the two-dimensional flow of a stratified fluid over an obstacle. *J. Fluid Mech.* **50**, 177–88. [p. 309.]

Tucker, G.B. (1989). Laboratory modelling of air flow in the Sofia valley. *Bulg. Geophys. J.* **15**, 92–100. [pp. 507, 509.]

Turner, J.S. (1973). *Buoyancy Effects in Fluids.* Cambridge University Press, 367 pp. [pp. 194, 247.]

Turner, J.S. (1986). Turbulent entrainment: the development of the entrainment assumption, and its application to geophysical flows. *J. Fluid Mech.* **173**, 431–471. [p. 185.]

Ungarish, M. & Hogg, A.J. (2018). Models of internal jumps and the fronts of gravity currents: unifying two-layer theories and deriving new results. *J. Fluid Mech.* **846**, 654–85. [p. 109.]

Ungarish, M. & Huppert, H.E. (2002). On gravity currents propagating at the base of a stratified ambient. *J. Fluid Mech.* **458**, 283–301. [p. 178.]

Ünülata, U., Oguz, T., Latif, M.A. & Ozsoy, E. (1990). On the physical oceanography of the Turkish straits. In *The Physical Oceanography of Sea Straits*, L. Pratt (ed), Kluwer, 25–60. [p. .]

Vanden-Broeck, J.-M. (1987). Free surface flow over an obstruction in a channel. *Phys. Fluids* **30**, 2315–17. [p. 77.]

Vanden-Broeck, J.-M. & Keller, J.B. (1987). Weir flows. *J. Fluid Mech.* **176**, 283–293. [p. 46.]

Vlasenko, V., Stashchuk, N. & Hutter, K. (2005). *Baroclinic Tides – Theoretical Modelling and Observational Evidence*. (Second Edition, 2012.) Cambridge University Press, 351 pp. [pp. 502, 504.]

Voight, B. & Sparks, R.S.J. (2010). Introduction to special section on the Eruption of Soufriére Hills Volcano, Montserrat, the CALIPSO Project, and the SEA-CALIPSO Arc-Crust Imaging Experiment. *Geophys. Res. Let.* **37**, L00E23, doi: 10.1029/2010GL044254. [p. 498.]

Vosper, S.B., Brown, A.R. & Webster, S. (2016). Orographic drag on islands in the NWP mountan grey zone. *Quart. J. Roy. Met. Soc.* **142**, 3128–37. [p. 519.]

Vosper, S.B., Ross, A.N., Renfrew, I.A., Sheridan, P., Elvidge, A.D. and Grubišić, V. (2018). Current challenges in orographic flow dynamics: turbulent exchange due to low-level gravity-wave processes. *Atmosphere* **361**, doi:10.3390/atmos9090361. [p. 519.]

Vreman, A.W., AL-Tarazi, M., Kuipers, J.A.M., Van Sint Annaland, M. & Bokhove, O. (2007). Supercritical shallow granular flow through a contraction: experiment, theory and simulation. *J. Fluid Mech.* **578**, 233–269. [p. 54.]

Wallace, J.M. & Hobbs, P.V. (1977). *Atmospheric Science – an Introductory Survey*. Academic Press, 467 pp. (Second edition, 2005.) [pp. 5, 419.]

Wallace, J.M., Tibaldi, S. & Simmons, A.J. (1983). Reduction of systematic forecast errors in the ECMWF model through the introduction of an envelope orography. *Quart. J. Roy. Met. Soc.* **109**, 683–717. [pp. 511, 512.]

Watson, G.N. (1966). *A Treatise on the Theory of Bessel Functions*. Cambridge University Press, 804 pp. [p. 207.]

Weber, M. (1919). Die Grundlagen der Ähnlichkeitsmechanik und ihre Verwertung bei Modellversuchen. *Jahrbuch der Schiffbautechnischen Gesellschaft Berlin*, Vol. 20, 355–477. [p. 14.]

Wei, T. & Smith, C.R. (1986). Secondary vortices in the wake of circular cylinders. *J. Fluid Mech.* **169**, 513–33. [pp. 467.]

Wei, S.N., Kao, T.W. & Pao, H.-P. (1975). Experimental study of upstream influence in the two-dimensional flow of a stratified fluid over an obstacle. *Geophys. Fluid Dyn.* **6**, 315–336. [p. 360.]

Weil, J.C., Traugott, S.C. & Wong, D.K. (1981). Stack plume interaction and flow characteristics for a notched ridge. *Rep. PPRP-61, Maryland Power Plant Siting Program*, Martin Marietta Corp., Baltimore, MD, 92 pp. [p. 470.]

Wells, M.G. & Wettlaufer, J.S. (2007). The long-term circulation driven by density currents in a two-layer stratified basin. *J. Fluid Mech.* **572**, 37–58. [p. 180.]

White, B.L. & Helfrich, K.R. (2008). Gravity currents and internal waves in a stratified fluid. *J. Fluid Mech.* **616**, 327–356. [pp. 177, 179.]

White, B.L. & Helfrich, K.R. (2012). A general description of a gravity current front propagating in a two-layer fluid. *J. Fluid Mech.* **711**, 545–575. [pp. 177, 178, 179.]

White, B.L. & Helfrich, K.R. (2014). A model for internal bores in continuous stratification. *J. Fluid Mech.* **761**, 282–304. [p. 336.]

Whitham, G.B. (1967). Variational methods and applications to water waves. *Proc. Roy. Soc. Lond. A* **229**, 6–25. [p. 120.]

Whitham, G.B. (1974). *Linear and Non-Linear Waves*. Wiley, 636 pp. [pp. 17, 31, 49, 50, 58, 59, 100, 121.]

Williams, R. & Armi, L. (1991). Two-layer hydraulics with comparable internal wave speeds. *J. Fluid Mech.* **230**, 667–691. [pp. 142, 144, 145, 145.]

Winant, C.D., Dorman, C.E., Friehe, C.A. & Beardsley, R.C. (1988). The maritime layer off Northern California: an example of supercritical flow. *J. Atmos. Sci.* **45**, 3588–605. [p. 141.]

Winters, K.B. (2016). The turbulent transition of a supercritical downslope flow: sensitivity to downstream conditions. *J. Fluid Mech.* **792**, 997–1012. [pp. 373, 387, 388, 388.]

Winters, K.B. & Armi, L. (2012). Hydraulic control of continuously stratified flow over an obstacle. *J. Fluid Mech.* **700**, 502–513. [p. 358.]

Winters, K.B. & Armi, L. (2014). Topographic control of stratified flows: upstream jets, blocking and isolating layers. *J. Fluid Mech.* **753**, 80–103. [pp. 325, 369, 370, 370, 371, 373.]

Winters, K.B. & D'Asaro, E.A. (1989). Two-dimensional instability of finite amplitude internal gravity wave packets near a critical level. *J. Geophys. Res.* **94**, 12709–19. [p. 232.]

Winters, K.B. & De La Fuente, A. (2012). Modelling rotating stratified flows at laboratory-scale using spectrally-based DNS. *Ocean Model.* **49–50**, 47–59. [p. 325.]

Winters, K.B. & Riley, J.J. (1992). Instabilities of internal waves near a critical level. *Dyn. Atmos. Oceans* **16**, 249–78. [p. 236.]

Winters, K.B. & Seim, H.E. (2000). The role of dissipation and mixing in exchange flow through a contracting channel. *J. Fluid Mech.* **407**, 265–290. [p. 171.]

Wolanski, E., Imberger, J. & Heron, M.L. (1984). Island wakes in shallow coastal waters. *J. Geophys. Res.* **89**, 10553–69. [p. 84.]

Wong, K.K. & Kao, T W. (1970). Stratified flow over extended obstacles and its application to topographic effect on ambient wind shear. *J. Atmos. Sci.* **27**, 884–9. [p. 285.]

Wood, I.R. (1968). Selective withdrawal from a stably stratified fluid. *J. Fluid Mech.*. **32**, 209–223. [pp. 129, 131, 144, 144, 147, 166, 166.]

Wood, I.R. (1970). A lock exchange flow. *J. Fluid Mech.* **42**, 671–87. [p. 158.]

Wood, I.R. & Simpson, J.E. (1984). Jumps in layered miscible fluids. *J. Fluid Mech.* **140**, 329–42. [p. 104.]

Woods, A.E. (2010). Turbulent plumes in nature. *Ann. Rev. Fluid Mech.* **42**, 391–412. [p. 499.]

Woods, A.W. (1988). The thermodynamics and fluid dynamics of eruption columns. *Bull. Volcanol.* **50**, 169–191. [p. 499.]

Woodson, C.B. (2018). The fate and impact of internal waves in nearshore ecosystems. *Ann. Rev. Marine Science* **10**, 421–441. [p. 503.]

Wu, T.Y. (1987). Generation of upstream advancing solitons by moving disturbances. *J. Fluid Mech.* **184**, 75–99. [p. 72.]

Wu, D.M. & Wu, T.Y. (1982). Three-dimensional nonlinear long waves due to moving surface pressure. In *Proc. 14th Symp. on Naval Hydrodynamics, Nat. Acad. Sci., Washington, DC*, pp. 103–25. [p. 72.]

Wurtele, M.G. (1957). The three-dimensional lee wave. *Beitr. Phys. Atmos.* **29**, 242–52. [p. 397.]

Wurtele, M.G., Sharman, R.D. & Datta, A. (1996). Atmospheric lee waves. *Ann. Rev. Fluid Mech.* **28**, 429–76. [p. 373.]

Yih, C-S. (1969). *Fluid Mechanics: a Concise Introduction to the Theory.* McGraw–Hill. 622 pp. [p. 147.]

Yih, C.S. & Guha, C.R. (1955). Hydraulic jump in a fluid system of two layers. *Tellus* **7**, 358–66. [pp. 104, 330.]

Yuan, C., Grimshaw, R., Johnson, E. & Chen, X. (2018). The propagation of internal solitary waves over variable topography in a horizontally two-dimensional framework. *J. Phys. Oceanog.* **48**, 283–300. [p. 94.]

Zaremba, L.J., Lawrence, G.A. & Pieters, R. (2003). Frictional two-layer flow. *J. Fluid Mech.* **474**, 339–354. [p. 171.]

Index

$1\frac{1}{2}$-layer model approximation, 97

Aden, Gulf of, 165
Airy's equation, 256
amplitude dispersion, 32, 333
anabatic flow, 192
anelastic approximation, 6
Antarctic Bottom Water, 504
Antarctic peninsula, 494, 507
Antarctic Slope Front, 504
atmospheric boundary-layer, 261

Bab al Mandab Strait, 165
baroclinic mode, 99
barotropic flow, 135
barotropic mode, 99
Benjamin–Ono (B–O) equation, 121
Bernoulli function, *see* total head function
Bernoulli integral, 10
Blasius boundary layer, 255
blocked flow drag, 517
blocking, 335
boundary-layer separation, 273
Boussinesq approximation, 6, 7
Boussinesq equation, 58
    linear, 57
breakdown of linear models, 424
Brunt–Väisälä frequency, *see* buoyancy frequency
Bulgaria, 507
buoyancy frequency, 4
buoyancy gradient, 209
buoyancy number, 181, 184
bushfire, 494

cascading in ocean, 503
caustic, 30
clean-air turbulence, 372
cnoidal wave, 60
collapse of streamlines, 429
columnar disturbance modes, 210
columnar modes, *see* columnar disturbance modes
continental shelves, 503
control points, 134

control section, 164
cowhorn eddy, 451
critical curve, 135
critical flow upstream, 334
critical layer, 204, 229
critical level, *see* critical layer
cross-flow separation, 439
Cumbria, 424

DEEPWAVE, 501
density current, *see* gravity current
dissipative process, 224
dividing stream-surface, 453
dividing streamline, 453
downslope flow
    ocean, 188
Drazin's model, 475

energy flux, 61
Ertel potential velocity, 11
estuary, 148
    inverse, 148
exchange
    maximal, 149
    submaximal, 149
exchange flow, 148
    lock, 157
expansion fan, 90
    Prandtl–Meyer, 92
external mode, 99

finite-depth boundary condition, 9
finite-depth equation, 121
flow
    critical, 18
    finite depth, 263
    infinite depth, 263
    one-dimensional, 18
    plunging, 144
    self-similar, 144, 166
    subcritical, 18
    supercritical, 18
    two-dimensional, 18

flow force, 61
flow-solve, 325
flow-splitting, 449, 482
focus, 436
foehn, 373, 495, 519
föhn, *see* foehn
free surface, 9
friction number, 87
Froude number, 13, 206
Froude number, F_0, 18
full vortex sheet model, FVS, 106

Garrett–Munk spectrum, 502
Gibraltar, 149
gravity current, 172
gravity-wave drag, 517
gravity-wave drag scheme, 515
Grimshaw–Yi equation, 323
group velocity, 25

half-saddle point, 439
half-singular point, 439
Hawaii, 489
Heaviside step function, 302
Hilo's band cloud, 490
Holmboe instability, 181, 248
horizontal energy flux, 13
horseshoe vortex, 482
hydraulic alternative, 328
hydraulic drag, 517
hydraulic drop, 15
hydraulic flow
    stratified, 325
hydraulic jump, 102, 333
    models, 328
hydrostatic approximation, 6, 7

infinite depth boundary condition, 9
instability
    Holmboe, 181, 248
    Kelvin–Helmholtz, 243
    Rayleigh–Taylor, 243
    triadic resonant, 260
intermediate-long-wave (ILW) equation, 121
internal mode, 99
internal wave beam, 260
intrinsic frequency, 216
isolated mixed regions, 274

Jan Mayen, 420, 423
jet, 489

Kadomtsev–Petviashvili (KP) equation, 92
Karman vortex sheet, 452
katabatic wind, 180
Kelvin ship-wake pattern, 29
Kelvin ship-wake wedge angle, 30
Kelvin wake, 397
Kelvin–Helmholtz instability, 243, 247
kinematic wave approximation, 49
Korteweg–de Vries equation, 59, 121
    extended (eKdV), 121, 322

forced (fKdV), 73, 322
    forced extended, (feKdV), 124

leaky mode, 216, 218
lee wave, 213
lee-side separation, 482
lift component, 393
Lighthill diagram, 28, 29, 213
Loch Ness monster topography, 386
lock exchange flow, 157
Loewe phenomenon, 190
Long's model, 288, 322
    finite depth, 297, 313
    infinite depth, 288

Mach angle, 21, 28, 89
Mach-stem reflection, 81
Macquarrie Island, 419
mass flux, 12
MATERHORN project, 195, 467, 498
maximal exchange solution, 157
Melbourne eddy, 507
Miles–Howard theorem, 244
modulation theory, 73
momentum flux, *see* flow force
momentum flux tensor, 12
momentum-source model, 304
Monin–Obukhov similarity theory, 468
Montserrat, 498
morning glory, 189

Nhu, 14, 267
node, 435
    simple, 438

ocean tides, 504
Olympic mountains, 492
orography, 1
over-reflection, 253
Owens Valley, 497

passenger mode, 335
plane of attachment, 435
plane of separation, 435
pliant boundary, *see* pliant surface, 15
pliant surface, 9
plume, 184
polynomial hill, 444, 446
post-wave separation, 273, 274
post-wave separation flow, 383
potential flow, 442
potential vorticity, 11
Prandtl number, 181
Prandtl–Meyer expansion fan, 92
Pyrenees, 495
PYREX, 495, 514

QRS framework, 56
quasi-biennial oscillation, 501

radiation condition, 9, 276
rarefaction, 113, 333
rarefied, 32

Rayleigh friction, 230
Rayleigh's criterion, 244
Rayleigh's equation, 106
Rayleigh's hydraulic equation, 35
Rayleigh–Taylor instability, 243
refractive index, 216
Reynolds number, 13, 181
Richardson number, 181, 244
    local, 207
Riemann invariants, 100
rigid list boundary condition, 9
ripple, 482
roll waves, 50
rotors, 274, 497, 519

saddle point, 435
    simple, 438
saturation, 419
saturation hypothesis, 238
Scorer's l^2, 217
self-similar flow, 144, 166
separatrices, 435
Sheppard criterion, 454, 468
Sierra Wave Project, 497
singular point
    classification, 435
singular solutions, 204
slowly varying approximation, 226, 232
solitary wave, 60
Soufriére Hills, 498
sound barrier, 25
South Georgia, 494
squashing, 360
St. Vincent, 491
stage
    gravity current, 176
    slumping, 176
    viscous, 176
stagnation point, 435
Strouhal number, 458
subcritical flow, 110
submaximal exchange solution, 157
subsurface motion, 504
supercritical flow, 110
supercritical leap, 15, 127
swirl, 395

T-REX, 497
Tasmania, 494
tidal motion, 504
Tollmien–Schlichting instability, 254
Tollmien–Schlichting wave, 265
topographic forcing, 124
topography, 1
topology of flow, 433
total head function, 10
total reflector, 199
Tramontana, 496
trapped mode, 216, 217

    inhomogeneities, 218
trapped motion, 216
triadic resonant instability, 260
triple deck, 263

valleys, 378
velocity gradient, 209
vertically integrated energy density, 12
virtual control point, 139
viscosity, 224, 254, 443
vortex wake model, 109
vorticity, 11
vorticity balance model, 330
vorticity equation, 11, 254

wake, 416, 418, 423
    classification, 456
    Kelvin, 416
    vortex-street-like, 467
wave action, 222
wave packet, 212
wavenumber surface, 213
White Island, 498
Witch of Agnesi, 268, 278, 292, 358
WKB approximation, *see* slowly varying
    approximation